100 Instructive Calculus-based PHYSICS Examples

Volume 2: Electricity and Magnetism

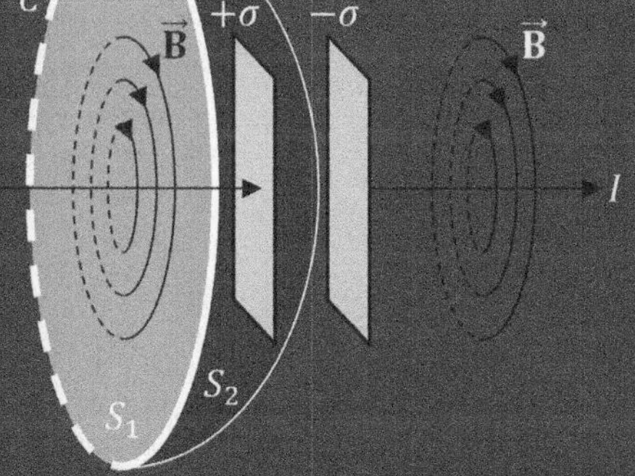

$$\oint_S \vec{\mathbf{E}} \cdot d\vec{\mathbf{A}} = \frac{q_{enc}}{\epsilon_0}$$

$$\oint_S \vec{\mathbf{B}} \cdot d\vec{\mathbf{A}} = 0$$

$$\oint_C \vec{\mathbf{E}} \cdot d\vec{\mathbf{s}} = -\frac{d\Phi_m}{dt}$$

$$\oint_C \vec{\mathbf{B}} \cdot d\vec{\mathbf{s}} = \mu_0 I_{enc} + \epsilon_0 \mu_0 \frac{d\Phi_e}{dt}$$

Chris McMullen, Ph.D.

Vol. 2

100 Instructive Calculus-based Physics Examples
Volume 2: Electricity and Magnetism
Fundamental Physics Problems

Chris McMullen, Ph.D.
Physics Instructor
Northwestern State University of Louisiana

www.monkeyphysicsblog.wordpress.com
www.improveyourmathfluency.com
www.chrismcmullen.wordpress.com

Zishka Publishing

ISBN: 978-1-941691-13-7

Textbooks > Science > Physics
Study Guides > Workbooks> Science

CONTENTS

INTRODUCTION

This book includes fully-solved examples with detailed explanations for 125 standard physics problems. There are also 16 math examples from integral calculus, which are essential toward mastering calculus-based physics. That makes a total of 141 problems.

Each example breaks the solution down into terms that make it easy to understand. The written explanations between the math help describe exactly what is happening, one step at a time. These examples are intended to serve as a helpful guide for solving similar standard physics problems from a textbook or course.

The best way to use this book is to write down the steps of the mathematical solution on a separate sheet of paper while reading through the example. Since writing is a valuable memory aid, this is an important step. In addition to writing down the solution, try to think your way through the solution. It may help to read through the solution at least two times: The first time, write it down and work it out on a separate sheet of paper as you solve it. The next time, think your way through each step as you read it.

Math and science books aren't meant to be read like novels. The best way to learn math and science is to think it through one step at a time. Read an idea, think about it, and then move on. Also write down the solutions and work them out on your own paper as you read. Students who do this tend to learn math and science better.

Note that these examples serve two purposes:

- They are primarily designed to help students understand how to solve standard physics problems. This can aid students who are struggling to figure out homework problems, or it can help students prepare for exams.
- These examples are also the solutions to the problems of the author's other book, *Essential Calculus-based Physics Study Guide Workbook*, ISBN 978-1-941691-11-3. That study guide workbook includes space on which to solve each problem.

1 COULOMB'S LAW

Coulomb's law
$F_e = k\frac{|q_1||q_2|}{R^2}$

Coulomb's Constant
$k = 8.99 \times 10^9 \ \frac{\text{N}\cdot\text{m}^2}{\text{C}^2} \approx 9.0 \times 10^9 \ \frac{\text{N}\cdot\text{m}^2}{\text{C}^2}$

Symbol	Name	SI Units
F_e	electric force	N
q	charge	C
R	separation	m
k	Coulomb's constant	$\frac{\text{N}\cdot\text{m}^2}{\text{C}^2}$ or $\frac{\text{kg}\cdot\text{m}^3}{\text{C}^2\cdot\text{s}^2}$

Elementary Charge
$e = 1.60 \times 10^{-19}$ C

Prefix	Name	Power of 10
m	milli	10^{-3}
μ	micro	10^{-6}
n	nano	10^{-9}
p	pico	10^{-12}

Note: The symbol μ is the lowercase Greek letter mu. When it is used as a metric prefix, it is called micro. For example, 32 μC is called 32 microCoulombs.

Example 1. A coin with a monkey's face has a net charge of −8.0 μC while a coin with a monkey's tail has a net charge of −3.0 μC. The coins are separated by 2.0 m. What is the electric force between the two coins? Is the force attractive or repulsive?

Solution. Begin by making a list of the known quantities:

- The coin with the monkey's face has a charge of $q_1 = -8.0\ \mu\text{C}$.
- The coin with the monkey's tail has a charge of $q_2 = -3.0\ \mu\text{C}$.
- The separation between the coins is $R = 2.0$ m.
- Coulomb's constant is $k = 9.0 \times 10^9\ \frac{\text{N}\cdot\text{m}^2}{\text{C}^2}$.

Convert the charges from microCoulombs (μC) to Coulombs (C). Recall that the metric prefix micro (μ) stands for one millionth: $\mu = 10^{-6}$.

$$q_1 = -8.0\ \mu\text{C} = -8.0 \times 10^{-6}\ \text{C}$$
$$q_2 = -3.0\ \mu\text{C} = -3.0 \times 10^{-6}\ \text{C}$$

Plug these values into Coulomb's law. It's convenient to suppress units and trailing zeroes (as in 8.0) until the end of the solution in order to avoid clutter. Note the absolute values (the reason for them is that the magnitude of the force is always positive).

$$F_e = k\frac{|q_1||q_2|}{R^2} = (9 \times 10^9)\frac{|-8 \times 10^{-6}||-3 \times 10^{-6}|}{(2)^2} = (9 \times 10^9)\frac{(8 \times 10^{-6})(3 \times 10^{-6})}{(2)^2}$$

If you're not using a calculator, it's convenient to separate the powers:

$$F_e = \frac{(9)(8)(3)}{(2)^2} \times 10^9 10^{-6} 10^{-6} = 54 \times 10^{-3} = 0.054\ \text{N}$$

Note that $10^9 10^{-6} 10^{-6} = 10^{9-6-6} = 10^{9-12} = 10^{-3}$ according to the rule $x^m x^n = x^{m+n}$. Also note that $R^2 = 2^2 = 4.0\ \text{m}^2$.

The answer is $F_e = 0.054$ N, which could also be expressed as 54×10^{-3} N, 5.4×10^{-2} N, or 54 mN (meaning milliNewtons, where the prefix milli, m, stands for 10^{-3}). The force is **repulsive** because two negative charges repel one another.

Example 2. A small glass rod and a small strand of monkey fur are each electrically neutral initially. When a monkey rubs the glass rod with the monkey fur, a charge of 800 nC is transferred between them. The monkey places the strand of fur 20 cm from the glass rod. What is the electric force between the fur and the rod? Is the force attractive or repulsive?

Solution. Begin by making a list of the known quantities:

- The small glass rod has a charge of $q_1 = 800\ \mu\text{C}$.
- The monkey fur has a charge of $q_2 = -800\ \mu\text{C}$.
- The separation between the objects is $R = 20$ cm.
- Coulomb's constant is $k = 9.0 \times 10^9\ \frac{\text{N}\cdot\text{m}^2}{\text{C}^2}$.

When the charge of 800 nC is transferred between the two objects, one object becomes positively charged and the other becomes negatively charged. (It doesn't matter which is positive and which is negative for the purpose of the calculation.) Convert the charges from nanoCoulombs (nC) to Coulombs (C). Recall that the metric prefix nano (n) stands for one billionth: $\text{n} = 10^{-9}$.

$$q_1 = 800 \text{ nC} = 800 \times 10^{-9} \text{ C} = 8.00 \times 10^{-7} \text{ C}$$
$$q_2 = -800 \text{ nC} = -800 \times 10^{-9} \text{ C} = -8.00 \times 10^{-7} \text{ C}$$

Note that $100 \times 10^{-9} = 10^2 \times 10^{-9} = 10^{2-9} = 10^{-7}$ according to the rule $x^m x^n = x^{m+n}$. Convert the distance from centimeters (cm) to meters (m).

$$R = 20 \text{ cm} = 0.20 \text{ m}$$

Plug these values into Coulomb's law. Note the absolute values (the reason for them is that the magnitude of the force is always positive).

$$F_e = k\frac{|q_1||q_2|}{R^2} = (9 \times 10^9)\frac{|8 \times 10^{-7}||-8 \times 10^{-7}|}{(0.2)^2} = (9 \times 10^9)\frac{(8 \times 10^{-7})(8 \times 10^{-7})}{(0.2)^2}$$

If you're not using a calculator, it's convenient to separate the powers:

$$F_e = \frac{(9)(8)(8)}{(0.2)^2} \times 10^9 10^{-7} 10^{-7} = 14{,}400 \times 10^{-5} = 0.144 \text{ N}$$

Note that $10^9 10^{-7} 10^{-7} = 10^{9-7-7} = 10^{9-14} = 10^{-5}$ according to the rule $x^m x^n = x^{m+n}$. Also note that $R^2 = 0.2^2 = 0.04 \text{ m}^2$.

The answer is $F_e = 0.144$ N, which could also be expressed as 144×10^{-3} N or 144 mN (meaning milliNewtons, where the prefix milli, m, stands for 10^{-3}). The force is **attractive** because oppositely charged objects attract one another. We know that the charges are opposite because when the charge was transferred (during the process of rubbing), one object gained electrons while the other lost electrons, making one object negative and the other object positive (since both were neutral prior to rubbing).

Example 3. A metal banana-shaped earring has a net charge of -2.0 μC while a metal apple-shaped earring has a net charge of 8.0 μC. The earrings are 3.0 m apart.

(A) What is the electric force between the two earrings? Is it attractive or repulsive?

Solution. Begin by making a list of the known quantities:

- The banana-shaped earring has a charge of $q_1 = -2.0$ μC.
- The apple-shaped earring has a charge of $q_2 = 8.0$ μC.
- The separation between the earrings is $R = 3.0$ m.
- Coulomb's constant is $k = 9.0 \times 10^9 \frac{\text{N}\cdot\text{m}^2}{\text{C}^2}$.

Convert the charges from microCoulombs (μC) to Coulombs (C). Recall that the metric prefix micro (μ) stands for one millionth: $\mu = 10^{-6}$.

$$q_1 = -2.0 \ \mu\text{C} = -2.0 \times 10^{-6} \text{ C}$$
$$q_2 = 8.0 \ \mu\text{C} = 8.0 \times 10^{-6} \text{ C}$$

Plug these values into Coulomb's law. Note the absolute values (the reason for them is that the magnitude of the force is always positive).

$$F_e = k\frac{|q_1||q_2|}{R^2} = (9 \times 10^9)\frac{|-2 \times 10^{-6}||8 \times 10^{-6}|}{(3)^2} = (9 \times 10^9)\frac{(2 \times 10^{-6})(8 \times 10^{-6})}{(3)^2}$$

If you're not using a calculator, it's convenient to separate the powers:

$$F_e = \frac{(9)(2)(8)}{(3)^2} \times 10^9 10^{-6} 10^{-6} = 16 \times 10^{-3} = 0.016 \text{ N}$$

Note that $10^9 10^{-6} 10^{-6} = 10^{9-6-6} = 10^{9-12} = 10^{-3}$ according to the rule $x^m x^n = x^{m+n}$. Also note that $R^2 = 3^2 = 9.0 \text{ m}^2$.

The answer is $F_e = 0.016$ N, which could also be expressed as 16×10^{-3} N, 1.6×10^{-2} N, or 16 mN (meaning milliNewtons, where the prefix milli, m, stands for 10^{-3}). The force is **attractive** because oppositely charged objects attract one another.

(B) The two earrings are brought together, touching one another for a few seconds, after which the earrings are once again placed 3.0 m apart. What is the electric force between the two earrings now? Is it attractive or repulsive?

Solution. To the extent possible, the charges would like to **neutralize** when contact is made. The -2.0 μC isn't enough to completely neutralize the $+8.0$ μC, so what will happen is that the -2.0 μC will pair up with $+2.0$ μC from the $+8.0$ μC, leaving a net excess charge of $+6.0$ μC. **One-half** of the **net excess charge**, $+6.0$ μC, resides on each earring. That is, after contact, each earring will have a charge of $q = +3.0$ μC. That's what happens conceptually. We can arrive at the same answer mathematically using the formula below.

$$q = \frac{q_1 + q_2}{2} = \frac{-2.0 \text{ μC} + 8.0 \text{ μC}}{2} = \frac{6.0 \text{ μC}}{2} = 3.0 \text{ μC}$$

Make a list of the known quantities:

- The charges both equal $q = 3.0 \text{ μC} = 3.0 \times 10^{-6}$ C.
- The separation between the earrings is $R = 3.0$ m.
- Coulomb's constant is $k = 9.0 \times 10^9 \frac{\text{N·m}^2}{\text{C}^2}$.

Set the two charges equal in Coulomb's law.

$$F_e = k\frac{q^2}{R^2} = (9 \times 10^9)\frac{(3 \times 10^{-6})^2}{(3)^2}$$

Apply the rule $(ab)^2 = a^2b^2$ from algebra: $(3 \times 10^{-6})^2 = 3^2 \times (10^{-6})^2 = 9 \times 10^{-12}$.

$$F_e = (9 \times 10^9)\frac{9 \times 10^{-12}}{(3)^2} = \frac{(9)(9)}{(3)^2} \times 10^9 10^{-6} 10^{-6} = 9.0 \times 10^{-3} = 0.0090 \text{ N}$$

The answer is $F_e = 0.0090$ N, which could also be expressed as 9.0×10^{-3} N or 9.0 mN (meaning milliNewtons, where the prefix milli, m, stands for 10^{-3}). The force is **repulsive** because two positive charges repel one another. (After they make contact, both earrings become **positively** charged.)

2 ELECTRIC FIELD

Relation between Electric Force and Electric Field	Electric Field Due to a Pointlike Charge
$\vec{F}_e = q\vec{E}$	$E = \frac{k\|q\|}{R^2}$

Coulomb's Constant
$k = 8.99 \times 10^9 \frac{N \cdot m^2}{C^2} \approx 9.0 \times 10^9 \frac{N \cdot m^2}{C^2}$

Symbol	Name	SI Units
E	electric field	N/C or V/m
F_e	electric force	N
q	charge	C
R	distance from the charge	m
k	Coulomb's constant	$\frac{N \cdot m^2}{C^2}$ or $\frac{kg \cdot m^3}{C^2 \cdot s^2}$

Prefix	Name	Power of 10
m	milli	10^{-3}
μ	micro	10^{-6}
n	nano	10^{-9}
p	pico	10^{-12}

Example 4. A small furball from a monkey has a net charge of $-300\ \mu\text{C}$. The furball is in the presence of an external electric field with a magnitude of 80,000 N/C. What is the magnitude of the electric force exerted on the furball?

Solution. Make a list of the known quantities and the desired unknown:

- The furball has a charge of $q = -300\ \mu\text{C} = -300 \times 10^{-6}\ \text{C} = -3.0 \times 10^{-4}\ \text{C}$. Recall that the metric prefix micro (μ) stands for 10^{-6}. Note that $100 \times 10^{-6} = 10^{-4}$.
- The electric **field** has a magnitude of $E = 80{,}000$ N/C.
- We are looking for the magnitude of the electric **force,** F_e.

Based on the list above, we should use the following equation.

$$F_e = |q|E = |-3.0 \times 10^{-4}| \times 80{,}000 = (3.0 \times 10^{-4}) \times 80{,}000 = 240{,}000 \times 10^{-4} = 24\ \text{N}$$

The magnitude of a vector is always positive (that's why we took the absolute value of the charge). The electric force exerted on the furball is $F_e = 24$ N.

Note that we used the equation $F_e = |q|E$ because we were finding the force exerted on a charge in the presence of an external electric field. We didn't use the equation $E = \frac{k|q|}{R^2}$ because the problem didn't require us to find an electric field created by the charge. (The problem didn't state what created the electric field, and it doesn't matter as it isn't relevant to solving the problem.)

Example 5. A gorilla's earring has a net charge of $800\ \mu\text{C}$. Find the magnitude of the electric field at a distance of 2.0 m from the earring.

Solution. Make a list of the known quantities and the desired unknown:

- The earring has a charge of $q = 800\ \mu\text{C} = 800 \times 10^{-6}\ \text{C} = 8.0 \times 10^{-4}\ \text{C}$. Recall that the metric prefix micro (μ) stands for 10^{-6}. Note that $100 \times 10^{-6} = 10^{-4}$.
- We wish to find the electric field at a distance of $R = 2.0$ m from the earring.
- We also know that Coulomb's constant is $k = 9.0 \times 10^9\ \frac{\text{N}\cdot\text{m}^2}{\text{C}^2}$.
- We are looking for the magnitude of the electric **field,** E.

Based on the list above, we should use the following equation.

$$E = \frac{k|q|}{R^2} = \frac{(9 \times 10^9)|8 \times 10^{-4}|}{(2)^2} = \frac{(9)(8)}{(2)^2} \times 10^9 10^{-4} = 18 \times 10^5\ \text{N/C} = 1.8 \times 10^6\ \text{N/C}$$

Note that $10^9 10^{-4} = 10^{9-4} = 10^5$ according to the rule $x^m x^{-n} = x^{m-n}$. Also note that $R^2 = 2^2 = 4.0\ \text{m}^2$ and that $18 \times 10^5 = 1.8 \times 10^6$. The magnitude of the electric field at a distance of 2.0 m from the earring is $E = 1.8 \times 10^6$ N/C.

Example 6. A monkey's earring experiences an electric force of 12 N in the presence of an external electric field of 30,000 N/C. The electric force is opposite to the electric field. What is the net charge of the earring?

Solution. Make a list of the known quantities and the desired unknown:

- The electric **force** has a magnitude of $F_e = 12$ N.
- The electric **field** has a magnitude of $E = 30{,}000$ N/C.
- We are looking for the net charge of the earring, q.

Based on the list above, we should use the following equation.

$$F_e = |q|E$$

Divide both sides of the equation by the magnitude of the electric field.

$$|q| = \frac{F_e}{E} = \frac{12}{30{,}000} = \frac{12}{3}\frac{1}{10{,}000} = 4.0 \times 10^{-4}\text{ C}$$

Note that $30{,}000 = (3)(10{,}000)$ and that $10{,}000 = 10^4$ such that $\frac{1}{10{,}000} = 10^{-4}$. The absolute value of the charge is $|q| = 4.0 \times 10^{-4}$ C. The charge must be **negative** since the force ($\vec{\mathbf{F}}_e$) is opposite to the electric field ($\vec{\mathbf{E}}$). The answer is $q = -4.0 \times 10^{-4}$ C, which can also be expressed as $q = -400$ μC, since the metric prefix micro (μ) stands for 10^{-6}.

Example 7. A small strand of monkey fur has a net charge of 80 μC. Where does the electric field created by the strand of fur equal 20,000 N/C?

Solution. Make a list of the known quantities and the desired unknown:

- The strand of fur has a charge of $q = 80\ \mu\text{C} = 80 \times 10^{-6}\text{ C} = 8.0 \times 10^{-5}$ C. Recall that the metric prefix micro (μ) stands for 10^{-6}. Note that $10 \times 10^{-6} = 10^{-5}$.
- The electric **field** has a magnitude of $E = 20{,}000$ N/C at the desired location.
- We also know that Coulomb's constant is $k = 9.0 \times 10^9\ \frac{\text{N}\cdot\text{m}^2}{\text{C}^2}$.
- We wish to find the distance, R, from the monkey fur for which the electric field has the specified magnitude.

Based on the list above, we should use the following equation.

$$E = \frac{k|q|}{R^2}$$

Multiply both sides of the equation by R^2 and divide both sides of the equation by E.

$$R^2 = \frac{k|q|}{E}$$

Squareroot both sides of the equation.

$$R = \sqrt{\frac{k|q|}{E}} = \sqrt{\frac{(9 \times 10^9)|8 \times 10^{-5}|}{20{,}000}} = \sqrt{\frac{(9)(8)}{2}\frac{10^9 10^{-5}}{10^4}} = \sqrt{36} = 6.0\text{ m}$$

Note that $20{,}000 = 2 \times 10^4$. Also note that $\frac{10^9 10^{-5}}{10^4} = \frac{10^4}{10^4} = 1$ according to the rules $x^m x^n = x^{m+n}$ and $\frac{x^m}{x^n} = x^{m-n}$. The answer is $R = 6.0$ m.

Example 8. A pear-shaped earring, shown as a dot (●) below, with a charge of 30 μC lies at the point (3.0 m, 6.0 m).

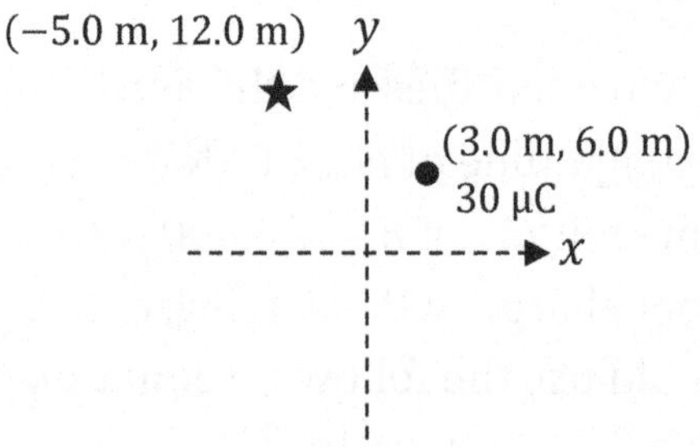

(A) What is the magnitude of the electric field at the point (−5.0 m, 12.0 m), which is marked as a star (★) above?

Solution. First, we need to find the distance between the charge and the point where we're trying to find the electric field. Apply the distance formula.

$$R = \sqrt{(x_2 - x_1)^2 + (y_2 - y_1)^2} = \sqrt{(-5 - 3)^2 + (12 - 6)^2}$$

$$R = \sqrt{(-8)^2 + 6^2} = \sqrt{64 + 36} = \sqrt{100} = 10 \text{ m}$$

Note that $-5 - 3 = -8$ (the star is 8.0 m to the left of the earring). Also note that the minus sign is squared: $(-8)^2 = +64$. Make a list of the known quantities and the desired unknown:

- The earring has a charge of $q = 30$ μC $= 30 \times 10^{-6}$ C $= 3.0 \times 10^{-5}$ C. Recall that the metric prefix micro (μ) stands for 10^{-6}. Note that $10 \times 10^{-6} = 10^{-5}$.
- We wish to find the electric field at a distance of $R = 10$ m from the earring.
- We also know that Coulomb's constant is $k = 9.0 \times 10^9 \, \frac{\text{N·m}^2}{\text{C}^2}$.
- We are looking for the magnitude of the electric **field**, E.

Based on the list above, we should use the following equation.

$$E = \frac{k|q|}{R^2} = \frac{(9 \times 10^9)|3 \times 10^{-5}|}{10^2} = (9)(3)\frac{10^9 10^{-5}}{10^2} = 27 \times 10^2 \text{ N/C} = 2.7 \times 10^3 \text{ N/C}$$

Note that $\frac{10^9 10^{-5}}{10^2} = \frac{10^{9-5}}{10^2} = \frac{10^4}{10^2} = 10^{4-2} = 10^2$ according to the rules $x^m x^n = x^{m+n}$ and $\frac{x^m}{x^n} = x^{m-n}$. The magnitude of the electric field at the specified point is $E = 2.7 \times 10^3$ N/C, which can also be expressed as $E = 2700$ N/C (except that scientific notation expresses the result properly to two significant figures).

(B) If a small strand of chimpanzee fur with a net charge of 500 μC is placed at the point (−5.0 m, 12.0 m), marked with a star (★), what force will be exerted on the strand of fur?

Solution. Use the result from part (A) with the following equation. Convert the charge from microCoulombs (μC) to Coulombs (C). Note that the metric prefix micro (μ) stands for 10^{-6}. The charge is $q_2 = 5.0 \times 10^{-4}$ C.

$$F_e = |q_2|E = |5.0 \times 10^{-4}| \times 2.7 \times 10^3 = 13.5 \times 10^{-1} = 1.35 \text{ N}$$

The electric force exerted on the strand of lemur fur is $F_e = 1.35$ N, which rounds up to $F_e = 1.4$ N when expressed with two significant figures.

3 SUPERPOSITION OF ELECTRIC FIELDS

Distance Formula

$$R_i = \sqrt{(x_2 - x_1)^2 + (y_2 - y_1)^2}$$

Direction of each Vector

$$\theta_i = \tan^{-1}\left(\frac{\Delta y}{\Delta x}\right)$$

Magnitude of each Electric Field or Electric Force Vector

$$E_1 = \frac{k|q_1|}{R_1^2} \quad , \quad E_2 = \frac{k|q_2|}{R_2^2} \quad , \quad E_3 = \frac{k|q_3|}{R_3^2} \quad \cdots$$

$$F_1 = \frac{k|q_1||q|}{R_1^2} \quad , \quad F_2 = \frac{k|q_2||q|}{R_2^2} \quad , \quad F_3 = \frac{k|q_3||q|}{R_3^2} \quad \cdots$$

Components of each Vector

$$E_{1x} = E_1 \cos\theta_1 \quad , \quad E_{1y} = E_1 \sin\theta_1 \quad , \quad E_{2x} = E_2 \cos\theta_2 \quad , \quad E_{2y} = E_2 \sin\theta_2 \quad \cdots$$

$$F_{1x} = F_1 \cos\theta_1 \quad , \quad F_{1y} = F_1 \sin\theta_1 \quad , \quad F_{2x} = F_2 \cos\theta_2 \quad , \quad F_{2y} = F_2 \sin\theta_2 \quad \cdots$$

Components of the Resultant Vector

$$E_x = E_{1x} + E_{2x} + \cdots + E_{Nx} \quad , \quad E_y = E_{1y} + E_{2y} + \cdots + E_{Ny}$$

$$F_x = F_{1x} + F_{2x} + \cdots + F_{Nx} \quad , \quad F_y = F_{1y} + F_{2y} + \cdots + F_{Ny}$$

Magnitude and Direction of the Resultant Vector

$$E = \sqrt{E_x^2 + E_y^2} \quad , \quad \theta_E = \tan^{-1}\left(\frac{E_y}{E_x}\right)$$

$$F_e = \sqrt{F_x^2 + F_y^2} \quad , \quad \theta_F = \tan^{-1}\left(\frac{F_y}{F_x}\right)$$

Coulomb's Constant
$k = 8.99 \times 10^9 \ \frac{\text{N}\cdot\text{m}^2}{\text{C}^2} \approx 9.0 \times 10^9 \ \frac{\text{N}\cdot\text{m}^2}{\text{C}^2}$

Symbol	Name	Units
E	magnitude of electric field	N/C or V/m
F_e	magnitude of electric force	N
θ	direction of $\vec{\mathbf{E}}$ or $\vec{\mathbf{F}}_e$	°
q	charge	C
R	distance from the charge	m
d	distance between two charges	m
k	Coulomb's constant	$\frac{\text{N}\cdot\text{m}^2}{\text{C}^2}$ or $\frac{\text{kg}\cdot\text{m}^3}{\text{C}^2\cdot\text{s}^2}$

Prefix	Name	Power of 10
m	milli	10^{-3}
μ	micro	10^{-6}
n	nano	10^{-9}
p	pico	10^{-12}

Example 9. A monkey-shaped earring with a charge of 6.0 μC lies at the point (0, 2.0 m). A banana-shaped earring with a charge of −6.0 μC lies at the point (0, −2.0 m). Determine the magnitude and direction of the net electric field at the point (2.0 m, 0).

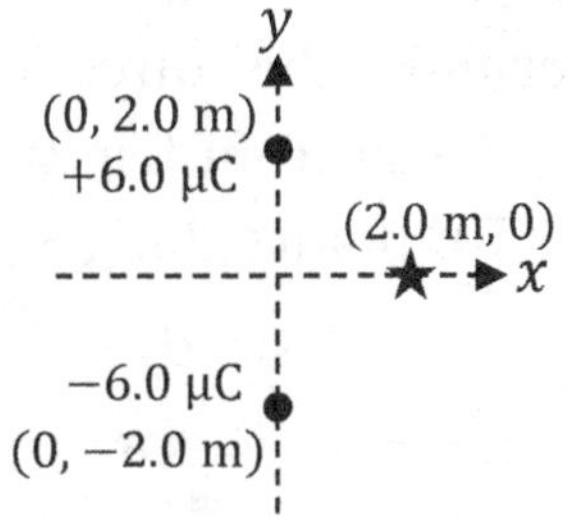

Solution. Begin by sketching the electric field vectors created by each of the charges. In order to do this, imagine a positive "test" charge at the point (2.0 m, 0), marked by a star (★) above. The point (2.0 m, 0) is called the **field point**: It's the point where we are determining the electric field. We label the monkey-shaped earring as charge 1 ($q_1 = 6.0$ μC) and the banana-shaped earring as charge 2 ($q_2 = -6.0$ μC).

- A positive "test" charge at (2.0 m, 0) would be repelled by $q_1 = 6.0$ μC. Thus, we draw $\vec{\mathbf{E}}_1$ directly away from $q_1 = 6.0$ μC (diagonally down and to the right).
- A positive "test" charge at (2.0 m, 0) would be attracted to $q_2 = -6.0$ μC. Thus, we draw $\vec{\mathbf{E}}_2$ towards $q_2 = -6.0$ μC (diagonally down and to the left).

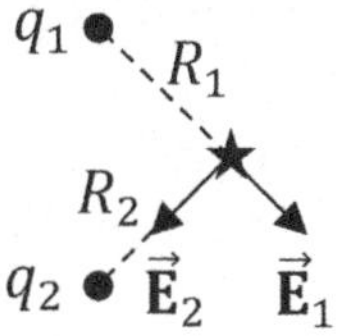

Find the distance between each charge (shown as a dot above) and the field point (shown as a star) where we're trying to find the net electric field. Compare the diagrams above and below. Note that $|\Delta x_1| = |\Delta x_2| = 2.0$ m and $|\Delta y_1| = |\Delta y_2| = 2.0$ m.

Apply the distance formula to find R_1 and R_2.

$$R_1 = \sqrt{\Delta x_1^2 + \Delta y_1^2} = \sqrt{2^2 + 2^2} = \sqrt{4+4} = \sqrt{8} = \sqrt{(4)(2)} = \sqrt{4}\sqrt{2} = 2\sqrt{2} \text{ m}$$

$$R_2 = \sqrt{\Delta x_2^2 + \Delta y_2^2} = \sqrt{2^2 + 2^2} = \sqrt{4+4} = \sqrt{8} = \sqrt{(4)(2)} = \sqrt{4}\sqrt{2} = 2\sqrt{2} \text{ m}$$

Next, find the direction of $\vec{\mathbf{E}}_1$ and $\vec{\mathbf{E}}_2$. First determine the reference angle from each right triangle. The reference angle is the smallest angle with the positive or negative x-axis.

$$\theta_{1ref} = \tan^{-1}\left|\frac{\Delta y_1}{\Delta x_1}\right| = \tan^{-1}\left(\frac{2}{2}\right) = \tan^{-1}(1) = 45°$$

$$\theta_{2ref} = \tan^{-1}\left|\frac{\Delta y_2}{\Delta x_2}\right| = \tan^{-1}\left(\frac{2}{2}\right) = \tan^{-1}(1) = 45°$$

Use the reference angles to determine the direction of each electric field vector counterclockwise from the $+x$-axis. Recall from trig that 0° points along $+x$, 90° points along $+y$, 180° points along $-x$, and 270° points along $-y$.

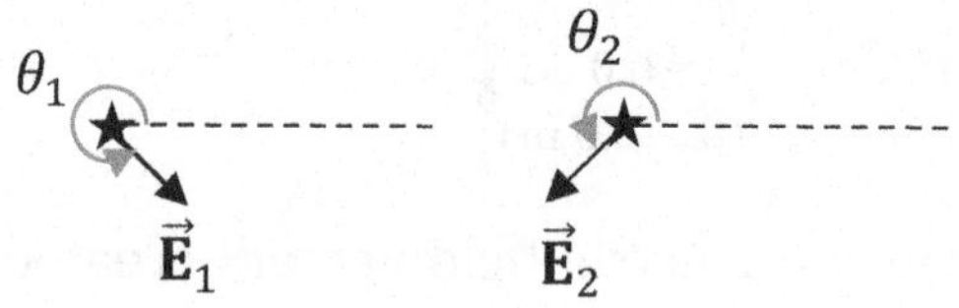

- $\vec{\mathbf{E}}_1$ lies in Quadrant IV (left diagram above):
$$\theta_1 = 360° - \theta_{1ref} = 360° - 45° = 315°$$
- $\vec{\mathbf{E}}_2$ lies in Quadrant III (right diagram above):
$$\theta_2 = 180° + \theta_{2ref} = 180° + 45° = 225°$$

Find the magnitude of the electric field created by each pointlike charge at the specified point. First convert the charges from μC to C: $q_1 = 6.0 \times 10^{-6}$ C and $q_2 = -6.0 \times 10^{-6}$ C. Use the values $R_1 = 2\sqrt{2}$ m and $R_2 = 2\sqrt{2}$ m, which we found previously.

$$E_1 = \frac{k|q_1|}{R_1^2} = \frac{(9 \times 10^9)|6 \times 10^{-6}|}{\left(2\sqrt{2}\right)^2} = \frac{(9)(6)}{8} \times 10^9 10^{-6} = 6.75 \times 10^3 \text{ N/C} = 6750 \text{ N/C}$$

$$E_2 = \frac{k|q_2|}{R_2^2} = \frac{(9 \times 10^9)|-6 \times 10^{-6}|}{\left(2\sqrt{2}\right)^2} = \frac{(9)(6)}{8} \times 10^9 10^{-6} = 6.75 \times 10^3 \text{ N/C} = 6750 \text{ N/C}$$

Note that $\left(2\sqrt{2}\right)^2 = (2)^2\left(\sqrt{2}\right)^2 = (4)(2) = 8$ since $\left(\sqrt{2}\right)^2 = 2$, and that $10^9 10^{-6} = 10^{9-6} = 10^3$ according to the rule $x^m x^{-n} = x^{m-n}$. Also note that the minus sign from $q_2 = -6.0 \times 10^{-6}$ C vanished with the absolute values (because the magnitude of an electric field vector can't be negative). Following is a summary of what we know thus far.

- $q_1 = 6.0 \times 10^{-6}$ C, $R_1 = 2\sqrt{2}$ m, $\theta_1 = 315°$, and $E_1 = 6750$ N/C.
- $q_2 = -6.0 \times 10^{-6}$ C, $R_2 = 2\sqrt{2}$ m, $\theta_2 = 225°$, and $E_2 = 6750$ N/C.

We are now prepared to add the electric field vectors. We will use the magnitudes and directions ($E_1 = 6750$ N/C, $E_2 = 6750$ N/C, $\theta_1 = 315°$, and $\theta_2 = 225°$) of the two electric field vectors to determine the magnitude and direction (E and θ_E) of the resultant vector. The first step of vector addition is to find the components of the known vectors. This step involves four equations (two for $\vec{\mathbf{E}}_1$ and two for $\vec{\mathbf{E}}_2$).

$$E_{1x} = E_1 \cos\theta_1 = 6750\cos(315°) = 6750\left(\frac{\sqrt{2}}{2}\right) = 3375\sqrt{2} \text{ N/C}$$

$$E_{1y} = E_1 \sin\theta_1 = 6750\sin(315°) = 6750\left(-\frac{\sqrt{2}}{2}\right) = -3375\sqrt{2} \text{ N/C}$$

$$E_{2x} = E_2 \cos\theta_2 = 6750\cos(225°) = 6750\left(-\frac{\sqrt{2}}{2}\right) = -3375\sqrt{2}\text{ N/C}$$

$$E_{2y} = E_2 \sin\theta_2 = 6750\sin(225°) = 6750\left(-\frac{\sqrt{2}}{2}\right) = -3375\sqrt{2}\text{ N/C}$$

It's a good idea to check the signs based on the previous diagram. For x, right is positive and left is negative. For y, up is positive and down is negative. (This is because we draw a coordinate system with $+x$ to the right and $+y$ up.)

- $E_{1x} = 3375\sqrt{2}$ N/C. It's positive because $\vec{\mathbf{E}}_1$ points to the right, not left.
- $E_{1y} = -3375\sqrt{2}$ N/C. It's negative because $\vec{\mathbf{E}}_1$ points downward, not upward.
- $E_{2x} = -3375\sqrt{2}$ N/C. It's negative because $\vec{\mathbf{E}}_2$ points to the left, not right.
- $E_{2y} = -3375\sqrt{2}$ N/C. It's negative because $\vec{\mathbf{E}}_2$ points downward, not upward.

The second step of vector addition is to add the respective components together.

$$E_x = E_{1x} + E_{2x} = 3375\sqrt{2} + \left(-3375\sqrt{2}\right) = 0$$

$$E_y = E_{1y} + E_{2y} = -3375\sqrt{2} + \left(-3375\sqrt{2}\right) = -6750\sqrt{2}\text{ N/C}$$

The final step of vector addition is to apply the Pythagorean theorem and inverse tangent to determine the magnitude and direction, respectively, of the resultant vector.

$$E = \sqrt{E_x^2 + E_y^2} = \sqrt{0^2 + \left(-6750\sqrt{2}\right)^2} = \sqrt{\left(6750\sqrt{2}\right)^2} = 6750\sqrt{2}\text{ N/C}$$

Note that $\sqrt{\left(6750\sqrt{2}\right)^2} = 6750\sqrt{2}$ according to the rule $\sqrt{x^2} = x$. Also note that the minus sign disappears since it is squared: $(-1)^2 = +1$. The magnitude of the net electric field at the specified point is $E = 6750\sqrt{2}$ N/C, which can also be expressed as $E = 6.75\sqrt{2} \times 10^3$ N/C. Apply the following inverse tangent in order to determine the direction of the net electric field at the field point.

$$\theta_E = \tan^{-1}\left(\frac{E_y}{E_x}\right) = \tan^{-1}\left(\frac{-6750\sqrt{2}}{0}\right) = 270°$$

Although $\frac{-6750\sqrt{2}}{0}$ is undefined (you can't divide by zero), the inverse tangent is defined. You can see this from the physics: Since $E_x = 0$ and $E_y < 0$ (find these values above), the answer lies on the negative y-axis, which corresponds to $\theta_E = 270°$.

The final answer is that the magnitude of the net electric field at the specified point equals $E = 6.75\sqrt{2} \times 10^3$ N/C and its direction is $\theta_E = 270°$. If you use a calculator, the magnitude of the electric field works out to $E = 9.5 \times 10^3$ N/C to two significant figures.

Example 10. A monkey places three charges on the vertices of an equilateral triangle, as illustrated below. Determine the magnitude and direction of the net electric force exerted on q_3.

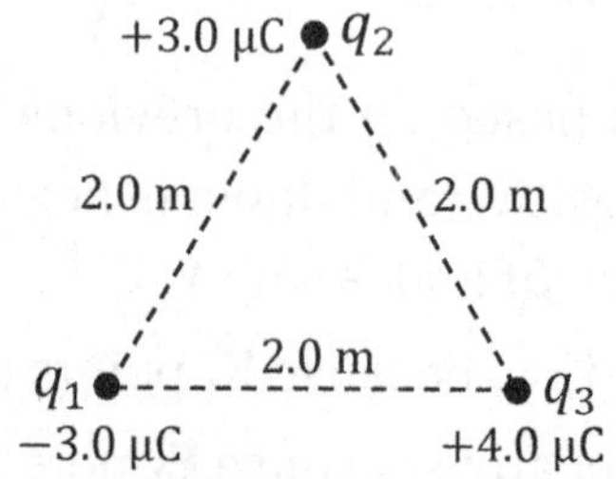

Solution. Begin by sketching the two electric forces exerted on q_3. Recall that opposite charges attract, while like charges repel.

- Since q_1 and q_3 have opposite signs, q_1 pulls q_3 to the left. Thus, we draw $\vec{\mathbf{F}}_1$ towards q_1 (to the left).
- Since q_2 and q_3 are both positive, q_2 pushes q_3 down and to the right. Thus, we draw $\vec{\mathbf{F}}_2$ directly away from q_2 (down and to the right).

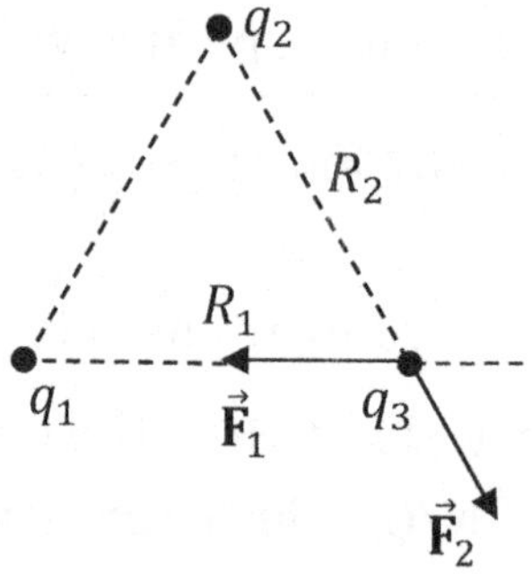

Find the distance between each charge. Since each charge lies on the vertex of an equilateral triangle, each R equals the edge length: $R_1 = R_2 = 2.0$ m. Next, find the direction of $\vec{\mathbf{F}}_1$ and $\vec{\mathbf{F}}_2$. Since $\vec{\mathbf{F}}_1$ points to the left, $\theta_1 = 180°$. (Recall from trig that 180° points along $-x$.) The angle θ_2 points in Quadrant IV (down and to the right). To determine the precise angle, first find the reference angle from the right triangle. The reference angle is the smallest angle with the positive or negative x-axis. The second vector lies $\theta_{2ref} = 60°$ from the horizontal. (**Equilateral** triangles have 60° angles.)

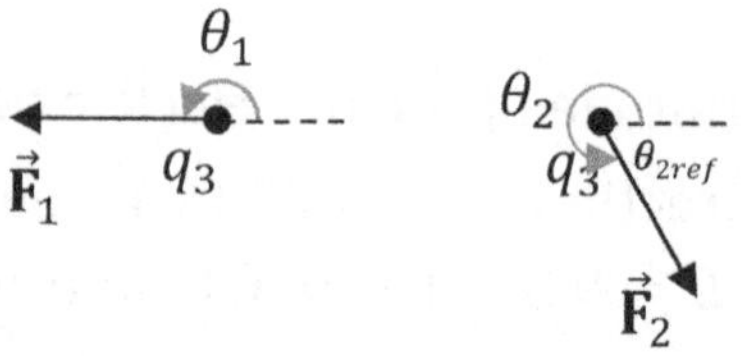

Relate the reference angle to θ_2 using geometry, where θ_2 is counterclockwise from the $+x$-axis. In Quadrant IV, the angle counterclockwise from the $+x$-axis equals 360° minus the reference angle.

$$\theta_2 = 360° - \theta_{2ref} = 360° - 60° = 300°$$

Find the magnitude of the electric force exerted on q_3 by each of the other charges. First convert the charges to SI units: $q_1 = -3.0 \times 10^{-6}$ C, $q_2 = 3.0 \times 10^{-6}$ C, and $q_3 = 4.0 \times 10^{-6}$ C. Use the values $R_1 = 2.0$ m and $R_2 = 2.0$ m, which we determined previously. Note that q_3 appears in both formulas below because we are finding the net force exerted on q_3.

$$F_1 = \frac{k|q_1||q_3|}{R_1^2} = \frac{(9 \times 10^9)|-3 \times 10^{-6}||4 \times 10^{-6}|}{(2)^2} = 27 \times 10^{-3} \text{ N} = 0.027 \text{ N}$$

$$F_2 = \frac{k|q_2||q_3|}{R_2^2} = \frac{(9 \times 10^9)|3 \times 10^{-6}||4 \times 10^{-6}|}{(2)^2} = 27 \times 10^{-3} \text{ N} = 0.027 \text{ N}$$

Note that $10^9 10^{-6} 10^{-6} = 10^{9-6-6} = 10^{-3}$ according to the rule $x^m x^n = x^{m+n}$. Also note that the minus signs vanished with the absolute values (because the magnitude of a force can't be negative). Following is a summary of what we know thus far.

- $q_1 = -3.0 \times 10^{-6}$ C, $R_1 = 2.0$ m, $\theta_1 = 180°$, and $F_1 = 0.027$ N.
- $q_2 = 3.0 \times 10^{-6}$ C, $R_2 = 2.0$ m, $\theta_2 = 300°$, and $F_2 = 0.027$ N.
- $q_3 = 4.0 \times 10^{-6}$ C.

We are now prepared to add the force vectors. We will use the magnitudes and directions ($F_1 = 0.027$ N, $F_2 = 0.027$ N, $\theta_1 = 180°$, and $\theta_2 = 300°$) of the two force vectors to determine the magnitude and direction (F and θ_F) of the resultant vector. The first step of vector addition is to find the components of the known vectors.

$$F_{1x} = F_1 \cos\theta_1 = (0.027)\cos(180°) = (0.027)(-1) = -0.027 \text{ N} = -27 \text{ mN}$$

$$F_{1y} = F_1 \sin\theta_1 = (0.027)\sin(180°) = (0.027)(0) = 0$$

$$F_{2x} = F_2 \cos\theta_2 = (0.027)\cos(300°) = (0.027)\left(\frac{1}{2}\right) = 0.0135 \text{ N} = \frac{27}{2} \text{ mN}$$

$$F_{2y} = F_2 \sin\theta_2 = (0.027)\sin(300°) = (0.027)\left(-\frac{\sqrt{3}}{2}\right) = -\frac{27}{2}\sqrt{3} \text{ mN}$$

We chose to express our intermediate answers in terms of milliNewtons (mN). Recall that the prefix milli (m) represents 10^{-3}, such that 0.027 N = 27 mN. It's a good idea to check the signs based on the previous diagram. For x, right is positive and left is negative. For y, up is positive and down is negative. (This is because we draw a coordinate system with $+x$ to the right and $+y$ up.)

- $F_{1x} = -27$ mN. It's negative because $\vec{\mathbf{F}}_1$ points to the left, not right.
- $F_{1y} = 0$. It's zero because $\vec{\mathbf{F}}_1$ is horizontal (it doesn't point up or down).
- $F_{2x} = \frac{27}{2}$ mN. It's positive because $\vec{\mathbf{F}}_2$ points to the right, not left.
- $F_{2y} = -\frac{27}{2}\sqrt{3}$ mN. It's negative because $\vec{\mathbf{F}}_2$ points downward, not upward.

The second step of vector addition is to add the respective components together.

$$F_x = F_{1x} + F_{2x} = -0.027 + 0.0135 = -0.0135 \text{ N} = -\frac{27}{2} \text{ mN}$$

$$F_y = F_{1y} + F_{2y} = 0 + \left(-0.0135\sqrt{3}\right) = -0.0135\sqrt{3} \text{ N} = -\frac{27}{2}\sqrt{3} \text{ mN}$$

The final step of vector addition is to apply the Pythagorean theorem and inverse tangent to determine the magnitude and direction of the resultant vector.

$$F = \sqrt{F_x^2 + F_y^2} = \sqrt{(-0.0135)^2 + \left(-0.0135\sqrt{3}\right)^2} = (0.0135)\sqrt{(-1)^2 + \left(-\sqrt{3}\right)^2}$$

We factored out 0.0135 to make the arithmetic simpler. The minus signs disappear since they are squared: $(-1)^2 = +1$ and $\left(-\sqrt{3}\right)^2 = +3$.

$$F = (0.0135)\sqrt{1+3} = (0.0135)\sqrt{4} = 0.0135(2) = 0.027 \text{ N} = 27 \text{ mN}$$

The magnitude of the net electric force is $F = 0.027$ N $= 27$ mN, where the prefix milli (m) stands for one-thousandth (10^{-3}). It can also be expressed as $F = 2.7 \times 10^{-2}$ N. Apply the following inverse tangent in order to determine the direction of the net electric force exerted on q_3.

$$\theta_F = \tan^{-1}\left(\frac{F_y}{F_x}\right) = \tan^{-1}\left(\frac{-0.0135\sqrt{3}}{-0.0135}\right) = \tan^{-1}\left(\sqrt{3}\right)$$

Note that -0.0135 cancels out. The reference angle for the answer is 60° since $\tan 60° = \sqrt{3}$. However, this isn't the answer because the answer doesn't lie in Quadrant I. Since $F_x < 0$ and $F_y < 0$ (find these values on the previous page), the answer lies in Quadrant III. Apply trig to determine θ_F from the reference angle: In Quadrant III, add the reference angle to 180°.

$$\theta_F = 180° + \theta_{ref} = 180° + 60° = 240°$$

The final answer is that the magnitude of the net electric force exerted on q_3 equals $F = 27$ mN and its direction is $\theta_F = 240°$.

Example 11. A monkey places two small spheres a distance of 9.0 m apart, as shown below. The left sphere has a charge of 25 μC, while the right sphere has a charge of 16 μC. Find the point where the net electric field equals zero.

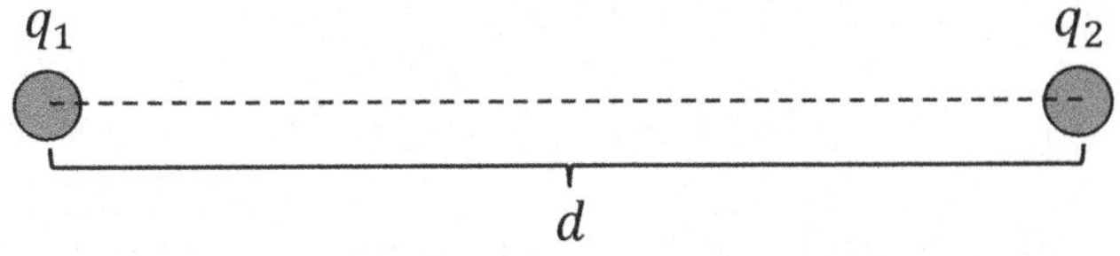

Solution. Consider each of the three regions shown below. Imagine placing a positive "test" charge in each region. Since q_1 and q_2 are both positive, the positive "test" charge would be repelled by both q_1 and q_2. Draw the electric fields away from q_1 and q_2 in each region.

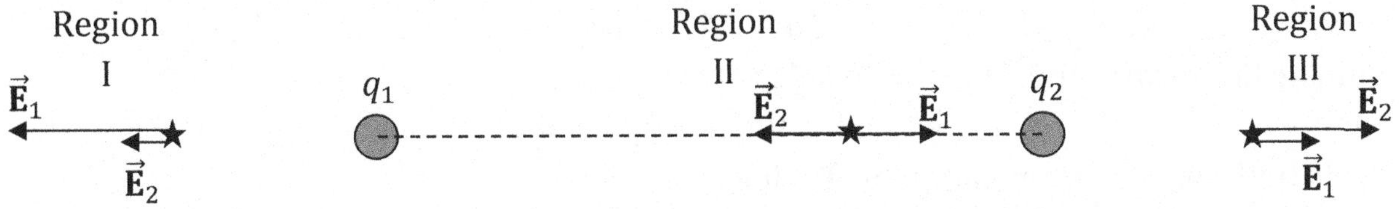

I. Region I is left of q_1. $\vec{\mathbf{E}}_1$ and $\vec{\mathbf{E}}_2$ both point left. They won't cancel here.
II. Region II is between q_1 and q_2. $\vec{\mathbf{E}}_1$ points right, while $\vec{\mathbf{E}}_2$ points left. They can cancel out in Region II.
III. Region III is right of q_2. $\vec{\mathbf{E}}_1$ and $\vec{\mathbf{E}}_2$ both point right. They won't cancel here.

We know that the answer lies in Region II, since that is the only place where the electric fields could cancel out. In order to find out exactly where in Region II the net electric field equals zero, set the magnitudes of the electric fields equal to one another.

$$E_1 = E_2$$

$$\frac{k|q_1|}{R_1^2} = \frac{k|q_2|}{R_2^2}$$

Divide both sides by k (it will cancel out) and **cross-multiply**.

$$|q_1|R_2^2 = |q_2|R_1^2$$

Plug in the values of the charges. Recall that the metric prefix micro (μ) stands for 10^{-6}.

$$|25 \times 10^{-6}|R_2^2 = |16 \times 10^{-6}|R_1^2$$

Divide both sides of the equation by 10^{-6} (it will cancel out).

$$25R_2^2 = 16R_1^2$$

Since we have two unknowns (R_1 and R_2), we need a second equation. Study the diagram below to relate R_1 and R_2 to the distance between the charges, d.

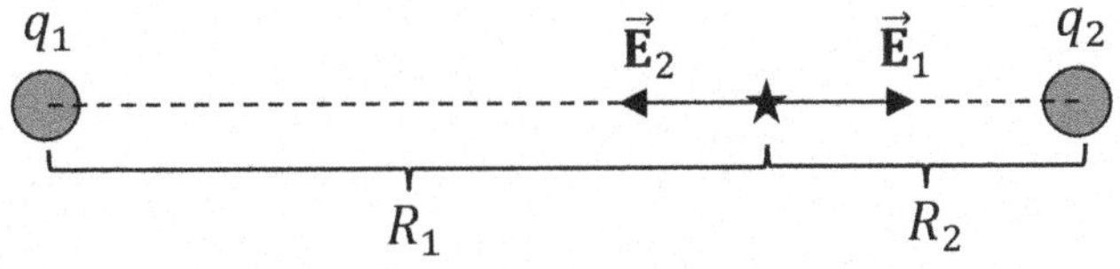

Observe that R_1 and R_2 add up to d (the distance between the charges).

$$R_1 + R_2 = d$$

Isolate R_2 in the previous equation: $R_2 = d - R_1$. Substitute this expression into the equation $25R_2^2 = 16R_1^2$, which we found previously.

$$25(d - R_1)^2 = 16R_1^2$$

Squareroot both sides of the equation and simplify.

$$\sqrt{25(d - R_1)^2} = \sqrt{16R_1^2}$$

Apply the rules $\sqrt{xy} = \sqrt{x}\sqrt{y}$ and $\sqrt{x^2} = x$ to write $\sqrt{25(d - R_1)^2} = \sqrt{25}\sqrt{(d - R_1)^2} = 5(d - R_1)$. Note that $\sqrt{16R_1^2} = 4R_1$.

$$5(d - R_1) = 4R_1$$

Distribute the 5 to both terms.

$$5d - 5R_1 = 4R_1$$

Combine like terms. Add $5R_1$ to both sides of the equation.

$$5d = 5R_1 + 4R_1 = 9R_1$$

Divide both sides of the equation by 9. Plug in $d = 9.0$ m.

$$R_1 = \frac{5d}{9} = \frac{(5)(9)}{9} = 5.0 \text{ m}$$

The net electric field is zero in Region II, a distance of $R_1 = 5.0$ m from the left charge (and therefore a distance of $R_2 = 4.0$ m from the right charge, since $R_1 + R_2 = d = 9.0$ m).

Example 12. A monkey places two small spheres a distance of 4.0 m apart, as shown below. The left sphere has a charge of 2.0 μC, while the right sphere has a charge of −8.0 μC. Find the point where the net electric field equals zero.

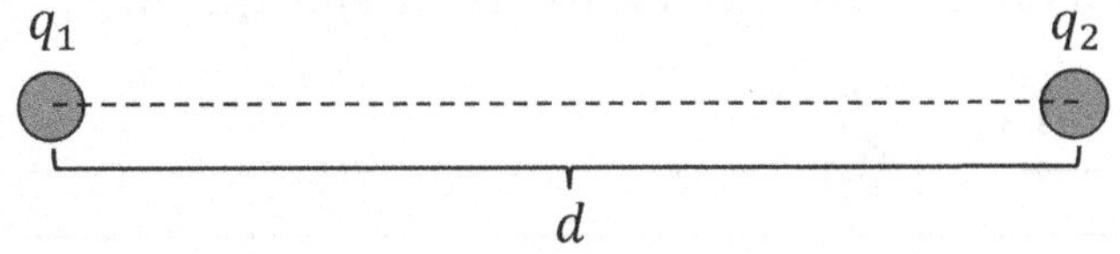

Solution. It's instructive to compare this problem to the previous problem. In this problem, the signs are opposite, which will make a significant difference in the solution.

Consider each of the three regions shown below. Imagine placing a positive "test" charge in each region. Since q_1 is positive, $\vec{\mathbf{E}}_1$ will point away from q_1 in each region because a positive "test" charge would be repelled by q_1. Since q_2 is negative, $\vec{\mathbf{E}}_2$ will point towards q_2 in each region because a positive "test" charge would be attracted to q_2. Draw the electric fields away from q_1 and towards q_2 in each region.

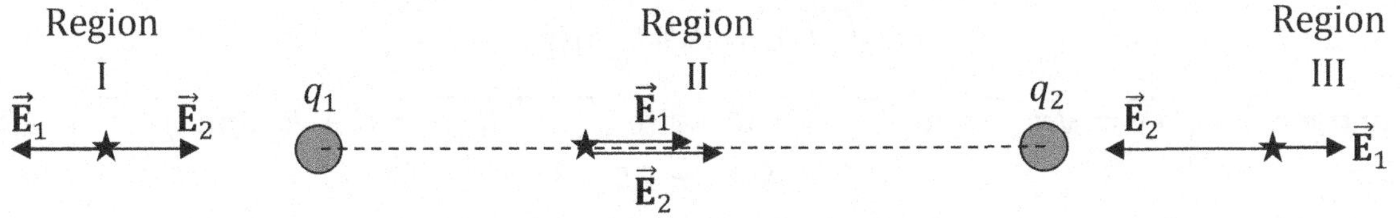

I. Region I is left of q_1. $\vec{\mathbf{E}}_1$ points left and $\vec{\mathbf{E}}_2$ points right. They can cancel here.
II. Region II is between q_1 and q_2. $\vec{\mathbf{E}}_1$ and $\vec{\mathbf{E}}_2$ both point right. They won't cancel out in Region II.
III. Region III is right of q_2. $\vec{\mathbf{E}}_1$ points right and $\vec{\mathbf{E}}_2$ points left. Although $\vec{\mathbf{E}}_1$ and $\vec{\mathbf{E}}_2$ point in opposite directions in Region III, they won't cancel in this region because it is closer to the stronger charge, such that $|E_2| > |E_1|$.

We know that the answer lies in Region I, since that is the only place where the electric fields could cancel out. In order to find out exactly where in Region I the net electric field equals zero, set the magnitudes of the electric fields equal to one another.

$$E_1 = E_2$$

$$\frac{k|q_1|}{R_1^2} = \frac{k|q_2|}{R_2^2}$$

Divide both sides by k (it will cancel out) and **cross-multiply**.

$$|q_1|R_2^2 = |q_2|R_1^2$$

Plug in the values of the charges. Recall that the metric prefix micro (μ) stands for 10^{-6}. Note the **absolute values.**

$$|2 \times 10^{-6}|R_2^2 = |-8 \times 10^{-6}|R_1^2$$

Divide both sides of the equation by 10^{-6} (it will cancel out).

$$2R_2^2 = 8R_1^2$$

Divide both sides of the equation by 2.

$$R_2^2 = 4R_1^2$$

Since we have two unknowns (R_1 and R_2), we need a second equation. Study the diagram below to relate R_1 and R_2 to the distance between the charges, d.

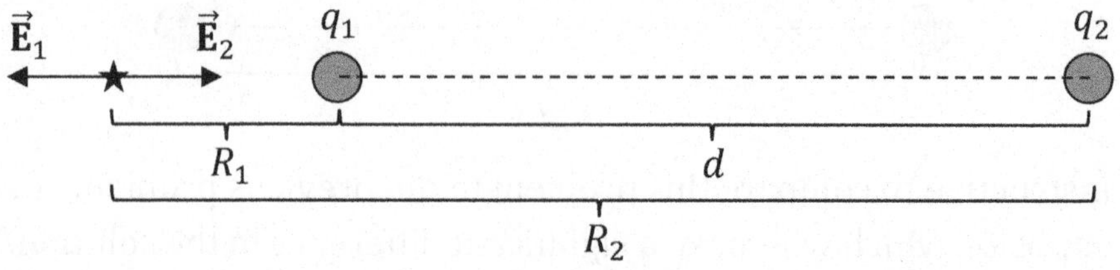

Observe that R_1 and d add up to R_2 in this example (which differs from the previous example).

$$R_1 + d = R_2$$

Substitute this expression into the equation $R_2^2 = 4R_1^2$, which we found previously.

$$(d + R_1)^2 = 4R_1^2$$

Squareroot both sides of the equation and simplify.

$$\sqrt{(d + R_1)^2} = \sqrt{4R_1^2}$$

Apply the rules $\sqrt{xy} = \sqrt{x}\sqrt{y}$ and $\sqrt{x^2} = x$ to write $\sqrt{(d + R_1)^2} = d + R_1$ and $\sqrt{4R_1^2} = 2R_1$.

$$d + R_1 = 2R_1$$

Combine like terms. Subtract R_1 from both sides of the equation.

$$d = 2R_1 - R_1$$

Note that $2R_1 - R_1 = R_1$.

$$R_1 = d = 4.0 \text{ m}$$

The net electric field is zero in Region I, a distance of $R_1 = 4.0$ m to the **left** of q_1 (and therefore a distance of $R_2 = R_1 + d = 4 + 4 = 8.0$ m from the right charge).

4 ELECTRIC FIELD MAPPING

Relation between Electric Force and Electric Field	Electric Field Due to a Pointlike Charge
$\vec{\mathbf{F}}_e = q\vec{\mathbf{E}}$	$E = \dfrac{k\|q\|}{R^2}$

Net Electric Field (Principle of Superposition)	Approximate Magnitude of Electric Field in a Map
$\vec{\mathbf{E}}_{net} = \vec{\mathbf{E}}_1 + \vec{\mathbf{E}}_2 + \cdots$	$E \approx \left\|\dfrac{\Delta V}{\Delta R}\right\|$

$$\vec{\mathbf{E}}_{net} = \vec{\mathbf{E}}_1 + \vec{\mathbf{E}}_2$$

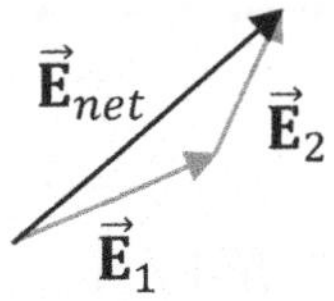

$$E \approx \left|\frac{\Delta V}{\Delta R}\right|$$

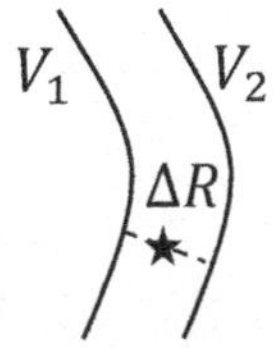

Symbol	Name	SI Units
E	magnitude of electric field	N/C or V/m
q	charge	C
k	Coulomb's constant	$\frac{\text{N}\cdot\text{m}^2}{\text{C}^2}$ or $\frac{\text{kg}\cdot\text{m}^3}{\text{C}^2\cdot\text{s}^2}$
R	distance from the charge	m
F_e	magnitude of electric force	N
V	electric potential	V
ΔV	potential difference	V
ΔR	distance between two equipotentials	m

Example 13. Sketch the electric field at points A, B, and C for the isolated positive sphere shown below.

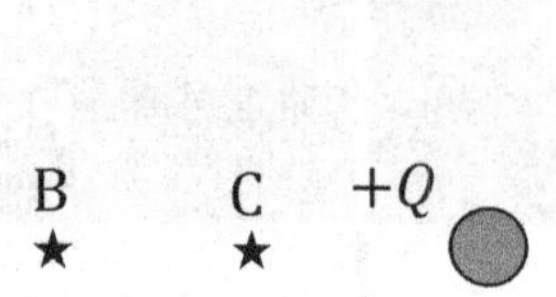

Solution. If a positive "test" charge were placed at points A, B, or C, it would be repelled by the positive sphere. Draw the electric field directly away from the positive sphere. Draw a shorter arrow at point B, since point B is further away from the positive sphere.

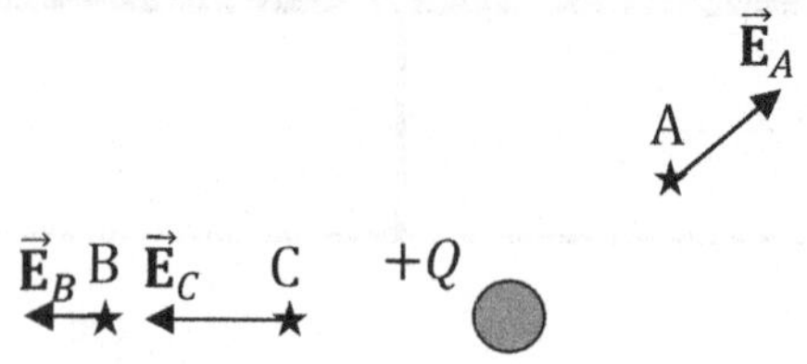

Example 14. Sketch the electric field at points D, E, and F for the isolated negative sphere shown below.

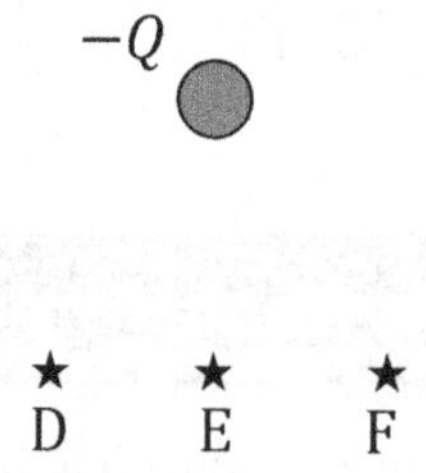

Solution. If a positive "test" charge were placed at points D, E, or F, it would be attracted to the negative sphere. Draw the electric field towards the negative sphere. Draw a longer arrow at point E, since point E is closer to the negative sphere.

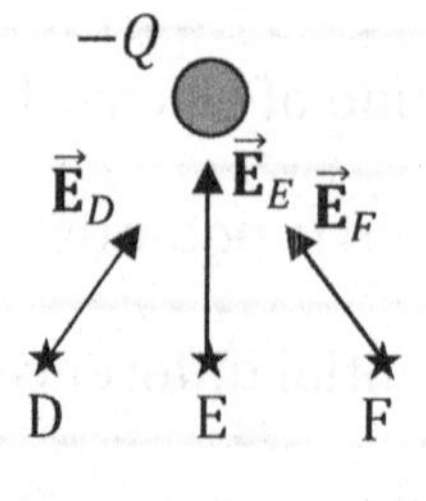

Example 15. Sketch the electric field at points G, H, I, J, K, and M for the pair of equal but oppositely charged spheres shown below. (We skipped L in order to avoid possible confusion in our solution, since we use $\vec{\mathbf{E}}_L$ for the electric field created by the left charge.)

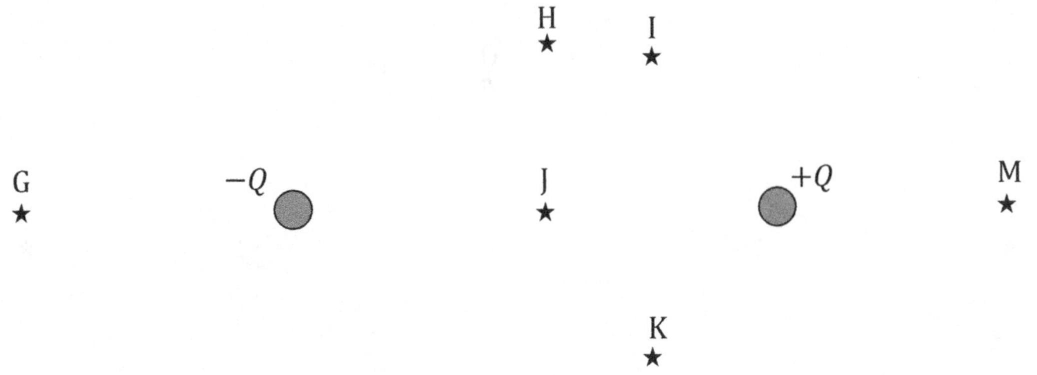

Solution. If a positive "test" charge were placed at any of these points, it would be repelled by the positive sphere and attracted to the negative sphere. First draw two separate electric fields (one for each sphere) and then draw the resultant of these two vectors for the net electric field. Move one of the arrows (we moved $\vec{\mathbf{E}}_R$) to join the two vectors ($\vec{\mathbf{E}}_L$ and $\vec{\mathbf{E}}_R$) tip-to-tail. The resultant vector, $\vec{\mathbf{E}}_{net}$, which is the net electric field at the specified point, begins at the tail of $\vec{\mathbf{E}}_L$ and ends at the tip of $\vec{\mathbf{E}}_R$. The length of $\vec{\mathbf{E}}_L$ or $\vec{\mathbf{E}}_R$ depends on how far the point is from the respective sphere.

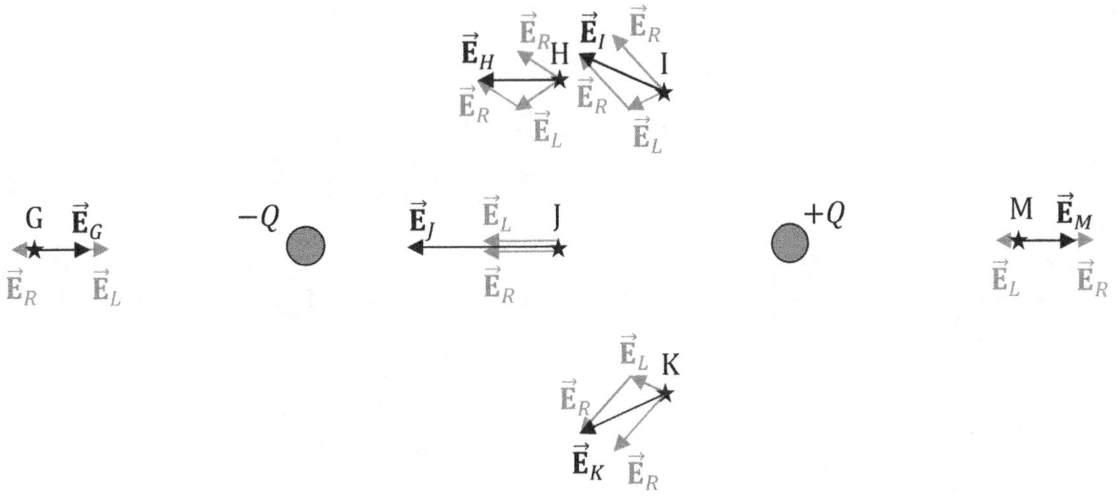

Example 16. Sketch the electric field at points N, O, P, S, T, and U for the pair of negative spheres shown below. (We skipped Q in order to avoid possible confusion with the charge. We also skipped R in order to avoid possible confusion in our solution, since we use $\vec{\mathbf{E}}_R$ for the electric field created by the right charge.)

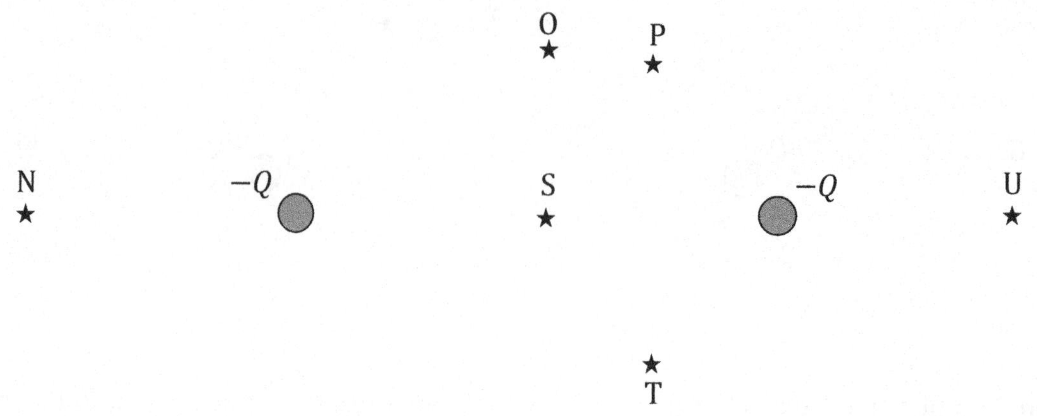

Solution. If a positive "test" charge were placed at any of these points, it would be attracted to each negative sphere. First draw two separate electric fields (one for each sphere) and then draw the resultant of these two vectors for the net electric field. Move one of the arrows (we moved $\vec{\mathbf{E}}_R$) to join the two vectors ($\vec{\mathbf{E}}_L$ and $\vec{\mathbf{E}}_R$) tip-to-tail. The resultant vector, $\vec{\mathbf{E}}_{net}$, which is the net electric field at the specified point, begins at the tail of $\vec{\mathbf{E}}_L$ and ends at the tip of $\vec{\mathbf{E}}_R$. The length of $\vec{\mathbf{E}}_L$ or $\vec{\mathbf{E}}_R$ depends on how far the point is from the respective sphere.

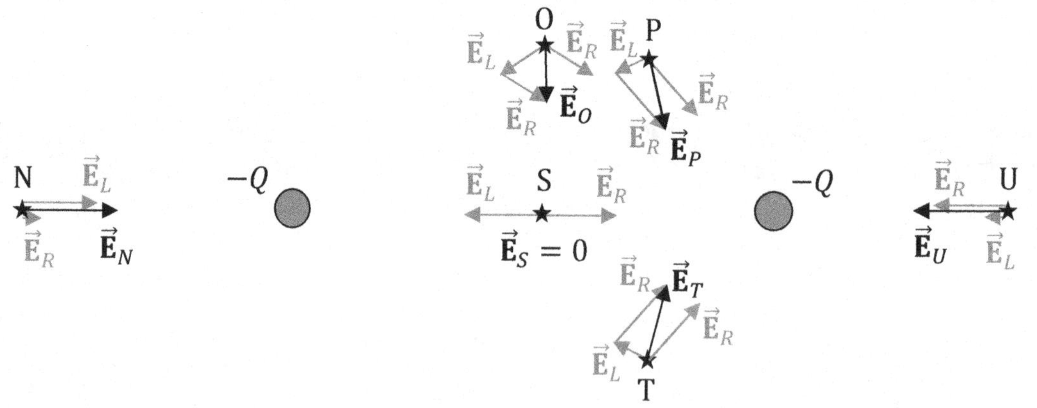

Example 17. Sketch the electric field at point V for the three charged spheres (two are negative, while the bottom left is positive) shown below, which form an equilateral triangle.

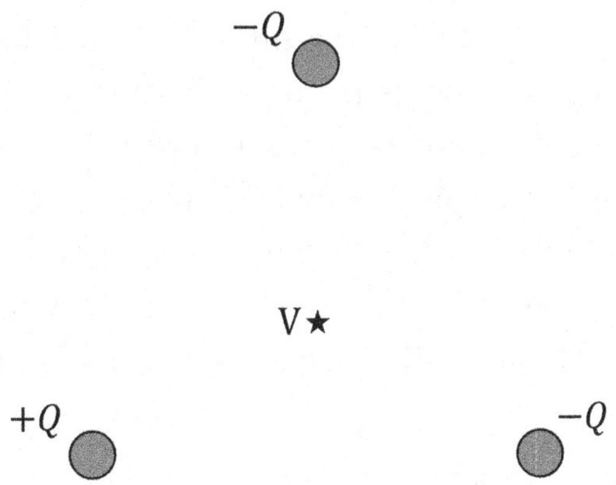

Solution. If a positive "test" charge were placed at point V, it would be repelled by the positive sphere and attracted to each negative sphere. First draw three separate electric fields (one for each sphere) and then draw the resultant of these three vectors for the net electric field. Move two of the arrows to join the three vectors tip-to-tail. The resultant vector, $\vec{\mathbf{E}}_{net}$, which is the net electric field at the specified point, begins at the tail of the first vector and ends at the tip of the last vector. Each of the three separate electric fields has the same length since point V is equidistant from the three charged spheres.

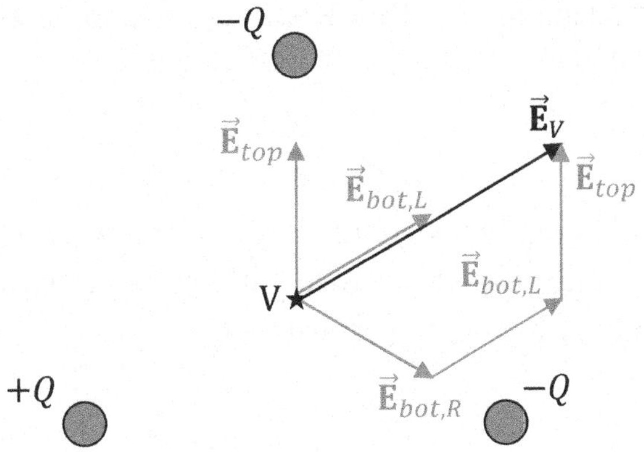

Isolated Positive Sphere Example: A Prelude to Example 18. Sketch an electric field map for the single isolated positive charge shown below.

Solution. No matter where you might place a positive "test" charge in the above diagram, the "test" charge would be repelled by the positive charge $+Q$. Therefore, the lines of force radiate outward away from the positive charge (since it is completely isolated, meaning that there aren't any other charges around). The equipotential surfaces must be perpendicular to the lines of force: In this case, the equipotential surfaces are concentric spheres centered around the positive charge. Although they are drawn as circles below, they are really spheres that extend in 3D space in front and behind the plane of the paper.

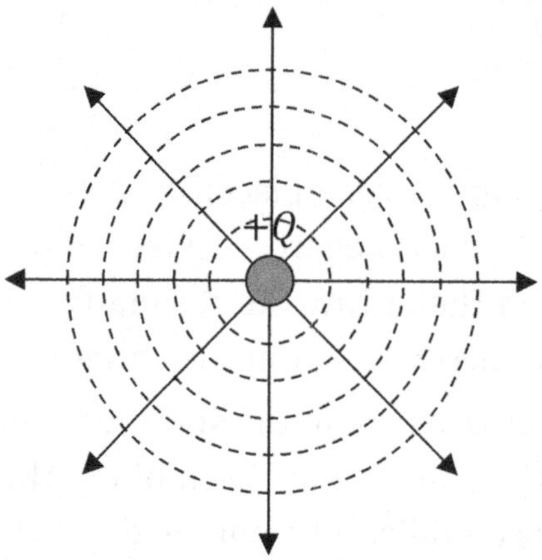

Isolated Negative Sphere Example: Another Prelude to Example 18. Sketch an electric field map for the single isolated negative charge shown below.

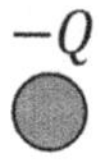

Solution. This electric field map is nearly identical to the previous electric field map: The only difference is that the lines radiate inward rather than outward, since a positive "test" charge would be attracted to the negative charge $-Q$.

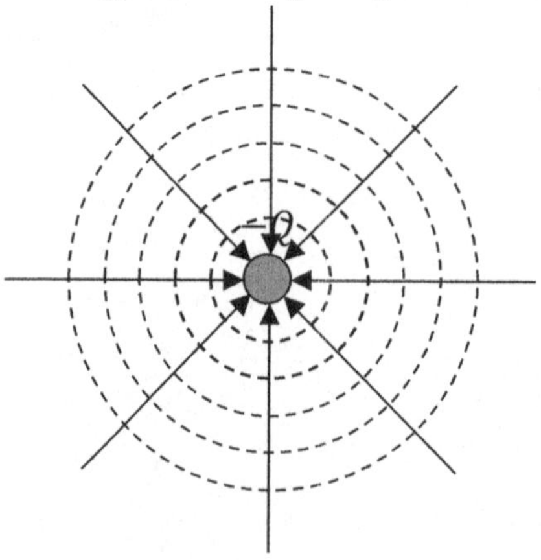

Electric Dipole Example: Another Prelude to Example 18. Sketch an electric field map for the equal but opposite charges (called an electric dipole) shown below.

Solution. Make this map one step at a time:

- First, sample the net electric field in a variety of locations using the superposition strategy that we applied in Examples 13-17. If you review Example 15, you will see that it featured the electric dipole (but note that this diagram is **reversed**):
  - The net electric field points to the **right** in the **center** of the diagram (where a positive "test" charge would be repelled by $+Q$ and attracted to $-Q$).
  - The net electric field points to the **right** anywhere along the **vertical line** that **bisects** the diagram, as the vertical components will cancel out there.
  - The net electric field points to the left along the horizontal line to the left of $+Q$ or to the right of $-Q$, but points to the right between the two charges.
  - The net electric field is somewhat **radial** (like the spokes of a bicycle wheel) **near either charge**, where the closer charge has the dominant effect.
- Beginning with the above features, draw smooth curves that leave $+Q$ and head into $-Q$ (except where the lines go beyond the scope of the diagram). The lines aren't perfectly radial near either charge, but curve so as to leave $+Q$ and reach $-Q$.
- Check several points in different regions: The **tangent** line at any point on a line of force should match the direction of the net electric field from **superposition**.
- Draw smooth curves for the equipotential surfaces. Wherever an equipotential intersects a line of force, the two curves must be perpendicular to one another.

In the diagram below, we used solid (—) lines for the lines of force (electric field lines) with arrows (→) indicating direction, and dashed lines (----) for the equipotentials.

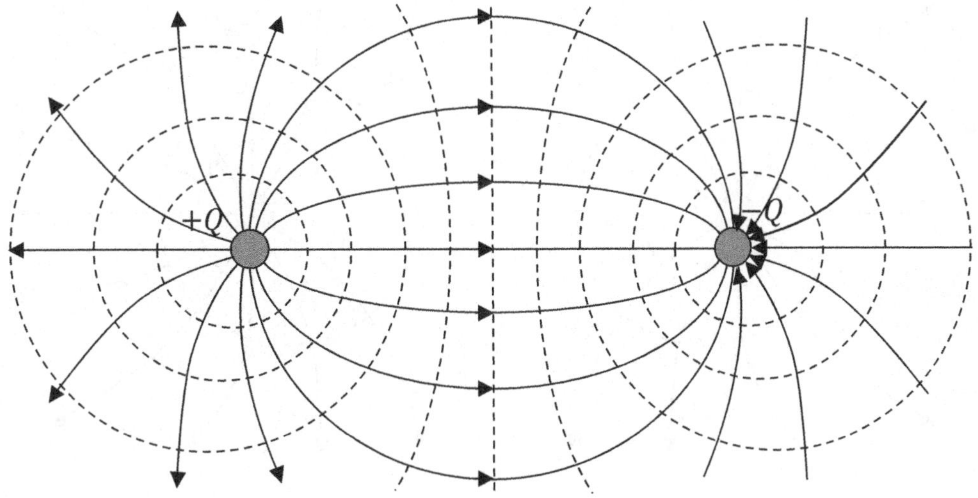

Example 18. Sketch an electric field map for the two positive charges shown below.

Solution. Make this map one step at a time:

- First, sample the net electric field in a variety of locations using the superposition strategy that we applied in the previous problems and examples.
  - The net electric field is **zero** in the **center** of the diagram (where a positive "test" charge would be repelled equally by both charged spheres). This point is called a **saddle point** in this diagram.
  - Along the horizontal line, the net electric field points to the left to the left of the left sphere and points to the right to the right of the right sphere.
  - The net electric field is somewhat **radial** (like the spokes of a bicycle wheel) **near either charge**, where the closer charge has the dominant effect.
- Beginning with the above features, draw smooth curves that leave each positive sphere. The lines aren't perfectly radial near either charge, but curve due to the influence of the other sphere. It's instructive to compare how the field lines curve in the diagram below versus how they curve for the electric dipole (see the previous page): They curve **backwards** in comparison.
- Check several points in different regions: The **tangent** line at any point on a line of force should match the direction of the net electric field from **superposition**. Compare the tangent lines at the points in the diagram from Example 16, but note that these arrows are **backwards** in comparison (since these are positive charges, whereas Example 16 has negative charges).
- Draw smooth curves for the equipotential surfaces. Wherever an equipotential intersects a line of force, the two curves must be perpendicular to one another.

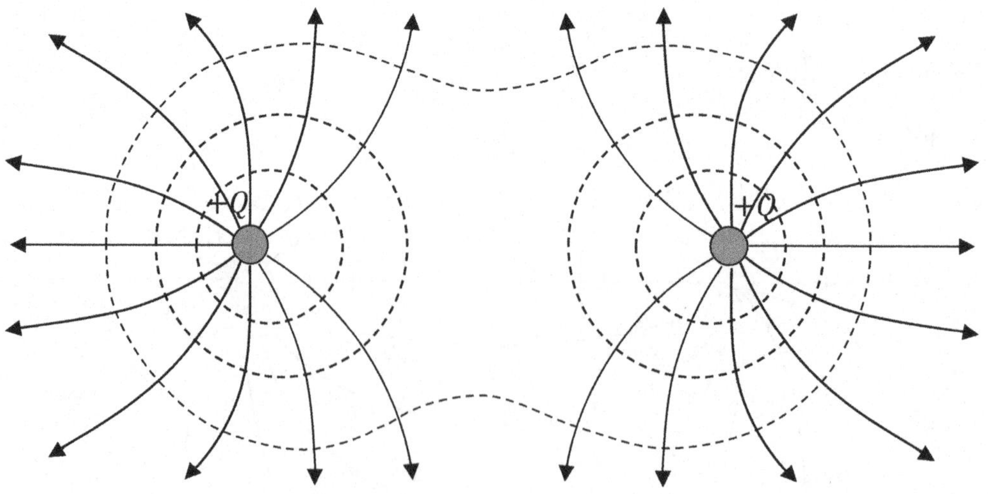

Example 19. Consider the map of equipotentials drawn below. Note that the diagram below shows equipotentials, **not** lines of force.

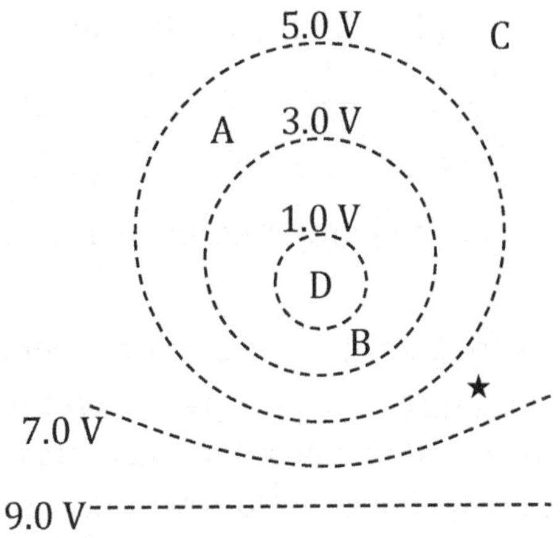

(A) Sketch the lines of force for the diagram above.

Solution. Draw smooth curves that are **perpendicular** to the equipotentials wherever the lines of force intersect the equipotentials. Include arrows showing that the lines of force travel from higher electric potential (in Volts) to lower electric potential.

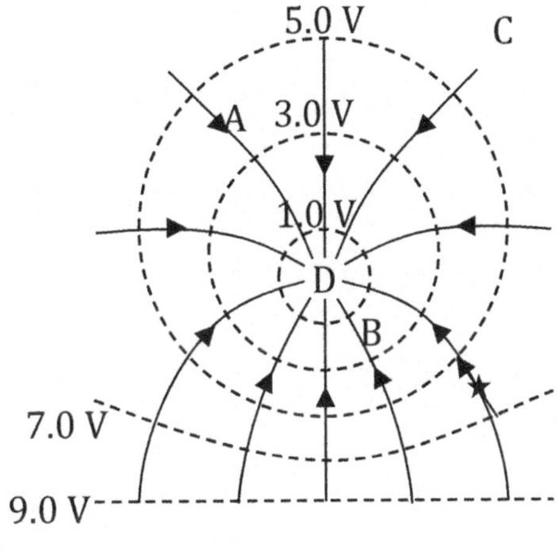

(B) At which point(s) is there a pointlike charge? Is the charge positive or negative?

Solution. There is a negative charge at point D where the lines of force **converge**.

(C) Rank the electric field strength at points A, B, and C.

Solution. Of these three points, the lines of force are more dense at point B and less dense at point C: $E_B > E_A > E_C$. (Note that the electric field is strongest near point D, where the field lines are most dense. However, point D isn't mentioned in the question.)

(D) Estimate the magnitude of the electric field at the star (★) in the previous diagram.
Solution. Draw an arrow through the star (★) that is on average roughly perpendicular to the neighboring equipotentials. The potential difference is $\Delta V = 7 - 5 = 2.0$ V. Measure the length of the line with a ruler: $\Delta R = 0.8$ cm $= 0.008$ m. Use the following formula.

$$E \approx \left|\frac{\Delta V}{\Delta R}\right| = \frac{2}{0.008} = 250\,\frac{\text{V}}{\text{m}} \text{ or } 250\,\frac{\text{N}}{\text{C}}$$

The magnitude of the electric field at the star (★) in the previous diagram is approximately $E \approx 250\frac{\text{V}}{\text{m}}$. Note that a Volt per meter (V/m) is equivalent to a Newton per Coulomb (N/C).

(E) If a negative "test" charge were placed at the star (★), which way would it be pushed?
Solution. It would be pushed **opposite** to the arrow drawn through the star in the previous diagram (because the arrow shows how a positive "test" charge would be pushed, and this question asks about a **negative** test charge). Note that this arrow isn't quite straight away from point D because there happens to be a positive horizontal rod at the bottom which has some influence. (You don't need to know anything about the positive horizontal rod in order to fully solve the problem. However, if you're wondering what could make an electric field map like the one shown in this diagram, it's a combination of a positive line charge and a positive point charge. You could make such an electric field map in a lab using conductive ink and carbon-impregnated paper, for example.)

5 ELECTROSTATIC EQUILIBRIUM

Electrostatic Equilibrium	
$\sum F_{1x} = 0$, $\sum F_{1y} = 0$, $\sum F_{2x} = 0$, $\sum F_{2y} = 0$	
Force in an Electric Field	**Force between Two Charges**
$F_e = \lvert q \rvert E$	$F_e = \frac{k\lvert q_1 \rvert \lvert q_2 \rvert}{R^2}$
Weight	**Static Friction Force**
$W = mg$	$f_s \leq \mu_s N$

Symbol	Name	SI Units
E	magnitude of electric field	N/C or V/m
F_e	magnitude of electric force	N
q	charge	C
R	distance between two charges	m
k	Coulomb's constant	$\frac{\text{N}\cdot\text{m}^2}{\text{C}^2}$ or $\frac{\text{kg}\cdot\text{m}^3}{\text{C}^2\cdot\text{s}^2}$
m	mass	kg
mg	weight	N
N	normal force	N
f	friction force	N
μ	coefficient of friction	unitless
T	tension	N

Example 20. A poor monkey spends his last nickel on two magic bananas. When he dangles the two $9\sqrt{3}$-kg (which is heavy because the magic bananas are made out of metal) bananas from cords, they "magically" spread apart as shown below. Actually, it's because they have equal electric charge.

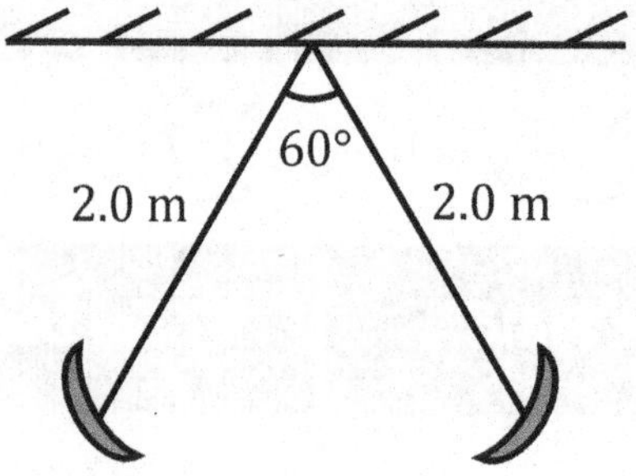

(A) What is the tension in each cord?

Solution. Draw and label a free-body diagram (FBD) for each banana.

- Weight (mg) pulls straight down.
- Tension (T) pulls along the cord.
- The two bananas repel one another with an electric force (F_e) via Coulomb's law. The electric force is repulsive because the (metallic) bananas have the same charge.

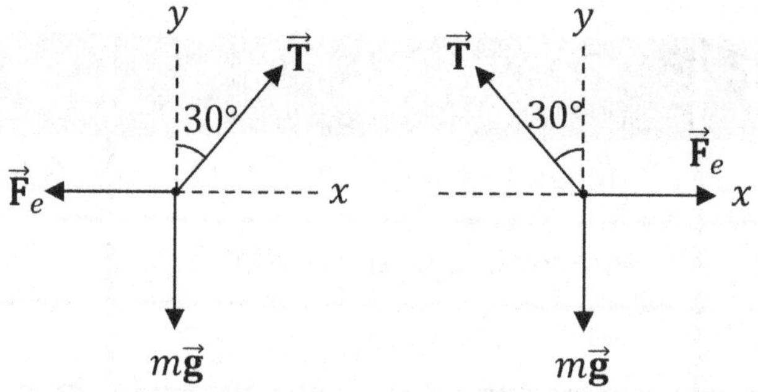

Note that the angle with the vertical is 30° (split the 60° angle at the top of the triangle in half to find this.) Apply Newton's second law. In **electrostatic equilibrium**, there is no acceleration (since the velocity, which equals zero, isn't changing): $a_x = 0$ and $a_y = 0$.

- Since tension doesn't lie on an axis, T appears in both the x- and y-sums with trig. In the FBD, since y is adjacent to 30°, cosine appears in the y-sum.
- Since the electric force is horizontal, F_e appears only the x-sums with no trig.
- Since weight is vertical, mg appears only in the y-sums with no trig.

$$\sum F_{1x} = ma_x \quad , \quad \sum F_{1y} = ma_y \quad , \quad \sum F_{2x} = ma_x \quad , \quad \sum F_{2y} = ma_y$$

$$T \sin 30° - F_e = 0 \quad , \quad T \cos 30° - mg = 0 \quad , \quad F_e - T \sin 30° = 0 \quad , \quad T \cos 30° - mg = 0$$

We can solve for tension in the equation from the y-sums.

$$T \cos 30° - mg = 0$$

Add weight (mg) to both sides of the equation.

$$T \cos 30° = mg$$

Divide both sides of the equation by cos 30°.

$$T = \frac{mg}{\cos 30°}$$

In this book, we will round $g = 9.81 \text{ m/s}^2$ to $g \approx 10 \text{ m/s}^2$ in order to show you how to obtain an approximate answer without using a calculator. (Feel free to use a calculator if you wish. It's a valuable skill to be able to estimate an answer without a calculator, which is the reason we will round 9.81 to 10.)

$$T = \frac{(9\sqrt{3})(9.81)}{\cos 30°} \approx \frac{(9\sqrt{3})(10)}{\frac{\sqrt{3}}{2}}$$

To divide by a fraction, multiply by its **reciprocal**. Note that the reciprocal of $\frac{\sqrt{3}}{2}$ is $\frac{2}{\sqrt{3}}$.

$$T \approx (9\sqrt{3})(10)\frac{2}{\sqrt{3}} = 180 \text{ N}$$

Note that the $\sqrt{3}$'s cancel since $\frac{\sqrt{3}}{\sqrt{3}} = 1$. The tension is $T \approx 180$ N. (If you don't round gravity, the answer for tension works out to $T = 177$ N.)

(B) What is the charge of each (metal) banana?

Solution. First, solve for the electric force in the equation from the x-sums from part (A).

$$T \sin 30° - F_e = 0$$

Add F_e to both sides of the equation.

$$T \sin 30° = F_e$$

Plug in the tension that we found in part (A).

$$F_e = T \sin 30° \approx 180 \sin 30° = 180\left(\frac{1}{2}\right) = 90 \text{ N}$$

The electric force is $F_e \approx 90$ N. Now apply Coulomb's law (Chapter 1).

$$F_e = k\frac{|q_1||q_2|}{R^2}$$

Since the charges are equal, $q_1 = q_2$, we will call them both q. If you set $q_1 = q$ and $q_2 = q$ in Coulomb's law (the previous equation), you get:

$$F_e = k\frac{q^2}{R^2}$$

Note that $qq = q^2$. Multiply both sides of the equation by R^2.

$$F_e R^2 = kq^2$$

Divide both sides of the equation by Coulomb's constant.

$$\frac{F_e R^2}{k} = q^2$$

Squareroot both sides of the equation. Note that $\sqrt{R^2} = R$. Recall that $k = 9.0 \times 10^9 \frac{\text{N·m}^2}{\text{C}^2}$.

$$q = \sqrt{\frac{F_e R^2}{k}} = R\sqrt{\frac{F_e}{k}}$$

Note that the distance between the (metal) bananas is $R = 2.0$ m because the bananas form an **equilateral** triangle with the vertex. Study the diagram on page 36.

$$q \approx 2\sqrt{\frac{90}{9 \times 10^9}} = 2\sqrt{\frac{10}{10^9}} = 2\sqrt{10^{-8}} = 2.0 \times 10^{-4} \text{ C}$$

Note that $\frac{90}{9} = 10$ and that $\frac{10}{10^9} = \frac{10^1}{10^9} = 10^{1-9} = 10^{-8}$ according to the rule $\frac{x^m}{x^n} = x^{m-n}$. Also note that $\sqrt{10^{-8}} = 10^{-4}$ since $(10^{-4})^2 = 10^{-4}10^{-4} = 10^{-4-4} = 10^{-8}$. The answer is $q \approx 2.0 \times 10^{-4}$ C, which can also be expressed as $q \approx 200$ μC using the metric prefix micro ($\mu = 10^{-6}$). Either both charges are positive ($q_1 = q_2 = +200$ μC) or both charges are negative ($q_1 = q_2 = -200$ μC). (Without more information, there is no way to determine whether they are both positive or both negative.)

Example 21. The 60-g banana-shaped earring illustrated below is suspended in midair in electrostatic equilibrium. The earring is connected to the floor by a thread that makes a 60° angle with the vertical. The earring has a charge of +300 μC and there is a uniform electric field directed 60° above the horizontal.

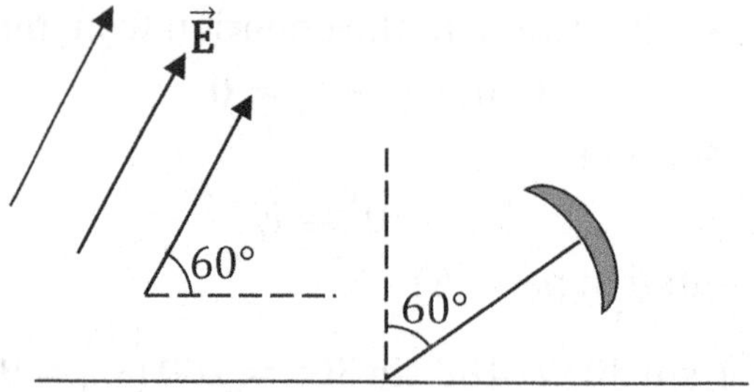

(A) Determine the magnitude of the electric field.
Solution. Draw and label a FBD for the earring.

- Weight (mg) pulls straight down.
- Tension (T) pulls along the thread.
- The electric field ($\vec{\mathbf{E}}$) exerts an electric force ($\vec{\mathbf{F}}_e = q\vec{\mathbf{E}}$) on the charge, which is parallel to the electric field since the charge is positive.

Note that the electric field is steeper than the tension. Note also the difference in how the two 60° angles are labeled in the diagrams: With electric field the 60° angle is with the horizontal, whereas with tension the 60° angle is with the vertical.

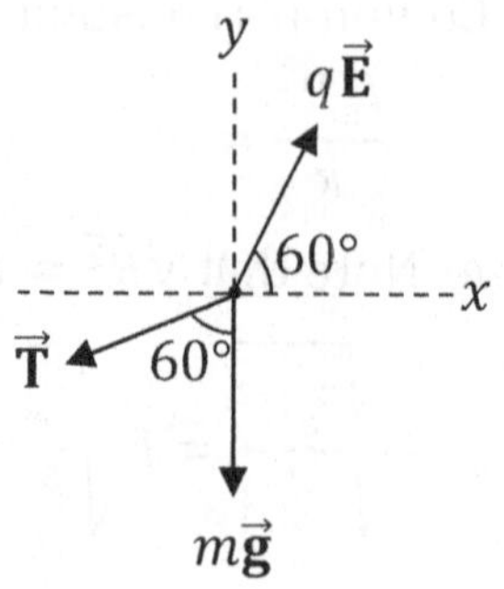

Apply Newton's second law. In **electrostatic equilibrium**, $a_x = 0$ and $a_y = 0$.

- Since tension doesn't lie on an axis, T appears in both the x- and y-sums with trig. In the FBD, since y is adjacent to $60°$ (for **tension**), cosine appears in the y-sum.
- Since the electric force doesn't lie on an axis, $|q|E$ appears in both the x- and y-sums with trig. Recall that $F_e = |q|E$ for a charge in an external electric field. In the FBD, since x is adjacent to $60°$ (for $q\vec{\mathbf{E}}$), cosine appears in the x-sum.
- Since weight is vertical, mg appears only in the y-sum with no trig.

$$\sum F_x = ma_x \quad , \quad \sum F_y = ma_y$$

$$|q|E\cos 60° - T\sin 60° = 0 \quad , \quad |q|E\sin 60° - T\cos 60° - mg = 0$$

In the left equation, add $T\sin 60°$ to both sides of the equation. In the right equation, add mg to both sides of the equation.

$$|q|E\cos 60° = T\sin 60° \quad , \quad |q|E\sin 60° - T\cos 60° = mg$$

Note that $\cos 60° = \frac{1}{2}$ and $\sin 60° = \frac{\sqrt{3}}{2}$. Substitute these into the previous equations.

$$\frac{|q|E}{2} = \frac{T\sqrt{3}}{2} \quad , \quad \frac{|q|E\sqrt{3}}{2} - \frac{T}{2} = mg$$

Each equation has two unknowns (E and T are both unknown). Make a substitution in order to solve for the unknowns. Isolate T in the first equation. Multiply both sides of the equation by 2 and divide both sides by $\sqrt{3}$.

$$T = \frac{|q|E}{\sqrt{3}}$$

Substitute $T = \frac{|q|E}{\sqrt{3}}$ into the equation $\frac{|q|E\sqrt{3}}{2} - \frac{T}{2} = mg$, which we obtained previously. In the first line below, we will copy the equation $\frac{|q|E\sqrt{3}}{2} - \frac{T}{2} = mg$, and in the line after that, we will rewrite the equation with T replaced by $\frac{|q|E}{\sqrt{3}}$. Compare the following two lines closely.

$$\frac{|q|E\sqrt{3}}{2} - \frac{T}{2} = mg$$

$$\frac{|q|E\sqrt{3}}{2} - \frac{|q|E}{2\sqrt{3}} = mg$$

The algebra is simpler if you multiply both sides of this equation by $\sqrt{3}$.

$$\frac{3|q|E}{2} - \frac{|q|E}{2} = mg\sqrt{3}$$

Note that $\sqrt{3}\sqrt{3} = 3$ and $\frac{\sqrt{3}}{\sqrt{3}} = 1$. **Factor** out the $|q|E$. The left-hand side becomes:

$$\frac{3|q|E}{2} - \frac{|q|E}{2} = \left(\frac{3}{2} - \frac{1}{2}\right)|q|E = |q|E$$

Since $\frac{3|q|E}{2} - \frac{|q|E}{2} = |q|E$ (because $\frac{3}{2} - \frac{1}{2} = 1$), the equation $\frac{3|q|E}{2} - \frac{|q|E}{2} = mg\sqrt{3}$ simplifies to:

$$|q|E = mg\sqrt{3}$$

Solve for the electric field. Divide both sides of the equation by the charge.

$$E = \frac{mg\sqrt{3}}{|q|}$$

Convert the mass from grams (g) to kilograms (kg): $m = 60 \text{ g} = 0.060 \text{ kg} = 6.0 \times 10^{-2}$ kg.
Convert the charge from microCoulombs (μC) to Coulombs: $q = 300\ \mu\text{C} = 3.00 \times 10^{-4}$ C.

$$E = \frac{mg\sqrt{3}}{|q|} = \frac{(6 \times 10^{-2})(9.81)\sqrt{3}}{3 \times 10^{-4}} \approx \frac{(6 \times 10^{-2})(10)\sqrt{3}}{3 \times 10^{-4}} = \frac{(6)(10)\sqrt{3}}{3}\frac{10^{-2}}{10^{-4}}$$

$$E \approx 20\sqrt{3} \times 10^2 \text{ N/C} = 2000\sqrt{3}\ \frac{\text{N}}{\text{C}}$$

Note that $\frac{10^{-2}}{10^{-4}} = 10^{-2-(-4)} = 10^{-2+4} = 10^2$ according to the rule $\frac{x^m}{x^n} = x^{m-n}$. The magnitude of the electric field is $E \approx 2000\sqrt{3}\ \frac{\text{N}}{\text{C}}$. If you don't round gravity, the answer works out to $E = 3400\ \frac{\text{N}}{\text{C}} = 3.4 \times 10^3\ \frac{\text{N}}{\text{C}}$ (to two significant figures).

(B) Determine the tension in the thread.

Solution. Plug $E = 2000\sqrt{3}\ \frac{\text{N}}{\text{C}}$ into $T = \frac{|q|E}{\sqrt{3}}$, which we found in part (A). See the middle of page 39.

$$T = \frac{|q|E}{\sqrt{3}} \approx \frac{(3 \times 10^{-4})(2000\sqrt{3})}{\sqrt{3}} = 6000 \times 10^{-4} = 0.60 \text{ N}$$

The tension equals $T \approx 0.60$ N, which can also be expressed as $T \approx \frac{3}{5}$ N. If you don't round gravity, the answer works out to $T = 0.59$ N.

6 INTEGRATION ESSENTIALS

Algebra Relations

$$\sqrt{x} = x^{1/2} \quad , \quad \sqrt{cx} = (cx)^{1/2} = c^{1/2}x^{1/2} \quad , \quad \frac{1}{x^n} = x^{-n}$$

Derivative of a Polynomial Term

$$\frac{d}{dx}(ax^b) = bax^{b-1}$$

Derivative of a Sum of Functions

$$\frac{d}{dx}(y_1 + y_2 + \cdots + y_N) = \frac{dy_1}{dx} + \frac{dy_2}{dx} + \cdots + \frac{dy_N}{dx}$$

Anti-derivative of a Polynomial Term

$$\int ax^b \, dx = \frac{ax^{b+1}}{b+1} \qquad (b \neq -1)$$

Definite Integral of a Polynomial Term

$$\int_{x=x_i}^{x_f} ax^b \, dx = \left[\frac{ax^{b+1}}{b+1}\right]_{x=x_i}^{x_f} = \frac{ax_f^{b+1}}{b+1} - \frac{ax_i^{b+1}}{b+1} \qquad (b \neq -1)$$

Integral of a Sum of Functions

$$\int (y_1 + y_2 + \cdots + y_N) \, dx = \int y_1 \, dx + \int y_2 \, dx + \cdots + \int y_N \, dx$$

Secant, Cosecant, and Cotangent
$\sec\theta = \frac{1}{\cos\theta}$, $\csc\theta = \frac{1}{\sin\theta}$, $\cot\theta = \frac{1}{\tan\theta}$
Trig Identities
$\sin^2 u + \cos^2 u = 1$, $\tan^2 u + 1 = \sec^2 u$, $1 + \cot^2 u = \csc^2 u$
$\cos^2 u = \frac{1+\cos 2u}{2}$, $\sin^2 u = \frac{1-\cos 2u}{2}$
Derivatives of Trig Functions
$\frac{d}{d\theta}\sin\theta = \cos\theta$, $\frac{d}{d\theta}\cos\theta = -\sin\theta$, $\frac{d}{d\theta}\tan\theta = \sec^2\theta$
$\frac{d}{d\theta}\sec\theta = \sec\theta\tan\theta$, $\frac{d}{d\theta}\csc\theta = -\csc\theta\cot\theta$, $\frac{d}{d\theta}\cot\theta = -\csc^2\theta$
Anti-derivatives of Trig Functions
$\int \sin\theta\, d\theta = -\cos\theta$, $\int \cos\theta\, d\theta = \sin\theta$, $\int \tan\theta\, d\theta = \ln\lvert\sec\theta\rvert$
$\int \sec\theta\, d\theta = \ln\lvert\sec\theta + \tan\theta\rvert$, $\int \csc\theta\, d\theta = -\ln\lvert\csc\theta + \cot\theta\rvert$, $\int \cot\theta\, d\theta = \ln\lvert\sin\theta\rvert$

Example 22. Perform the following definite integral.

$$\int_{x=2}^{3} 6x^2\, dx$$

Solution. Compare $6x^2$ with ax^b to see that $a = 6$ and $b = 2$. The anti-derivative is $\frac{ax^{b+1}}{b+1}$.

$$\int_{x=2}^{3} 6x^2\, dx = \left[\frac{6x^{2+1}}{2+1}\right]_{x=2}^{3} = \left[\frac{6x^3}{3}\right]_{x=2}^{3} = [2x^3]_{x=2}^{3}$$

The anti-derivative of $6x^2$ is $2x^3$, but we're not finished yet. The notation $[2x^3]_{x=2}^{3}$ means to evaluate the function $2x^3$ when $x = 3$, then evaluate the function $2x^3$ when $x = 2$, and finally to subtract the two results: $2(3)^3 - 2(2)^3$.

$$\int_{x=2}^{3} 6x^2\, dx = [2x^3]_{x=2}^{3} = 2(3)^3 - 2(2)^3 = 2(27) - 2(8) = 54 - 16 = 38$$

The answer to the definite integral is 38.

Example 23. Perform the following definite integral.

$$\int_{x=1}^{2} \frac{8\,dx}{x^3}$$

Solution. Since x appears in the denominator, apply the rule $\frac{1}{x^n} = x^{-n}$ to write $\frac{8}{x^3}$ as $8x^{-3}$. Compare $8x^{-3}$ with ax^b to see that $a = 8$ and $b = -3$. The anti-derivative is $\frac{ax^{b+1}}{b+1}$.

$$\int_{x=1}^{2} \frac{8\,dx}{x^3} = \int_{x=1}^{2} 8x^{-3}\,dx = \left[\frac{8x^{-3+1}}{-3+1}\right]_{x=1}^{2} = \left[\frac{8x^{-2}}{-2}\right]_{x=1}^{2} = [-4x^{-2}]_{x=1}^{2} = \left[-\frac{4}{x^2}\right]_{x=1}^{2}$$

Note that $-3 + 1 = -2$ and that $\frac{8}{-2} = -4$. In the last step, we used the rule that $x^{-2} = \frac{1}{x^2}$. The notation $\left[-\frac{4}{x^2}\right]_{x=1}^{2}$ means to evaluate the function $-\frac{4}{x^2}$ when $x = 2$, then evaluate the function $-\frac{4}{x^2}$ when $x = 1$, and finally to subtract the two results: $-\frac{4}{2^2} - \left(-\frac{4}{1^2}\right)$.

$$\int_{x=1}^{2} \frac{8\,dx}{x^3} = \left[-\frac{4}{x^2}\right]_{x=1}^{2} = -\frac{4}{2^2} - \left(-\frac{4}{1^2}\right) = -\frac{4}{4} + \frac{4}{1} = -1 + 4 = 3$$

Recall that two minus signs make a plus sign. The answer to the definite integral is 3.

Example 24. Perform the following definite integral.

$$\int_{x=9}^{16} 3\sqrt{x}\,dx$$

Solution. The trick to this problem is to recall the rule from algebra that $\sqrt{x} = x^{1/2}$ to write $3\sqrt{x}$ as $3x^{1/2}$. Compare $3x^{1/2}$ with ax^b to see that $a = 3$ and $b = \frac{1}{2}$. The anti-derivative is $\frac{ax^{b+1}}{b+1}$. Note that $b + 1 = \frac{1}{2} + 1 = \frac{1}{2} + \frac{2}{2} = \frac{1+2}{2} = \frac{3}{2}$ (to add fractions, first find a **common denominator**). Also note that $\frac{a}{b+1} = \frac{3}{3/2} = 3\left(\frac{2}{3}\right) = 2$ (to divide by a fraction, multiply by its **reciprocal**: note that the reciprocal of $\frac{3}{2}$ is $\frac{2}{3}$).

$$\int_{x=9}^{16} 3\sqrt{x}\,dx = \int_{x=9}^{16} 3x^{1/2}\,dx = \left[\frac{3x^{1/2+1}}{\frac{1}{2}+1}\right]_{x=4}^{9} = \left[\frac{3x^{3/2}}{\frac{3}{2}}\right]_{x=4}^{9} = [2x^{3/2}]_{x=9}^{16}$$

The notation $[2x^{3/2}]_{x=9}^{16}$ means to evaluate the function $2x^{3/2}$ when $x = 16$, then evaluate the function $2x^{3/2}$ when $x = 9$, and finally to subtract the two results: $2(16)^{3/2} - 2(9)^{3/2}$.

$$\int_{x=9}^{16} 3\sqrt{x}\,dx = [2x^{3/2}]_{x=9}^{16} = 2(16)^{3/2} - 2(9)^{3/2} = 2(64) - 2(27) = 128 - 54 = 74$$

Note that $(16)^{3/2} = \left(\sqrt{16}\right)^3 = (4)^3 = 64$ and $(9)^{3/2} = \left(\sqrt{9}\right)^3 = (3)^3 = 27$. (You can try entering 16^1.5 on your calculator, for example. Note that $\frac{3}{2} = 1.5$.) The answer to the definite integral is 74.

Example 25. Perform the following definite integral.

$$\int_{x=1}^{3} (8x^3 - 6x)\, dx$$

Solution. The rule $\int (y_1 + y_2)\, dx = \int y_1\, dx + \int y_2\, dx$ means that we can integrate the function $(8x^3 - 6x)$ by finding the anti-derivative of each term separately.

- The first term is $8x^3$. Compare $8x^3$ with ax^b to see that $a = 8$ and $b = 3$.

$$\int 8x^3\, dx = \frac{8x^{3+1}}{3+1} = \frac{8x^4}{4} = 2x^4$$

- The second term is $-6x$. Compare $-6x$ with ax^b to see that $a = -6$ and $b = 1$ (since a power of 1 is implied when you don't see a power).

$$\int -6x\, dx = \frac{-6x^{1+1}}{1+1} = -\frac{6x^2}{2} = -3x^2$$

Add these two anti-derivatives together and evaluate the definite integral over the limits:

$$\int_{x=1}^{3} (8x^3 - 6x)\, dx = [2x^4 - 3x^2]_{x=1}^{3} = [2(3)^4 - 3(3)^2] - [2(1)^4 - 3(1)^2]$$

Distribute the minus sign. Two minus signs make a plus sign.

$$\int_{x=1}^{3} (8x^3 - 6x)\, dx = 2(3)^4 - 3(3)^2 - 2(1)^4 + 3(1)^2 = 2(81) - 3(9) - 2(1) + 3(1)$$

$$= 162 - 27 - 2 + 3 = 136$$

The answer to the definite integral is 136.

Example 26. Perform the following definite integral.

$$\int_{\theta=30°}^{90°} \cos\theta\, d\theta$$

Solution. First find the anti-derivative of $\cos\theta$. See page 42.

$$\int_{\theta=30°}^{90°} \cos\theta\, d\theta = [\sin\theta]_{\theta=30°}^{90°} = \sin(90°) - \sin(30°) = 1 - \frac{1}{2} = \frac{2}{2} - \frac{1}{2} = \frac{2-1}{2} = \frac{1}{2}$$

Subtract fractions with a **common denominator**. The answer to the definite integral is $\frac{1}{2}$.

Example 27. Perform the following definite integral.

$$\int_{\theta=45^\circ}^{135^\circ} \sin\theta \, d\theta$$

Solution. First find the anti-derivative of $\sin\theta$. See page 42.

$$\int_{\theta=45^\circ}^{135^\circ} \sin\theta \, d\theta = [-\cos\theta]_{\theta=45^\circ}^{135^\circ} = -\cos(135^\circ) - [-\cos(45^\circ)] = -\cos(135^\circ) + \cos(45^\circ)$$

$$= -\left(-\frac{\sqrt{2}}{2}\right) + \frac{\sqrt{2}}{2} = \frac{\sqrt{2}}{2} + \frac{\sqrt{2}}{2} = \frac{2\sqrt{2}}{2} = \sqrt{2}$$

Note that two minus signs make a plus sign. The answer to the definite integral is $\sqrt{2}$.

Example 28. Perform the following definite integral.

$$\int_{x=0}^{8} \left(\frac{x}{4} + 1\right)^3 dx$$

Solution. Make the substitution $u = \frac{x}{4} + 1$. Take an implicit derivative of $u = \frac{x}{4} + 1$:

- On the left-hand side: $\frac{d}{du}(u) = 1$. Multiply by du to get $1\,du = du$.
- On the right-hand side: $\frac{d}{dx}\left(\frac{x}{4} + 1\right) = \frac{1}{4}$. Multiply by dx to get $\frac{dx}{4}$.

The implicit derivative of $u = \frac{x}{4} + 1$ is therefore $du = \frac{dx}{4}$. Solve for dx to get $dx = 4du$.

$$u = \frac{x}{4} + 1 \quad , \quad dx = 4du$$

Now we must adjust the limits. Plug each limit into $u = \frac{x}{4} + 1$:

$$u(x = 0) = \frac{0}{4} + 1 = 0 + 1 = 1$$

$$u(x = 8) = \frac{8}{4} + 1 = 2 + 1 = 3$$

The integral becomes:

$$\int_{x=0}^{8} \left(\frac{x}{4} + 1\right)^3 dx = \int_{u=1}^{3} 4u^3 \, du$$

Now we can integrate over u.

$$\int_{u=1}^{3} 4u^3 \, du = \left[\frac{4u^{3+1}}{3+1}\right]_{u=1}^{3} = \left[\frac{4u^4}{4}\right]_{u=1}^{3} = [u^4]_{u=1}^{3} = (3)^4 - (1)^4 = 81 - 1 = 80$$

The answer to both the old and new integrals is 80.

$$\int_{x=0}^{8} \left(\frac{x}{4} + 1\right)^3 dx = \int_{u=1}^{3} 4u^3 \, du = 80$$

Example 29. Perform the following definite integral.

$$\int_{x=1}^{5} 3\sqrt{2x-1}\,dx$$

Solution. Make the substitution $u = 2x - 1$. (Unlike the next two examples, this integral does __not__ involve a trig substitution even though it involves a squareroot. The reason is that the variable x is __not__ squared in this example. Compare this integral with the following two examples closely.) Take an implicit derivative of $u = 2x - 1$:

- On the left-hand side: $\frac{d}{du}(u) = 1$. Multiply by du to get $1\,du = du$.
- On the right-hand side: $\frac{d}{dx}(2x-1) = 2$. Multiply by dx to get $2dx$.

The implicit derivative of $u = 2x - 1$ is therefore $du = 2dx$. Solve for dx to get $dx = \frac{du}{2}$.

$$u = 2x - 1 \quad , \quad dx = \frac{du}{2}$$

Now we must adjust the limits. Plug each limit into $u = 2x - 1$:

$$u(x=1) = 2(1) - 1 = 2 - 1 = 1$$
$$u(x=5) = 2(5) - 1 = 10 - 1 = 9$$

The integral becomes:

$$\int_{x=1}^{5} 3\sqrt{2x-1}\,dx = \int_{u=1}^{9} \frac{3}{2}\sqrt{u}\,du = \int_{u=1}^{9} \frac{3}{2}u^{1/2}\,du = \frac{3}{2}\int_{u=1}^{9} u^{1/2}\,du$$

Note that $u^{1/2} = \sqrt{u}$ (see Example 24). Now we can integrate over u.

$$\frac{3}{2}\int_{u=1}^{9} u^{1/2}\,du = \frac{3}{2}\left[\frac{u^{1/2+1}}{1/2+1}\right]_{u=1}^{9} = \frac{3}{2}\left[\frac{u^{3/2}}{3/2}\right]_{u=1}^{9} = \frac{3}{2}\left[\frac{2u^{3/2}}{3}\right]_{u=1}^{9} = \left[u^{3/2}\right]_{u=1}^{9}$$

Note that $\frac{1}{2} + 1 = \frac{3}{2}$. To divide by a fraction, multiply by its **reciprocal**. Note that the reciprocal of $\frac{3}{2}$ is $\frac{2}{3}$. This is why $\frac{3}{2}\left[\frac{u^{3/2}}{3/2}\right]_{u=1}^{9} = \frac{3}{2}\left[\frac{2u^{3/2}}{3}\right]_{u=1}^{9} = \left[u^{3/2}\right]_{u=1}^{9}$. Note also that $\frac{3}{2}\frac{2}{3} = 1$.

$$\left[u^{3/2}\right]_{u=1}^{9} = (9)^{3/2} - (1)^{3/2} = \left(\sqrt{9}\right)^3 - \left(\sqrt{1}\right)^3 = 3^3 - 1^3 = 27 - 1 = 26$$

Note that $(9)^{3/2} = \left(\sqrt{9}\right)^3 = (3)^3 = 27$. You can verify on your calculator, for example, that 9^1.5 = 27. (Note that $\frac{3}{2} = 1.5$.) The answer to both the old and new integrals is 26.

$$\int_{x=1}^{5} 3\sqrt{2x-1}\,dx = \int_{u=1}^{9} \frac{3}{2}\sqrt{u}\,du = 26$$

Example 30. Perform the following definite integral.

$$\int_{x=0}^{3} 4\sqrt{9-x^2}\,dx$$

Solution. One way to integrate a function of the form $\sqrt{a^2-x^2}$ is to make the substitution $x = a\sin u$. The reason for this is that it will allow us to take advantage of the trig identity $\sin^2 u + \cos^2 u = 1$ (as we will see later in the solution). Write $9 - x^2$ as $3^2 - x^2$. Now it looks like $a^2 - x^2$ with $a = 3$. Since $a = 3$ in this example, we make the substitution $x = 3\sin u$. Take an implicit derivative:

- On the left-hand side: $\frac{d}{dx}(x) = 1$. Multiply by dx to get $1\,dx = dx$.
- On the right-hand side: $\frac{d}{du}(3\sin u) = 3\cos u$. Multiply by du to get $3\cos u\,du$. Recall from calculus that $\frac{d}{du}\sin u = \cos u$ (you can find trig derivatives on page 42).

The implicit derivative of $x = 3\sin u$ is therefore $dx = 3\cos u\,du$. We will make the following pair of substitutions in the original integral:

$$x = 3\sin u \quad , \quad dx = 3\cos u\,du$$

Now we must adjust the limits. Solve for u to obtain $u = \sin^{-1}\left(\frac{x}{3}\right)$:

$$u(x=0) = \sin^{-1}\left(\frac{0}{3}\right) = 0°$$

$$u(x=3) = \sin^{-1}\left(\frac{3}{3}\right) = \sin^{-1}(1) = 90°$$

The integral becomes:

$$\int_{x=0}^{3} 4\sqrt{9-x^2}\,dx = \int_{u=0}^{90°} 4\sqrt{9-9\sin^2 u}\;3\cos u\,du = 12\int_{u=0}^{90°} \sqrt{9-9\sin^2 u}\,\cos u\,du$$

Factor out the 9 to write $\sqrt{9-9\sin^2 u} = \sqrt{9(1-\sin^2 u)} = 3\sqrt{1-\sin^2 u}$.

$$\int_{x=0}^{3} 4\sqrt{9-x^2}\,dx = 12\int_{u=0}^{90°} 3\sqrt{1-\sin^2 u}\,\cos u\,du = 36\int_{u=0}^{90°} \sqrt{1-\sin^2 u}\,\cos u\,du$$

Use the trig identity $\sin^2 u + \cos^2 u = 1$ to replace $\sqrt{1-\sin^2 u}$ with $\sqrt{\cos^2 u} = \cos u$.

$$\int_{x=0}^{3} 4\sqrt{9-x^2}\,dx = 36\int_{u=0}^{90°} \cos u\,\cos u\,du = 36\int_{u=0}^{90°} \cos^2 u\,du$$

Apply the trig identity $\cos^2 u = \frac{1+\cos 2u}{2}$. (Page 42 lists a few handy trig identities.)

$$36\int_{u=0}^{90°} \cos^2 u\,du = 36\int_{u=0}^{90°} \frac{1+\cos 2u}{2}\,du = 36\int_{u=0}^{90°} \frac{1}{2}\,du + 36\int_{u=0}^{90°} \frac{\cos 2u}{2}\,du$$

$$18\int_{u=0}^{90°} du + 18\int_{u=0}^{90°} \cos 2u\,du$$

The first integral will only work in **radians**. (You can get away with degrees when **all** of the angles appear in the arguments of trig functions. The first integral has an angle that is **not** in an argument of a trig function, so we **must** use radians.) Convert to radians using $360° = 2\pi$ rad. The new limits are from $u = 0$ to $u = \frac{\pi}{2}$ radians.

$$18 \int_{u=0}^{90°} du + 18 \int_{u=0}^{90°} \cos 2u \, du = 18 \int_{u=0}^{\pi/2} du + 18 \int_{u=0}^{\pi/2} \cos 2u \, du$$

$$= 18[u]_{u=0}^{\pi/2} + 18 \left[\frac{\sin 2u}{2}\right]_{u=0}^{\pi/2} = 18[u]_{u=0}^{\pi/2} + 9[\sin 2u]_{u=0}^{\pi/2}$$

Note that $\int \cos 2u \, du = \frac{\sin 2u}{2}$ because $\frac{d}{du}(\sin 2u) = 2\cos 2u$.

$$18[u]_{u=0}^{\pi/2} + 9[\sin 2u]_{u=0}^{\pi/2} = 18\left(\frac{\pi}{2} - 0\right) + 9\left[\sin\left[2\left(\frac{\pi}{2}\right)\right] - \sin[2(0)]\right]$$

$$= 9\pi - 0 + 9[\sin \pi - \sin 0] = 9\pi + 9(0 - 0) = 9\pi$$

The answer to the definite integral is 9π.

Example 31. Perform the following definite integral.

$$\int_{x=0}^{4} \frac{dx}{\sqrt{16 + x^2}}$$

Solution. One way to integrate a function of the form $\sqrt{a^2 + x^2}$ is to make the substitution $x = a \tan u$. The reason for this is that it will allow us to take advantage of the trig identity $1 + \tan^2 u = \sec^2 u$ (as we will see later in the solution). Write $16 + x^2$ as $4^2 + x^2$. Now it looks like $a^2 + x^2$ with $a = 4$. Since $a = 4$ in this example, we make the substitution $x = 4 \tan u$. Take an implicit derivative:

- On the left-hand side: $\frac{d}{dx}(x) = 1$. Multiply by dx to get $1\,dx = dx$.
- On the right-hand side: $\frac{d}{du}(4 \tan u) = 4 \sec^2 u$. Multiply by du to get $4 \sec^2 u \, du$. Recall from calculus that $\frac{d}{du} \tan u = \sec^2 u$ (you can find trig derivatives on page 42).

The implicit derivative of $x = 4 \tan u$ is therefore $dx = 4 \sec^2 u \, du$. We will make the following pair of substitutions in the original integral:

$$x = 4 \tan u \quad , \quad dx = 4 \sec^2 u \, du$$

Now we must adjust the limits. Solve for u to obtain $u = \tan^{-1}\left(\frac{x}{4}\right)$:

$$u(x = 0) = \tan^{-1}\left(\frac{0}{4}\right) = 0°$$

$$u(x = 4) = \tan^{-1}\left(\frac{4}{4}\right) = \tan^{-1}(1) = 45°$$

The integral becomes:

$$\int_{x=0}^{4} \frac{dx}{\sqrt{16+x^2}} = \int_{u=0°}^{45°} \frac{4\sec^2 u}{\sqrt{16+16\tan^2 u}} du$$

Factor out the 16 to write $\sqrt{16+16\tan^2 u} = \sqrt{16(1+\tan^2 u)} = 4\sqrt{1+\tan^2 u}$

$$\int_{u=0°}^{45°} \frac{4\sec^2 u}{\sqrt{16+16\tan^2 u}} du = \int_{u=0°}^{45°} \frac{4\sec^2 u}{4\sqrt{1+\tan^2 u}} du = \int_{u=0°}^{45°} \frac{\sec^2 u}{\sqrt{1+\tan^2 u}} du$$

Note that the 4 cancels out. Use the trig identity $1+\tan^2 u = \sec^2 u$ to replace $\sqrt{1+\tan^2 u}$ with $\sqrt{\sec^2 u}$. (Page 42 lists a few handy trig identities. Note that $1+\tan^2 u = \tan^2 u + 1$.)

$$\int_{u=0°}^{45°} \frac{\sec^2 u}{\sqrt{1+\tan^2 u}} du = \int_{u=0°}^{45°} \frac{\sec^2 u\, du}{\sqrt{\sec^2 u}} = \int_{u=0°}^{45°} \frac{\sec^2 u\, du}{\sec u} = \int_{u=0°}^{45°} \sec u\, du$$

Look up the integral of secant (on page 42).

$$\int_{u=0°}^{45°} \sec u\, du = [\ln|\sec u + \tan u|]_{u=0°}^{45°} = \ln|\sec 45° + \tan 45°| - \ln|\sec 0° + \tan 0°|$$

$$\int_{u=0°}^{45°} \sec u\, du = \ln|\sqrt{2}+1| - \ln|1+0| = \ln|\sqrt{2}+1| - \ln|1| = \ln|\sqrt{2}+1| - 0 = \ln|\sqrt{2}+1|$$

We used the logarithm identity $\ln(1) = 0$. The answer to the definite integral is $\ln|\sqrt{2}+1|$.

Example 32. Perform the following double integral.

$$\int_{x=0}^{3} \int_{y=0}^{x} 10xy^2\, dxdy$$

Solution. Since the upper y-limit depends on x, we must carry out the y-integration first.

$$\int_{x=0}^{3} \int_{y=0}^{x} 10xy^2\, dxdy = \int_{x=0}^{3} \left(\int_{y=0}^{x} 10xy^2\, dy \right) dx$$

When integrating over y, we treat the independent variable x like any other constant. Therefore, we can factor $10x$ out of the y integral (but be careful not to pull x out of the x integral). We're applying the same concept as $\int cf(y)\, dy = c \int f(y)\, dy$.

$$\int_{x=0}^{3} \left(\int_{y=0}^{x} 10xy^2\, dy \right) dx = \int_{x=0}^{3} 10x \left(\int_{y=0}^{x} y^2\, dy \right) dx$$

To help make this clear, we will carry out the complete definite integral over y in parentheses before proceeding.

$$\int_{x=0}^{3} 10x \left(\int_{y=0}^{x} y^2\, dy \right) dx = \int_{x=0}^{3} 10x \left(\left[\frac{y^3}{3} \right]_{y=0}^{y=x} \right) dx = \int_{x=0}^{3} 10x \left(\frac{x^3}{3} - \frac{0^3}{3} \right) dx$$

$$= \int_{x=0}^{3} 10x\left(\frac{x^3}{3}\right) dx = \int_{x=0}^{3} \frac{10x^4}{3} dx = \frac{10}{3}\int_{x=0}^{3} x^4\, dx$$

Now we have a single integral over x.

$$\frac{10}{3}\int_{x=0}^{3} x^4\, dx = \frac{10}{3}\left[\frac{x^5}{5}\right]_{x=0}^{x=3} = \frac{2}{3}[x^5]_{x=0}^{x=3} = \frac{2}{3}(3^5 - 0^5) = \frac{2}{3}(243 - 0) = 162$$

The final answer to the double integral is 162.

Example 33. Perform the following double integral.

$$\int_{x=0}^{y}\int_{y=0}^{3} 10xy^2\, dxdy$$

Solution. Since the x-limit depends on y, we must carry out the x-integration first.

$$\int_{x=0}^{y}\int_{y=0}^{3} 10xy^2\, dxdy = \int_{y=0}^{3}\left(\int_{x=0}^{y} 10xy^2\, dx\right) dy$$

When integrating over x, we treat the independent variable y like any other constant. We can factor $10y^2$ out of the x integral (but be careful not to pull y^2 out of the y integral). We're applying the same concept as $\int cf(x)\, dx = c\int f(x)\, dx$.

$$\int_{y=0}^{3}\left(\int_{x=0}^{y} 10xy^2\, dx\right) dy = \int_{y=0}^{3} 10y^2\left(\int_{x=0}^{y} x\, dx\right) dy$$

To help make this clear, we will carry out the complete definite integral over x in parentheses before proceeding.

$$\int_{y=0}^{3} 10y^2\left(\int_{x=0}^{y} x\, dx\right) dy = \int_{y=0}^{3} 10y^2\left(\left[\frac{x^2}{2}\right]_{x=0}^{x=y}\right) dy = \int_{y=0}^{3} 10y^2\left(\frac{y^2}{2} - \frac{0^2}{2}\right) dy$$

$$= \int_{y=0}^{3} 10y^2\left(\frac{y^2}{2}\right) dy = \int_{y=0}^{3} 5y^4\, dy$$

Now we have a single integral over y.

$$\int_{y=0}^{3} 5y^4\, dy = \left[\frac{5y^5}{5}\right]_{y=0}^{y=3} = [y^5]_{y=0}^{y=3} = 3^5 - 0^5 = 243$$

The final answer to the double integral is 243.

Example 34. Perform the following double integral.

$$\int_{x=0}^{2}\int_{y=0}^{3} x^3y^2\,dxdy$$

Solution. Since all of the limits are constants, in this example we are free to perform the integration in any order. We choose to integrate over y first.

$$\int_{x=0}^{2}\int_{y=0}^{3} x^3y^2\,dxdy = \int_{x=0}^{2}\left(\int_{y=0}^{3} x^3y^2\,dy\right)dx$$

When integrating over y, we treat the independent variable x like any other constant. Therefore, we can factor x^3 out of the y integral (but be careful not to pull x^3 out of the x integral). We're applying the same concept as $\int cf(y)\,dy = c\int f(y)\,dy$.

$$\int_{x=0}^{2}\left(\int_{y=0}^{3} x^3y^2\,dy\right)dx = \int_{x=0}^{2} x^3\left(\int_{y=0}^{3} y^2\,dy\right)dx$$

To help make this clear, we will carry out the complete definite integral over y in parentheses before proceeding.

$$\int_{x=0}^{2} x^3\left(\int_{y=0}^{3} y^2\,dy\right)dx = \int_{x=0}^{2} x^3\left(\left[\frac{y^3}{3}\right]_{y=0}^{y=3}\right)dx = \int_{x=0}^{2} x^3\left(\frac{3^3}{3}-\frac{0^3}{3}\right)dx$$

$$\int_{x=0}^{2} x^3\left(\frac{27}{3}-\frac{0}{3}\right)dx = \int_{x=0}^{2} x^3\left(\frac{27}{3}\right)dx = 9\int_{x=0}^{2} x^3\,dx$$

Now we have a single integral over x.

$$9\int_{x=0}^{2} x^3\,dx = 9\left[\frac{x^4}{4}\right]_{x=0}^{x=2} = 9\left(\frac{2^4}{4}-\frac{0^4}{4}\right) = 9\left(\frac{16}{4}-\frac{0}{4}\right) = 9(4-0) = 9(4) = 36$$

The final answer to the double integral is 36.

Example 35. Perform the following triple integral.

$$\int_{x=0}^{4}\int_{y=0}^{x}\int_{z=0}^{y} 6z\,dx\,dy\,dz$$

Solution. Since the y- and z-limits both involve variables, we must perform the y and z integrals first. Furthermore, in this example we must perform the z integral **before** performing the y integral. Why? Because there is a y in the upper limit of the z integral, we won't be able to integrate over y until we perform the z-integration. Thus we begin with the z-integration.

$$\int_{x=0}^{4}\int_{y=0}^{x}\int_{z=0}^{y} 6z\,dx\,dy\,dz = \int_{x=0}^{4}\int_{y=0}^{x}\left(\int_{z=0}^{y} 6z\,dz\right)dy\,dx$$

To help make this clear, we will carry out the complete definite integral over z in parentheses before proceeding.

$$\int_{x=0}^{4}\int_{y=0}^{x}\left(\int_{z=0}^{y} 6z\,dz\right)dy\,dx = \int_{x=0}^{4}\int_{y=0}^{x}\left(\left[\frac{6z^2}{2}\right]_{z=0}^{z=y}\right)dy\,dx = \int_{x=0}^{4}\int_{y=0}^{x}\left([3z^2]_{z=0}^{z=y}\right)dy\,dx$$

$$= \int_{x=0}^{4}\int_{y=0}^{x}[3y^2 - 3(0)^2]dy\,dx = \int_{x=0}^{4}\int_{y=0}^{x} 3y^2\,dy\,dx$$

Now we have a double integral similar to the previous examples. We perform the y-integration next because there is a variable in the upper limit of the y integral.

$$\int_{x=0}^{4}\int_{y=0}^{x} 3y^2\,dy\,dx = \int_{x=0}^{4}\left(\int_{y=0}^{x} 3y^2\,dy\right)dx = \int_{x=0}^{4}\left(\left[\frac{3y^3}{3}\right]_{y=0}^{x}\right)dx = \int_{x=0}^{4}\left([y^3]_{y=0}^{x}\right)dx$$

$$= \int_{x=0}^{4}(x^3 - 0)\,dx = \int_{x=0}^{4} x^3\,dx = \left[\frac{x^4}{4}\right]_{x=0}^{4} = \frac{4^4}{4} - \frac{0^4}{4} = 64 - 0 = 64$$

Cartesian Coordinates
(x, y, z)

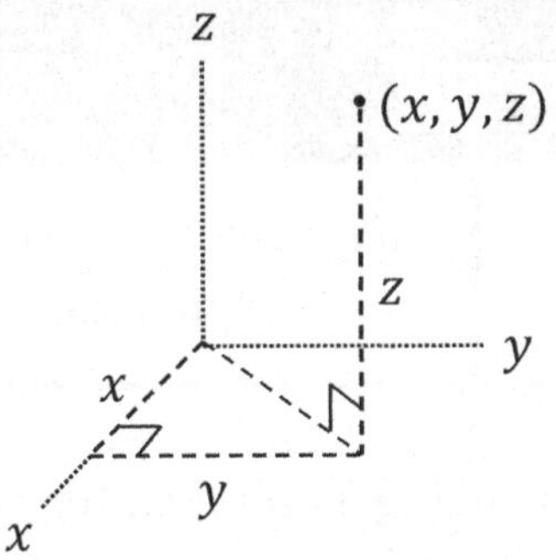

2D Polar Coordinates
(r, θ)

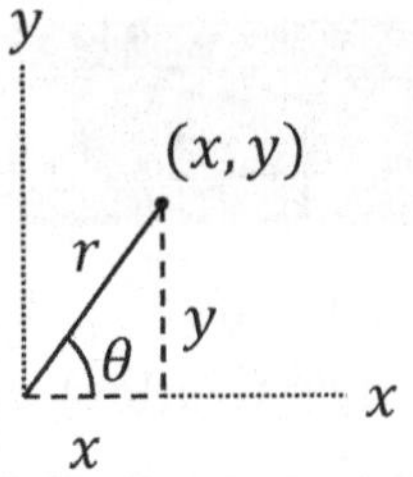

$$x = r\cos\theta \quad , \quad r = \sqrt{x^2 + y^2}$$
$$y = r\sin\theta \quad , \quad \theta = \tan^{-1}\left(\frac{y}{x}\right)$$

Cylindrical Coordinates
(r_c, θ, z)

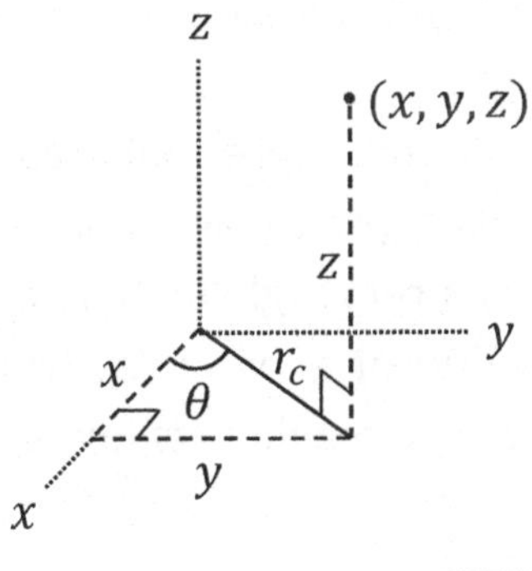

$$x = r_c\cos\theta \quad , \quad r_c = \sqrt{x^2 + y^2}$$
$$y = r_c\sin\theta \quad , \quad \theta = \tan^{-1}\left(\frac{y}{x}\right)$$

Spherical Coordinates
(r, θ, φ)

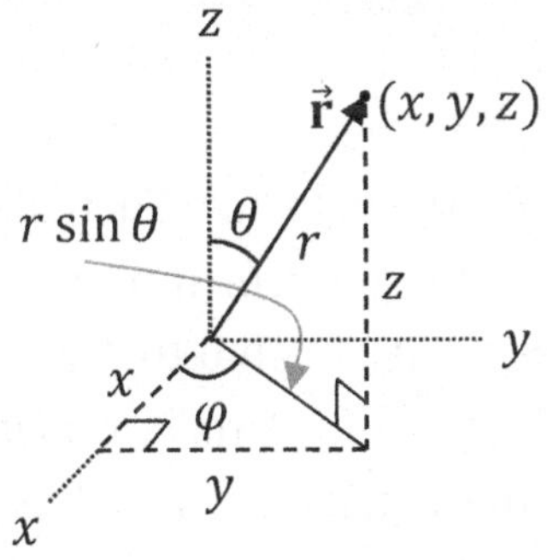

$$x = r\sin\theta\cos\varphi \quad , \quad r = \sqrt{x^2 + y^2 + z^2}$$
$$y = r\sin\theta\sin\varphi \quad , \quad \theta = \cos^{-1}\left(\frac{z}{r}\right)$$
$$z = r\cos\theta \quad , \quad \varphi = \tan^{-1}\left(\frac{y}{x}\right)$$

Differential Arc Length		
$ds = dx$ (along x)	$ds = dy$ or dz (along y or z)	$ds = ad\theta$ (circular arc of radius a)
Differential Area Element		
$dA = dxdy$ (polygon in xy plane)	$dA = rdrd\theta$ (pie slice, disc, thick ring)	$dA = a^2 \sin\theta\, d\theta d\varphi$ (sphere of radius a)
Differential Volume Element		
$dV = dxdydz$ (bounded by flat sides)	$dV = r_c dr_c d\theta dz$ (cylinder or cone)	$dV = r^2 \sin\theta\, drd\theta d\varphi$ (spherical)

Example 36. Find the area of the triangle illustrated below using a double integral.

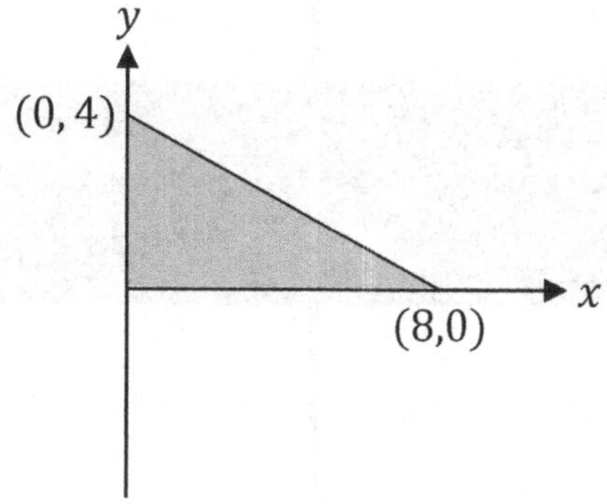

Solution. For a triangle, use Cartesian coordinates: Express the differential area element as $dA = dxdy$. This requires integrating over both x and y. We **don't** want to let x vary from 0 to 8 and let y vary from 0 to 4 because that would give us a rectangle instead of a triangle. If we let x vary from 0 to 8, observe that for a given value of x, y varies from 0 to the hypotenuse of the triangle, which is less than 4 (except when x reaches its lower limit). See the diagram below.

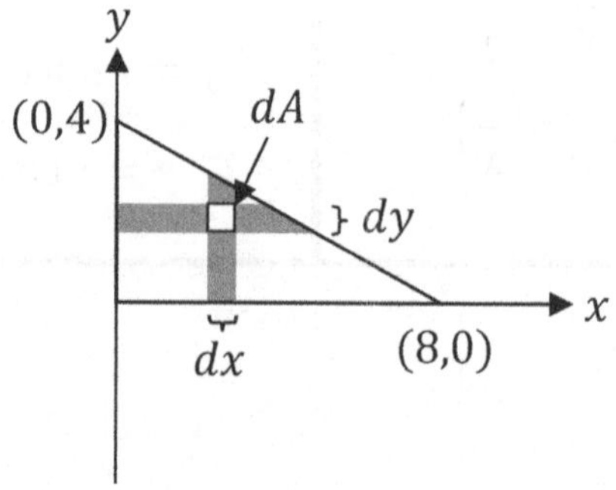

To find the proper upper limit for y, we need the equation of the line that serves as the triangle's hypotenuse. That line has a slope equal to $\frac{y_2-y_1}{x_2-x_1} = \frac{0-4}{8-0} = -\frac{4}{8} = -\frac{1}{2}$ and a y-intercept of 4. Since the general equation for a line is $y = mx + b$, the equation for this line is $y = -\frac{1}{2}x + 4$. For a given value of x, y will vary from 0 to $y = -\frac{1}{2}x + 4$, where the upper limit came from the equation for the line (of the hypotenuse). See the vertical gray band in the previous diagram. Now we have the integration limits.

$$A = \int dA = \int_{x=0}^{8} \int_{y=0}^{-\frac{x}{2}+4} dy\, dx$$

We must perform the y-integration first because it has x in its upper limit.

$$A = \int_{x=0}^{8} \left(\int_{y=0}^{-\frac{x}{2}+4} dy \right) dx = \int_{x=0}^{8} \left([y]_{y=0}^{y=-\frac{x}{2}+4} \right) dx = \int_{x=0}^{8} \left(-\frac{x}{2} + 4 - 0 \right) dx = \int_{x=0}^{8} \left(-\frac{x}{2} + 4 \right) dx$$

$$= \left[-\frac{x^2}{4} + 4x \right]_{x=0}^{x=8} = -\frac{(8)^2}{4} + 4(8) - \left[-\frac{(0)^2}{4} + 4(0) \right] = -\frac{64}{4} + 32 = -16 + 32 = 16$$

The final answer to the double integral is $A = 16$. Of course, we don't need calculus to find the area of a triangle. We could just use the formula one-half base times height: $A = \frac{1}{2}bh = \frac{1}{2}(8)(4) = 16$. However, there are some integrals that can only be done with calculus. (On a similar note, while area can be calculated as a single integral, there are some physics integrals that can only be done as double or triple integrals, so multi-integration is a necessary skill.) We will see examples of integrals of this form in physics problems in Chapters 7, 8, and other chapters in this book.

Example 37. A thick circular ring has an inner radius of 2.0 m and an outer radius of 4.0 m. Find the area of this ring using a double integral.

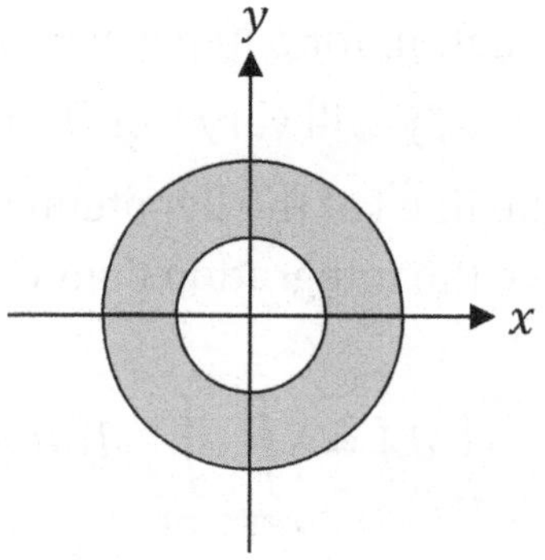

Solution. For a solid circular ring, use $dA = r dr d\theta$. This requires integrating over both r and θ. Unlike in the previous example, r and θ each have constant limits:

$$2.0 \text{ m} \leq r \leq 4.0 \text{ m}$$

$$0 \leq \theta \leq 2\pi \quad \text{(full circle)}$$

For any value of θ, the variable r ranges between 2.0 m and 4.0 m (the inner and outer radii of the thick ring). Contrast this with the previous example where the upper limit of y depended on the value of x. Substitute $dA = r dr d\theta$ into the area integral.

$$A = \int dA = \int_{r=2}^{4} \int_{\theta=0}^{2\pi} r dr d\theta$$

Since all of the limits are constant, we can do these integrals in any order.

$$A = \int_{r=2}^{4} \left(\int_{\theta=0}^{2\pi} r d\theta \right) dr$$

When integrating over θ, treat the independent variable r as a constant. This means that you can pull r out of the θ integral (but be careful not to pull r out of the r integral).

$$A = \int_{r=2}^{4} r \left(\int_{\theta=0}^{2\pi} d\theta \right) dr = \int_{r=2}^{4} r[\theta]_{\theta=0}^{2\pi} dr = \int_{r=2}^{4} r(2\pi - 0) dr = \int_{r=2}^{4} 2\pi r \, dr$$

$$A = \int_{r=2}^{4} 2\pi r \, dr = 2\pi \int_{r=2}^{4} r \, dr = 2\pi \left[\frac{r^2}{2}\right]_{r=2}^{4} = 2\pi \left(\frac{4^2}{2} - \frac{2^2}{2}\right) = 2\pi(8 - 2) = 12\pi$$

The final answer to the double integral is $A = 12\pi \text{ m}^2$. If you use a calculator, the area works out to $A = 37 \text{ m}^2$.

7 ELECTRIC FIELD INTEGRALS

Electric Field Integral		
$\vec{\mathbf{E}} = k \int \frac{\hat{\mathbf{R}}}{R^2} dq$	$\hat{\mathbf{R}} = \frac{\vec{\mathbf{R}}}{R}$	$Q = \int dq$
Differential Charge Element		
$dq = \lambda ds$ (line or thin arc)	$dq = \sigma dA$ (surface area)	$dq = \rho dV$ (volume)
Relation Among Coordinate Systems and Unit Vectors		
$x = r \cos\theta$ $y = r \sin\theta$ $\hat{\mathbf{r}} = \hat{\mathbf{x}} \cos\theta + \hat{\mathbf{y}} \sin\theta$ (2D polar)	$x = r_c \cos\theta$ $y = r_c \sin\theta$ $\hat{\mathbf{r}}_c = \hat{\mathbf{x}} \cos\theta + \hat{\mathbf{y}} \sin\theta$ (cylindrical)	$x = r \sin\theta \cos\varphi$ $y = r \sin\theta \sin\varphi$ $z = r \cos\theta$ $\hat{\mathbf{r}} = \hat{\mathbf{x}} \cos\varphi \sin\theta$ $+\hat{\mathbf{y}} \sin\varphi \sin\theta + \hat{\mathbf{z}} \cos\theta$ (spherical)
Differential Arc Length		
$ds = dx$ (along x)	$ds = dy$ or dz (along y or z)	$ds = a d\theta$ (circular arc of radius a)
Differential Area Element		
$dA = dxdy$ (polygon in xy plane)	$dA = rdrd\theta$ (pie slice, disc, thick ring)	$dA = a^2 \sin\theta \, d\theta d\varphi$ (sphere of radius a)
Differential Volume Element		
$dV = dxdydz$ (bounded by flat sides)	$dV = r_c dr_c d\theta dz$ (cylinder or cone)	$dV = r^2 \sin\theta \, drd\theta d\varphi$ (spherical)

Symbol	Name	SI Units
dq	differential charge element	C
Q	total charge of the object	C
$\vec{\mathbf{E}}$	electric field	N/C or V/m
k	Coulomb's constant	$\frac{\text{N}\cdot\text{m}^2}{\text{C}^2}$ or $\frac{\text{kg}\cdot\text{m}^3}{\text{C}^2\cdot\text{s}^2}$
$\vec{\mathbf{R}}$	a vector from each dq to the field point	m
$\hat{\mathbf{R}}$	a unit vector from each dq toward the field point	unitless
R	the distance from each dq to the field point	m
x, y, z	Cartesian coordinates of dq	m, m, m
$\hat{\mathbf{x}}, \hat{\mathbf{y}}, \hat{\mathbf{z}}$	unit vectors along the $+x$-, $+y$-, $+z$-axes	unitless
r, θ	2D polar coordinates of dq	m, rad
r_c, θ, z	cylindrical coordinates of dq	m, rad, m
r, θ, φ	spherical coordinates of dq	m, rad, rad
$\hat{\mathbf{r}}, \hat{\boldsymbol{\theta}}, \hat{\boldsymbol{\varphi}}$	unit vectors along spherical coordinate axes	unitless
$\hat{\mathbf{r}}_c$	a unit vector pointing away from the $+z$-axis	unitless
λ	linear charge density (for an arc)	C/m
σ	surface charge density (for an area)	C/m^2
ρ	volume charge density	C/m^3
ds	differential arc length	m
dA	differential area element	m^2
dV	differential volume element	m^3

Note: The symbols λ, σ, and ρ are the lowercase Greek letters lambda, sigma, and rho.

Example 38. A uniformly charged rod of length L lies on the y-axis with one end at the origin, as illustrated below. The positively charged rod has total charge Q. Derive an equation for the electric field at the point $(0, -a)$, where a is a constant, in terms of k, Q, a, L, and appropriate unit vectors.

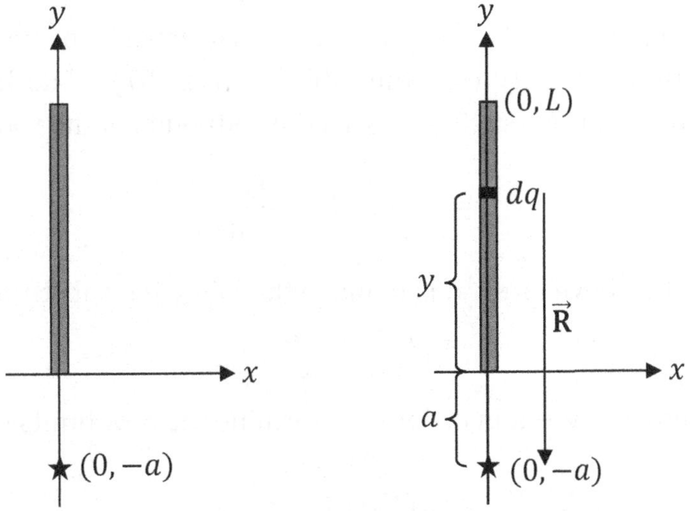

Solution. Begin with a labeled diagram: See the diagram above on the right. Draw a representative dq. Draw $\vec{\mathbf{R}}$ from the source, dq, to the field point $(0, -a)$. When we perform the integration, we effectively integrate over every dq that makes up the rod. Start the math with the electric field integral.

$$\vec{\mathbf{E}} = k \int \frac{\hat{\mathbf{R}}}{R^2} dq$$

Examine the right diagram above:

- The vector $\vec{\mathbf{R}}$ extends from the source (each dq that makes up the rod) to the field point $(0, -a)$. Note that $\vec{\mathbf{R}}$ has a different length for each dq that makes up the rod. Therefore, its magnitude, R, can't come out of the integral. From the diagram above on the right, you should see that $R = y + a$ (since dq is y units to the origin and another a units to the field point).
- The unit vector $\hat{\mathbf{R}}$ points one unit in the direction of $\vec{\mathbf{R}}$. From the diagram above on the right, you should see that $\vec{\mathbf{R}}$ points in the $-y$-direction. Therefore, $\hat{\mathbf{R}} = -\hat{\mathbf{y}}$ (since $\hat{\mathbf{y}}$ is a unit vector that points along the y-axis).

Substitute the expressions for R and $\hat{\mathbf{R}}$ into the electric field integral.

$$\vec{\mathbf{E}} = k \int \frac{\hat{\mathbf{R}}}{R^2} dq = k \int \frac{-\hat{\mathbf{y}}}{(y + a)^2} dq$$

For a thin rod, we write $dq = \lambda ds$ (see the table on page 57). Note that $-\hat{\mathbf{y}}$ is a constant and may come out of the integral, since $\hat{\mathbf{y}}$ is exactly one unit long and always points along the $+y$-axis. In contrast, the symbol y is **not** a constant and may **not** come out of the integral.

$$\vec{\mathbf{E}} = -k\hat{\mathbf{y}} \int \frac{dq}{(y+a)^2} = -k\hat{\mathbf{y}} \int \frac{\lambda ds}{(y+a)^2}$$

Since the rod has **uniform** charge density, we may pull λ out of the integral.

$$\vec{\mathbf{E}} = -k\lambda\hat{\mathbf{y}} \int \frac{ds}{(y+a)^2}$$

For a rod lying along the y-axis, we work with Cartesian coordinates and write the differential arc length as $ds = dy$ (see the table on page 57). The limits of integration correspond to the length of the rod: $0 \leq y \leq L$ (the endpoints of the rod).

$$\vec{\mathbf{E}} = -k\lambda\hat{\mathbf{y}} \int_{y=0}^{L} \frac{dy}{(y+a)^2}$$

One way to perform the above integral is to make the following substitution (Chapter 6).

$$u = y + a$$

$$du = dy$$

Plug the limits of y into $u = y + a$ in order to determine the new limits of integration.

$$u(0) = a$$

$$u(L) = L + a$$

Substitute the expressions and limits into the previous integral.

$$\vec{\mathbf{E}} = -k\lambda\hat{\mathbf{y}} \int_{u=a}^{L+a} \frac{du}{u^2}$$

Apply the rule from algebra that $\frac{1}{u^2} = u^{-2}$.

$$\vec{\mathbf{E}} = -k\lambda\hat{\mathbf{y}} \int_{u=a}^{L+a} u^{-2}\,du = -k\lambda\hat{\mathbf{y}} \left[\frac{u^{-2+1}}{-2+1}\right]_{u=a}^{L+a} = -k\lambda\hat{\mathbf{y}}[-u^{-1}]_{u=a}^{L+a} = k\lambda\hat{\mathbf{y}}[u^{-1}]_{u=a}^{L+a}$$

Note that $-2 + 1 = -1$. The two minus signs combine to make a plus sign: $(-1)(-1) = +1$. Apply the rule from algebra that $\frac{1}{u} = u^{-1}$.

$$\vec{\mathbf{E}} = k\lambda\hat{\mathbf{y}} \left[\frac{1}{u}\right]_{u=a}^{L+a} = k\lambda\hat{\mathbf{y}} \left(\frac{1}{L+a} - \frac{1}{a}\right)$$

In order to subtract the fractions, make a **common denominator** of $a(L + a)$. Multiply $\frac{1}{L+a}$ by $\frac{a}{a}$ to make $\frac{a}{a(L+a)}$ and multiply $\frac{1}{a}$ by $\frac{L+a}{L+a}$ to make $\frac{L+a}{a(L+a)}$.

$$\vec{\mathbf{E}} = k\lambda\hat{\mathbf{y}} \left[\frac{a}{a(L+a)} - \frac{L+a}{a(L+a)}\right] = k\lambda\hat{\mathbf{y}} \left[\frac{a-(L+a)}{a(L+a)}\right]$$

Distribute the minus sign: $a - (L + a) = a - L - a = -L$.

$$\vec{\mathbf{E}} = k\lambda\hat{\mathbf{y}} \left[\frac{-L}{a(L+a)}\right] = \frac{-k\lambda L}{a(L+a)}\hat{\mathbf{y}}$$

We're not finished yet. According to the directions, we must eliminate the constant λ from our answer. The way to do this is to integrate over dq to find the total charge of the rod, Q. A convenient feature of this integral is that we may use the same substitutions from before.

$$Q = \int \lambda\, ds = \lambda \int_{y=0}^{L} dy = \lambda L$$

Divide both sides of the equation by L.

$$\lambda = \frac{Q}{L}$$

Substitute this expression for λ into our previous expression for $\vec{\mathbf{E}}$.

$$\vec{\mathbf{E}} = \frac{-k\lambda L}{a(L+a)}\hat{\mathbf{y}} = \frac{-kL}{a(L+a)}\left(\frac{Q}{L}\right)\hat{\mathbf{y}} = \frac{-kQ}{a(L+a)}\hat{\mathbf{y}}$$

The electric field at the point $(0, -a)$ is $\vec{\mathbf{E}} = \frac{-kQ}{a(L+a)}\hat{\mathbf{y}}$. It has a magnitude of $E = \frac{kQ}{a(L+a)}$ and a direction of $-\hat{\mathbf{y}}$. From the diagrams at the beginning of the solution, it should be clear that a positive test charge placed at $(0, -a)$ would be pushed downward along $-\hat{\mathbf{y}}$.

Tip: Check the units of your answer for consistency, as this can help to catch mistakes. Any expression for electric field should include the units of k times the unit of charge divided by the unit of length squared. Our answer, $E = \frac{kQ}{a(L+a)}$, meets this criteria, since $a(L+a)$ has units of length squared (a is in meters and $L + a$ is also in meters).

Example 39. A rod has non-uniform charge density $\lambda = \beta|x|$ (note the absolute values, which make λ nonnegative), where β is a constant, and endpoints at $\left(-\frac{L}{2}, 0\right)$ and $\left(\frac{L}{2}, 0\right)$, where L is a constant, as illustrated below. The positively charged rod has total charge Q. Derive an equation for the electric field at the point $(0, p)$, where p is a constant, in terms of k, Q, L, p, and appropriate unit vectors.

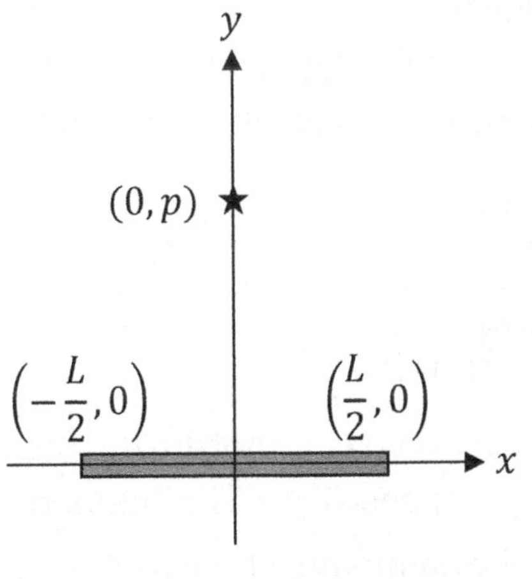

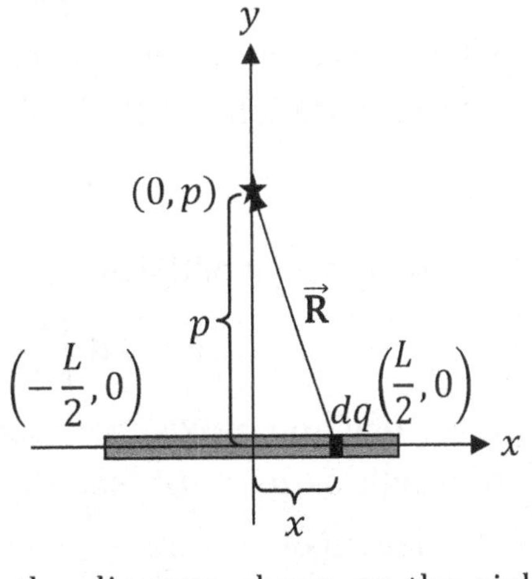

Solution. Begin with a labeled diagram. See the diagram above on the right. Draw a representative dq. Draw $\vec{\mathbf{R}}$ from the source, dq, to the field point $(0, p)$. When we perform the integration, we effectively integrate over every dq that makes up the rod. Start the math with the electric field integral.

$$\vec{\mathbf{E}} = k\int \frac{\hat{\mathbf{R}}}{R^2}dq$$

Examine the right diagram on the previous page:

- The vector $\vec{\mathbf{R}}$ extends from the source (each dq that makes up the rod) to the field point $(0, p)$. Note that $\vec{\mathbf{R}}$ has a different length for each dq that makes up the rod. Therefore, its magnitude, R, can't come out of the integral. What we can do is express R in terms of the constant p and the variable x using the Pythagorean theorem: $R = \sqrt{x^2 + p^2}$.
- The unit vector $\hat{\mathbf{R}}$ points one unit in the direction of $\vec{\mathbf{R}}$. Since $\vec{\mathbf{R}}$ points in a different direction for each dq that makes up the rod, its direction, $\hat{\mathbf{R}}$, can't come out of the integral. $\vec{\mathbf{R}}$ extends x units to the left (along $-\hat{\mathbf{x}}$)* and p units up (along $\hat{\mathbf{y}}$). Therefore, $\vec{\mathbf{R}} = -x\hat{\mathbf{x}} + p\hat{\mathbf{y}}$. Find the unit vector $\hat{\mathbf{R}}$ by dividing $\vec{\mathbf{R}}$ by R.†

$$\hat{\mathbf{R}} = \frac{\vec{\mathbf{R}}}{R} = \frac{-x\hat{\mathbf{x}} + p\hat{\mathbf{y}}}{\sqrt{x^2 + p^2}}$$

Substitute the expressions for R and $\hat{\mathbf{R}}$ into the electric field integral. We will write $\frac{\hat{\mathbf{R}}}{R^2}$ as $\hat{\mathbf{R}}\frac{1}{R^2}$ in order to make these substitutions clearly.

$$\vec{\mathbf{E}} = k\int \frac{\hat{\mathbf{R}}}{R^2}dq = k\int \hat{\mathbf{R}}\frac{1}{R^2}dq = k\int \frac{-x\hat{\mathbf{x}} + p\hat{\mathbf{y}}}{\sqrt{x^2 + p^2}}\frac{1}{x^2 + p^2}\,dq = k\int \frac{-x\hat{\mathbf{x}} + p\hat{\mathbf{y}}}{(x^2 + p^2)^{3/2}}\,dq$$

Note that $R^2 = \left(\sqrt{x^2 + p^2}\right)^2 = x^2 + p^2$ and that $\sqrt{x^2 + p^2}(x^2 + p^2) = (x^2 + p^2)^{3/2}$, where we applied the rule from algebra that $\sqrt{u}\,u = u^{1/2}u^1 = u^{1/2+1} = u^{3/2}$, where $u = x^2 + p^2$. For a thin rod, we write $dq = \lambda ds$ (see the table on page 57).

$$\vec{\mathbf{E}} = k\int \frac{-x\hat{\mathbf{x}} + p\hat{\mathbf{y}}}{(x^2 + p^2)^{3/2}}\lambda ds$$

Since the rod has non-uniform charge density, we may **not** pull λ out of the integral. Instead, make the substitution $\lambda = \beta|x|$ using the expression given in the problem.

$$\vec{\mathbf{E}} = k\int \frac{-x\hat{\mathbf{x}} + p\hat{\mathbf{y}}}{(x^2 + p^2)^{3/2}}\beta|x|ds$$

The constant β may come out of the integral.

$$\vec{\mathbf{E}} = k\beta\int \frac{-x\hat{\mathbf{x}} + p\hat{\mathbf{y}}}{(x^2 + p^2)^{3/2}}|x|ds$$

For a rod lying along the x-axis, we work with Cartesian coordinates and write the differential arc length as $ds = dx$ (see the table on page 57). The limits of integration correspond to the length of the rod: $-\frac{L}{2} \leq x < \frac{L}{2}$ (the endpoints of the rod).

* When x is negative, this is self-correcting: The minus sign in x (when it's negative) will make $-x\hat{\mathbf{x}}$ positive (since two minus signs make a plus sign), meaning that $\hat{\mathbf{R}}$ points up and to the right (instead of up and to the left) when x is negative.

† The vector $\vec{\mathbf{R}}$ equals its magnitude (R) times its direction ($\hat{\mathbf{R}}$): $\vec{\mathbf{R}} = R\hat{\mathbf{R}}$. Solving for $\hat{\mathbf{R}}$, we get $\hat{\mathbf{R}} = \frac{\vec{\mathbf{R}}}{R}$.

$$\vec{\mathbf{E}} = k\beta \int_{y=-L/2}^{L/2} \frac{-x\hat{\mathbf{x}} + p\hat{\mathbf{y}}}{(x^2 + p^2)^{3/2}} |x| dx$$

Split this into two separate integrals using the rule $\int (f_1 + f_2)\, dx = \int f_1\, dx + \int f_2\, dx$. Note that $\frac{-x\hat{\mathbf{x}}+p\hat{\mathbf{y}}}{(x^2+p^2)^{3/2}} = \frac{-x\hat{\mathbf{x}}}{(x^2+p^2)^{3/2}} + \frac{p\hat{\mathbf{y}}}{(x^2+p^2)^{3/2}}$.

$$\vec{\mathbf{E}} = -k\beta \int_{y=-L/2}^{L/2} \frac{x\hat{\mathbf{x}}}{(x^2 + p^2)^{3/2}} |x| dx + k\beta \int_{y=-L/2}^{L/2} \frac{p\hat{\mathbf{y}}}{(x^2 + p^2)^{3/2}} |x| dx$$

The first integral is zero. One way to see this (without doing all of the math!) is to look at the picture. By symmetry, the horizontal component of the electric field will vanish. Therefore, the $\hat{\mathbf{x}}$-integral is zero: $\vec{\mathbf{E}}$ will be along $\hat{\mathbf{y}}$ (straight up). You may pull $p\hat{\mathbf{y}}$ out of the second integral since they are constants.

$$\vec{\mathbf{E}} = 0 + k\beta p\hat{\mathbf{y}} \int_{y=-L/2}^{L/2} \frac{|x| dx}{(x^2 + p^2)^{3/2}}$$

The integrand is an **even** function of x because $\frac{|x|}{(x^2+p^2)^{3/2}}$ is the same whether you plug in $+x$ or $-x$, since $|-x| = |x|$ and since $(-x)^2 = x^2$. Therefore, we would obtain the same result for the integral from $x = -\frac{L}{2}$ to 0 as we would for the integral from $x = 0$ to $\frac{L}{2}$. This means that we can change the limits to $x = 0$ to $\frac{L}{2}$ and add a factor of 2 in front of the integral. Now we may drop the absolute values from $|x|$, since x won't be negative when we integrate from $x = 0$ to $\frac{L}{2}$.

$$\vec{\mathbf{E}} = 2k\beta p\hat{\mathbf{y}} \int_{y=0}^{L/2} \frac{x dx}{(x^2 + p^2)^{3/2}}$$

This integral can be performed via the following trigonometric substitution (Chapter 6):

$$x = p \tan\theta$$
$$dx = p \sec^2\theta\, d\theta$$

It's easier to ignore the new limits for now and deal with them later. Note that the denominator of the integral simplifies as follows, using the trig identity $\tan^2\theta + 1 = \sec^2\theta$:

$$(x^2 + p^2)^{3/2} = [(p \tan\theta)^2 + p^2]^{3/2} = [p^2(\tan^2\theta + 1)]^{3/2} = (p^2 \sec^2\theta)^{3/2} = p^3 \sec^3\theta$$

In the last step, we applied the rule from algebra that $(a^2x^2)^{3/2} = (a^2)^{3/2}(x^2)^{3/2} = a^3x^3$. Substitute the above expressions for x, dx, and $(a^2 + x^2)^{3/2}$ into the previous integral.

$$\vec{\mathbf{E}} = 2k\beta p\hat{\mathbf{y}} \int \frac{(p\tan\theta)(p\sec^2\theta\, d\theta)}{p^3\sec^3\theta} = 2k\beta p\hat{\mathbf{y}} \int \frac{\tan\theta\, d\theta}{p\sec\theta} = 2k\beta\hat{\mathbf{y}} \int \frac{\tan\theta\, d\theta}{\sec\theta}$$

Recall from trig that $\tan\theta = \frac{\sin\theta}{\cos\theta}$ and $\sec\theta = \frac{1}{\cos\theta}$. Therefore, it follows that:

$$\frac{\tan\theta}{\sec\theta} = \frac{\sin\theta}{\cos\theta} \div \frac{1}{\cos\theta} = \frac{\sin\theta}{\cos\theta} \times \frac{\cos\theta}{1} = \sin\theta$$

The electric field integral becomes:

$$\vec{\mathbf{E}} = 2k\beta\hat{\mathbf{y}}\int \sin\theta \, d\theta = 2k\beta\hat{\mathbf{y}}[-\cos\theta]_{\theta_0}^{\theta}$$

We still need to evaluate the anti-derivative, $-\cos\theta$, over the limits. Instead of figuring out what the limits over θ are, it's simpler to draw a right triangle to express $\cos\theta$ in terms of x, which will allow us to evaluate the function over the old limits. Since $x = p\tan\theta$, which means that $\tan\theta = \frac{x}{p}$, draw a right triangle with x opposite to θ and with p adjacent to θ.

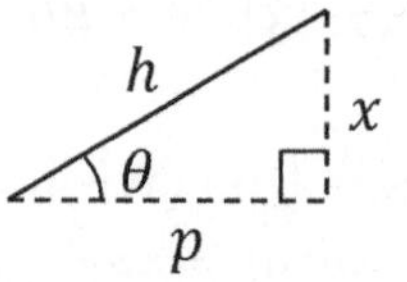

Apply the Pythagorean theorem to find the hypotenuse.

$$h = \sqrt{x^2 + p^2}$$

From the right triangle, you should see that $\cos\theta = \frac{p}{h} = \frac{p}{\sqrt{x^2+p^2}}$. Substitute this into the previous expression for electric field and evaluate $-\cos\theta = -\frac{p}{\sqrt{x^2+p^2}}$ over the limits from $x = 0$ to $\frac{L}{2}$.

$$\vec{\mathbf{E}} = 2k\beta\hat{\mathbf{y}}[-\cos\theta]_{\theta_0}^{\theta} = 2k\beta\hat{\mathbf{y}}\left[-\frac{p}{\sqrt{x^2+p^2}}\right]_{x=0}^{L/2}$$

$$\vec{\mathbf{E}} = 2k\beta\hat{\mathbf{y}}\left(-\frac{p}{\sqrt{\frac{L^2}{4}+p^2}} + \frac{p}{p}\right) = 2k\beta\hat{\mathbf{y}}\left(-\frac{p}{\sqrt{\frac{L^2}{4}+p^2}} + 1\right) = 2k\beta\hat{\mathbf{y}}\left(1 - \frac{p}{\sqrt{\frac{L^2}{4}+p^2}}\right)$$

Note that $-\frac{p}{\sqrt{\frac{L^2}{4}+p^2}} + 1 = 1 - \frac{p}{\sqrt{\frac{L^2}{4}+p^2}}$.

We're not finished yet. According to the directions, we must eliminate the constant β from our answer. The way to do this is to integrate over dq to find the total charge of the rod, Q. We use the same substitutions from before.

$$Q = \int \lambda \, ds = \int \beta|x| \, ds = \beta\int_{x=-L/2}^{L/2} |x| \, dx = 2\beta\int_{x=0}^{L/2} x \, dx = 2\beta\left[\frac{x^2}{2}\right]_{x=0}^{L/2} = \beta[x^2]_{x=0}^{L/2} = \frac{\beta L^2}{4}$$

Multiply both sides of the equation by 4 and divide both sides by L^2.

$$\beta = \frac{4Q}{L^2}$$

Substitute this expression for β into our previous expression for $\vec{\mathbf{E}}$.

$$\vec{\mathbf{E}} = 2k\beta\hat{\mathbf{y}}\left(1 - \frac{p}{\sqrt{\frac{L^2}{4}+p^2}}\right) = 2k\left(\frac{4Q}{L^2}\right)\hat{\mathbf{y}}\left(1 - \frac{p}{\sqrt{\frac{L^2}{4}+p^2}}\right) = \frac{8kQ}{L^2}\hat{\mathbf{y}}\left(1 - \frac{p}{\sqrt{\frac{L^2}{4}+p^2}}\right)$$

Uniformly Charged Ring Example: A Prelude to Example 40. A thin wire is bent into the shape of the semicircle illustrated below. The radius of the semicircle is denoted by the symbol a and the negatively charged wire has total charge $-Q$ and uniform charge density. Derive an equation for the electric field at the origin in terms of k, Q, a, and appropriate unit vectors.

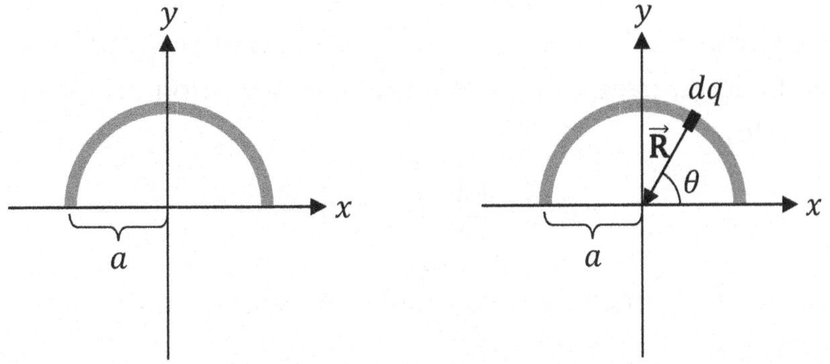

Solution. Begin with a labeled diagram. See the diagram above on the right. Draw a representative dq: Since this is a very thin semicircular wire (and <u>not</u> a solid semicircle – that is, it's <u>not</u> half of a solid disc, nor is it a thick ring), dq must lie on the circumference and not inside the semicircle. Draw $\vec{\mathbf{R}}$ from the source, dq, to the field point (at the origin). When we perform the integration, we effectively integrate over every dq that makes up the semicircle. Start the math with the electric field integral.

$$\vec{\mathbf{E}} = k \int \frac{\hat{\mathbf{R}}}{R^2} dq$$

Examine the right diagram above:

- The vector $\vec{\mathbf{R}}$ extends from the source (each dq that makes up the semicircle) to the field point $(0,0)$. The vector $\vec{\mathbf{R}}$ has the same length for each dq that makes up the semicircle. Its magnitude, R, equals the radius of the semicircle: $R = a$. (Note that this would <u>not</u> be the case for a thick ring or a solid semicircle.)
- The unit vector $\hat{\mathbf{R}}$ points one unit in the direction of $\vec{\mathbf{R}}$. For each dq that makes up the semicircle, $\vec{\mathbf{R}}$ points in a different direction, so its direction, $\hat{\mathbf{R}}$, can't come out of the integral. From the diagram above on the right, you should see that $\vec{\mathbf{R}}$ points inward (toward the origin). Since the unit vector $\hat{\mathbf{r}}$ of 2D polar coordinates points **outward**, we can write $\hat{\mathbf{R}} = -\hat{\mathbf{r}}$ because $\hat{\mathbf{R}}$ points **inward**.

Substitute the expressions $R = a$ and $\hat{\mathbf{R}} = -\hat{\mathbf{r}}$ into the electric field integral.

$$\vec{\mathbf{E}} = k \int \frac{\hat{\mathbf{R}}}{R^2} dq = k \int \frac{-\hat{\mathbf{r}}}{a^2} dq$$

We may pull $-\frac{1}{a^2}$ out of the integral, since the radius (a) is constant.

$$\vec{\mathbf{E}} = -\frac{k}{a^2} \int \hat{\mathbf{r}}\, dq$$

For a thin semicircular wire, we write $dq = \lambda ds$ (see the table on page 57).

$$\vec{\mathbf{E}} = -\frac{k}{a^2}\int \hat{\mathbf{r}}\,\lambda ds$$

Since the semicircle has **uniform** charge density, we may pull λ out of the integral.

$$\vec{\mathbf{E}} = -\frac{k\lambda}{a^2}\int \hat{\mathbf{r}}\,ds$$

For a semicircle, we work with 2D polar coordinates and write the differential arc length as $ds = a\,d\theta$ (see the table on page 57). The limits of integration are from $\theta = 0$ to $\theta = \pi$ **radians** (for half a circle).

$$\vec{\mathbf{E}} = -\frac{k\lambda}{a^2}\int_{\theta=0}^{\pi} \hat{\mathbf{r}}\,(a\,d\theta)$$

Since the radius (a) of the semicircle is constant, we may pull it out of the integral. Note that $\frac{1}{a^2}(a) = \frac{a}{a^2} = \frac{1}{a}$.

$$\vec{\mathbf{E}} = -\frac{k\lambda}{a}\int_{\theta=0}^{\pi} \hat{\mathbf{r}}\,d\theta$$

Since $\hat{\mathbf{r}}$ points in a different direction for each value of θ, it is not constant and therefore may **not** come out of the integral. What we need to do is express $\hat{\mathbf{r}}$ in terms of θ. You can find such an equation on page 57. For this problem, we need the expression for $\hat{\mathbf{r}}$ in 2D polar coordinates (not the $\hat{\mathbf{r}}$ of spherical coordinates).‡

$$\hat{\mathbf{r}} = \hat{\mathbf{x}}\cos\theta + \hat{\mathbf{y}}\sin\theta$$

Substitute this expression into the electric field integral.

$$\vec{\mathbf{E}} = -\frac{k\lambda}{a}\int_{\theta=0}^{\pi} (\hat{\mathbf{x}}\cos\theta + \hat{\mathbf{y}}\sin\theta)\,d\theta$$

Separate this integral into two integrals – one for each term.

$$\vec{\mathbf{E}} = -\frac{k\lambda}{a}\int_{\theta=0}^{\pi} \hat{\mathbf{x}}\cos\theta\,d\theta - \frac{k\lambda}{a}\int_{\theta=0}^{\pi} \hat{\mathbf{y}}\sin\theta\,d\theta$$

Unlike the 2D polar unit vector $\hat{\mathbf{r}}$, the Cartesian unit vectors $\hat{\mathbf{x}}$ and $\hat{\mathbf{y}}$ are constants (for example, $\hat{\mathbf{x}}$ always points one unit along the $+x$-axis). Therefore, we may factor $\hat{\mathbf{x}}$ and $\hat{\mathbf{y}}$ out of the integrals.

‡ Here is where this equation comes from. First, any 2D vector can be expressed in terms of unit vectors in the form $\vec{\mathbf{A}} = A_x\hat{\mathbf{x}} + A_y\hat{\mathbf{y}}$. Next, the position vector is special in that its x- and y-components are the x- and y-coordinates: $\vec{\mathbf{r}} = x\hat{\mathbf{x}} + y\hat{\mathbf{y}}$. An alternative way to express the position vector is $\vec{\mathbf{r}} = r\hat{\mathbf{r}}$. Going x units along $\hat{\mathbf{x}}$ and then going y units along $\hat{\mathbf{y}}$ is equivalent to going r units outward along $\hat{\mathbf{r}}$. (See the 2D polar coordinate diagram from Chapter 6 on page 53, and note that $\hat{\mathbf{r}}$ is along the hypotenuse of the right triangle.) Set these two equations equal to one another: $x\hat{\mathbf{x}} + y\hat{\mathbf{y}} = r\hat{\mathbf{r}}$. Divide both sides of the equation by r: $\hat{\mathbf{r}} = \frac{x}{r}\hat{\mathbf{x}} + \frac{y}{r}\hat{\mathbf{y}}$. Recall that $x = r\cos\theta$ and $y = r\sin\theta$, such that $\frac{x}{r} = \cos\theta$ and $\frac{y}{r} = \sin\theta$. Therefore, $\hat{\mathbf{r}} = \hat{\mathbf{x}}\cos\theta + \hat{\mathbf{y}}\sin\theta$.

$$\vec{\mathbf{E}} = -\frac{k\lambda}{a}\hat{\mathbf{x}} \int_{\theta=0}^{\pi} \cos\theta \, d\theta - \frac{k\lambda}{a}\hat{\mathbf{y}} \int_{\theta=0}^{\pi} \sin\theta \, d\theta$$

$$\vec{\mathbf{E}} = -\frac{k\lambda}{a}\hat{\mathbf{x}}[\sin\theta]_{\theta=0}^{\pi} - \frac{k\lambda}{a}\hat{\mathbf{y}}[-\cos\theta]_{\theta=0}^{\pi}$$

$$\vec{\mathbf{E}} = -\frac{k\lambda}{a}\hat{\mathbf{x}}(\sin\pi - \sin 0) - \frac{k\lambda}{a}\hat{\mathbf{y}}(-\cos\pi + \cos 0)$$

$$\vec{\mathbf{E}} = -\frac{k\lambda}{a}\hat{\mathbf{x}}(0 - 0) - \frac{k\lambda}{a}\hat{\mathbf{y}}(1 + 1) = -\frac{k\lambda}{a}\hat{\mathbf{y}}(2) = -\frac{2k\lambda}{a}\hat{\mathbf{y}}$$

Note that π radians $= 180°$. According to the directions, we must eliminate the constant λ from our answer. The way to do this is to integrate over dq to find the total charge of the semicircle, $-Q$. This integral involves the same substitutions from before. Note that the integral equals $-Q$ because the problem states that the semicircle is **negative** and has a total charge of $-Q$. We **<u>must</u>** use **radians** (instead of degrees) for the limits of θ (unless **<u>every</u>** θ appears in the argument of a trig function, which is **<u>not</u>** the case here).

$$-Q = \int dq = \int \lambda \, ds = \lambda \int ds = \lambda \int a \, d\theta = a\lambda \int_{\theta=0}^{\pi} d\theta = a\lambda[\theta]_{\theta=0}^{\pi} = a\lambda(\pi - 0) = \pi a\lambda$$

Multiply both sides of the equation by -1 in order to solve for Q. In our notation, the symbol Q is positive, such that we can write the total charge of the semicircle as $-Q$, which is negative. Also, the symbol λ is negative (so that the two minus signs in the equation below make Q positive, such that $-Q$ is negative).

$$Q = -\pi a\lambda$$

The signs get a little tricky with a negatively charged object. However, there is a simpler alternative: You could instead solve the problem for a positively charged object, and simply change the direction of the final answer at the end of the solution to deal with the negative charge. (We didn't do it that way, but we could have. You would get the same final answer.)

Solve for λ in terms of Q: Divide both sides of the equation by πa.

$$\lambda = \frac{-Q}{\pi a}$$

Substitute this expression for λ into our previous expression for $\vec{\mathbf{E}}$.

$$\vec{\mathbf{E}} = -\frac{2k\lambda}{a}\hat{\mathbf{y}} = -\frac{2k}{a}\left(\frac{-Q}{\pi a}\right)\hat{\mathbf{y}} = \frac{2kQ}{\pi a^2}\hat{\mathbf{y}}$$

The electric field at the origin is $\vec{\mathbf{E}} = \frac{2kQ}{\pi a^2}\hat{\mathbf{y}}$. It has a magnitude of $E = \frac{2kQ}{\pi a^2}$ and a direction of $\hat{\mathbf{E}} = +\hat{\mathbf{y}}$. If you study the picture, by symmetry it should make sense that the electric field points straight **up** at the origin, along the $+y$-axis: A positive "test" charge placed at the origin would be attracted upward towards the negatively charged semicircle. Note that the horizontal component of the electric field cancels out.

Example 40. Two thin wires are bent into the shape of semicircles, as illustrated below. Each semicircle has the same radius (a) and is uniformly charged. The top semicircle has charge $+Q$, while the bottom semicircle has charge $-Q$. Derive an equation for the electric field at the origin in terms of k, Q, a, and appropriate unit vectors.

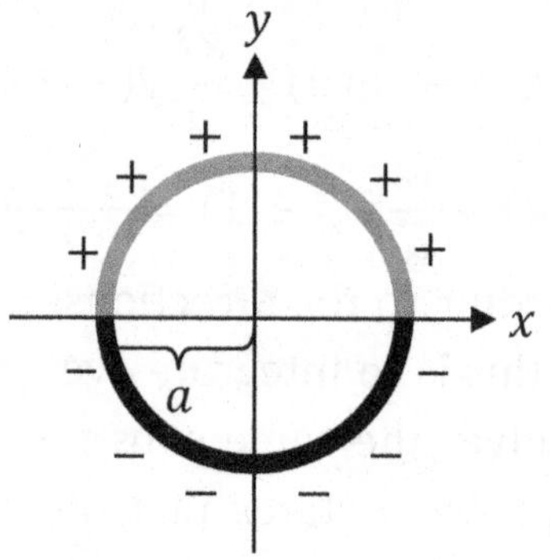

Solution. Apply the principle of superposition (see Chapter 3). First, ignore the negative semicircle (as in the left diagram below) and find the electric field at the origin due to the positive semicircle. Call this $\vec{\mathbf{E}}_1$. This part of the solution is nearly identical to the previous example. The only difference is that here the electric field has the opposite direction since this semicircle is **positive**, whereas it was negative in the example. Therefore, $\vec{\mathbf{E}}_1$ is the same as the expression on the bottom of the previous page, except for an overall negative sign. A positive "test" charge placed at the origin would be repelled downward away from the positively charged semicircle.

$$\vec{\mathbf{E}}_1 = -\frac{2kQ}{\pi a^2}\hat{\mathbf{y}}$$

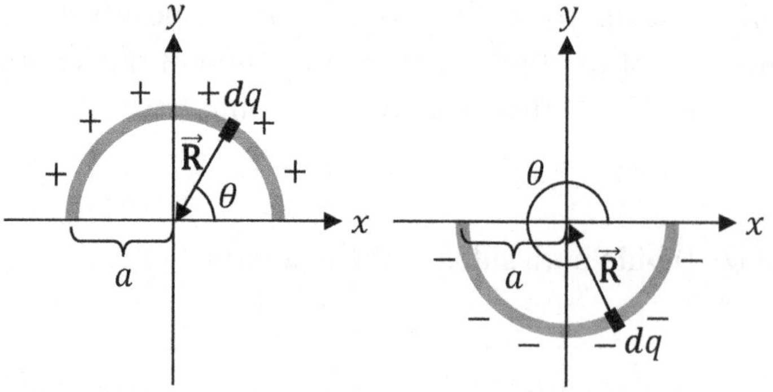

Next, ignore the positive semicircle (as in the right diagram above) and find the electric field at the origin due to the negative semicircle. Call this $\vec{\mathbf{E}}_2$. The answer is identical to the previous step. A positive "test" charge placed at the origin would be attracted downward towards the negatively charged semicircle.

$$\vec{\mathbf{E}}_2 = -\frac{2kQ}{\pi a^2}\hat{\mathbf{y}}$$

Note that $\vec{\mathbf{E}}_1$ and $\vec{\mathbf{E}}_2$ are both equal and both point downward: A positive "test" charge at the origin would be repelled downward by the positive semicircle, and would also be attracted downward to the negative semicircle. That's why $\vec{\mathbf{E}}_1$ and $\vec{\mathbf{E}}_2$ both point downward. Add the vectors $\vec{\mathbf{E}}_1$ and $\vec{\mathbf{E}}_2$ to find the net electric field, $\vec{\mathbf{E}}_{net}$, at the origin. Since $\vec{\mathbf{E}}_1$ and $\vec{\mathbf{E}}_2$ are both equal (in magnitude and direction), the answer is simply two times $\vec{\mathbf{E}}_1$. The final answer for the electric field created at the origin by the composite object is:

$$\vec{\mathbf{E}}_{net} = \vec{\mathbf{E}}_1 + \vec{\mathbf{E}}_2 = -\frac{2kQ}{\pi a^2}\hat{\mathbf{y}} - -\frac{2kQ}{\pi a^2}\hat{\mathbf{y}} = -\frac{4kQ}{\pi a^2}\hat{\mathbf{y}}$$

Example 41. A very thin uniformly charged ring lies in the xy plane, centered about the origin as illustrated below. The radius of the ring is denoted by the symbol a and the positively charged ring has total charge Q. Derive an equation for the electric field at the point $(0,0,p)$ in terms of k, Q, a, p, and appropriate unit vectors.

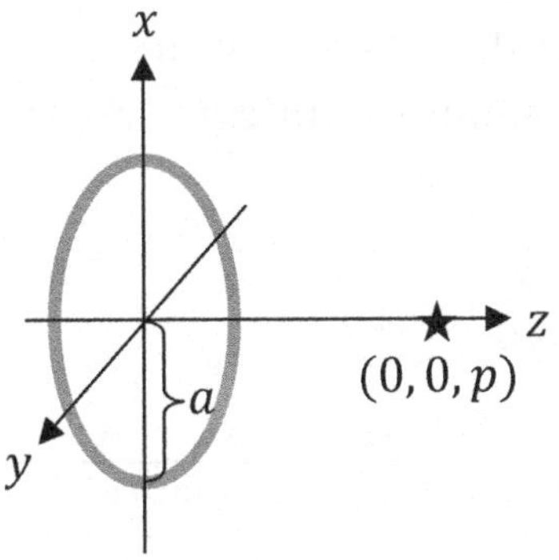

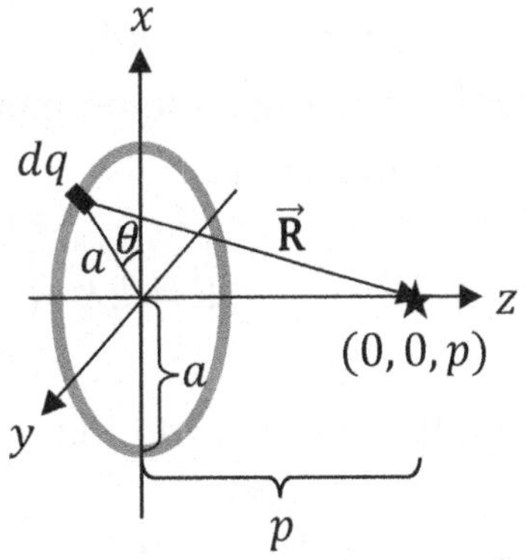

Solution. Begin with a labeled diagram. See the diagram above on the right. Draw a representative dq on the thin ring. Draw $\vec{\mathbf{R}}$ from the source, dq, to the field point $(0,0,p)$. When we perform the integration, we effectively integrate over every dq that makes up the thin ring. Start the math with the electric field integral.

$$\vec{\mathbf{E}} = k\int \frac{\hat{\mathbf{R}}}{R^2}dq$$

Examine the right diagram above:

- The vector $\vec{\mathbf{R}}$ extends from the source (each dq that makes up the ring) to the field point $(0,0,p)$. Express the magnitude of $\vec{\mathbf{R}}$, denoted by R, in terms of the radius a and the constant p using the Pythagorean theorem.

$$R = \sqrt{a^2 + p^2}$$

- The unit vector $\hat{\mathbf{R}}$ points one unit in the direction of $\vec{\mathbf{R}}$. Since $\vec{\mathbf{R}}$ points in a different direction for each dq that makes up the thin ring, its direction, $\hat{\mathbf{R}}$, can't come out of the integral. Note that $\vec{\mathbf{R}}$ extends a units inward, towards the z-axis (along $-\hat{\mathbf{r}}_c$), and p units along the z-axis (along $\hat{\mathbf{z}}$). Therefore, $\vec{\mathbf{R}} = -a\hat{\mathbf{r}}_c + p\hat{\mathbf{z}}$. Note that this is the $\hat{\mathbf{r}}_c$ of **cylindrical** coordinates (Chapter 6). Find the unit vector $\hat{\mathbf{R}}$ by dividing $\vec{\mathbf{R}}$ by R.

$$\hat{\mathbf{R}} = \frac{\vec{\mathbf{R}}}{R} = \frac{-a\hat{\mathbf{r}}_c + p\hat{\mathbf{z}}}{\sqrt{a^2 + p^2}}$$

Substitute the previous expressions for R and $\hat{\mathbf{R}}$ into the electric field integral. We will write $\frac{\hat{\mathbf{R}}}{R^2}$ as $\hat{\mathbf{R}}\frac{1}{R^2}$ in order to make these substitutions clearly. Note that $R^2 = \left(\sqrt{a^2+p^2}\right)^2 = a^2+p^2$ and note that $(a^2+p^2)\sqrt{a^2+p^2} = (a^2+p^2)^1(a^2+p^2)^{1/2} = (a^2+p^2)^{3/2}$.

$$\vec{\mathbf{E}} = k\int \frac{\hat{\mathbf{R}}}{R^2}dq = k\int \hat{\mathbf{R}}\frac{1}{R^2}dq = k\int \frac{-a\hat{\mathbf{r}}_c + p\hat{\mathbf{z}}}{\sqrt{a^2+p^2}}\frac{1}{a^2+p^2}dq = k\int \frac{-a\hat{\mathbf{r}}_c + p\hat{\mathbf{z}}}{(a^2+p^2)^{3/2}}dq$$

For a very thin ring (circular arc), we write $dq = \lambda ds$ (see the table on page 57).

$$\vec{\mathbf{E}} = k\int \frac{-a\hat{\mathbf{r}}_c + p\hat{\mathbf{z}}}{(a^2+p^2)^{3/2}}(\lambda ds)$$

Since the thin ring has **uniform** charge density, we may pull λ out of the integral.

$$\vec{\mathbf{E}} = k\lambda\int \frac{-a\hat{\mathbf{r}}_c + p\hat{\mathbf{z}}}{(a^2+p^2)^{3/2}}ds$$

For a very thin ring, we work with 2D polar coordinates and write the differential arc length as $ds = a\,d\theta$ (see the table on page 57). The limits of integration are from $\theta = 0$ to $\theta = 2\pi$ __radians__ (for full a circle).

$$\vec{\mathbf{E}} = k\lambda\int_{\theta=0}^{2\pi} \frac{-a\hat{\mathbf{r}}_c + p\hat{\mathbf{z}}}{(a^2+p^2)^{3/2}}(a\,d\theta)$$

Split the integral into two parts, noting that $\frac{-a\hat{\mathbf{r}}_c+p\hat{\mathbf{z}}}{(a^2+p^2)^{3/2}} = \frac{-a\hat{\mathbf{r}}_c}{(a^2+p^2)^{3/2}} + \frac{p\hat{\mathbf{z}}}{(a^2+p^2)^{3/2}}$.

$$\vec{\mathbf{E}} = -k\lambda\int_{\theta=0}^{2\pi} \frac{a\hat{\mathbf{r}}_c}{(a^2+p^2)^{3/2}}(a\,d\theta) + k\lambda\int_{\theta=0}^{2\pi} \frac{p\hat{\mathbf{z}}}{(a^2+p^2)^{3/2}}(a\,d\theta)$$

Since a, p, and $\hat{\mathbf{z}}$ are constants (since $\hat{\mathbf{z}}$ is one unit long and points along the $+z$-axis), we may pull them out of the integrals. However, $\hat{\mathbf{r}}_c$ is __not__ a constant, as it points in a different direction for each dq that makes up the thin ring, so $\hat{\mathbf{r}}_c$ may __not__ come out of the integral.

$$\vec{\mathbf{E}} = -\frac{ka^2\lambda}{(a^2+p^2)^{3/2}}\int_{\theta=0}^{2\pi} \hat{\mathbf{r}}_c\,d\theta + \frac{kap\lambda\hat{\mathbf{z}}}{(a^2+p^2)^{3/2}}\int_{\theta=0}^{2\pi} d\theta$$

Study the diagram at the beginning of this example. It should be clear from symmetry that a positive test charge placed at the field point $(0,0,p)$ would be pushed directly to the right, along the $+z$-axis. Therefore, the first integral, which contains $\hat{\mathbf{r}}_c$, will equal zero (though you can do the extra work§ if you like), and we only need to perform the second integral.

$$\vec{\mathbf{E}} = \frac{kap\lambda\hat{\mathbf{z}}}{(a^2+p^2)^{3/2}}\int_{\theta=0}^{2\pi} d\theta = \frac{2\pi kap\lambda\hat{\mathbf{z}}}{(a^2+p^2)^{3/2}}$$

We're not finished yet. According to the directions, we must eliminate the constant λ from our answer. The way to do this is to integrate over dq to find the total charge of the ring, Q.

§ This extra work is straightforward: Make the substitution $\hat{\mathbf{r}}_c = \hat{\mathbf{x}}\cos\theta + \hat{\mathbf{y}}\sin\theta$ (see page 57). Then the integral $\int_{\theta=0}^{2\pi}\hat{\mathbf{r}}_c\,d\theta$ becomes $\hat{\mathbf{x}}\int_{\theta=0}^{2\pi}\cos\theta\,d\theta + \hat{\mathbf{y}}\int_{\theta=0}^{2\pi}\sin\theta\,d\theta$ (since $\hat{\mathbf{x}}$ and $\hat{\mathbf{y}}$ are constants, unlike $\hat{\mathbf{r}}_c$). The integral of either $\cos\theta$ or $\sin\theta$ over one full cycle equals zero.

We use the same substitutions from before.

$$Q = \int dq = \int \lambda\, ds = \lambda \int ds = \lambda \int_{\theta=0}^{2\pi} a\, d\theta = a\lambda \int_{\theta=0}^{2\pi} d\theta = 2\pi a\lambda$$

Divide both sides of the equation by $2\pi a$.

$$\lambda = \frac{Q}{2\pi a}$$

Substitute this expression for λ into our previous expression for $\vec{\mathbf{E}}$.

$$\vec{\mathbf{E}} = \frac{2\pi k a p \lambda \hat{\mathbf{z}}}{(a^2+p^2)^{3/2}} = \frac{2\pi k a p}{(a^2+p^2)^{3/2}}\left(\frac{Q}{2\pi a}\right)\hat{\mathbf{z}} = \frac{kQp}{(a^2+p^2)^{3/2}}\hat{\mathbf{z}}$$

The electric field at the point $(0,0,p)$ is $\vec{\mathbf{E}} = \frac{kQp}{(a^2+p^2)^{3/2}}\hat{\mathbf{z}}$. It has a magnitude of $E = \frac{kQp}{(a^2+p^2)^{3/2}}$ and a direction of $\hat{\mathbf{z}}$.

Example 42. A solid disc with non-uniform charge density $\sigma = \beta r_c$, where β is a constant, lies in the xy plane, centered about the origin as illustrated below. The radius of the disc is denoted by the symbol a and the positively charged disc has total charge Q. Derive an equation for the electric field at the point $(0,0,p)$ in terms of k, Q, a, p, and appropriate unit vectors.

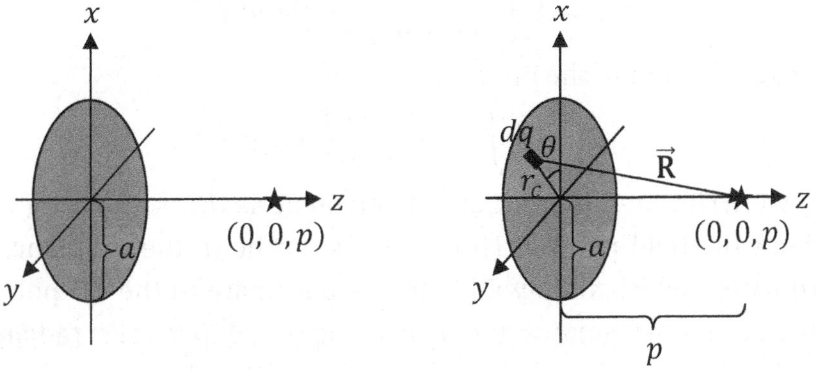

Solution. Begin with a labeled diagram. See the diagram above on the right. Draw a representative dq: Since this is a solid disc (unlike the previous example), most of the dq's lie within the area of the disc, so we drew dq inside the disc (not on its circumference). Draw $\vec{\mathbf{R}}$ from the source, dq, to the field point $(0,0,p)$. When we perform the integration, we effectively integrate over every dq that makes up the solid disc. Start the math with the electric field integral.

$$\vec{\mathbf{E}} = k \int \frac{\hat{\mathbf{R}}}{R^2}\, dq$$

Examine the right diagram above:

- The vector $\vec{\mathbf{R}}$ extends from the source (each dq that makes up the disc) to the field point $(0,0,p)$. Express the magnitude of $\vec{\mathbf{R}}$, denoted by R, in terms of the variable r_c and the constant p using the Pythagorean theorem.

$$R = \sqrt{r_c^2 + p^2}$$

- The unit vector $\hat{\mathbf{R}}$ points one unit in the direction of $\vec{\mathbf{R}}$. Note that $\vec{\mathbf{R}}$ extends r_c units towards the origin (along $-\hat{\mathbf{r}}_c$) and p units along the $+z$-axis (along $\hat{\mathbf{z}}$). Therefore, $\vec{\mathbf{R}} = -r_c\,\hat{\mathbf{r}}_c + p\,\hat{\mathbf{z}}$. Note that this is the $\hat{\mathbf{r}}_c$ of **cylindrical** coordinates (Chapter 6). Find the unit vector $\hat{\mathbf{R}}$ by dividing $\vec{\mathbf{R}}$ by R.

$$\hat{\mathbf{R}} = \frac{\vec{\mathbf{R}}}{R} = \frac{-r_c\,\hat{\mathbf{r}}_c + p\,\hat{\mathbf{z}}}{\sqrt{r_c^2 + p^2}}$$

Substitute the previous expressions for R and $\hat{\mathbf{R}}$ into the electric field integral. We will write $\frac{\hat{\mathbf{R}}}{R^2}$ as $\hat{\mathbf{R}}\frac{1}{R^2}$ in order to make these substitutions clearly. Note that $R^2 = \left(\sqrt{r_c^2 + p^2}\right)^2 = r_c^2 + p^2$ and that $\sqrt{r_c^2 + p^2}(r_c^2 + p^2) = (r_c^2 + p^2)^{1/2+1} = (r_c^2 + p^2)^{3/2}$.

$$\vec{\mathbf{E}} = k\int \frac{\hat{\mathbf{R}}}{R^2}dq = k\int \hat{\mathbf{R}}\frac{1}{R^2}dq = k\int \frac{-r_c\,\hat{\mathbf{r}}_c + p\,\hat{\mathbf{z}}}{\sqrt{r_c^2 + p^2}}\frac{1}{r_c^2 + p^2}dq = k\int \frac{-r_c\,\hat{\mathbf{r}}_c + p\,\hat{\mathbf{z}}}{(r_c^2 + p^2)^{3/2}}dq$$

For a solid disc, we write $dq = \sigma dA$ (see the table on page 57).

$$\vec{\mathbf{E}} = k\int \frac{-r_c\,\hat{\mathbf{r}}_c + p\,\hat{\mathbf{z}}}{(r_c^2 + p^2)^{3/2}}(\sigma dA)$$

Since the solid disc has non-uniform charge density, we may <u>**not**</u> pull σ out of the integral. Instead, make the substitution $\sigma = \beta r_c$ using the expression given in the problem.

$$\vec{\mathbf{E}} = k\int \frac{-r_c\,\hat{\mathbf{r}}_c + p\,\hat{\mathbf{z}}}{(r_c^2 + p^2)^{3/2}}(\beta r_c)dA$$

The constant β may come out of the integral.

$$\vec{\mathbf{E}} = k\beta\int \frac{-r_c\,\hat{\mathbf{r}}_c + p\,\hat{\mathbf{z}}}{(r_c^2 + p^2)^{3/2}}(r_c)dA$$

For a solid disc, we write the differential area element as $dA = r_c dr_c d\theta$ (see the table on page 57), and since the field point at $(0,0,p)$ does not lie in the xy plane, we work with cylindrical coordinates (which simply adds the z-coordinate to the 2D polar coordinates). The limits of integration are from $r_c = 0$ to $r_c = a$ and $\theta = 0$ to $\theta = 2\pi$ <u>**radians**</u>.

$$\vec{\mathbf{E}} = k\beta \int_{r_c=0}^{a}\int_{\theta=0}^{2\pi} \frac{-r_c\,\hat{\mathbf{r}}_c + p\,\hat{\mathbf{z}}}{(r_c^2 + p^2)^{3/2}}(r_c)(r_c dr_c d\theta) = k\beta \int_{r_c=0}^{a}\int_{\theta=0}^{2\pi} \frac{-r_c\,\hat{\mathbf{r}}_c + p\,\hat{\mathbf{z}}}{(r_c^2 + p^2)^{3/2}} r_c^2 dr_c d\theta$$

Note that $r_c r_c = r_c^2$. Split the integral into two parts, noting that $\frac{-r_c\,\hat{\mathbf{r}}_c + p\,\hat{\mathbf{z}}}{(r_c^2+p^2)^{3/2}} = \frac{-r_c\,\hat{\mathbf{r}}_c}{(r_c^2+p^2)^{3/2}} + \frac{p\,\hat{\mathbf{z}}}{(r_c^2+p^2)^{3/2}}$.

$$\vec{\mathbf{E}} = -k\beta \int_{r_c=0}^{a}\int_{\theta=0}^{2\pi} \frac{r_c\hat{\mathbf{r}}_c r_c^2 dr_c d\theta}{(r_c^2 + p^2)^{3/2}} + k\beta \int_{r_c=0}^{a}\int_{\theta=0}^{2\pi} \frac{p\hat{\mathbf{z}} r_c^2 dr_c d\theta}{(r_c^2 + p^2)^{3/2}}$$

Since p and $\hat{\mathbf{z}}$ are constants (since $\hat{\mathbf{z}}$ is one unit long and points along the $+z$-axis), we may pull them out of the second integral. However, r_c and $\hat{\mathbf{r}}_c$ are <u>**not**</u> constants, as they differ for each dq that makes up the solid disc, so r_c and $\hat{\mathbf{r}}_c$ may <u>**not**</u> come out of the integrals. Note that $r_c r_c^2 = r_c^3$.

$$\vec{\mathbf{E}} = -k\beta \int_{r_c=0}^{a} \int_{\theta=0}^{2\pi} \frac{\hat{\mathbf{r}}_c r_c^3 dr_c d\theta}{(r_c^2 + p^2)^{3/2}} + k\beta p\hat{\mathbf{z}} \int_{r_c=0}^{a} \int_{\theta=0}^{2\pi} \frac{r_c^2 dr_c d\theta}{(r_c^2 + p^2)^{3/2}}$$

Study the diagram at the beginning of this example. It should be clear from symmetry that a positive test charge placed at the field point $(0, 0, p)$ would be pushed directly to the right, along the $+z$-axis. Therefore, the first integral, which contains $\hat{\mathbf{r}}_c$, will equal zero (though you can do the extra work if you like), and we only need to perform the second integral.

$$\vec{\mathbf{E}} = k\beta p\hat{\mathbf{z}} \int_{r_c=0}^{a} \int_{\theta=0}^{2\pi} \frac{r_c^2 dr_c d\theta}{(r_c^2 + p^2)^{3/2}}$$

When we integrate over θ, we treat the independent variable r_c as a constant. In fact, nothing in the integrand depends on the variable θ, so we can factor everything out of the θ-integration.

$$\vec{\mathbf{E}} = k\beta p\hat{\mathbf{z}} \int_{r_c=0}^{a} \frac{r_c^2 dr_c}{(r_c^2 + p^2)^{3/2}} \int_{\theta=0}^{2\pi} d\theta = k\beta p\hat{\mathbf{z}} \int_{r_c=0}^{a} \frac{r_c^2 dr_c}{(r_c^2 + p^2)^{3/2}} [\theta]_{\theta=0}^{2\pi}$$

$$\vec{\mathbf{E}} = k\beta p\hat{\mathbf{z}} \int_{r_c=0}^{a} \frac{r_c^2 dr_c}{(r_c^2 + p^2)^{3/2}} (2\pi - 0) = 2\pi k\beta p\hat{\mathbf{z}} \int_{r_c=0}^{a} \frac{r_c^2 dr_c}{(r_c^2 + p^2)^{3/2}}$$

The remaining integral can be performed via the following trigonometric substitution:

$$r_c = p \tan\psi$$

$$dr_c = p \sec^2\psi \, d\psi$$

Solving for ψ, we get $\theta = \tan^{-1}\left(\frac{r_c}{p}\right)$, which shows that the new limits of integration are from $\psi = \tan^{-1}(0) = 0°$ to $\psi = \tan^{-1}\left(\frac{a}{p}\right)$, which for now we will simply call ψ_{max}. Note that the denominator of the integral simplifies as follows, using the trig identity $\tan^2\psi + 1 = \sec^2\psi$:

$$(r_c^2 + p^2)^{3/2} = [(p\tan\psi)^2 + p^2]^{3/2} = [p^2(\tan^2\psi + 1)]^{3/2} = (p^2\sec^2\psi)^{3/2} = p^3\sec^3\psi$$

In the last step, we applied the rule from algebra that $(p^2x^2)^{3/2} = (p^2)^{3/2}(x^2)^{3/2} = p^3x^3$. Substitute the above expressions for r_c, dr_c, and $(r_c^2 + p^2)^{3/2}$ into the previous integral.

$$\vec{\mathbf{E}} = 2\pi k\beta p\hat{\mathbf{z}} \int_{r_c=0}^{a} \frac{r_c^2 dr_c}{(r_c^2 + p^2)^{3/2}} = 2\pi k\beta p\hat{\mathbf{z}} \int_{\psi=0°}^{\psi_{max}} \frac{(p^2\tan^2\psi)(p\sec^2\psi \, d\psi)}{p^3\sec^3\psi}$$

Note that $\frac{p^2 p}{p^3} = 1$ and $\frac{\sec^2\psi}{\sec^3\psi} = \frac{1}{\sec\psi}$.

$$\vec{\mathbf{E}} = 2\pi k\beta p\hat{\mathbf{z}} \int_{\psi=0°}^{\psi_{max}} \frac{\tan^2\psi \, d\psi}{\sec\psi}$$

Recall from trig that $\sec\psi = \frac{1}{\cos\psi}$ and $\tan\psi = \frac{\sin\psi}{\cos\psi}$. Therefore, it follows that:

$$\frac{\tan^2\psi}{\sec\psi} = \frac{\sin^2\psi}{\cos^2\psi} \div \frac{1}{\cos\psi} = \frac{\sin^2\psi}{\cos^2\psi} \times \frac{\cos\psi}{1} = \frac{\sin^2\psi}{\cos\psi}$$

To divide by a fraction, multiply by its **reciprocal**. Also recall from trig that $\sin^2\psi + \cos^2\psi = 1$, such that $\sin^2\psi = 1 - \cos^2\psi$. Substitute this into the previous equation.

$$\frac{\tan^2\psi}{\sec\psi} = \frac{\sin^2\psi}{\cos\psi} = \frac{1-\cos^2\psi}{\cos\psi} = \frac{1}{\cos\psi} - \cos\psi = \sec\psi - \cos\psi$$

The electric field integral becomes:

$$\vec{\mathbf{E}} = 2\pi k\beta p\hat{\mathbf{z}} \int_{\psi=0°}^{\psi_{max}} \frac{\tan^2\psi\, d\psi}{\sec\psi} = 2\pi k\beta p\hat{\mathbf{z}} \int_{\psi=0°}^{\psi_{max}} \sec\psi\, d\psi - 2\pi k\beta p\hat{\mathbf{z}} \int_{\psi=0°}^{\psi_{max}} \cos\psi\, d\psi$$

You can look up these anti-derivatives in Chapter 6 (page 42).

$$\vec{\mathbf{E}} = 2\pi k\beta p\hat{\mathbf{z}}[\ln|\sec\psi + \tan\psi| - \sin\psi]_{\psi=0°}^{\psi_{max}}$$

Evaluate the anti-derivative over the limits.

$$\vec{\mathbf{E}} = 2\pi k\beta p\hat{\mathbf{z}}(\ln|\sec\psi_{max} + \tan\psi_{max}| - \sin\psi_{max} - \ln|\sec 0 + \tan 0| + \sin 0)$$

Recall from trig that $\sec 0 = 1$, $\tan 0 = 0$, and $\sin 0 = 0$.

$$\vec{\mathbf{E}} = 2\pi k\beta p\hat{\mathbf{z}}(\ln|\sec\psi_{max} + \tan\psi_{max}| - \sin\psi_{max} - \ln|1|)$$

Apply the rule for logarithms that $\ln(1) = 0$.

$$\vec{\mathbf{E}} = 2\pi k\beta p\hat{\mathbf{z}}(\ln|\sec\psi_{max} + \tan\psi_{max}| - \sin\psi_{max})$$

Recall that what we called ψ_{max} is given by $\psi_{max} = \tan^{-1}\left(\frac{a}{p}\right)$, such that $\cos\psi_{max}$ is the complicated looking expression $\cos\left(\tan^{-1}\left(\frac{a}{p}\right)\right)$. When you find yourself taking the cosine of an inverse tangent, you can find a simpler way to write it by drawing a right triangle and applying the Pythagorean theorem. Since $\psi_{max} = \tan^{-1}\left(\frac{a}{p}\right)$, it follows that $\tan\psi_{max} = \frac{a}{p}$. We can make a right triangle from this: Since the tangent of ψ_{max} equals the opposite over the adjacent, we draw a right triangle with a opposite and p adjacent to ψ_{max}.

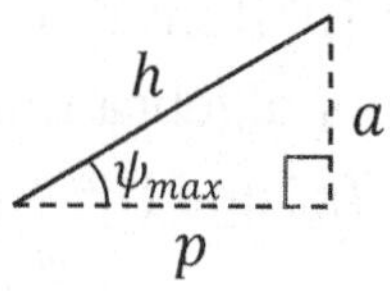

Find the hypotenuse, h, of the right triangle from the Pythagorean theorem.

$$h = \sqrt{p^2 + a^2}$$

Use the right triangle to write expressions for the secant, tangent, and sine of ψ_{max}.

$$\sin\psi_{max} = \frac{a}{h} = \frac{a}{\sqrt{p^2+a^2}} \quad , \quad \sec\psi = \frac{h}{p} = \frac{\sqrt{p^2+a^2}}{p} \quad , \quad \tan\psi = \frac{a}{p}$$

Substitute these expressions into the previous equation for electric field.

$$\vec{\mathbf{E}} = 2\pi k\beta p\hat{\mathbf{z}}\left(\ln\left|\frac{\sqrt{p^2+a^2}}{p} + \frac{a}{p}\right| - \frac{a}{\sqrt{p^2+a^2}}\right) = 2\pi k\beta p\hat{\mathbf{z}}\left(\ln\left|\frac{a+\sqrt{p^2+a^2}}{p}\right| - \frac{a}{\sqrt{p^2+a^2}}\right)$$

Note that $\frac{\sqrt{p^2+a^2}}{p} + \frac{a}{p} = \frac{\sqrt{p^2+a^2}+a}{p} = \frac{a+\sqrt{p^2+a^2}}{p}$. We're not finished yet. According to the directions, we must eliminate the constant β from our answer. The way to do this is to integrate over dq to find the total charge of the disc, Q. We use the same substitutions from before.

$$Q = \int dq = \int \sigma\, dA = \int \beta r_c\, dA = \beta \int_{r_c=0}^{a} \int_{\theta=0}^{2\pi} r_c (r_c dr_c d\theta) = \beta \int_{r_c=0}^{a} r_c^2 dr_c \int_{\theta=0}^{2\pi} d\theta$$

$$Q = \beta \left[\frac{r_c^3}{3}\right]_{r_c=0}^{a} [\theta]_{\theta=0}^{2\pi} = \beta\left(\frac{a^3}{3} - \frac{0^3}{3}\right)(2\pi - 0) = \beta\left(\frac{a^3}{3}\right)(2\pi) = \frac{2\pi\beta a^3}{3}$$

Multiply both sides of the equation by 3 and divide by $2\pi a^3$.

$$\beta = \frac{3Q}{2\pi a^3}$$

Substitute this expression for σ into our previous expression for $\vec{\mathbf{E}}$.

$$\vec{\mathbf{E}} = 2\pi k\left(\frac{3Q}{2\pi a^3}\right) p\hat{\mathbf{z}}\left(\ln\left|\frac{a+\sqrt{p^2+a^2}}{p}\right| - \frac{a}{\sqrt{p^2+a^2}}\right)$$

$$\vec{\mathbf{E}} = \frac{3kQp}{a^3}\hat{\mathbf{z}}\left(\ln\left|\frac{a+\sqrt{p^2+a^2}}{p}\right| - \frac{a}{\sqrt{p^2+a^2}}\right)$$

The electric field at the point $(0, 0, p)$ is $\vec{\mathbf{E}} = \frac{3kQp}{a^3}\hat{\mathbf{z}}\left(\ln\left|\frac{a+\sqrt{p^2+a^2}}{p}\right| - \frac{a}{\sqrt{p^2+a^2}}\right)$. It has a magnitude of $E = \frac{3kQp}{a^3}\left(\ln\left|\frac{a+\sqrt{p^2+a^2}}{p}\right| - \frac{a}{\sqrt{p^2+a^2}}\right)$ and a direction of $\hat{\mathbf{z}}$.

Example 43. A thick semicircular ring with non-uniform charge density $\sigma = \beta r$, where β is a constant, lies in the xy plane, centered about the origin as illustrated below. The thick ring has inner radius $\frac{a}{2}$, outer radius a, and the positively charged ring has total charge Q. Derive an equation for the electric field at the origin in terms of k, Q, a and appropriate unit vectors.

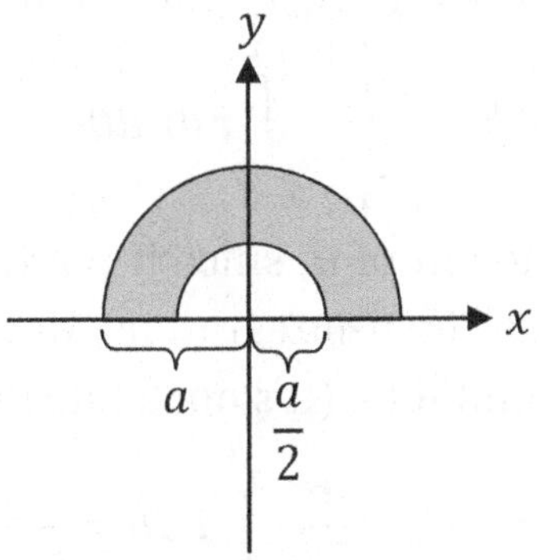

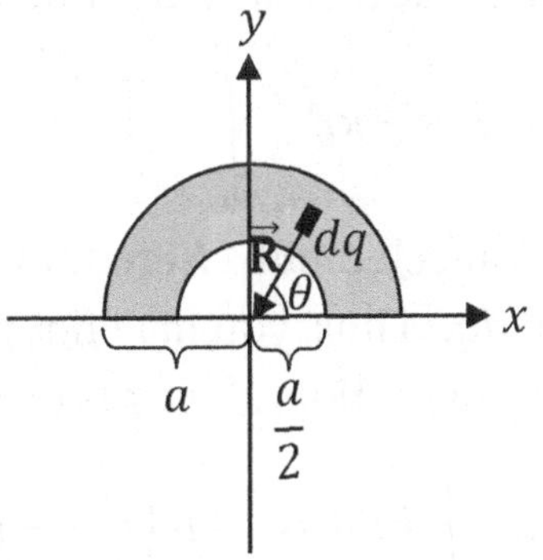

Solution. Begin with a labeled diagram. See the diagram above on the right. Draw a representative dq: Since this is a thick ring (unlike a very thin ring), most of the dq's lie within the area of the ring, so we drew dq inside of the area of the ring's thickness (not on its circumference). Draw $\vec{\mathbf{R}}$ from the source, dq, to the field point $(0,0)$. When we perform the integration, we effectively integrate over every dq that makes up the thick ring. Start the math with the electric field integral.

$$\vec{\mathbf{E}} = k \int \frac{\hat{\mathbf{R}}}{R^2} dq$$

Examine the right diagram on the previous page:

- The vector $\vec{\mathbf{R}}$ extends from the source (each dq that makes up the thick ring) to the field point $(0,0)$. Note that $\vec{\mathbf{R}}$ extends r units inward. Therefore, $R = r$ (using 2D polar coordinates).
- The unit vector $\hat{\mathbf{R}}$ points one unit in the direction of $\vec{\mathbf{R}}$. Note that $\vec{\mathbf{R}}$ points **inward**, towards the origin (along $-\hat{\mathbf{r}}$). Therefore, $\hat{\mathbf{R}} = -\hat{\mathbf{r}}$, since the $\hat{\mathbf{r}}$ from 2D polar coordinates points outward.

Substitute the expressions $R = r$ and $\hat{\mathbf{R}} = -\hat{\mathbf{r}}$ into the electric field integral.

$$\vec{\mathbf{E}} = k \int \frac{\hat{\mathbf{R}}}{R^2} dq = -k \int \frac{\hat{\mathbf{r}}}{r^2} dq$$

For a thick ring (unlike a very thin ring), we write $dq = \sigma dA$ (see the table on page 57).

$$\vec{\mathbf{E}} = -k \int \frac{\hat{\mathbf{r}}}{r^2} (\sigma dA)$$

Since the thick ring has non-uniform charge density, we may <u>**not**</u> pull σ out of the integral. Instead, make the substitution $\sigma = \beta r$ using the expression given in the problem.

$$\vec{\mathbf{E}} = -k \int \frac{\hat{\mathbf{r}}}{r^2} \beta r dA$$

Note that $\frac{1}{r^2} r = \frac{1}{r}$. The constant β may come out of the integral.

$$\vec{\mathbf{E}} = -k\beta \int \frac{\hat{\mathbf{r}}}{r} dA$$

For a thick solid ring (unlike a very thin ring), we write the differential area element as $dA = r dr d\theta$ (see the table on page 57). The limits of integration are from $r = \frac{a}{2}$ to $r = a$ (the inner and outer radii of the ring) and $\theta = 0$ to $\theta = \pi$ <u>**radians**</u> (for a semicircle).

$$\vec{\mathbf{E}} = -k\beta \int_{r_c=a/2}^{a} \int_{\theta=0}^{\pi} \frac{\hat{\mathbf{r}}}{r} (r dr d\theta) = -k\beta \int_{r_c=a/2}^{a} \int_{\theta=0}^{\pi} \hat{\mathbf{r}}\, dr\, d\theta$$

Note that the r's canceled out. Note that $\hat{\mathbf{r}}$ is a function of θ, since it points in a different direction for each dq. Thus, you may <u>**not**</u> pull $\hat{\mathbf{r}}$ out of the θ-integration. However, you may pull $\hat{\mathbf{r}}$ out of the r-integration, since $\hat{\mathbf{r}}$ is exactly one unit long (it's not a function of r).

$$\vec{\mathbf{E}} = -k\beta \int_{r=\frac{a}{2}}^{a} dr \int_{\theta=0}^{\pi} \hat{\mathbf{r}}\, d\theta = -k\beta\left(a - \frac{a}{2}\right) \int_{\theta=0}^{\pi} \hat{\mathbf{r}}\, d\theta = -k\beta\left(\frac{a}{2}\right) \int_{\theta=0}^{\pi} \hat{\mathbf{r}}\, d\theta = -\frac{k\beta a}{2} \int_{\theta=0}^{\pi} \hat{\mathbf{r}}\, d\theta$$

Note that $\int_{r=a/2}^{a} dr = a - \frac{a}{2} = \frac{2a}{2} - \frac{a}{2} = \frac{2a-a}{2} = \frac{a}{2}$. Apply the equation $\hat{\mathbf{r}} = \hat{\mathbf{x}} \cos\theta + \hat{\mathbf{y}} \sin\theta$

from page 57.

$$\vec{\mathbf{E}} = -\frac{k\beta a}{2}\int_{\theta=0}^{\pi} (\hat{\mathbf{x}}\cos\theta + \hat{\mathbf{y}}\sin\theta)\, d\theta$$

Separate the integral into two terms.

$$\vec{\mathbf{E}} = -\frac{k\beta a}{2}\int_{\theta=0}^{\pi} \hat{\mathbf{x}}\cos\theta\, d\theta - \frac{k\beta a}{2}\int_{\theta=0}^{\pi} \hat{\mathbf{y}}\sin\theta\, d\theta$$

We may pull $\hat{\mathbf{x}}$ and $\hat{\mathbf{y}}$ out of the integrals since they are constants: They are exactly one unit long and each always points in the same direction, unlike $\hat{\mathbf{r}}$.

$$\vec{\mathbf{E}} = -\frac{k\beta a}{2}\hat{\mathbf{x}}\int_{\theta=0}^{\pi} \cos\theta\, d\theta - \frac{k\beta a}{2}\hat{\mathbf{y}}\int_{\theta=0}^{\pi} \sin\theta\, d\theta$$

You can find the anti-derivatives of $\cos\theta$ and $\sin\theta$ in Chapter 6 (page 42).

$$\vec{\mathbf{E}} = -\frac{k\beta a}{2}\hat{\mathbf{x}}[\sin\theta]_{\theta=0}^{\pi} - \frac{k\beta a}{2}\hat{\mathbf{y}}[-\cos\theta]_{\theta=0}^{\pi}$$

Evaluate the anti-derivatives over the limits.

$$\vec{\mathbf{E}} = -\frac{k\beta a}{2}\hat{\mathbf{x}}(\sin\pi - \sin 0) - \frac{k\beta a}{2}\hat{\mathbf{y}}(-\cos\pi + \cos 0)$$

$$\vec{\mathbf{E}} = -\frac{k\beta a}{2}\hat{\mathbf{x}}(0 - 0) - \frac{k\beta a}{2}\hat{\mathbf{y}}[-(-1) + 1] = 0 - \frac{k\beta a}{2}\hat{\mathbf{y}}(1 + 1)$$

$$\vec{\mathbf{E}} = -\frac{k\beta a}{2}\hat{\mathbf{y}}(2) = -k\beta a\hat{\mathbf{y}}$$

We're not finished yet. According to the directions, we must eliminate the constant β from our answer. The way to do this is to integrate over dq to find the total charge of the thick ring, Q. We use the same substitutions from before.

$$Q = \int dq = \int \sigma\, dA = \int \beta r\, dA = \beta \int_{r=a/2}^{a}\int_{\theta=0}^{\pi} r(r dr d\theta) = \beta \int_{r=a/2}^{a} r^2 dr \int_{\theta=0}^{\pi} d\theta$$

$$Q = \beta\left[\frac{r^3}{3}\right]_{r_c=\frac{a}{2}}^{a}[\theta]_{\theta=0}^{\pi} = \beta\left[\frac{a^3}{3} - \frac{\left(\frac{a}{2}\right)^3}{3}\right](\pi - 0)$$

$$Q = \beta\left(\frac{a^3}{3} - \frac{1}{3}\frac{a^3}{8}\right)(\pi) = \pi\beta\left(\frac{a^3}{3} - \frac{a^3}{24}\right)$$

Subtract fractions by finding a **common denominator**. Note that $\frac{a^3}{3} = \frac{a^3}{3}\frac{8}{8} = \frac{8a^3}{24}$.

$$Q = \pi\beta\left(\frac{8a^3}{24} - \frac{a^3}{24}\right) = \pi\beta\left(\frac{8a^3 - a^3}{24}\right) = \pi\beta\left(\frac{7a^3}{24}\right) = \frac{7\pi\beta a^3}{24}$$

Multiply both sides of the equation by 24 and divide by $7\pi a^3$.

$$\beta = \frac{24Q}{7\pi a^3}$$

Substitute this expression for σ into our previous expression for $\vec{\mathbf{E}}$.

$$\vec{\mathbf{E}} = -k\beta a\hat{\mathbf{y}} = -k\left(\frac{24Q}{7\pi a^3}\right)a\hat{\mathbf{y}} = -\frac{24kQ}{7\pi a^2}\hat{\mathbf{y}}$$

The electric field at the origin is $\vec{\mathbf{E}} = -\frac{24kQ}{7\pi a^2}\hat{\mathbf{y}}$. It has a magnitude of $E = \frac{24kQ}{7\pi a^2}$ and a direction of $-\hat{\mathbf{y}}$.

8 GAUSS'S LAW

Gauss's Law		
$\oint_S \vec{\mathbf{E}} \cdot d\vec{\mathbf{A}} = \frac{q_{enc}}{\epsilon_0}$		
Permittivity of Free Space		
$\epsilon_0 = \frac{1}{4\pi k}$		
Charge Enclosed		
$q_{enc} = \int dq$		
Differential Charge Element		
$dq = \lambda ds$ (line or thin arc)	$dq = \sigma dA$ (surface area)	$dq = \rho dV$ (volume)
Differential Arc Length		
$ds = dx$ (along x)	$ds = dy$ or dz (along y or z)	$ds = a d\theta$ (circular arc of radius a)
Differential Area Element		
$dA = dxdy$ (polygon in xy plane)	$dA = rdrd\theta$ (pie slice, disc, thick ring)	$dA = a^2 \sin\theta \, d\theta d\varphi$ (sphere of radius a)
Differential Volume Element		
$dV = dxdydz$ (bounded by flat sides)	$dV = r_c dr_c d\theta dz$ (cylinder or cone)	$dV = r^2 \sin\theta \, drd\theta d\varphi$ (spherical)

Symbol	Name	SI Units
dq	differential charge element	C
q_{enc}	the charge enclosed by the Gaussian surface	C
Q	total charge of the object	C
$\vec{\mathbf{E}}$	electric field	N/C or V/m
Φ_e	electric flux	$\frac{\text{N}\cdot\text{m}^2}{\text{C}}$ or $\frac{\text{kg}\cdot\text{m}^3}{\text{C}\cdot\text{s}^2}$
Φ_e^{net}	the net electric flux through a closed surface	$\frac{\text{N}\cdot\text{m}^2}{\text{C}}$ or $\frac{\text{kg}\cdot\text{m}^3}{\text{C}\cdot\text{s}^2}$
ϵ_0	permittivity of free space	$\frac{\text{C}^2}{\text{N}\cdot\text{m}^2}$ or $\frac{\text{C}^2\cdot\text{s}^2}{\text{kg}\cdot\text{m}^3}$
x, y, z	Cartesian coordinates of dq	m, m, m
$\hat{\mathbf{x}}, \hat{\mathbf{y}}, \hat{\mathbf{z}}$	unit vectors along the $+x$-, $+y$-, $+z$-axes	unitless
r, θ	2D polar coordinates of dq	m, rad
r_c, θ, z	cylindrical coordinates of dq	m, rad, m
r, θ, φ	spherical coordinates of dq	m, rad, rad
$\hat{\mathbf{r}}, \hat{\boldsymbol{\theta}}, \hat{\boldsymbol{\varphi}}$	unit vectors along spherical coordinate axes	unitless
$\hat{\mathbf{r}}_c$	a unit vector pointing away from the $+z$-axis	unitless
λ	linear charge density (for an arc)	C/m
σ	surface charge density (for an area)	C/m^2
ρ	volume charge density	C/m^3
ds	differential arc length	m
dA	differential area element	m^2
dV	differential volume element	m^3

Note: The symbol Φ is the uppercase Greek letter phi* and ϵ is epsilon.

* Cool physics note: If you rotate uppercase phi (Φ) sideways, it resembles Darth Vader's™ spaceship.

Conceptual Gauss's Law Example: A Prelude to Examples 44-50. Consider the electric field map for the electric dipole (Chapter 4) shown below. The four dashed curves marked A, B, C, and D represent closed surfaces in three-dimensional space: For example, A and D are really spheres, though they are drawn as circles. Describe how each of the closed surfaces A, B, C, and D relates to Gauss's law.

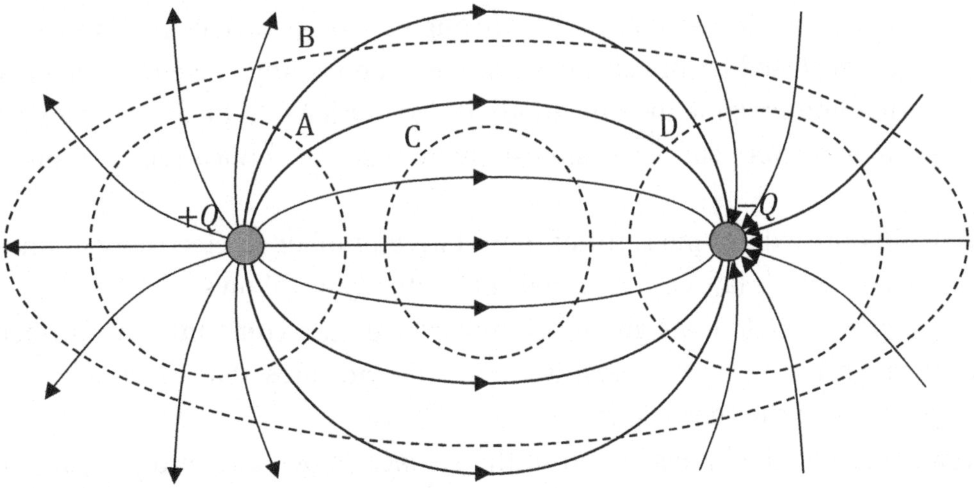

Solution. We begin with the equation for Gauss's law.

$$\oint_S \vec{\mathbf{E}} \cdot d\vec{\mathbf{A}} = \frac{q_{enc}}{\epsilon_0}$$

According to **Gauss's law**, the **net** electric flux ($\oint_S \vec{\mathbf{E}} \cdot d\vec{\mathbf{A}}$) through a **closed** surface (often called a Gaussian surface) is proportional to the charge enclosed (q_{enc}) by the surface. The symbol $\oint$ is called a **closed** integral: It represents that the integral is over a **closed** surface. The constant ϵ_0 is called the **permittivity of free space** and equals $\epsilon_0 = \frac{1}{4\pi k}$, where k is Coulomb's constant (see Chapter 1).

In order to solve this example, we need to know how Gauss's law relates to the concept of **electric flux** (Φ_e). The equation for electric flux involves the scalar product between electric field ($\vec{\mathbf{E}}$) and the differential area element ($d\vec{\mathbf{A}}$), where the direction of $d\vec{\mathbf{A}}$ is **perpendicular to the surface**. The symbol S below stands for the surface over which area is integrated.

$$\Phi_e = \int_S \vec{\mathbf{E}} \cdot d\vec{\mathbf{A}}$$

Recall the scalar product from Volume 1 (the equation is given below), where θ is the angle between $\vec{\mathbf{E}}$ and $d\vec{\mathbf{A}}$. Note that $\vec{\mathbf{E}} \cdot d\vec{\mathbf{A}}$ is maximum when $\theta = 0°$ and zero when $\theta = 90°$.

$$\vec{\mathbf{E}} \cdot d\vec{\mathbf{A}} = E \cos\theta \, dA$$

Since this is a conceptual example, we need to interpret electric flux ($\Phi_e = \int_S \vec{\mathbf{E}} \cdot d\vec{\mathbf{A}}$) conceptually. Electric flux provides a measure of the relative number of electric field lines passing through a surface. (Electric field lines were discussed in Chapter 4.) Gauss's law involves the **net** electric flux (Φ_e^{net}) through a **closed** surface. Since the net electric flux equals $\Phi_e^{net} = \oint_S \vec{\mathbf{E}} \cdot d\vec{\mathbf{A}}$, Gauss's law can be expressed as:

$$\Phi_e^{net} = \frac{q_{enc}}{\epsilon_0}$$

This shows that the **net electric flux** (Φ_e^{net}) through a closed surface is proportional to the **net charge** (q_{enc}) **enclosed** by the surface. This is true for any closed surface (even if it's just a surface that you imagine in your mind: it doesn't have to be a physical surface with walls – in fact, an imaginary surface is almost always used in applications of Gauss's law).

Since electric flux provides a measure of the relative number of electric field lines passing through a surface, Gauss's law can be stated conceptually as follows:

- If more electric field lines go out of the surface than come into the surface, the net flux through the surface is **positive**, and this indicates that there is a net positive charge inside of the closed surface.
- If fewer electric field lines go out of the surface than come into the surface, the net flux through the surface is **negative**, and this indicates that there is a net negative charge inside of the closed surface.
- If the same number of electric field lines go out of the surface as come into the surface, the net flux through the surface is zero. This indicates that the net charge inside of the closed surface is **zero**.

We will now apply the above conceptual interpretation of Gauss's law to the diagram given at the beginning of this example:

- The net flux through closed surface A is positive (every field line comes out of the surface – no field lines come back into it) because it encloses a **positive** charge.
- The net flux through closed surface B is **zero** (the same number of field lines enter the surface as exit the surface). The positive and negative charges inside cancel out (the net charge enclosed is zero).
- The net flux through closed surface C is **zero** (every field line that goes into the surface also comes back out of it). There aren't any charges enclosed.
- The net flux through closed surface D is negative (every field line goes into the surface – no field lines come back out of it) because it encloses a **negative** charge.

Uniform Sphere Example: A Prelude to Example 44. A solid spherical insulator† centered about the origin with positive charge Q has radius a and uniform charge density ρ. Derive an expression for the electric field both inside and outside of the sphere.

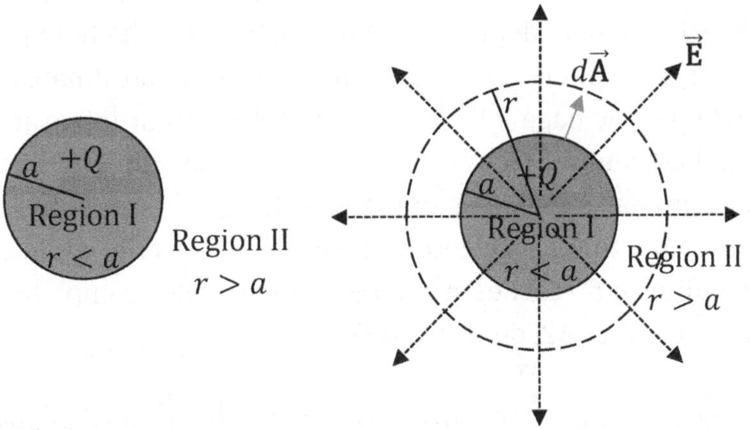

First sketch the electric field lines for the positive sphere. Regardless of where a positive test charge might be placed, it would be repelled directly away from the center of the sphere. Therefore, the electric field lines radiate outward, as shown above on the right (but realize that the field lines really radiate outward in three dimensions, and we are really working with spheres, not circles). We choose our Gaussian surface to be a sphere (shown as a dashed circle above) concentric with the charged sphere such that $\vec{\mathbf{E}}$ and $d\vec{\mathbf{A}}$ will be parallel and the magnitude of $\vec{\mathbf{E}}$ will be constant over the Gaussian surface (since every point on the Gaussian sphere is equidistant from the center of the positive sphere). Write the formula for Gauss's law.

$$\oint_S \vec{\mathbf{E}} \cdot d\vec{\mathbf{A}} = \frac{q_{enc}}{\epsilon_0}$$

Recall (from Volume 1) that the scalar product is $\vec{\mathbf{E}} \cdot d\vec{\mathbf{A}} = E \cos\theta \, dA$, and $\theta = 0°$ since the electric field lines radiate outward, perpendicular to the surface (and therefore parallel to $d\vec{\mathbf{A}}$, which is always perpendicular to the surface).

$$\oint_S E \cos 0° \, dA = \frac{q_{enc}}{\epsilon_0}$$

Recall from trig that $\cos 0° = 1$.

$$\oint_S E \, dA = \frac{q_{enc}}{\epsilon_0}$$

The magnitude of the electric field is constant over the Gaussian sphere, since every point on the sphere is equidistant from the center of the positive sphere. Therefore, we may pull E out of the integral.

† If it were a conductor, all of the excess charge would move to the surface due to Gauss's law. We'll discuss why that's the case as part of the example with the conducting shell (pages 93-95).

$$E \oint_S dA = \frac{q_{enc}}{\epsilon_0}$$

This area integral is over the surface of the Gaussian sphere of radius r, where r is a variable because the electric field depends on how close or far the field point is away from the charged sphere. For a sphere, we work with spherical coordinates (Chapter 6) and write $dA = r^2 \sin\theta \, d\theta d\varphi$ (see page 79). This is a purely angular integration (over the two spherical angles θ and φ), and we treat r as a constant in this integral because every point on the surface of the Gaussian sphere has the same value of r. For a sphere, φ varies from 0 to 2π while θ varies from 0 to π: φ sweeps out a horizontal circle, while a given value of θ sweeps out a cone and once θ reaches π the cone traces out a complete sphere (this was discussed in Volume 1, Chapter 32, on page 348).

$$A = \oint_S dA = \int_{\theta=0}^{\pi} \int_{\varphi=0}^{2\pi} r^2 \sin\theta \, d\theta \, d\varphi = r^2 \int_{\theta=0}^{\pi} \int_{\varphi=0}^{2\pi} \sin\theta \, d\theta \, d\varphi$$

When integrating over φ, we treat the independent variable θ as a constant. Therefore, we may pull θ out of the φ-integration.

$$A = r^2 \int_{\theta=0}^{\pi} \sin\theta \left(\int_{\varphi=0}^{2\pi} d\varphi \right) d\theta = r^2 \int_{\theta=0}^{\pi} \sin\theta \, [\varphi]_{\varphi=0}^{2\pi} \, d\theta = r^2 \int_{\theta=0}^{\pi} \sin\theta \, (2\pi) \, d\theta$$

$$A = 2\pi r^2 \int_{\theta=0}^{\pi} \sin\theta \, d\theta = 2\pi r^2 [-\cos\theta]_{\theta=0}^{\pi} = 2\pi r^2(-\cos\pi + \cos 0) = 2\pi r^2(1+1) = 4\pi r^2$$

Although we could have simply looked up the formula for the surface area of a sphere and found that it was $A = 4\pi r^2$, there are some problems where you need to know how to perform the integration. Substitute this expression for surface area into the previous equation for electric field.

$$E \oint_S dA = EA = E 4\pi r^2 = \frac{q_{enc}}{\epsilon_0}$$

Isolate the magnitude of the electric field by dividing both sides of the equation by $4\pi r^2$.

$$E = \frac{q_{enc}}{4\pi\epsilon_0 r^2}$$

Now we need to determine how much charge is enclosed by the Gaussian surface. For a solid charged sphere, we write $dq = \rho dV$ (see page 79). Since the sphere has uniform charge density, we may pull ρ out of the integral.

$$q_{enc} = \int dq = \int \rho \, dV = \rho \int dV$$

For a sphere, we work with spherical coordinates and write the differential volume element as $dV = r^2 \sin\theta \, dr d\theta d\varphi$ (see page 79). Since we are now integrating over volume (not surface area), r is a variable and we will have a triple integral.

We must consider two different regions:

- The Gaussian sphere could be smaller than the charged sphere. This will help us find the electric field in region I.
- The Gaussian surface could be larger than the charged sphere. This will help us find the electric field in region II.

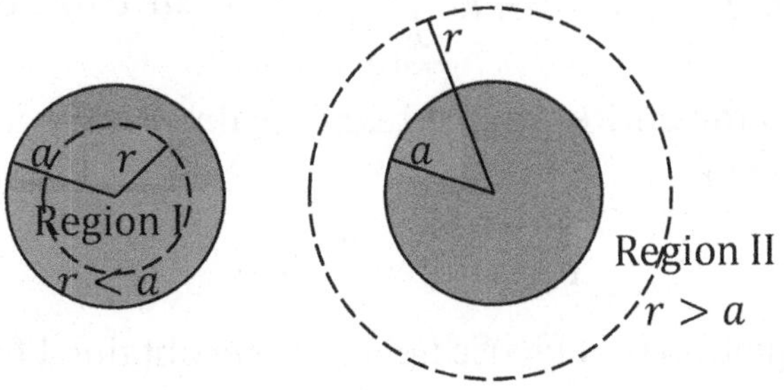

Region I: $r < a$.

Inside of the charged sphere, only a fraction of the sphere's charge is enclosed by the Gaussian sphere. In this region, the upper limit of the r-integration is the variable r: The larger the Gaussian sphere, the more charge it encloses, up to a maximum radius of a (the radius of the charged sphere).

$$q_{enc} = \rho \int dV = \rho \int_{r=0}^{r} \int_{\theta=0}^{\pi} \int_{\varphi=0}^{2\pi} r^2 \sin\theta \, dr \, d\theta \, d\varphi$$

This integral is separable:

$$q_{enc} = \rho \int_{r=0}^{r} r^2 \, dr \int_{\theta=0}^{\pi} \sin\theta \, d\theta \int_{\varphi=0}^{2\pi} d\varphi = \rho \left[\frac{r^3}{3}\right]_{r=0}^{r} [-\cos\theta]_{\theta=0}^{\pi} [\varphi]_{\varphi=0}^{2\pi}$$

$$q_{enc} = \rho \left(\frac{r^3}{3}\right)(-\cos\pi + \cos 0)(2\pi) = \rho \left(\frac{r^3}{3}\right)(1+1)(2\pi) = \frac{4\pi\rho r^3}{3}$$

You should recognize that $\frac{4\pi r^3}{3}$ is the volume of a sphere. For a uniformly charged sphere, charge equals ρ times volume. (For a non-uniform sphere, you would need to integrate: Then you couldn't just multiply ρ times volume.) Substitute this expression for the charge enclosed into the previous equation for electric field.

$$E_I = \frac{q_{enc}}{4\pi\epsilon_0 r^2} = \frac{4\pi\rho r^3}{3} \div 4\pi\epsilon_0 r^2 = \frac{4\pi\rho r^3}{3} \times \frac{1}{4\pi\epsilon_0 r^2} = \frac{\rho r}{3\epsilon_0}$$

Since the electric field points outward (away from the center of the sphere), we can include a direction with the electric field by adding on the spherical unit vector $\hat{\mathbf{r}}$.

$$\vec{\mathbf{E}}_I = \frac{\rho r}{3\epsilon_0} \hat{\mathbf{r}}$$

The answer is different outside of the charged sphere. We will explore that next.

Region II: $r > a$.

Outside of the charged sphere, 100% of the charge is enclosed by the Gaussian surface. This changes the upper limit of the r-integration to a (since all of the charge lies inside a sphere of radius a).

$$q_{enc} = Q = \rho \int dV = \rho \int_{r=0}^{a} \int_{\theta=0}^{\pi} \int_{\varphi=0}^{2\pi} r^2 \sin\theta \, dr \, d\theta \, d\varphi$$

We don't need to work out the entire triple integral again. We'll get the same expression as before, but with a in place of r.

$$q_{enc} = Q = \frac{4\pi\rho a^3}{3}$$

Substitute this into the equation for electric field that we obtained from Gauss's law.

$$E_{II} = \frac{q_{enc}}{4\pi\epsilon_0 r^2} = \frac{4\pi\rho a^3}{3} \div 4\pi\epsilon_0 r^2 = \frac{4\pi\rho a^3}{3} \times \frac{1}{4\pi\epsilon_0 r^2} = \frac{\rho a^3}{3\epsilon_0 r^2}$$

We can turn this into a vector by including the appropriate unit vector.

$$\vec{\mathbf{E}}_{II} = \frac{\rho a^3}{3\epsilon_0 r^2}\hat{\mathbf{r}}$$

Alternate forms of the answers in regions I and II.

There are multiple ways to express our answers for regions I and II. For example, we could use the equation $\epsilon_0 = \frac{1}{4\pi k}$ to work with Coulomb's constant (k) instead of the permittivity of free space (ϵ_0). Since the total charge of the sphere is $Q = \frac{4\pi\rho a^3}{3}$ (we found this equation for region II above), we can express the electric field in terms of the total charge (Q) of the sphere instead of the charge density (ρ).

Region I: $r < a$.

$$\vec{\mathbf{E}}_{I} = \frac{\rho r}{3\epsilon_0}\hat{\mathbf{r}} = \frac{4\pi k\rho r}{3}\hat{\mathbf{r}} = \frac{Qr}{4\pi\epsilon_0 a^3}\hat{\mathbf{r}} = \frac{kQr}{a^3}\hat{\mathbf{r}}$$

Region II: $r > a$.

$$\vec{\mathbf{E}}_{II} = \frac{\rho a^3}{3\epsilon_0 r^2}\hat{\mathbf{r}} = \frac{4\pi k\rho a^3}{3r^2}\hat{\mathbf{r}} = \frac{Q}{4\pi\epsilon_0 r^2}\hat{\mathbf{r}} = \frac{kQ}{r^2}\hat{\mathbf{r}}$$

Note that the electric field in region II is identical to the electric field created by a pointlike charge (see Chapter 2). Note also that the expressions for the electric field in the two different regions both agree at the boundary: That is, in the limit that r approaches a, both expressions approach $\frac{kQ}{a^2}\hat{\mathbf{r}}$.

Non-uniform Sphere Example: Another Prelude to Example 44. A solid spherical insulator centered about the origin with positive charge Q has radius a and non-uniform charge density $\rho = \beta r$, where β is a positive constant. Derive an expression for the electric field both inside and outside of the sphere.

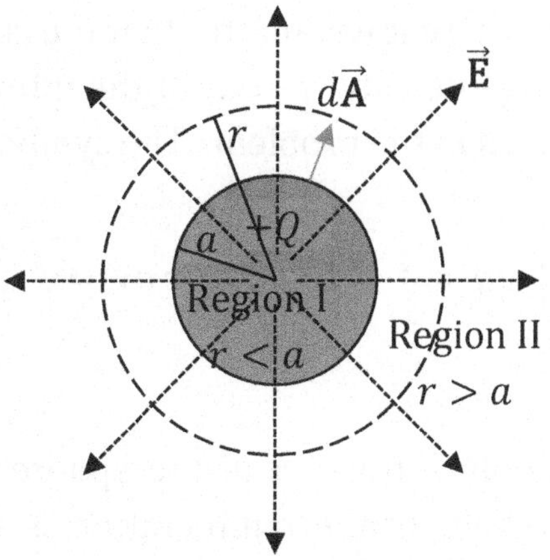

This problem is just like the previous example, except that the charge is non-uniform. The electric field lines and Gaussian surface are the same as before. To save time, we'll simply repeat the steps that are identical to the previous example, and pick up from where this solution deviates from the earlier one. It would be a good exercise to see if you can understand each step (if not, review the previous example for the explanation).

$$\oint_S \vec{\mathbf{E}} \cdot d\vec{\mathbf{A}} = \frac{q_{enc}}{\epsilon_0}$$

$$\oint_S E \cos 0° \, dA = \frac{q_{enc}}{\epsilon_0}$$

$$\oint_S E \, dA = \frac{q_{enc}}{\epsilon_0}$$

$$E \oint_S dA = \frac{q_{enc}}{\epsilon_0}$$

$$A = \oint_S dA = \int_{\theta=0}^{\pi} \int_{\varphi=0}^{2\pi} r^2 \sin\theta \, d\theta \, d\varphi = r^2 \int_{\theta=0}^{\pi} \int_{\varphi=0}^{2\pi} \sin\theta \, d\theta \, d\varphi$$

$$A = r^2 \int_{\theta=0}^{\pi} \sin\theta \left(\int_{\varphi=0}^{2\pi} d\varphi \right) d\theta = r^2 \int_{\theta=0}^{\pi} \sin\theta \, [\varphi]_{\varphi=0}^{2\pi} \, d\theta = r^2 \int_{\theta=0}^{\pi} \sin\theta \, (2\pi) \, d\theta$$

$$A = 2\pi r^2 \int_{\theta=0}^{\pi} \sin\theta \, d\theta = 2\pi r^2 [-\cos\theta]_{\theta=0}^{\pi} = 2\pi r^2(-\cos\pi + \cos 0) = 2\pi r^2(1+1) = 4\pi r^2$$

$$E\oint_S dA = EA = E4\pi r^2 = \frac{q_{enc}}{\epsilon_0}$$

$$E = \frac{q_{enc}}{4\pi\epsilon_0 r^2}$$

Now we have reached the point where it will matter that the charge density is non-uniform. This time, we can't pull the charge density (ρ) out of the integral. Instead, we must apply the equation $\rho = \beta r$ that was given in the problem. The symbol β, however, is constant and may come out of the integral.

$$q_{enc} = \int dq = \int \rho\, dV = \int \beta r\, dV = \beta \int r\, dV$$

<u>Region I</u>: $r < a$.

Inside of the charged sphere, only a fraction of the sphere's charge is enclosed by the Gaussian sphere, so the upper limit of the r-integration is the variable r. In spherical coordinates, we write the differential volume element as $dV = r^2 \sin\theta\, dr d\theta d\varphi$.

$$q_{enc} = \beta \int r\, dV = \beta \int_{r=0}^{r}\int_{\theta=0}^{\pi}\int_{\varphi=0}^{2\pi} r(r^2 \sin\theta\, dr\, d\theta\, d\varphi) = \beta \int_{r=0}^{r}\int_{\theta=0}^{\pi}\int_{\varphi=0}^{2\pi} r^3 \sin\theta\, dr\, d\theta\, d\varphi$$

This integral is separable:

$$q_{enc} = \beta \int_{r=0}^{r} r^3\, dr \int_{\theta=0}^{\pi} \sin\theta\, d\theta \int_{\varphi=0}^{2\pi} d\varphi = \beta\left[\frac{r^4}{4}\right]_{r=0}^{r} [-\cos\theta]_{\theta=0}^{\pi} [\varphi]_{\varphi=0}^{2\pi}$$

$$q_{enc} = \beta\left(\frac{r^4}{4}\right)(-\cos\pi + \cos 0)(2\pi) = \beta\left(\frac{r^4}{4}\right)(1+1)(2\pi) = \pi\beta r^4$$

As usual with problems that feature non-uniform densities, we will eliminate the constant of proportionality (in this case, β). Note that if we integrate all of the way to $r = a$, we would obtain the total charge of the sphere, Q. We would also get the same expression as above, except with r replaced by a:

$$Q = \pi\beta a^4$$

Isolate β by dividing both sides of the equation by πa^4.

$$\beta = \frac{Q}{\pi a^4}$$

Substitute this expression into the previous equation for the charge enclosed.

$$q_{enc} = \pi\beta r^4 = \pi\left(\frac{Q}{\pi a^4}\right) r^4 = \frac{Q r^4}{a^4}$$

Now substitute the charge enclosed into the previous equation for electric field.

$$E_I = \frac{q_{enc}}{4\pi\epsilon_0 r^2} = \frac{Q r^4}{a^4} \div 4\pi\epsilon_0 r^2 = \frac{Q r^4}{a^4} \times \frac{1}{4\pi\epsilon_0 r^2} = \frac{Q r^2}{4\pi\epsilon_0 a^4}$$

Recall that $\frac{1}{4\pi\epsilon_0} = k$.

$$E_I = \frac{kQr^2}{a^4}$$

We can add a unit vector to indicate the direction of the electric field vector.

$$\vec{\mathbf{E}}_I = \frac{kQr^2}{a^4}\hat{\mathbf{r}}$$

Region II: $r > a$.

Outside of the charged sphere, 100% of the charge is enclosed by the Gaussian surface.

$$q_{enc} = Q$$

Substitute this into the equation for electric field which we found from Gauss's law.

$$E_{II} = \frac{q_{enc}}{4\pi\epsilon_0 r^2} = \frac{Q}{4\pi\epsilon_0 r^2}$$

Apply the equation $\frac{1}{4\pi\epsilon_0} = k$ and add a unit vector to turn the magnitude (E) of the electric field into a vector ($\vec{\mathbf{E}}$).

$$\vec{\mathbf{E}}_{II} = \frac{kQ}{r^2}\hat{\mathbf{r}}$$

Example 44. A solid spherical insulator centered about the origin with positive charge Q has radius a and non-uniform charge density $\rho = \beta r^2$, where β is a positive constant. Derive an expression for the electric field both inside and outside of the sphere.

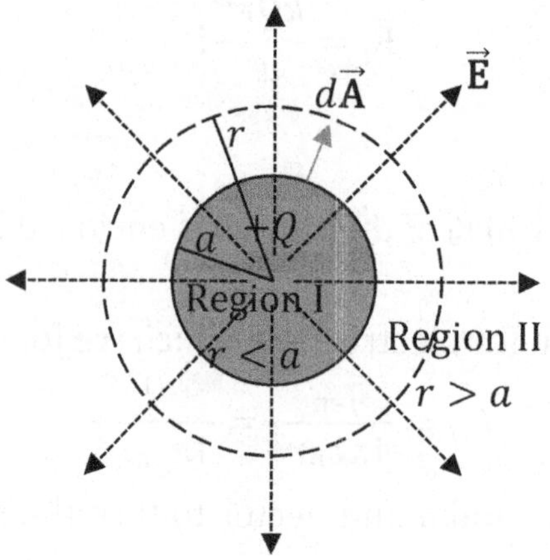

Solution. This problem is just like the previous examples, except for the charge density. The electric field lines and Gaussian surface are the same as before. To save time, we'll simply repeat the steps that are identical to the example on pages 83-86, and pick up from where this solution deviates from the earlier example. It would be a good exercise to see if you can understand each step (if not, review pages 83-86 for the explanation).

$$\oint_S \vec{\mathbf{E}} \cdot d\vec{\mathbf{A}} = \frac{q_{enc}}{\epsilon_0}$$

$$\oint_S E \cos 0° \, dA = \frac{q_{enc}}{\epsilon_0}$$

$$\oint_S E \, dA = \frac{q_{enc}}{\epsilon_0}$$

$$E \oint_S dA = \frac{q_{enc}}{\epsilon_0}$$

$$A = \oint_S dA = \int_{\theta=0}^{\pi} \int_{\varphi=0}^{2\pi} r^2 \sin\theta \, d\theta \, d\varphi = r^2 \int_{\theta=0}^{\pi} \int_{\varphi=0}^{2\pi} \sin\theta \, d\theta \, d\varphi$$

$$A = r^2 \int_{\theta=0}^{\pi} \sin\theta \left(\int_{\varphi=0}^{2\pi} d\varphi \right) d\theta = r^2 \int_{\theta=0}^{\pi} \sin\theta \, [\varphi]_{\varphi=0}^{2\pi} \, d\theta = r^2 \int_{\theta=0}^{\pi} \sin\theta \, (2\pi) \, d\theta$$

$$A = 2\pi r^2 \int_{\theta=0}^{\pi} \sin\theta \, d\theta = 2\pi r^2 [-\cos\theta]_{\theta=0}^{\pi} = 2\pi r^2(-\cos\pi + \cos 0) = 2\pi r^2(1+1) = 4\pi r^2$$

$$E \oint_S dA = EA = E 4\pi r^2 = \frac{q_{enc}}{\epsilon_0}$$

$$E = \frac{q_{enc}}{4\pi\epsilon_0 r^2}$$

Now we have reached the point where this example deviates from the previous example. We must apply the equation $\rho = \beta r^2$ that was given in the problem. The symbol β is constant and may come out of the integral.

$$q_{enc} = \int dq = \int \rho \, dV = \int \beta r^2 \, dV = \beta \int r^2 \, dV$$

<u>Region I</u>: $r < a$.

Inside of the charged sphere, only a fraction of the sphere's charge is enclosed by the Gaussian sphere, so the upper limit of the r-integration is the variable r. In spherical coordinates, we write the differential volume element as $dV = r^2 \sin\theta \, dr d\theta d\varphi$.

$$q_{enc} = \beta \int r^2 \, dV = \beta \int_{r=0}^{r} \int_{\theta=0}^{\pi} \int_{\varphi=0}^{2\pi} r^2 (r^2 \sin\theta \, dr \, d\theta \, d\varphi) = \beta \int_{r=0}^{r} \int_{\theta=0}^{\pi} \int_{\varphi=0}^{2\pi} r^4 \sin\theta \, dr \, d\theta \, d\varphi$$

This integral is separable:

$$q_{enc} = \beta \int_{r=0}^{r} r^4 \, dr \int_{\theta=0}^{\pi} \sin\theta \, d\theta \int_{\varphi=0}^{2\pi} d\varphi = \beta \left[\frac{r^5}{5}\right]_{r=0}^{r} [-\cos\theta]_{\theta=0}^{\pi} [\varphi]_{\varphi=0}^{2\pi}$$

$$q_{enc} = \beta \left(\frac{r^5}{5}\right)(-\cos\pi + \cos 0)(2\pi) = \beta \left(\frac{r^5}{5}\right)(1+1)(2\pi) = \frac{4\pi\beta r^5}{5}$$

As usual with problems that feature non-uniform densities, we will eliminate the constant of proportionality (in this case, β). Note that if we integrate all of the way to $r = a$, we would obtain the total charge of the sphere, Q. We would also get the same expression as above, except with r replaced by a:

$$Q = \frac{4\pi\beta a^5}{5}$$

Isolate β by multiplying both sides of the equation by 5 and dividing by $4\pi a^5$.

$$\beta = \frac{5Q}{4\pi a^5}$$

Substitute this expression into the previous equation for the charge enclosed.

$$q_{enc} = \frac{4\pi\beta r^5}{5} = \frac{4\pi}{5}\left(\frac{5Q}{4\pi a^5}\right) r^5 = \frac{Qr^5}{a^5}$$

Now substitute the charge enclosed into the previous equation for electric field.

$$E_I = \frac{q_{enc}}{4\pi\epsilon_0 r^2} = \frac{Qr^5}{a^5} \div 4\pi\epsilon_0 r^2 = \frac{Qr^5}{a^5} \times \frac{1}{4\pi\epsilon_0 r^2} = \frac{Qr^3}{4\pi\epsilon_0 a^5}$$

Recall that $\frac{1}{4\pi\epsilon_0} = k$.

$$E_I = \frac{kQr^3}{a^5}$$

We can add a unit vector to indicate the direction of the electric field vector.

$$\vec{\mathbf{E}}_I = \frac{kQr^3}{a^5}\hat{\mathbf{r}}$$

Region II: $r > a$.

Outside of the charged sphere, 100% of the charge is enclosed by the Gaussian surface.

$$q_{enc} = Q$$

Substitute this into the equation for electric field which we found from Gauss's law.

$$E_{II} = \frac{q_{enc}}{4\pi\epsilon_0 r^2} = \frac{Q}{4\pi\epsilon_0 r^2}$$

Apply the equation $\frac{1}{4\pi\epsilon_0} = k$ and add a unit vector to turn the magnitude (E) of the electric field into a vector ($\vec{\mathbf{E}}$).

$$\vec{\mathbf{E}}_{II} = \frac{kQ}{r^2}\hat{\mathbf{r}}$$

Spherical Shell Example: A Prelude to Example 45. A solid spherical insulator centered about the origin with positive charge $5Q$ has radius a and uniform charge density ρ. Concentric with the spherical insulator is a thick spherical conducting shell of inner radius b, outer radius c, and total charge $3Q$. Derive an expression for the electric field in each region.

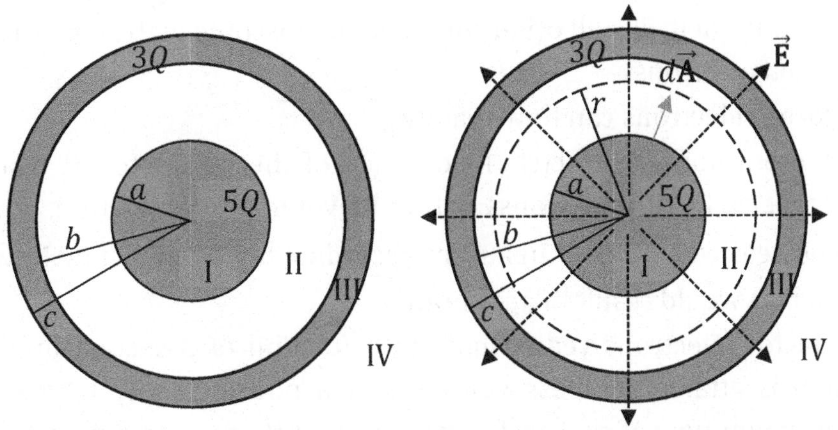

Solution. This problem is very similar to the previous examples, except that there are four regions, and there is a little "trick" to working with the conducting shell for region III. The same Gaussian sphere from the previous example applies here, and the math for Gauss's law works much the same way here. The real difference is in figuring out the charge enclosed in each region.

<u>Region I</u>: $r < a$.
The conducting shell does **<u>not</u>** matter for region I, since none of its charge will reside in a Gaussian sphere with $r < a$. Thus, we will obtain the same result as for region I of the example on pages 83-86, except that the total charge of the sphere is now $5Q$ (this was stated in the problem). We'll use one of the alternate forms of the equation for electric field that involves the total charge of the inner sphere (see the bottom of page 86), except that where we previously had Q for the total charge, we'll change that to $5Q$ for this problem.

$$\vec{\mathbf{E}}_I = \frac{5kQr}{a^3}\hat{\mathbf{r}}$$

<u>Region II</u>: $a < r < b$.
The conducting shell also does **<u>not</u>** matter for region II, since again none of its charge will reside in a Gaussian sphere with $r < b$. (Recall that Gauss's law involves the charge enclosed, q_{enc}, by the Gaussian sphere.) We obtain the same result as for region II of the previous examples (see the bottom of page 86), with Q replaced by $5Q$ for this problem.

$$\vec{\mathbf{E}}_{II} = \frac{5kQ}{r^2}\hat{\mathbf{r}}$$

Region III: $b < r < c$.

Now the conducting shell matters. The electric field inside of the conducting shell equals zero once electrostatic equilibrium is attained (which just takes a fraction of a second).

$$\vec{\mathbf{E}}_{III} = 0$$

Following are the reasons that the electric field must be zero in region III:

- The conducting shell, like all other forms of macroscopic matter, consists of protons, neutrons, and electrons.
- In a **conductor**, electrons can flow readily.
- If there were a nonzero electric field inside of the conductor, it would cause the charges (especially, the electrons) within its volume to accelerate. This is because the electric field would result in a force according to $\vec{\mathbf{F}} = q\vec{\mathbf{E}}$, while Newton's second law ($\sum \vec{\mathbf{F}} = m\vec{\mathbf{a}}$) would result in acceleration.
- In a conductor, electrons redistribute in a fraction of a second until **electrostatic equilibrium** is attained (unless you connect a power supply to the conductor to create a constant flow of charge, for example, but there is no power supply involved in this problem).
- Once electrostatic equilibrium is attained, the charges won't be moving, and therefore the electric field within the conducting shell must be zero.

Because the electric field is zero in region III, we can reason how the total charge of the conducting shell (which equals $3Q$ according to the problem) is distributed. According to Gauss's law, $\oint_S \vec{\mathbf{E}} \cdot d\vec{\mathbf{A}} = \frac{q_{enc}}{\epsilon_0}$. Since $\vec{\mathbf{E}}_{III} = 0$ in region III, the right-hand side of Gauss's law must also equal zero, meaning that the charge enclosed by a Gaussian surface in region III must equal zero: $q_{enc}^{III} = 0$.

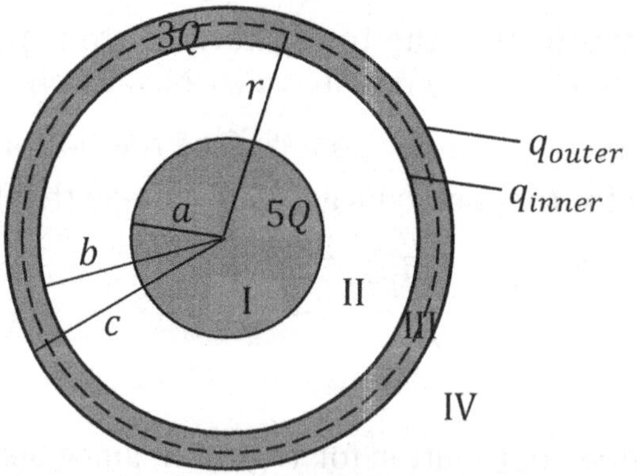

A Gaussian sphere in region III ($b < r < c$) encloses the inner sphere plus the inner surface of the conducting shell (see the dashed curve in the diagram above).

$$q_{enc}^{III} = 5Q + q_{inner} = 0$$

Since the charge enclosed in region III is zero, we can conclude that $q_{inner} = -5Q$. The total charge of the conducting shell equals $3Q$ according to the problem. Therefore, if we add the charge of the inner surface and outer surface of the conducting shell together, we must get $3Q$.

$$q_{inner} + q_{outer} = 3Q$$
$$-5Q + q_{outer} = 3Q$$

Add $5Q$ to both sides of the equation.

$$q_{outer} = 3Q + 5Q = 8Q$$

The conducting shell has a charge of $-5Q$ on its inner surface and a charge of $8Q$ on its outer surface, for a total charge of $3Q$.

It's important to note that "inner sphere" and "inner surface of the conducting shell" are two different things. In our notation, q_{inner} represents the charge on the "inner surface of the conducting shell." Note that q_{inner} does **not** refer to the "inner sphere."

In this example:

- The inner sphere has a charge of $q_{sphere} = 5Q$.
- The inner surface of the conducting shell has a charge of $q_{inner} = -5Q$.
- The outer surface of the conducting shell has a charge of $q_{outer} = 8Q$.
- The total charge of the conducting shell is $q_{shell} = q_{inner} + q_{outer} = 3Q$.

Note that the excess charge residing on the conducting shell can only reside on its inner or outer surfaces. (Specifically, $-5Q$ resides on its inner surface and $8Q$ resides on its outer surface.) The net charge within the thickness of its volume must be zero (meaning that the protons and electrons in that region will be equal in number so that they cancel out). This follows from the fact that the net electric field must be zero in region III ($\vec{\mathbf{E}}_{III} = 0$): Since Gauss's law states that $\oint_S \vec{\mathbf{E}} \cdot d\vec{\mathbf{A}} = \frac{q_{enc}}{\epsilon_0}$, in region III Gauss's law requires that the net charge enclosed within the solid thickness of the conducting shell be zero. This is true in general of conductors in electrostatic equilibrium: Any excess charge that the conductor has (from either an excess or deficiency of electrons compared to protons) must reside on its surface.

Region IV: $r < c$.

A Gaussian sphere in region IV encloses a total charge of $8Q$ (the $5Q$ from the inner sphere plus the $3Q$ from the conducting shell). The formula for the electric field is the same as for region II, except for changing the total charge enclosed to $8Q$.

$$\vec{\mathbf{E}}_{IV} = \frac{8kQ}{r^2}\hat{\mathbf{r}}$$

Example 45. A solid spherical insulator centered about the origin with negative charge $-6Q$ has radius a and non-uniform charge density $\rho = -\beta r$, where β is a positive constant. Concentric with the spherical insulator is a thick spherical conducting shell of inner radius b, outer radius c, and total charge $+4Q$. Derive an expression for the electric field in each region.

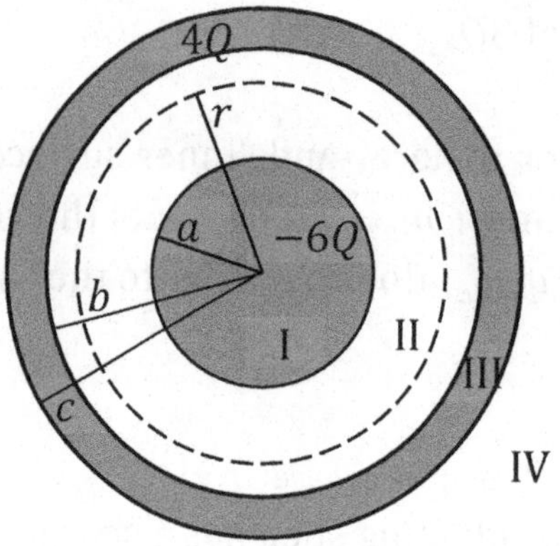

Solution. This problem is very similar to the previous example, except that the charge density is non-uniform.

Region I: $r < a$.

- The conducting shell doesn't matter in region I. We solved the problem with a non-uniformly charged sphere with charge density $\rho = \beta r$ in the example on pages 87-89. In that example, the total charge of the sphere was Q and we found that the electric field was $\frac{kQr^2}{a^4}\hat{\mathbf{r}}$ **inside** of the sphere. See the top of page 89.
- The only differences with this problem are that the charge density is negative ($\rho = -\beta r$) and the total charge of the sphere is $-6Q$. The only difference this will make in the final answer is that Q will be replaced by $-6Q$. Thus, $\vec{\mathbf{E}}_I = -\frac{6kQr^2}{a^4}\hat{\mathbf{r}}$.

Region II: $a < r < b$.

- The conducting shell also doesn't matter in region II. Review the example on pages 87-89 which had a sphere with charge density $\rho = \beta r$, where the total charge of the sphere was Q and we found that the electric field was $\frac{kQ}{r^2}\hat{\mathbf{r}}$ **outside** of the sphere.
- The only differences with this problem are that the charge density is negative ($\rho = -\beta r$) and the total charge of the sphere is $-6Q$. The only difference this will make in the final answer is that Q will be replaced by $-6Q$. Thus, $\vec{\mathbf{E}}_{II} = -\frac{6kQ}{r^2}\hat{\mathbf{r}}$.

Region III: $b < r < c$.

- Inside the conducting shell, $\vec{\mathbf{E}}_{III} = 0$ for the same reasons discussed in the previous example. See page 94.

<u>Region IV</u>: $r > c$.

- Outside of all of the spheres, the electric field is the same as that of a pointlike charge: It equals $\frac{kQ_{tot}}{r^2}\hat{\mathbf{r}}$ (see that bottom of page 86), except that in this problem the total charge isn't Q. The total charge enclosed by a Gaussian sphere in region IV is $-6Q + 4Q = -2Q$. Thus, we replace Q_{tot} with $-2Q$ to obtain $\vec{\mathbf{E}}_{IV} = -\frac{2kQ}{r^2}\hat{\mathbf{r}}$.

Infinite Line Charge Example: A Prelude to Example 46. An infinite line of positive charge lies on the z-axis and has uniform charge density λ. Derive an expression for the electric field created by the infinite line charge.

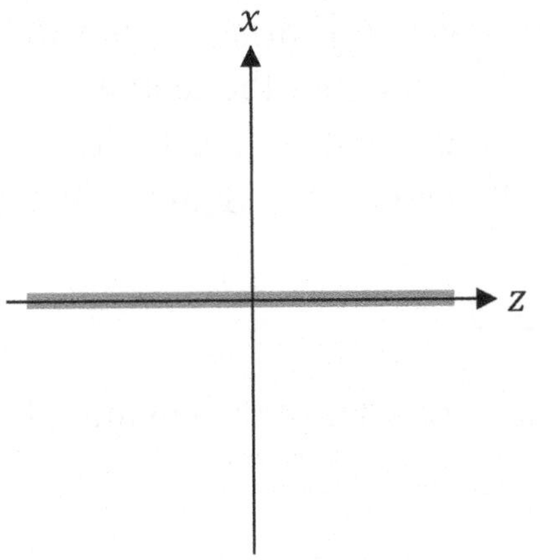

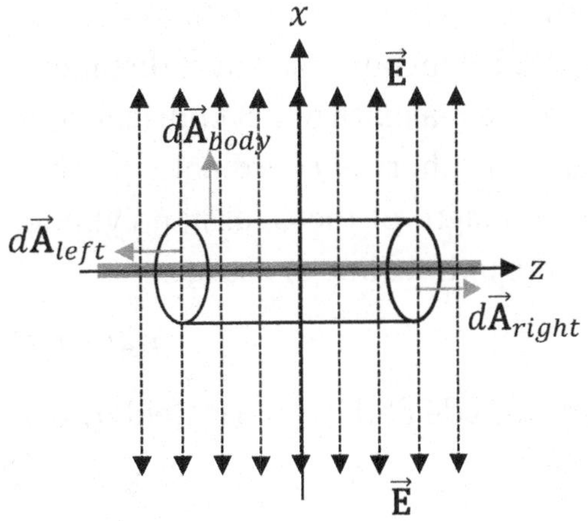

Solution. First sketch the electric field lines for the infinite line charge. Wherever a positive test charge might be placed, it would be repelled directly away from the z-axis. Therefore, the electric field lines radiate outward, as shown above on the right (but realize that the field lines really radiate outward in three dimensions). We choose our Gaussian surface to be a cylinder (see the right diagram above) coaxial with the line charge such that $\vec{\mathbf{E}}$ and $d\vec{\mathbf{A}}$ will be parallel along the body of the cylinder (and $\vec{\mathbf{E}}$ and $d\vec{\mathbf{A}}$ will be perpendicular along the ends). Write the formula for Gauss's law.

$$\oint_S \vec{\mathbf{E}} \cdot d\vec{\mathbf{A}} = \frac{q_{enc}}{\epsilon_0}$$

The scalar product is $\vec{\mathbf{E}} \cdot d\vec{\mathbf{A}} = E \cos\theta \, dA$. Study the direction of $\vec{\mathbf{E}}$ and $d\vec{\mathbf{A}}$ at each end and the body of the cylinder in the previous diagram.

- For the ends, $\theta = 90°$ because $\vec{\mathbf{E}}$ and $d\vec{\mathbf{A}}$ are perpendicular.
- For the body, $\theta = 0°$ because $\vec{\mathbf{E}}$ and $d\vec{\mathbf{A}}$ are parallel.

$$\int_{left} E \cos 90° \, dA + \int_{body} E \cos 0° \, dA + \int_{right} E \cos 90° \, dA = \frac{q_{enc}}{\epsilon_0}$$

Recall from trig that $\cos 0° = 1$ and $\cos 90° = 0$.

$$0 + \int_{body} E\, dA + 0 = \frac{q_{enc}}{\epsilon_0}$$

The magnitude of the electric field is constant over the body of the Gaussian cylinder, since every point on the body of the cylinder is equidistant from the axis of the line charge. Therefore, we may pull E out of the integral.

$$E \oint_{body} dA = \frac{q_{enc}}{\epsilon_0}$$

$$EA = \frac{q_{enc}}{\epsilon_0}$$

The area is the surface area of the Gaussian cylinder of radius r_c (which is a variable, since the electric field depends upon the distance from the line charge). Note that r_c is the distance from the z-axis (from cylindrical coordinates), and **not** the distance to the origin (which we use in spherical coordinates). The area of the body of a cylinder is $A = 2\pi r_c L$, where L is the length of the Gaussian cylinder. (We choose the Gaussian cylinder to be finite, unlike the infinite line of charge.)

$$E 2\pi r_c L = \frac{q_{enc}}{\epsilon_0}$$

Isolate the magnitude of the electric field by dividing both sides of the equation by $2\pi r_c L$.

$$E = \frac{q_{enc}}{2\pi\epsilon_0 r_c L}$$

Now we need to determine how much charge is enclosed by the Gaussian cylinder. For a line charge, we write $dq = \lambda ds$ (see page 79). Since the line charge has uniform charge density, we may pull λ out of the integral.

$$q_{enc} = \int dq = \int \lambda\, ds = \lambda \int ds = \lambda \int dz = \lambda L$$

The charge enclosed by the Gaussian cylinder is $q_{enc} = \lambda L$, where L is the length of the Gaussian cylinder. Substitute this into the previous equation for electric field.

$$E = \frac{q_{enc}}{2\pi\epsilon_0 r_c L} = \frac{\lambda L}{2\pi\epsilon_0 r_c L} = \frac{\lambda}{2\pi\epsilon_0 r_c}$$

We can turn this into a vector by including the appropriate unit vector.

$$\vec{\mathbf{E}} = \frac{\lambda}{2\pi\epsilon_0 r_c}\hat{\mathbf{r}}_c$$

Note that $\hat{\mathbf{r}}_c$ is directed away from the z-axis.

Infinite Conducting Cylinder Example: Another Prelude to Example 46. An infinite solid cylindrical conductor coaxial with the z-axis has radius a and uniform charge density σ. Derive an expression for the electric field both inside and outside of the conductor.

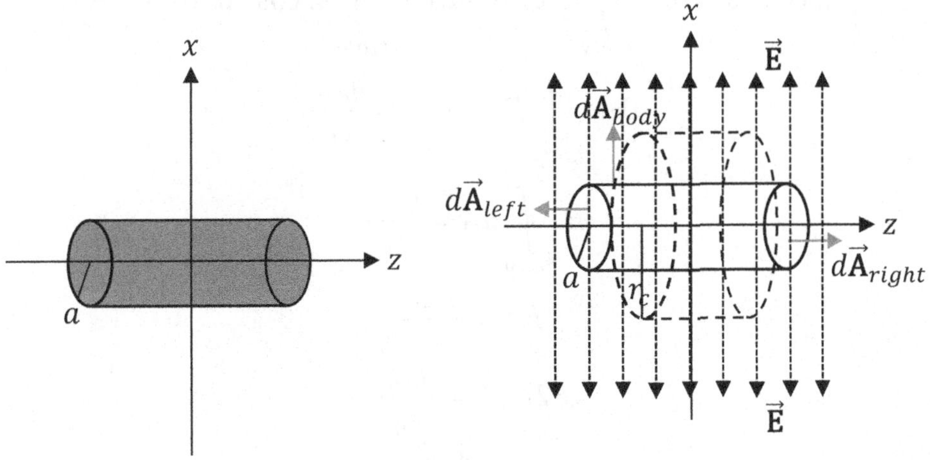

Solution. The electric field lines for this infinite charged cylinder radiate away from the z-axis just like the electric field lines for the infinite line charge in the previous example. As with the previous example, we will apply a Gaussian cylinder coaxial with the z-axis, and the math will be virtually the same as in the previous example. One difference is that this problem involves two different regions, and another difference is that this problem involves a surface charge density (σ) instead of a linear charge density (λ).

<u>Region I</u>: $r_c < a$.
Since the cylinder is a conductor, the electric field will be zero in region I.

$$\vec{\mathbf{E}}_I = 0$$

The reasoning is the same as it was a few examples back when we had a problem with a conducting shell. If there were a nonzero electric field inside the conducting cylinder, it would cause the charged particles inside of it (namely the valence electrons) to accelerate. Within a fraction of a second, the electrons in the conducting cylinder will attain electrostatic equilibrium, after which point the electric field will be zero inside the volume of the conducting cylinder. Since $\vec{\mathbf{E}}_I = 0$ inside the conducting cylinder and since $\oint_S \vec{\mathbf{E}} \cdot d\vec{\mathbf{A}} = \frac{q_{enc}}{\epsilon_0}$ according to Gauss's law, the net charge within the conducting cylinder must also be zero. Therefore, the net charge that resides on the conducting cylinder must reside on its surface.

<u>Region II</u>: $r_c > a$.
Since the electric field lines for this problem closely resemble the electric field lines of the previous example, the math for Gauss's law will start out the same. We will repeat those equations to save time, and then pick up our discussion where this solution deviates from the previous one. Again, it would be wise to try to follow along, and review the previous example if necessary.

$$\oint_S \vec{\mathbf{E}} \cdot d\vec{\mathbf{A}} = \frac{q_{enc}}{\epsilon_0}$$

$$\int_{left} E \cos 90^\circ \, dA + \int_{body} E \cos 0^\circ \, dA + \int_{right} E \cos 90^\circ \, dA = \frac{q_{enc}}{\epsilon_0}$$

$$0 + \int_{body} E \, dA + 0 = \frac{q_{enc}}{\epsilon_0}$$

$$E \oint_{body} dA = \frac{q_{enc}}{\epsilon_0}$$

$$EA = \frac{q_{enc}}{\epsilon_0}$$

$$E 2\pi r_c L = \frac{q_{enc}}{\epsilon_0}$$

$$E = \frac{q_{enc}}{2\pi\epsilon_0 r_c L}$$

This is the point where the solution is different for the infinite conducting cylinder than for the infinite line charge. When we determine how much charge is enclosed by the Gaussian cylinder, we will work with $dq = \sigma dA$ (see page 79). For a solid cylinder, we would normally use ρdV, but since this is a **conducting** cylinder, as we reasoned in region I, all of the charge resides on its surface, not within its volume. (If this had been an insulator instead of a conductor, then we would use ρdV for a solid cylinder.) Since the cylinder has uniform charge density, we may pull σ out of the integral.

$$q_{enc} = \int dq = \int \sigma \, dA = \sigma \int dA = \sigma A$$

The surface area of the body of a cylinder is $A = 2\pi a L$. Here, we use the radius of the conducting cylinder (a), not the radius of the Gaussian cylinder (r_c), because the charge lies on a cylinder of radius a (we're finding the charge enclosed). Plug this expression into the equation for the charge enclosed.

$$q_{enc} = \sigma A = \sigma 2\pi a L$$

Substitute this into the previous equation for electric field.

$$E_{II} = \frac{q_{enc}}{2\pi\epsilon_0 r_c L} = \frac{\sigma 2\pi a L}{2\pi\epsilon_0 r_c L} = \frac{\sigma a}{\epsilon_0 r_c}$$

We can turn this into a vector by including the appropriate unit vector.

$$\vec{\mathbf{E}}_{II} = \frac{\sigma a}{\epsilon_0 r_c} \hat{\mathbf{r}}_c$$

It's customary to write this in terms of **charge per unit length**, $\lambda = \frac{Q}{L}$, rather than charge per unit area, $\sigma = \frac{Q}{A} = \frac{Q}{2\pi a L}$, in which case the answer is identical to the previous example:

$$\vec{\mathbf{E}}_{II} = \frac{\lambda}{2\pi\epsilon_0 r_c} \hat{\mathbf{r}}_c$$

Example 46. An infinite solid cylindrical insulator coaxial with the z-axis has radius a and uniform positive charge density ρ. Derive an expression for the electric field both inside and outside of the charged insulator.

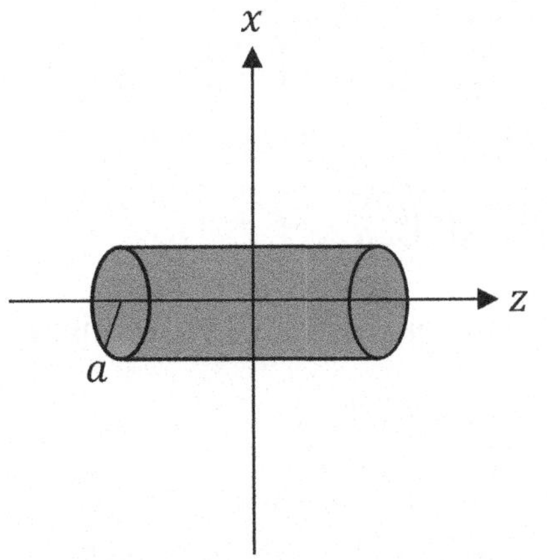

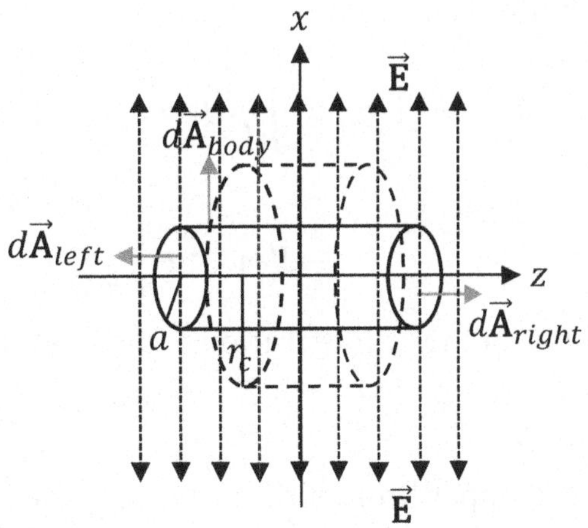

Solution. This problem is similar to the previous example. Since the electric field lines for this problem closely resemble‡ the electric field lines of the previous example, the math for Gauss's law will start out the same. We will repeat those equations to save time, and then pick up our discussion where this solution deviates from the previous one. Again, it would be wise to try to follow along, and review the example on pages 97-98 if necessary.

$$\oint_S \vec{\mathbf{E}} \cdot d\vec{\mathbf{A}} = \frac{q_{enc}}{\epsilon_0}$$

$$\int_{left} E \cos 90^\circ \, dA + \int_{body} E \cos 0^\circ \, dA + \int_{right} E \cos 90^\circ \, dA = \frac{q_{enc}}{\epsilon_0}$$

$$0 + \int_{body} E \, dA + 0 = \frac{q_{enc}}{\epsilon_0}$$

$$E \oint_{body} dA = \frac{q_{enc}}{\epsilon_0}$$

$$EA = \frac{q_{enc}}{\epsilon_0}$$

$$E 2\pi r_c L = \frac{q_{enc}}{\epsilon_0}$$

$$E = \frac{q_{enc}}{2\pi\epsilon_0 r_c L}$$

This is the point where the solution is different for the infinite conducting cylinder than for the infinite cylindrical insulator. When we determine how much charge is enclosed by the

‡ The electric field lines are the same outside of the cylinder, but inside of the cylinder is a different matter. For the conducting cylinder, there are no electric field lines inside of the cylinder, but for this cylindrical insulator, the electric field lines radiate away from the z-axis both inside and outside of the cylinder.

Gaussian cylinder, we will work with $dq = \rho dV$ and $dV = r_c dr_c d\theta dz$ (see page 79). Since the cylinder has uniform charge density, we may pull ρ out of the integral. The upper limit of the r_c-integration is r_c in region I ($r_c < a$) and a in region II ($r_c > a$). The length of the Gaussian cylinder is L.

$$q_{enc} = \int dq = \int \rho\, dV = \rho \int dV = \rho \int_{r_c=0}^{r_c \text{ or } a} r_c\, dr_c \int_{\theta=0}^{2\pi} d\theta \int_{z=0}^{L} dz$$

$$q_{enc} = \rho \left[\frac{r_c^2}{2}\right]_{r_c=0}^{r_c \text{ or } a} [\theta]_{\theta=0}^{2\pi} [z]_{z=0}^{L}$$

The upper limit of the r_c-integration is different in each region, as we will consider below.

<u>**Region I**</u>: $r_c < a$.

In Region I, the upper limit of the r_c-integration is r_c.

$$q_{enc} = \rho \left[\frac{r_c^2}{2}\right]_{r_c=0}^{r_c} [\theta]_{\theta=0}^{2\pi} [z]_{z=0}^{L} = \rho\left(\frac{r_c^2}{2}\right)(2\pi)(L) = \pi\rho r_c^2 L$$

Note that $\pi r_c^2 L$ is the volume of the Gaussian cylinder. Substitute this into the previous equation for electric field.

$$E_I = \frac{q_{enc}}{2\pi\epsilon_0 r_c L} = \frac{\pi\rho r_c^2 L}{2\pi\epsilon_0 r_c L} = \frac{\rho r_c}{2\epsilon_0}$$

We can turn this into a vector by including the appropriate unit vector.

$$\vec{\mathbf{E}}_I = \frac{\rho r_c}{2\epsilon_0}\hat{\mathbf{r}}_c$$

<u>**Region II**</u>: $r_c > a$.

In Region I, the upper limit of the r_c-integration is a.

$$q_{enc} = \rho \left[\frac{r_c^2}{2}\right]_{r_c=0}^{a} [\theta]_{\theta=0}^{2\pi} [z]_{z=0}^{L} = \rho\left(\frac{a^2}{2}\right)(2\pi)(L) = \pi\rho a^2 L$$

Substitute this into the previous equation for electric field.

$$E_{II} = \frac{q_{enc}}{2\pi\epsilon_0 r_c L} = \frac{\pi\rho a^2 L}{2\pi\epsilon_0 r_c L} = \frac{\rho a^2}{2\epsilon_0 r_c}$$

We can turn this into a vector by including the appropriate unit vector.

$$\vec{\mathbf{E}}_{II} = \frac{\rho a^2}{2\epsilon_0 r_c}\hat{\mathbf{r}}_c$$

It's customary to write this in terms of **charge per unit length**, $\lambda = \frac{Q}{L}$, rather than charge per unit volume, $\rho = \frac{Q}{V} = \frac{Q}{\pi a^2 L}$, in which case the answer is identical to the infinite line charge:

$$\vec{\mathbf{E}} = \frac{\lambda}{2\pi\epsilon_0 r_c}\hat{\mathbf{r}}_c$$

Infinite Plane Example: A Prelude to Examples 47-48. A very thin, infinitely§ large sheet of charge lies in the xy plane. The positively charged sheet has uniform charge density σ. Derive an expression for the electric field on either side of the infinite sheet.

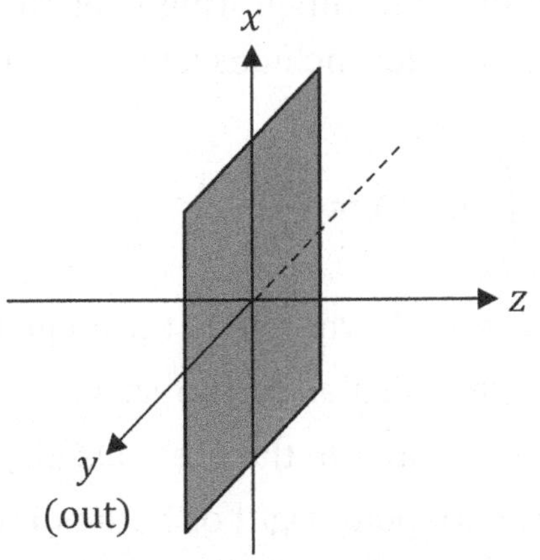

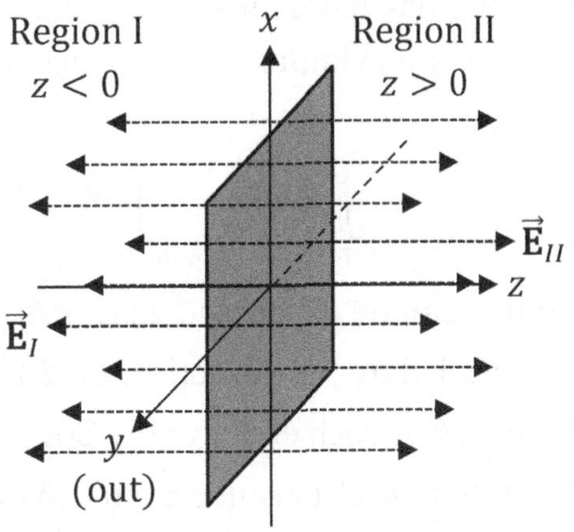

Solution. First sketch the electric field lines for the infinite sheet of positive charge. If a positive test charge were placed to the right of the sheet, it would be repelled to the right. If a positive test charge were placed to the left of the sheet, it would be reprelled to the left. Therefore, the electric field lines are directed away from the sheet, as shown above on the right. We choose a right-circular cylinder to serve as our Gaussian surface, as illustrated below. The reasons for this choice are:

- $\vec{\mathbf{E}}$ is parallel to $d\vec{\mathbf{A}}$ at the ends of the cylinder (both are horizontal).
- $\vec{\mathbf{E}}$ is perpendicular to $d\vec{\mathbf{A}}$ over the body of the cylinder ($\vec{\mathbf{E}}$ is horizontal while $d\vec{\mathbf{A}}$ is not).
- The magnitude of $\vec{\mathbf{E}}$ is constant over either end of the cylinder, since every point on the end is equidistant from the infinite sheet.

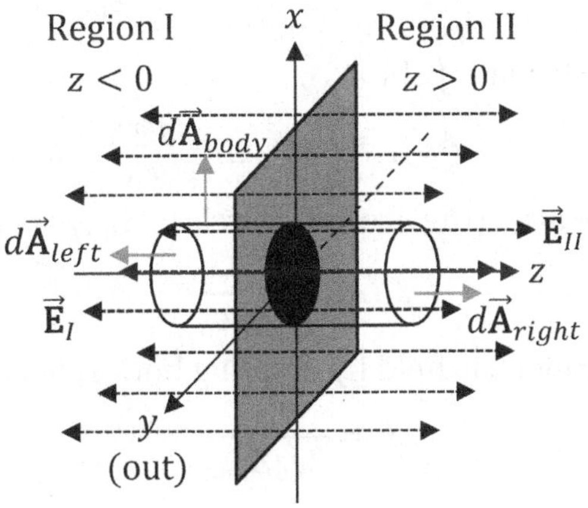

§ In practice, this result for the "infinite" sheet applies to a finite sheet when you're calculating the electric field at a distance from the sheet that is very small compared to the dimensions of the sheet and which is not near the edges of the sheet. A common example encountered in the laboratory is the parallel-plate capacitor (Chapter 12), for which the separation between the plates is very small compared to the size of the plates.

Write the formula for Gauss's law.

$$\oint_S \vec{\mathbf{E}} \cdot d\vec{\mathbf{A}} = \frac{q_{enc}}{\epsilon_0}$$

The net flux on the left-hand side of the equation involves integrating over the complete surface of the Gaussian cylinder. The surface of the cylinder includes a left end, a body, and a right end.

$$\int_{left} \vec{\mathbf{E}} \cdot d\vec{\mathbf{A}} + \int_{body} \vec{\mathbf{E}} \cdot d\vec{\mathbf{A}} + \int_{right} \vec{\mathbf{E}} \cdot d\vec{\mathbf{A}} = \frac{q_{enc}}{\epsilon_0}$$

The scalar product can be expressed as $\vec{\mathbf{E}} \cdot d\vec{\mathbf{A}} = E \cos\theta \, dA$, where θ is the angle between $\vec{\mathbf{E}}$ and $d\vec{\mathbf{A}}$. Also recall that the direction of $d\vec{\mathbf{A}}$ is perpendicular to the surface. Study the direction of $\vec{\mathbf{E}}$ and $d\vec{\mathbf{A}}$ at each end and the body of the cylinder in the previous diagram.

- For the ends, $\theta = 0°$ because $\vec{\mathbf{E}}$ and $d\vec{\mathbf{A}}$ either both point right or both point left.
- For the body, $\theta = 90°$ because $\vec{\mathbf{E}}$ and $d\vec{\mathbf{A}}$ are perpendicular.

$$\int_{left} E \cos 0° \, dA + \int_{body} E \cos 90° \, dA + \int_{right} E \cos 0° \, dA = \frac{q_{enc}}{\epsilon_0}$$

Recall from trig that $\cos 0° = 1$ and $\cos 90° = 0$.

$$\int_{left} E \, dA + 0 + \int_{right} E \, dA = \frac{q_{enc}}{\epsilon_0}$$

Over either end, the magnitude of the electric field is constant, since every point on one end is equidistant from the infinite sheet. Therefore, we may pull E out of the integrals. (We choose our Gaussian surface to be centered about the infinite sheet such that the value of E must be the same** at both ends.)

$$E \int_{left} dA + E \int_{right} dA = \frac{q_{enc}}{\epsilon_0}$$

The remaining integrals are trivial: $\int dA = A$.

$$EA_{left} + EA_{right} = \frac{q_{enc}}{\epsilon_0}$$

The two ends have the same area (the area of a circle): $A_{left} = A_{right} = A_{end}$.

$$2EA_{end} = \frac{q_{enc}}{\epsilon_0}$$

Isolate the magnitude of the electric field by dividing both sides of the equation by $2A_{end}$.

$$E = \frac{q_{enc}}{2\epsilon_0 A_{end}}$$

** Once we reach our final answer, we will see that this doesn't matter: It turns out that the electric field is independent of the distance from the infinite charged sheet.

Now we need to determine how much charge is enclosed by the Gaussian surface. We find this the same way that we calculated Q in the previous chapter.

$$q_{enc} = \int dq$$

This integral isn't over the Gaussian surface itself: Rather, this integral is over the region of the sheet enclosed by the Gaussian surface. This region is the circle shaded in black on the previous diagram. For a sheet of charge, we write $dq = \sigma dA$ (see page 79).

$$q_{enc} = \int \sigma \, dA$$

Since the sheet has uniform charge density, we may pull σ out of the integral.

$$q_{enc} = \sigma \int dA = \sigma A_{end}$$

This area, which is the area of the sheet enclosed by the Gaussian cylinder, and which is shown as a black circle on the previous diagram, is the same as the area of either end of the cylinder. Substitute this expression for the charge enclosed into the previous equation for electric field.

$$E = \frac{q_{enc}}{2\epsilon_0 A_{end}} = \frac{\sigma A_{end}}{2\epsilon_0 A_{end}} = \frac{\sigma}{2\epsilon_0}$$

The magnitude of the electric field is $E = \frac{\sigma}{2\epsilon_0}$, which is a constant. Thus, the electric field created by an infinite sheet of charge is uniform. We can use a unit vector to include the direction of the electric field with our answer: $\vec{\mathbf{E}} = \frac{\sigma}{2\epsilon_0}\hat{\mathbf{z}}$ for $z > 0$ (to the right of the sheet) and $\vec{\mathbf{E}} = -\frac{\sigma}{2\epsilon_0}\hat{\mathbf{z}}$ for $z < 0$ (to the left of the sheet), since $\hat{\mathbf{z}}$ points one unit along the $+z$-axis. There is a clever way to combine both results into a single equation: We can simply write $\vec{\mathbf{E}} = \frac{\sigma}{2\epsilon_0}\frac{z}{|z|}\hat{\mathbf{z}}$, since $\frac{z}{|z|} = +1$ if $z > 0$ and $\frac{z}{|z|} = -1$ if $z < 0$.

Example 47. Two very thin, infinitely large sheets of charge have equal and opposite uniform charge densities $+\sigma$ and $-\sigma$. The positive sheet lies in the xy plane at $z = 0$ while the negative sheet is parallel to the first at $z = d$. Derive an expression for the electric field in each of the three regions.

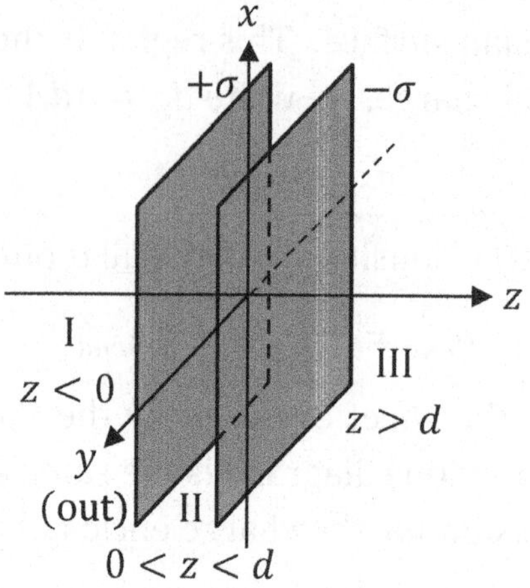

Solution. Apply the principle of superposition (see Chapter 3). First, ignore the negative plane and find the electric field in each region due to the positive plane. Call this $\vec{\mathbf{E}}_1$. This is identical to the previous example, where we found that that the magnitude of the electric field is $E_1 = \frac{\sigma}{2\epsilon_0}$.

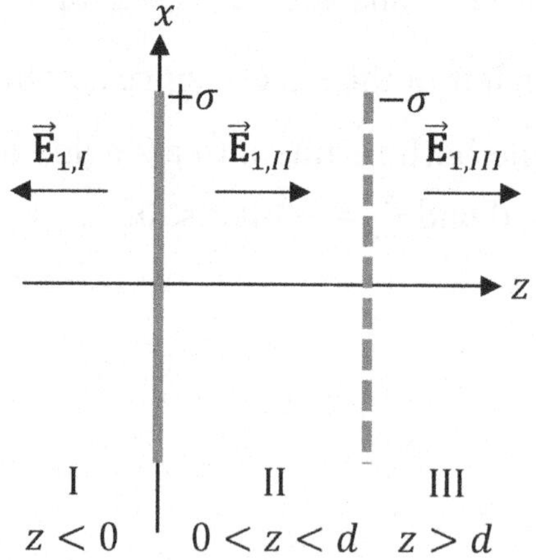

The direction of the electric field is different in each region. A positive "test" charge would be **repelled** by the positive plane.

- To the left of the positive plane (region I), $\vec{\mathbf{E}}_1$ points to the left.

$$\vec{\mathbf{E}}_{1,I} = -\frac{\sigma}{2\epsilon_0}\hat{\mathbf{z}}$$

- To the right of the positive plane (regions II and III), $\vec{\mathbf{E}}_1$ points to the right.

$$\vec{\mathbf{E}}_{1,II} = \vec{\mathbf{E}}_{1,III} = \frac{\sigma}{2\epsilon_0}\hat{\mathbf{z}}$$

The direction of $\vec{\mathbf{E}}_1$ in each region is illustrated above with **solid** arrows.

Next, ignore the positive plane and find the electric field in each region due to the negative plane. Call this $\vec{\mathbf{E}}_2$. The magnitude of the electric field is the same: $E_2 = \frac{\sigma}{2\epsilon_0}$. However, the direction is different.

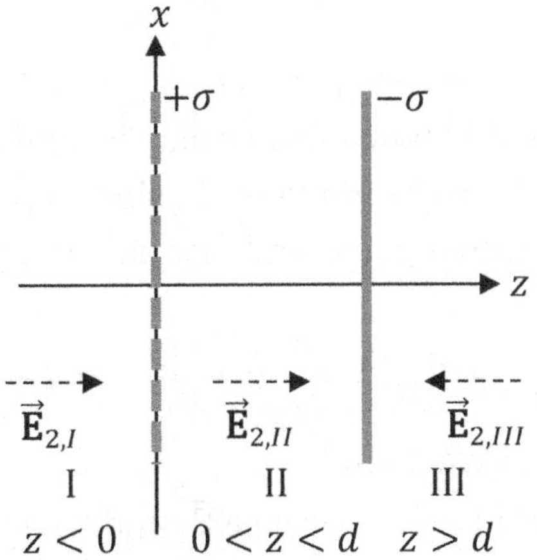

A positive "test" charge would be **attracted** to the negative plane.

- To the left of the negative plane (regions I and II), $\vec{\mathbf{E}}_2$ points to the right.

$$\vec{\mathbf{E}}_{2,I} = \vec{\mathbf{E}}_{2,II} = \frac{\sigma}{2\epsilon_0}\hat{\mathbf{z}}$$

- To the right of the negative plane (region I), $\vec{\mathbf{E}}_2$ points to the left.

$$\vec{\mathbf{E}}_{2,III} = -\frac{\sigma}{2\epsilon_0}\hat{\mathbf{z}}$$

The direction of $\vec{\mathbf{E}}_2$ in each region is illustrated above with **dashed** arrows.

Add the vectors $\vec{\mathbf{E}}_1$ and $\vec{\mathbf{E}}_2$ to find the net electric field, $\vec{\mathbf{E}}_{net}$, in each region. Note that $\vec{\mathbf{E}}_1$ and $\vec{\mathbf{E}}_2$ are both equal in magnitude. They also have the same direction in region II, but opposite directions in regions I and III, as illustrated below.

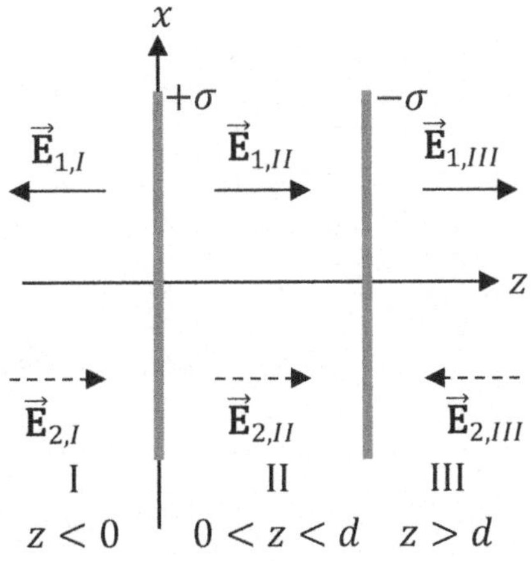

<u>Region I</u>: $z < 0$ (left of the positive plane).

In Region I, the net electric field is zero because $\vec{\mathbf{E}}_1$ and $\vec{\mathbf{E}}_2$ are equal and opposite.

$$\vec{\mathbf{E}}_{net,I} = \vec{\mathbf{E}}_{1,I} + \vec{\mathbf{E}}_{2,I} = -\frac{\sigma}{2\epsilon_0}\hat{\mathbf{z}} + \frac{\sigma}{2\epsilon_0}\hat{\mathbf{z}} = 0$$

<u>Region II</u>: $0 < z < d$ (in between the two planes).

In Region II, the net electric field is twice $\vec{\mathbf{E}}_1$ because $\vec{\mathbf{E}}_1$ and $\vec{\mathbf{E}}_2$ are equal and point in the same direction (both point to the right in region II, since a positive "test" charge would be repelled to the **right** by the positive plane and also attracted to the **right** by the negative plane).

$$\vec{\mathbf{E}}_{net,II} = \vec{\mathbf{E}}_{1,II} + \vec{\mathbf{E}}_{2,II} = \frac{\sigma}{2\epsilon_0}\hat{\mathbf{z}} + \frac{\sigma}{2\epsilon_0}\hat{\mathbf{z}} = 2\frac{\sigma}{2\epsilon_0}\hat{\mathbf{z}} = \frac{\sigma}{\epsilon_0}\hat{\mathbf{z}}$$

<u>Region III</u>: $z > d$ (left of the positive plane).

In Region III, the net electric field is zero because $\vec{\mathbf{E}}_1$ and $\vec{\mathbf{E}}_2$ are equal and opposite.

$$\vec{\mathbf{E}}_{net,III} = \vec{\mathbf{E}}_{1,III} + \vec{\mathbf{E}}_{2,III} = \frac{\sigma}{2\epsilon_0}\hat{\mathbf{z}} - \frac{\sigma}{2\epsilon_0}\hat{\mathbf{z}} = 0$$

Example 48. Two very thin, infinitely large sheets of charge have equal and opposite uniform charge densities $+\sigma$ and $-\sigma$. The two sheets are perpendicular to one another, with the positive sheet lying in the xy plane and with the negative sheet lying in the yz plane. Derive an expression for the electric field in the octant where x, y, and z are all positive.

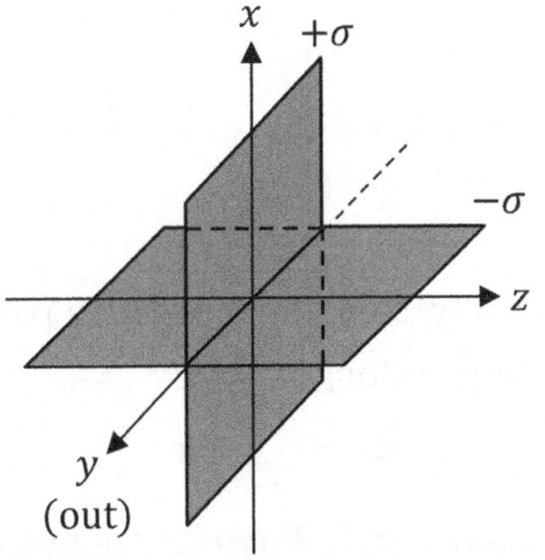

Solution. Apply the principle of superposition (see Chapter 3), much like we did in the previous example. First, ignore the negative plane and find the electric field in the desired region due to the positive plane. Call this $\vec{\mathbf{E}}_1$. This is similar to the previous examples, where we found that that the magnitude of the electric field is $E_1 = \frac{\sigma}{2\epsilon_0}$.

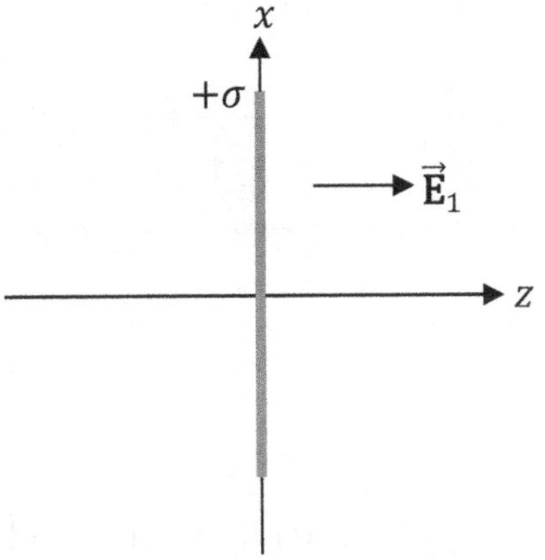

A positive "test" charge would be **repelled** by the positive plane. To the right of the positive plane (where z is positive), $\vec{\mathbf{E}}_1$ points to the right (along $+z$).

$$\vec{\mathbf{E}}_1 = \frac{\sigma}{2\epsilon_0}\hat{\mathbf{z}}$$

The direction of $\vec{\mathbf{E}}_1$ in the specified region is illustrated above with a **<u>solid</u>** arrow.

Next, ignore the positive plane and find the electric field in the desired region due to the negative plane. Call this $\vec{\mathbf{E}}_2$. The magnitude of the electric field is the same: $E_2 = \frac{\sigma}{2\epsilon_0}$. However, the direction is different.

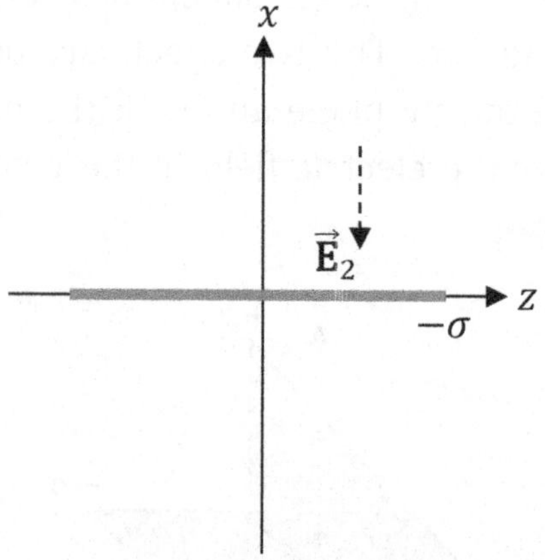

A positive "test" charge would be **attracted** to the negative plane. Above the negative plane (where x is positive), $\vec{\mathbf{E}}_2$ points down (along $-x$).

$$\vec{\mathbf{E}}_2 = -\frac{\sigma}{2\epsilon_0}\hat{\mathbf{x}}$$

The direction of $\vec{\mathbf{E}}_2$ in the specified region is illustrated above with a **dashed** arrow.

Add the vectors $\vec{\mathbf{E}}_1$ and $\vec{\mathbf{E}}_2$ to find the net electric field, $\vec{\mathbf{E}}_{net}$, in the desired region. Note that $\vec{\mathbf{E}}_1$ and $\vec{\mathbf{E}}_2$ are both equal in magnitude. They are also perpendicular, as illustrated below.

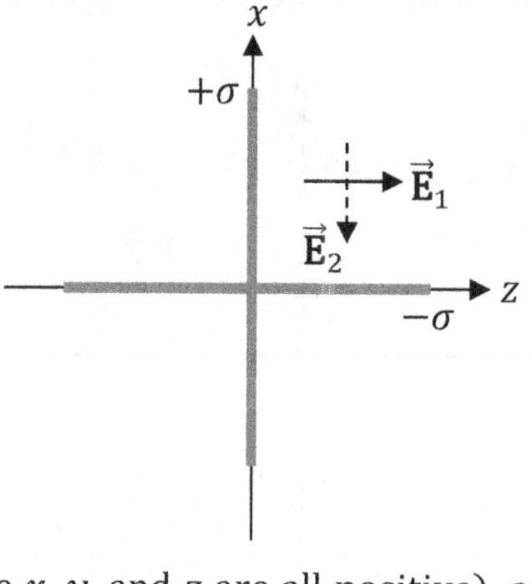

In the specified region (where x, y, and z are all positive), add the electric field vectors, $\vec{\mathbf{E}}_1$ and $\vec{\mathbf{E}}_2$, to find the net electric field, $\vec{\mathbf{E}}_{net}$.

$$\vec{\mathbf{E}}_{net} = \vec{\mathbf{E}}_1 + \vec{\mathbf{E}}_2 = \frac{\sigma}{2\epsilon_0}\hat{\mathbf{z}} - \frac{\sigma}{2\epsilon_0}\hat{\mathbf{x}} = \frac{\sigma}{2\epsilon_0}(\hat{\mathbf{z}} - \hat{\mathbf{x}})$$

Apply the Pythagorean theorem to find the magnitude of the net electric field, E_{net}.

$$E_{net} = \sqrt{E_1^2 + E_2^2} = \sqrt{\left(\frac{\sigma}{2\epsilon_0}\right)^2 + \left(\frac{\sigma}{2\epsilon_0}\right)^2} = \sqrt{\left(\frac{\sigma}{2\epsilon_0}\right)^2 2} = \frac{\sigma}{2\epsilon_0}\sqrt{2}$$

We applied the rules from algebra that $\sqrt{xy} = \sqrt{x}\sqrt{y}$ and $\sqrt{x^2} = x$. The net electric field has a magnitude of $E_{net} = \frac{\sigma}{2\epsilon_0}\sqrt{2}$ and points diagonally between $\vec{\mathbf{E}}_1$ and $\vec{\mathbf{E}}_2$.

Infinite Slab Example: A Prelude to Example 49. An infinite slab is like a very thick infinite sheet: It's basically a rectangular box with two infinite dimensions and one finite thickness. The infinite nonconducting slab with thickness T shown below is parallel to the xy plane, centered about $z = 0$, and has uniform positive charge density ρ. Derive an expression for the electric field in each region.

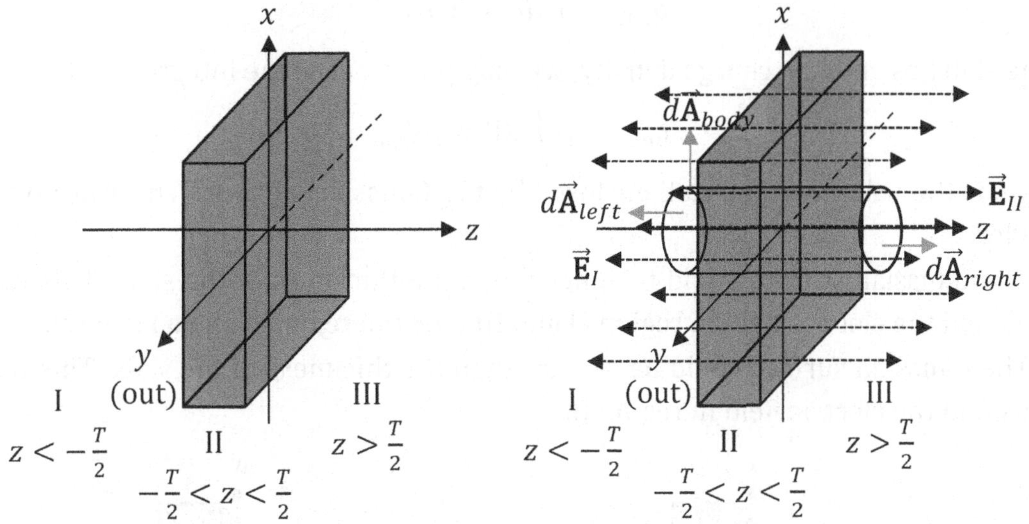

Solution. This problem is similar to the infinite plane example on pages 103-105. The difference is that the infinite slab has thickness, unlike the infinite plane. The electric field lines look the same, and we choose the same Gaussian surface: a right-circular cylinder. The math will also start out the same with Gauss's law. To save time, we'll simply repeat the steps that are identical to the example on pages 103-105, and pick up from where this solution deviates from that example. It would be a good exercise to see if you can understand each step (if not, review pages 103-105 for the explanation).

$$\oint_S \vec{\mathbf{E}} \cdot d\vec{\mathbf{A}} = \frac{q_{enc}}{\epsilon_0}$$

$$\int_{left} \vec{\mathbf{E}} \cdot d\vec{\mathbf{A}} + \int_{body} \vec{\mathbf{E}} \cdot d\vec{\mathbf{A}} + \int_{right} \vec{\mathbf{E}} \cdot d\vec{\mathbf{A}} = \frac{q_{enc}}{\epsilon_0}$$

$$\int_{left} E \cos 0° \, dA + \int_{body} E \cos 90° \, dA + \int_{right} E \cos 0° \, dA = \frac{q_{enc}}{\epsilon_0}$$

$$\int_{left} E \, dA + 0 + \int_{right} E \, dA = \frac{q_{enc}}{\epsilon_0}$$

$$E \int_{left} dA + E \int_{right} dA = \frac{q_{enc}}{\epsilon_0}$$

$$2EA_{end} = \frac{q_{enc}}{\epsilon_0}$$

$$E = \frac{q_{enc}}{2\epsilon_0 A_{end}}$$

What's different now is the charge enclosed by the Gaussian surface. Since the slab has thickness, the Gaussian surface encloses a volume of charge, so we use $dq = \rho dV$ (see page 79) instead of σdA in the integral to find q_{enc}.

$$q_{enc} = \int dq = \int \rho\, dV$$

Since the slab has uniform charge density, we may pull ρ out of the integral.

$$q_{enc} = \rho \int dV = \rho V_{enc}$$

Here, V_{enc} is the volume of the slab enclosed by the Gaussian surface. There are two cases to consider:

- The Gaussian surface could be longer than the thickness of the slab. This will help us find the electric field in regions I and III (see the regions labeled below).
- The Gaussian surface could be shorter than the thickness of the slab. This will help us find the electric field in region II.

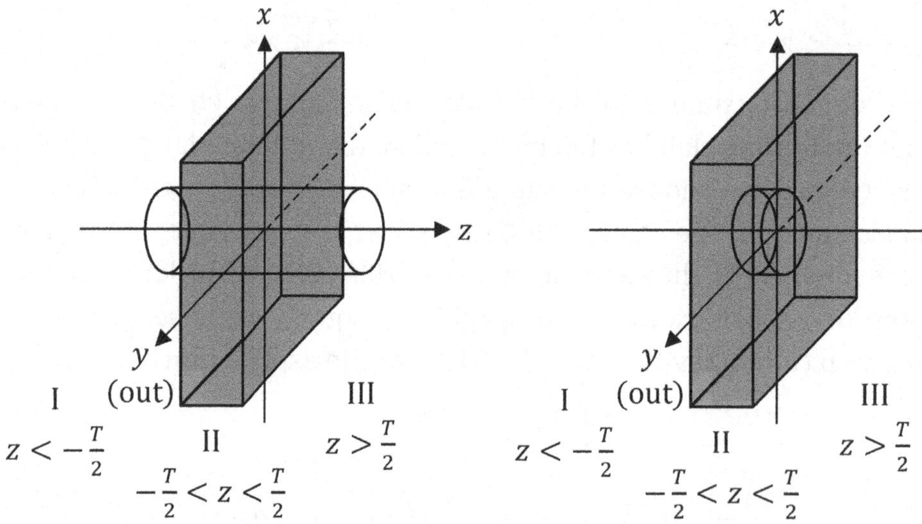

<u>**Regions I and III**</u>: $z < -\frac{T}{2}$ and $z > \frac{T}{2}$.

When the Gaussian cylinder is longer than the thickness of the slab, the volume of charge enclosed equals the intersection of the cylinder and the slab: It is a cylinder with a length equal to the thickness of the slab. The volume of this cylinder equals the thickness of the slab times the area of the circular end.

$$V_{enc} = TA_{end}$$

In this case, the charge enclosed is:

$$q_{enc} = \rho V_{enc} = \rho T A_{end}$$

Substitute this expression into the previous equation for electric field.

$$E_I = E_{III} = \frac{q_{enc}}{2\epsilon_0 A_{end}} = \frac{\rho T A_{end}}{2\epsilon_0 A_{end}} = \frac{\rho T}{2\epsilon_0}$$

The answer will be different in region II.

<u>Region II</u>: $-\frac{T}{2} < z < \frac{T}{2}$.

When the Gaussian cylinder is shorter than the thickness of the slab, the volume of charge enclosed equals the volume of the Gaussian cylinder. The length of the Gaussian cylinder is $2|z|$ (since the Gaussian cylinder extends from $-z$ to $+z$), where $-\frac{T}{2} < z < \frac{T}{2}$. The shorter the Gaussian cylinder, the less charge it encloses. The volume of the Gaussian cylinder equals $2|z|$ times the area of the circular end.

$$V_{enc} = 2|z|A_{end}$$

In this case, the charge enclosed is:

$$q_{enc} = \rho V_{enc} = \rho 2|z|A_{end}$$

Substitute this expression into the previous equation for electric field.

$$E_{II} = \frac{q_{enc}}{2\epsilon_0 A_{end}} = \frac{\rho 2|z|A_{end}}{2\epsilon_0 A_{end}} = \frac{\rho|z|}{\epsilon_0}$$

Note that the two expressions (for the different regions) agree at the boundary: That is, if you plug in $z = \frac{T}{2}$ into the expression for the electric field in region II, you get the same answer as in regions I and III. It is true in general that the electric field is continuous across a boundary (**except** when working with a **conductor**): If you remember this, you can use it to help check your answers to problems that involve multiple regions.

Example 49. An infinite nonconducting slab with thickness T is parallel to the xy plane, centered about $z = 0$, and has non-uniform charge density $\rho = \beta|z|$, where β is a positive constant. Derive an expression for the electric field in each region.

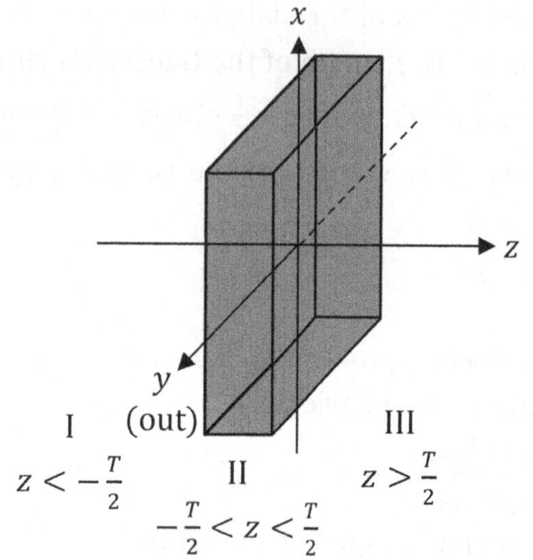

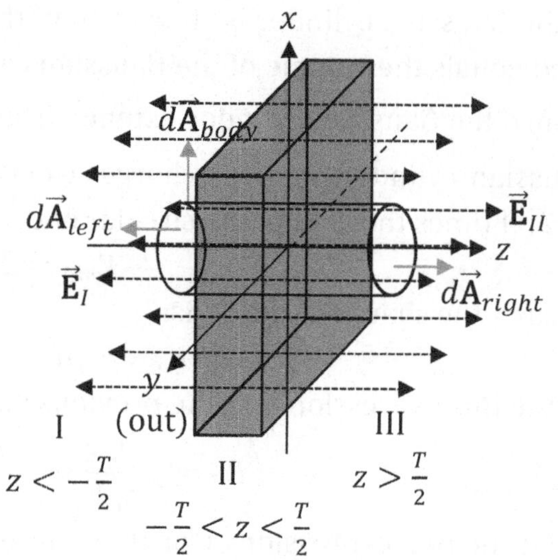

Solution. This problem is similar to the previous example. The difference is that this slab has non-uniform charge density. The electric field lines look the same, and we choose the same Gaussian surface: a right-circular cylinder. The math will also start out the same with Gauss's law. To save time, we'll simply repeat the steps that are identical to the example on pages 103-105, and pick up from where this solution deviates from the previous one. It would be a good exercise to see if you can understand each step (if not, review pages 103-105 for the explanation).

$$\oint_S \vec{\mathbf{E}} \cdot d\vec{\mathbf{A}} = \frac{q_{enc}}{\epsilon_0}$$

$$\int_{left} \vec{\mathbf{E}} \cdot d\vec{\mathbf{A}} + \int_{body} \vec{\mathbf{E}} \cdot d\vec{\mathbf{A}} + \int_{right} \vec{\mathbf{E}} \cdot d\vec{\mathbf{A}} = \frac{q_{enc}}{\epsilon_0}$$

$$\int_{left} E \cos 0° \, dA + \int_{body} E \cos 90° \, dA + \int_{right} E \cos 0° \, dA = \frac{q_{enc}}{\epsilon_0}$$

$$\int_{left} E \, dA + 0 + \int_{right} E \, dA = \frac{q_{enc}}{\epsilon_0}$$

$$E \int_{left} dA + E \int_{right} dA = \frac{q_{enc}}{\epsilon_0}$$

$$2EA_{end} = \frac{q_{enc}}{\epsilon_0}$$

$$E = \frac{q_{enc}}{2\epsilon_0 A_{end}}$$

What's different now is the charge enclosed by the Gaussian surface, since this slab has non-uniform charge density. Since the slab has thickness, the Gaussian surface encloses a volume of charge, so we use $dq = \rho dV$ (see page 79).

$$q_{enc} = \int dq = \int \rho \, dV$$

Since this slab has non-uniform charge density, we may **not** pull ρ out of the integral. Instead, we must apply the equation $\rho = \beta|z|$ that was given in the problem. The symbol β, however, is constant and may come out of the integral.

$$q_{enc} = \int \beta|z| \, dV = \beta \int |z| \, dV$$

For a Gaussian cylinder, $dV = r_c dr_c d\theta dz$ (see page 79). To deal with the absolute values, multiply by 2 and begin the z integral from 0. (We dealt with absolute values in a similar fashion in Example 39 in Chapter 7.) The limits of the z-integration are from 0 to z in region II ($|z| < \frac{T}{2}$) and from 0 to $\frac{T}{2}$ for regions I and III ($|z| > \frac{T}{2}$). The radius of the Gaussian cylinder is a.

$$q_{enc} = 2\beta \int_{r_c=0}^{a} r_c \, dr_c \int_{\theta=0}^{2\pi} d\theta \int_{z=0}^{z \text{ or } T/2} z \, dz$$

The upper limit of the z-integration is different in each region, as we will consider on the next page. Note that the double integral over r_c and θ equals the area of the end, which we have already called A_{end}. It will equal $A_{end} = \pi a^2$ (the area of a circle), but since it will cancel with A_{end} in a future step, there isn't any benefit to writing it as πa^2.

$$q_{enc} = 2\beta \, A_{end} \int_{z=0}^{z \text{ or } T/2} z \, dz$$

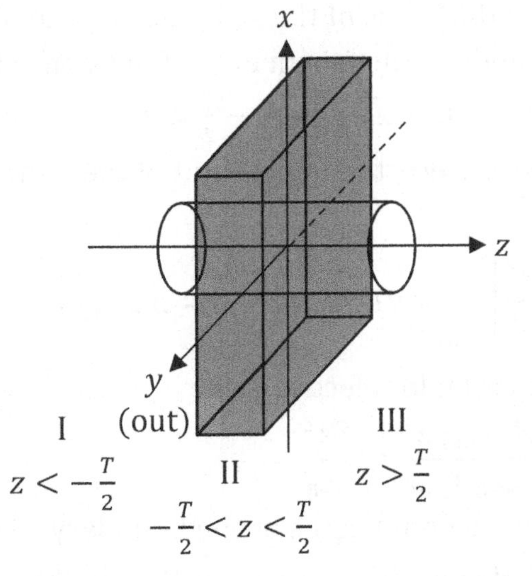

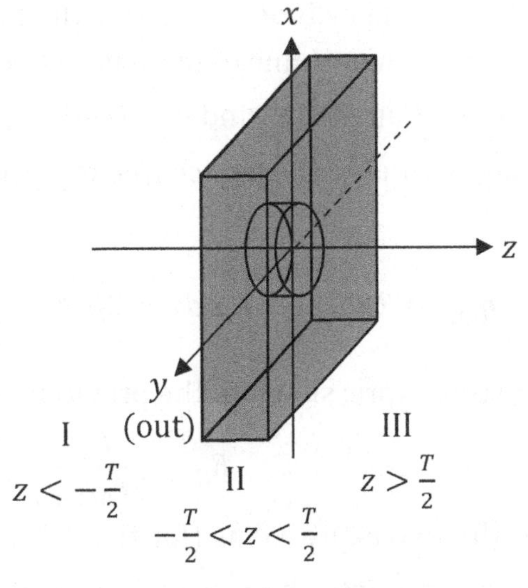

There are two cases to consider:

- The Gaussian surface could be longer than the thickness of the slab. This will help us find the electric field in regions I and III (see the regions labeled below).
- The Gaussian surface could be shorter than the thickness of the slab. This will help us find the electric field in region II.

Regions I and III: $z < -\frac{T}{2}$ and $z > \frac{T}{2}$.

When the Gaussian cylinder is longer than the thickness of the slab, the volume of charge enclosed equals the intersection of the cylinder and the slab: It is a cylinder with a length equal to the thickness of the slab. Here, the upper limit of the z-integration is $\frac{T}{2}$.

$$q_{enc} = 2\beta\, A_{end} \int_{z=0}^{T/2} z\, dz = 2\beta\, A_{end} \left[\frac{z^2}{2}\right]_{z=0}^{T/2} = 2\beta\, A_{end} \left(\frac{T^2/4}{2}\right) = 2\beta\, A_{end} \left(\frac{T^2}{8}\right)$$

Note that $(T/2)^2 = T^2/4$ and that $\frac{T^2/4}{2} = \frac{T^2}{4} \div 2 = \frac{T^2}{4} \times \frac{1}{2} = \frac{T^2}{8}$. To divide by 2, multiply by its **reciprocal**. Note that the reciprocal of 2 is $\frac{1}{2}$. Also note that $2 \times \frac{1}{8} = \frac{2}{8} = \frac{1}{4}$.

$$q_{enc} = \beta\, A_{end} \frac{T^2}{4}$$

Substitute this expression into the previous equation for electric field.

$$E_I = E_{III} = \frac{q_{enc}}{2\epsilon_0 A_{end}} = \frac{\beta\, A_{end}\, T^2/4}{2\epsilon_0 A_{end}} = \frac{\beta T^2}{8\epsilon_0}$$

Note that $\frac{1}{4} \div 2 = \frac{1}{4} \times \frac{1}{2} = \frac{1}{8}$. The answer will be different in region II.

Region II: $-\frac{T}{2} < z < \frac{T}{2}$.

When the Gaussian cylinder is shorter than the thickness of the slab, the volume of charge enclosed equals the volume of the Gaussian cylinder. The length of the Gaussian cylinder is $2|z|$ (since the Gaussian cylinder extends from $-z$ to $+z$), where $-\frac{T}{2} < z < \frac{T}{2}$. The shorter the Gaussian cylinder, the less charge it encloses. Here, the upper limit of the z-integration is z.

$$q_{enc} = 2\beta\, A_{end} \int_{z=0}^{z} z\, dz = 2\beta\, A_{end} \left[\frac{z^2}{2}\right]_{z=0}^{z} = 2\beta\, A_{end} \left(\frac{z^2}{2}\right) = \beta\, A_{end}\, z^2$$

Substitute this expression into the previous equation for electric field.

$$E_{II} = \frac{q_{enc}}{2\epsilon_0 A_{end}} = \frac{\beta\, A_{end}\, z^2}{2\epsilon_0 A_{end}} = \frac{\beta z^2}{2\epsilon_0}$$

Note that the two expressions (for the different regions) agree at the boundary: That is, if you plug in $z = \frac{T}{2}$ into the expression for the electric field in region II, you get the same answer as in regions I and III.

Spherical Cavity Example: A Prelude to Example 50. A solid spherical insulator centered about the origin has radius a and uniform positive charge density ρ. The solid sphere has a spherical cavity with radius b. The distance between the center of the charged sphere and the center of the cavity is d, for which $d + b < a$. Derive an expression for the electric field in the region inside the cavity.

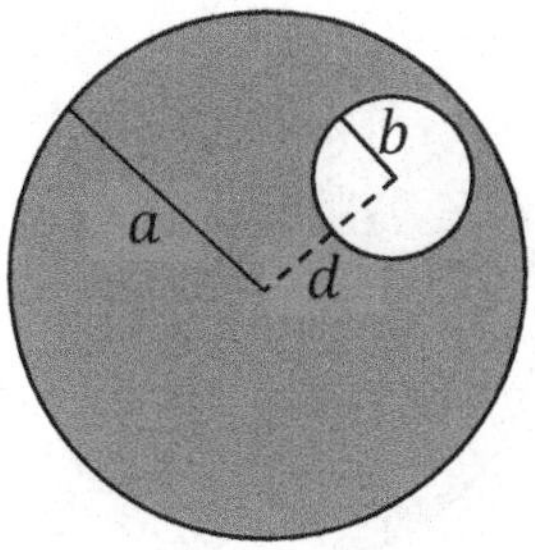

Solution. The "trick" to cavity problems is to apply the principle of **superposition** (Chapter 3). Although these aren't pointlike charges, the same principle of adding electric field vectors still applies. Geometrically, we can visualize the complete sphere as the sum of the object shown above and the small sphere that is missing from the object shown above.

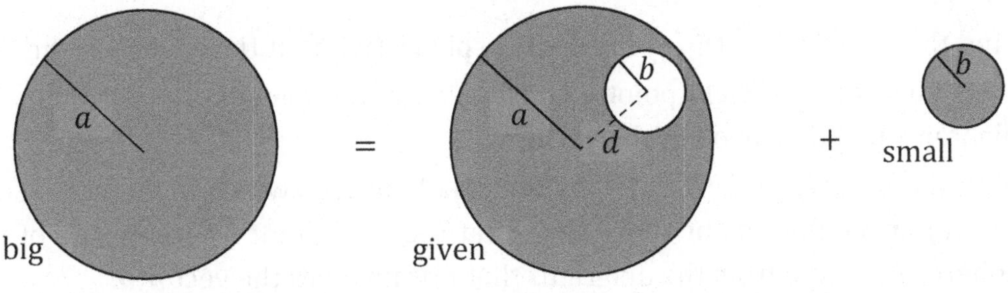

Our goal is to calculate the electric field at a point inside the cavity of the given shape. We marked such a point in the diagram below with a star (★). The electric field at the field point (★) due to the given shape (the original diagram that includes the cavity) **plus** the electric field at the field point (★) due to the small sphere (which is the same size as the cavity) **equals** the electric field at the field point (★) due to the big sphere (which doesn't have a cavity). We just wrote an equation in words, and now we will write the same equation with symbols:

$$\vec{\mathbf{E}}_{big} = \vec{\mathbf{E}}_{given} + \vec{\mathbf{E}}_{small}$$

Now we will draw the same equation with a picture:

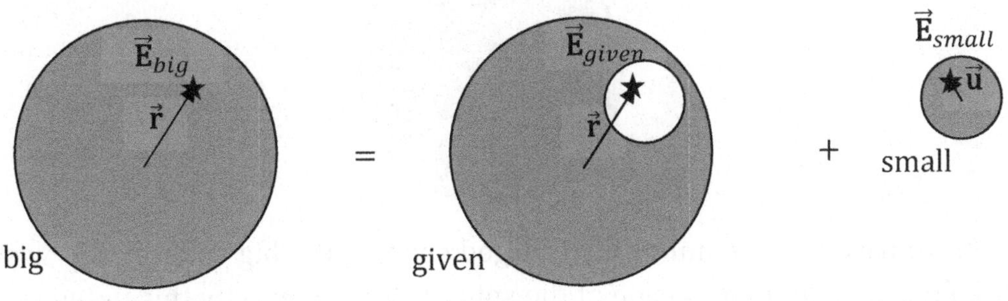

Consider the three vectors illustrated below:

- The vector $\vec{\mathbf{r}}$ extends from the center of the **big** sphere to the field point (★).
- The vector $\vec{\mathbf{u}}$ extends from the center of the **small** cavity to the field point (★).
- The vector $\vec{\mathbf{d}}$ extends from the center of the **big** sphere to the center of the **small** cavity.

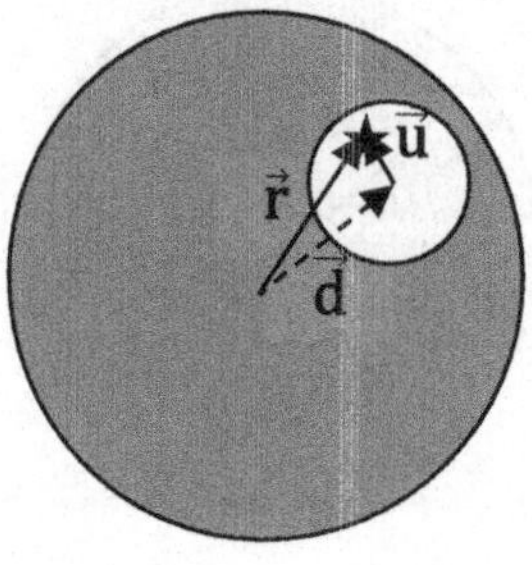

Since $\vec{\mathbf{d}}$ and $\vec{\mathbf{u}}$ join tip-to-tail to form $\vec{\mathbf{r}}$, recall from Volume 1 (vector addition) that $\vec{\mathbf{r}}$ is the resultant vector:

$$\vec{\mathbf{r}} = \vec{\mathbf{d}} + \vec{\mathbf{u}}$$

Here is our plan:

1. Find the electric field ($\vec{\mathbf{E}}_{big}$) for the big sphere (without the cavity) at the field point (★). Note that the field point (★) is inside of the big sphere. Also note from the diagrams that this involves the vector $\vec{\mathbf{r}}$.
2. Find the electric field ($\vec{\mathbf{E}}_{small}$) for the small sphere (which is the same size as the cavity) at the field point (★). Note that the field point (★) is inside of the small sphere. Also note from the diagrams that this involves the vector $\vec{\mathbf{u}}$.
3. Subtract our answers for the first two steps to obtain the electric field ($\vec{\mathbf{E}}_{given}$) for the given shape at the field point (★). Recall the equation $\vec{\mathbf{E}}_{big} = \vec{\mathbf{E}}_{given} + \vec{\mathbf{E}}_{small}$ from the previous page. We're subtracting $\vec{\mathbf{E}}_{small}$ from both sides in order to solve for $\vec{\mathbf{E}}_{given}$.

$$\vec{\mathbf{E}}_{given} = \vec{\mathbf{E}}_{big} - \vec{\mathbf{E}}_{small}$$

<u>**Step 1**</u>:

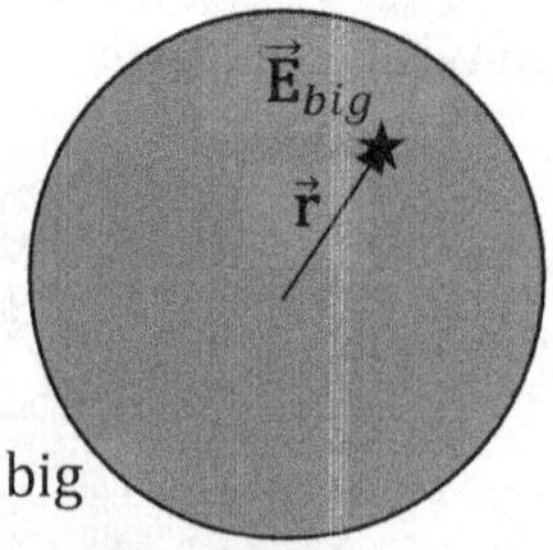

In this step, we have a single uniformly charged sphere (the big sphere) and wish to find the electric field at a point (★) inside of the sphere. We did exactly this in the example on pages 83-86. Rather than repeat that example, let us borrow an expression from region I ($r < a$) of that example (see the bottom of page 86).

$$\vec{\mathbf{E}}_{big} = \frac{\rho r}{3\epsilon_0}\hat{\mathbf{r}}$$

<u>Step 2</u>:

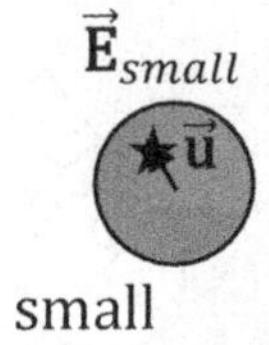

In this step, we have a single uniformly charged sphere (the small sphere, which is the same size as the cavity) and wish to find the electric field at a point (★) inside of the sphere. This is identical to what we did in Step 1, except that this step involves the vector $\vec{\mathbf{u}}$ instead of the vector $\vec{\mathbf{r}}$. We will obtain the same answer with r replaced by u and $\hat{\mathbf{r}}$ replaced by $\hat{\mathbf{u}}$.

$$\vec{\mathbf{E}}_{small} = \frac{\rho u}{3\epsilon_0}\hat{\mathbf{u}}$$

<u>Step 3</u>:

As discussed earlier, we will simply subtract electric field vectors to obtain our answer for the electric field at the field point (★) due to the given shape.

$$\vec{\mathbf{E}}_{given} = \vec{\mathbf{E}}_{big} - \vec{\mathbf{E}}_{small}$$

$$\vec{\mathbf{E}}_{given} = \frac{\rho r}{3\epsilon_0}\hat{\mathbf{r}} - \frac{\rho u}{3\epsilon_0}\hat{\mathbf{u}}$$

We can **factor** out the $\frac{\rho}{3\epsilon_0}$.

$$\vec{\mathbf{E}}_{given} = \frac{\rho}{3\epsilon_0}(r\hat{\mathbf{r}} - u\hat{\mathbf{u}})$$

Recall that any vector can be expressed as its magnitude times its direction. For example, $\vec{\mathbf{A}} = A\hat{\mathbf{A}}$. Therefore, $\vec{\mathbf{r}} = r\hat{\mathbf{r}}$ and $\vec{\mathbf{u}} = u\hat{\mathbf{u}}$.

$$\vec{\mathbf{E}}_{given} = \frac{\rho}{3\epsilon_0}(\vec{\mathbf{r}} - \vec{\mathbf{u}})$$

Earlier, we showed that $\vec{\mathbf{r}} = \vec{\mathbf{d}} + \vec{\mathbf{u}}$. Subtracting $\vec{\mathbf{u}}$ from both sides, we get $\vec{\mathbf{r}} - \vec{\mathbf{u}} = \vec{\mathbf{d}}$.

$$\vec{\mathbf{E}}_{given} = \frac{\rho}{3\epsilon_0}\vec{\mathbf{d}}$$

Recall that $\vec{\mathbf{d}}$ is a vector joining the center of the big sphere to the center of the cavity: It is a constant. Therefore, the electric field inside of the cavity equals $\vec{\mathbf{E}}_{given} = \frac{\rho}{3\epsilon_0}\vec{\mathbf{d}}$ and is <u>uniform</u> in this region.

Example 50. An infinite solid cylindrical insulator coaxial with the z-axis has radius a and uniform positive charge density ρ. The solid cylinder has a cylindrical cavity with radius b. The distance between the axis of the charged insulator and the axis of the cavity is d, for which $d + b < a$. Derive an expression for the electric field in the region inside the cavity.

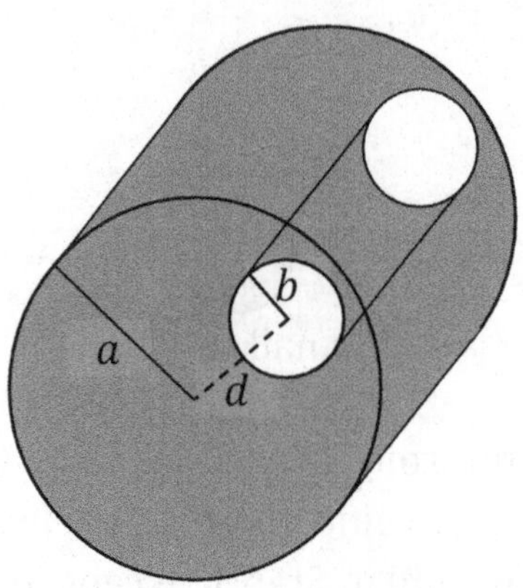

Solution. This problem is very similar to the previous example, except that the shape is a cylinder instead of a sphere. The strategy will be the same. The difference is that we will use the equation for a cylinder in place of the equation for a sphere. As in the previous example, we will visualize the complete cylinder as the sum of the object shown above and the small cylinder that is missing from the object shown above.

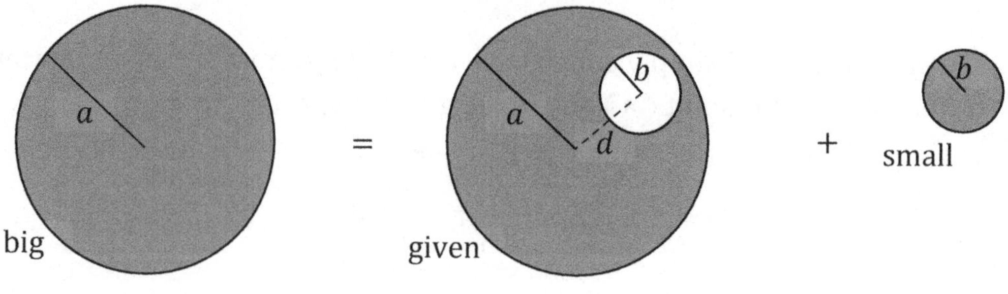

The electric field at the field point (★) due to the given shape (the original diagram that includes the cavity) **plus** the electric field at the field point (★) due to the small cylinder (which is the same size as the cavity) **equals** the electric field at the field point (★) due to the big cylinder (which doesn't have a cavity):

$$\vec{\mathbf{E}}_{big} = \vec{\mathbf{E}}_{given} + \vec{\mathbf{E}}_{small}$$

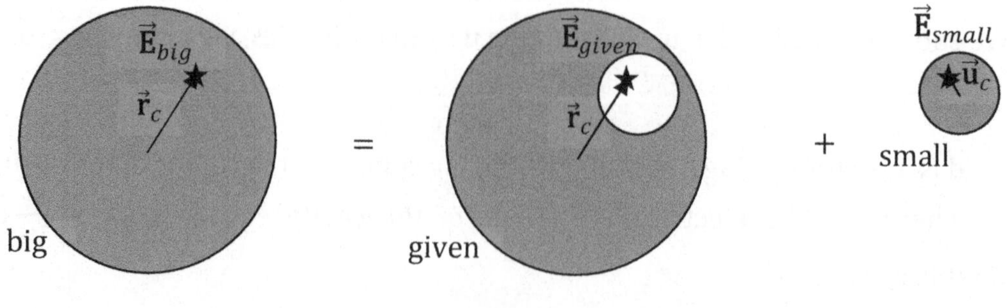

Consider the three vectors illustrated below:

- The vector $\vec{\mathbf{r}}_c$ extends from the axis of the **big** cylinder to the field point (★).
- The vector $\vec{\mathbf{u}}_c$ extends from the axis of the **small** cavity to the field point (★).
- The vector $\vec{\mathbf{d}}$ extends from the axis of the **big** cylinder to the axis of the **small** cavity.

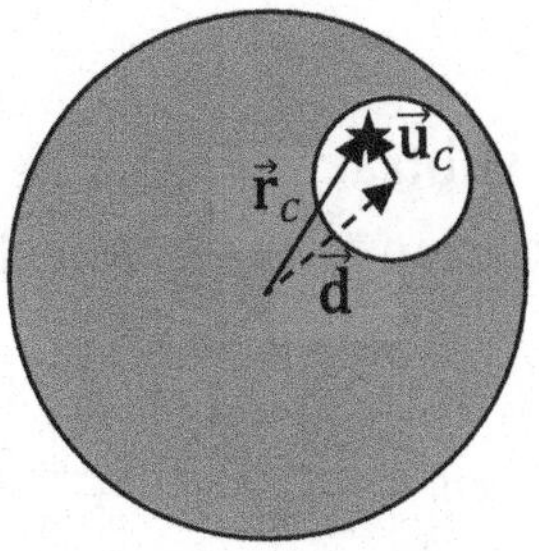

Since $\vec{\mathbf{d}}$ and $\vec{\mathbf{u}}_c$ join tip-to-tail to form $\vec{\mathbf{r}}_c$, recall from Volume 1 (vector addition) that $\vec{\mathbf{r}}_c$ is the resultant vector:

$$\vec{\mathbf{r}}_c = \vec{\mathbf{d}} + \vec{\mathbf{u}}_c$$

Step 1:

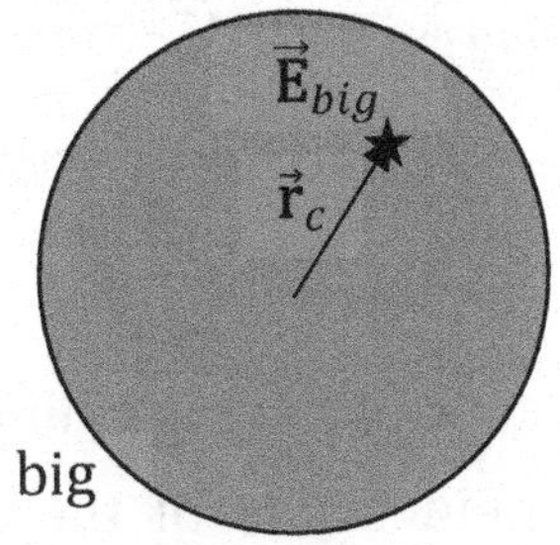

In this step, we have a single uniformly charged infinite cylinder (the big cylinder) and wish to find the electric field at a point (★) inside of the cylinder. We did exactly this in Example 46 on pages 101-102. Rather than repeat that example, let us borrow an expression from region I ($r_c < a$) of that example (see page 102).

$$\vec{\mathbf{E}}_{big} = \frac{\rho r_c}{2\epsilon_0}\hat{\mathbf{r}}_c$$

Step 2:

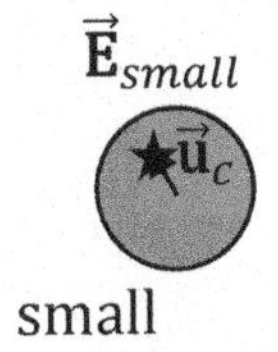

In this step, we have a single uniformly charged infinite cylinder (the small cylinder, which is the same size as the cavity) and wish to find the electric field at a point (★) inside of the cylinder. This is identical to what we did in Step 1, except that this step involves the vector

$\vec{\mathbf{u}}_c$ instead of the vector $\vec{\mathbf{r}}_c$. We will obtain the same answer with r_c replaced by u_c and $\hat{\mathbf{r}}_c$ replaced by $\hat{\mathbf{u}}_c$.

$$\vec{\mathbf{E}}_{small} = \frac{\rho u_c}{2\epsilon_0}\hat{\mathbf{u}}_c$$

<u>Step 3</u>:

As discussed earlier, we will simply subtract electric field vectors to obtain our answer for the electric field at the field point (★) due to the given shape.

$$\vec{\mathbf{E}}_{given} = \vec{\mathbf{E}}_{big} - \vec{\mathbf{E}}_{small}$$

$$\vec{\mathbf{E}}_{given} = \frac{\rho r}{2\epsilon_0}\hat{\mathbf{r}}_c - \frac{\rho u}{2\epsilon_0}\hat{\mathbf{u}}_c$$

We can **factor** out the $\frac{\rho}{2\epsilon_0}$.

$$\vec{\mathbf{E}}_{given} = \frac{\rho}{2\epsilon_0}(r_c\hat{\mathbf{r}}_c - u_c\hat{\mathbf{u}}_c)$$

Recall that any vector can be expressed as its magnitude times its direction. For example, $\vec{\mathbf{A}} = A\hat{\mathbf{A}}$. Therefore, $\vec{\mathbf{r}}_c = r_c\hat{\mathbf{r}}_c$ and $\vec{\mathbf{u}}_c = u_c\hat{\mathbf{u}}_c$.

$$\vec{\mathbf{E}}_{given} = \frac{\rho}{2\epsilon_0}(\vec{\mathbf{r}}_c - \vec{\mathbf{u}}_c)$$

Earlier, we showed that $\vec{\mathbf{r}}_c = \vec{\mathbf{d}} + \vec{\mathbf{u}}_c$. Subtracting $\vec{\mathbf{r}}_c$ from both sides, we get $\vec{\mathbf{r}}_c - \vec{\mathbf{u}}_c = \vec{\mathbf{d}}$.

$$\vec{\mathbf{E}}_{given} = \frac{\rho}{2\epsilon_0}\vec{\mathbf{d}}$$

Recall that $\vec{\mathbf{d}}$ is a vector joining the axis of the big cylinder to the axis of the cavity: It is a constant. Therefore, the electric field inside of the cavity equals $\vec{\mathbf{E}}_{given} = \frac{\rho}{3\epsilon_0}\vec{\mathbf{d}}$ and is <u>**uniform**</u> in this region.

9 ELECTRIC POTENTIAL

Distance Formula
$R_i = \sqrt{(x_2 - x_1)^2 + (y_2 - y_1)^2}$
Electric Potential due to a Single Pointlike Charge
$V = \frac{kq}{R}$
Net Electric Potential due to a System of Pointlike Charges
$V_{net} = V_1 + V_2 + \cdots + V_N = \frac{kq_1}{R_1} + \frac{kq_2}{R_2} + \cdots + \frac{kq_N}{R_N}$
Relationship between Electric Potential Energy and Electric Potential
$PE_e = qV$
Electric Potential Energy for two Pointlike Charges
$PE_e = k\frac{q_1 q_2}{R}$

Potential Difference	Electrical Work
$\Delta V = V_f - V_i$	$W_e = q\Delta V$

Symbol	Name	SI Units
V	electric potential	V
ΔV	potential difference	V
PE_e	electric potential energy	J
W_e	electrical work	J
q	charge	C
R	distance from the charge	m
k	Coulomb's constant	$\frac{\text{N}\cdot\text{m}^2}{\text{C}^2}$ or $\frac{\text{kg}\cdot\text{m}^3}{\text{C}^2\cdot\text{s}^2}$

Coulomb's Constant
$k = 8.99 \times 10^9 \ \frac{\text{N}\cdot\text{m}^2}{\text{C}^2} \approx 9.0 \times 10^9 \ \frac{\text{N}\cdot\text{m}^2}{\text{C}^2}$

Prefix	Name	Power of 10
m	milli	10^{-3}
μ	micro	10^{-6}
n	nano	10^{-9}
p	pico	10^{-12}

Example 51. A monkey-shaped earring with a charge of $7\sqrt{2}$ μC lies at the point (0, 1.0 m). A banana-shaped earring with a charge of $-3\sqrt{2}$ μC lies at the point (0, −1.0 m). Determine the net electric potential at the point (1.0 m, 0).

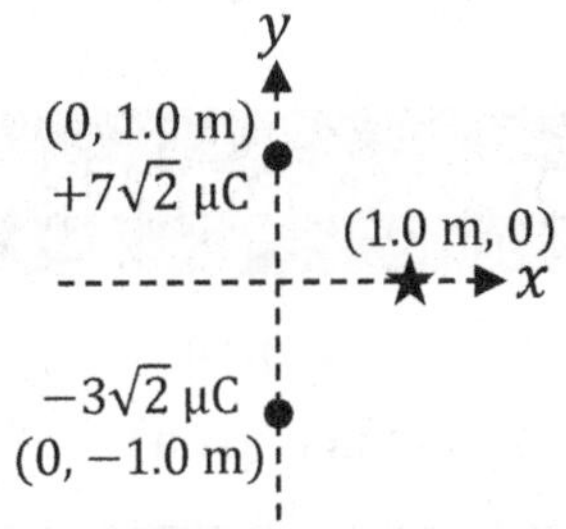

Solution. First we need to determine how far each charge is from the field point at (1.0 m, 0), which is marked with a star (★). We choose to call $q_1 = 7\sqrt{2}$ μC and $q_2 = -3\sqrt{2}$ μC. The distances R_1 and R_2 are illustrated below.

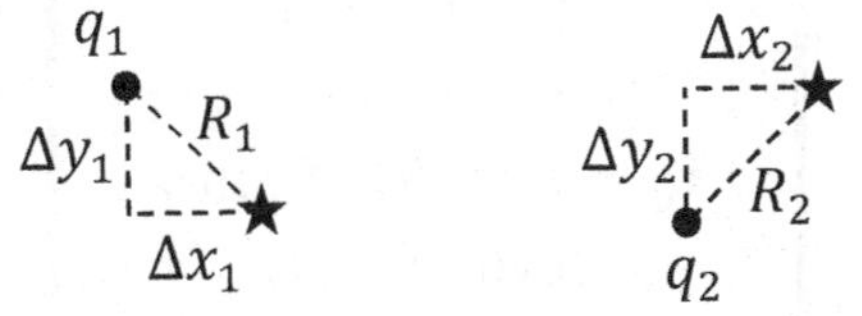

Apply the distance formula to determine these distances. Note that $|\Delta x_1| = |\Delta x_2| = 1.0$ m and $|\Delta y_1| = |\Delta y_2| = 1.0$ m.

$$R_1 = \sqrt{\Delta x_1^2 + \Delta y_1^2} = \sqrt{1^2 + 1^2} = \sqrt{2}\text{ m}$$

$$R_2 = \sqrt{\Delta x_2^2 + \Delta y_2^2} = \sqrt{1^2 + 1^2} = \sqrt{2}\text{ m}$$

The net electric potential equals the sum of the electric potentials for each pointlike charge.

$$V_{net} = \frac{kq_1}{R_1} + \frac{kq_2}{R_2}$$

Convert the charges to SI units: $q_1 = 7\sqrt{2}\ \mu\text{C} = 7\sqrt{2} \times 10^{-6}$ C and $q_2 = -3\sqrt{2}\ \mu\text{C} = -3\sqrt{2} \times 10^{-6}$ C. Recall that the metric prefix micro (μ) stands for one millionth: $\mu = 10^{-6}$. Factor out Coulomb's constant.

$$V_{net} = \frac{kq_1}{R_1} + \frac{kq_2}{R_2} = k\left(\frac{q_1}{R_1} + \frac{q_2}{R_2}\right) = (9 \times 10^9)\left(\frac{7\sqrt{2} \times 10^{-6}}{\sqrt{2}} + \frac{-3\sqrt{2} \times 10^{-6}}{\sqrt{2}}\right)$$

The $\sqrt{2}$'s cancel out.

$$V_{net} = (9 \times 10^9)(7 \times 10^{-6} - 3 \times 10^{-6})$$

Factor out the 10^{-6}.

$$V_{net} = (9 \times 10^9)(10^{-6})(7 - 3) = (9 \times 10^9)(10^{-6})(4) = 36 \times 10^3\text{ V} = 3.6 \times 10^4\text{ V}$$

Note that $10^9 10^{-6} = 10^{9-6} = 10^3$ according to the rule $x^m x^{-n} = x^{m-n}$. Also note that $36 \times 10^3 = 3.6 \times 10^4$. The answer is $V_{net} = 3.6 \times 10^4$ V, which is the same as 36×10^3 V, and can also be expressed as $V_{net} = 36$ kV, since the metric prefix kilo (k) stands for $10^3 = 1000$.

Electric Potential Integrals		
$V = k\int \frac{dq}{R}$	$\Delta V = -\int \vec{\mathbf{E}} \cdot d\vec{\mathbf{s}}$	$Q = \int dq$
Differential Charge Element		
$dq = \lambda ds$ (line or thin arc)	$dq = \sigma dA$ (surface area)	$dq = \rho dV$ (volume)
Relation Among Coordinate Systems and Unit Vectors		
$x = r\cos\theta$ $y = r\sin\theta$ $\hat{\mathbf{r}} = \hat{\mathbf{x}}\cos\theta + \hat{\mathbf{y}}\sin\theta$ (2D polar)	$x = r_c\cos\theta$ $y = r_c\sin\theta$ $\hat{\mathbf{r}}_c = \hat{\mathbf{x}}\cos\theta + \hat{\mathbf{y}}\sin\theta$ (cylindrical)	$x = r\sin\theta\cos\varphi$ $y = r\sin\theta\sin\varphi$ $z = r\cos\theta$ $\hat{\mathbf{r}} = \hat{\mathbf{x}}\cos\varphi\sin\theta$ $+\hat{\mathbf{y}}\sin\varphi\sin\theta + \hat{\mathbf{z}}\cos\theta$ (spherical)
Differential Arc Length		
$ds = dx$ (along x)	$ds = dy$ or dz (along y or z)	$ds = a d\theta$ (circular arc of radius a)
Differential Area Element		
$dA = dxdy$ (polygon in xy plane)	$dA = rdrd\theta$ (pie slice, disc, thick ring)	$dA = a^2\sin\theta\, d\theta d\varphi$ (sphere of radius a)
Differential Volume Element		
$dV = dxdydz$ (bounded by flat sides)	$dV = r_c dr_c d\theta dz$ (cylinder or cone)	$dV = r^2\sin\theta\, drd\theta d\varphi$ (spherical)

Symbol	Name	SI Units
dq	differential charge element	C
Q	total charge of the object	C
V	electric potential	V
$\vec{\mathbf{E}}$	electric field	N/C or V/m
k	Coulomb's constant	$\frac{\text{N}\cdot\text{m}^2}{\text{C}^2}$ or $\frac{\text{kg}\cdot\text{m}^3}{\text{C}^2\cdot\text{s}^2}$
$\vec{\mathbf{R}}$	a vector from each dq to the field point	m
R	the distance from each dq to the field point	m
$d\vec{\mathbf{s}}$	differential displacement vector (see Chapter 21 of Volume 1)	m
x, y, z	Cartesian coordinates of dq	m, m, m
r, θ	2D polar coordinates of dq	m, rad
r_c, θ, z	cylindrical coordinates of dq	m, rad, m
r, θ, φ	spherical coordinates of dq	m, rad, rad
λ	linear charge density (for an arc)	C/m
σ	surface charge density (for an area)	C/m^2
ρ	volume charge density	C/m^3
ds	differential arc length	m
dA	differential area element	m^2
dV	differential volume element	m^3

Note: The symbols λ, σ, and ρ are the lowercase Greek letters lambda, sigma, and rho.

Finite Rod Example: A Prelude to Examples 52-53. A rod lies on the y-axis with endpoints at the origin and $(0, L)$, as illustrated below. The positively charged rod has uniform charge density λ and total charge Q. Derive an equation for the electric potential at the point $(L, 0)$ in terms of k, Q, and L.

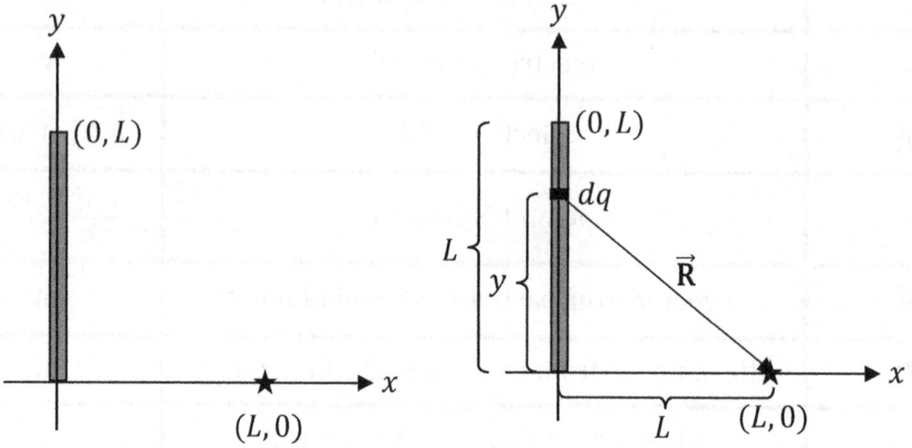

Solution. Begin with a labeled diagram. See the diagram above on the right. Draw a representative dq. Draw $\vec{\mathbf{R}}$ from the source, dq, to the field point $(L, 0)$. When we perform the integration, we effectively integrate over every dq that makes up the rod. Start the math with the electric potential integral.

$$V = k \int \frac{dq}{R}$$

Examine the right diagram above. The vector $\vec{\mathbf{R}}$ has a different length for each dq that makes up the rod. Therefore, its magnitude, R, can't come out of the integral. What we can do is express R in terms of the constant L and the variable y using the Pythagorean theorem: $R = \sqrt{L^2 + y^2}$. Substitute the expression for R into the electric potential integral.

$$V = k \int \frac{dq}{\sqrt{L^2 + y^2}}$$

For a thin rod, we write $dq = \lambda ds$ (see page 126).

$$V = k \int \frac{\lambda ds}{\sqrt{L^2 + y^2}}$$

Since the rod has uniform charge density, we may pull λ out of the integral.

$$V = k\lambda \int \frac{ds}{\sqrt{L^2 + y^2}}$$

For a rod lying along the y-axis, we work with Cartesian coordinates and write the differential arc length as $ds = dy$. The limits of integration correspond to the length of the rod: $0 \leq y \leq L$.

$$V = k\lambda \int_{y=0}^{L} \frac{dy}{\sqrt{L^2 + y^2}}$$

This integral can be performed via the following trigonometric substitution:

$$y = L \tan \theta$$
$$dy = L \sec^2 \theta \, d\theta$$

Solving for θ, we get $\theta = \tan^{-1}\left(\frac{y}{L}\right)$, which shows that the new limits of integration are from $\theta = \tan^{-1}(0) = 0°$ to $\theta = \tan^{-1}\left(\frac{L}{L}\right) = \tan^{-1}(1) = 45°$. Note that the denominator of the integral simplifies as follows, using the trig identity $1 + \tan^2 \theta = \sec^2 \theta$:

$$\sqrt{L^2 + y^2} = \sqrt{L^2 + (L \tan \theta)^2} = \sqrt{L^2(1 + \tan^2 \theta)} = \sqrt{L^2 \sec^2 \theta} = L \sec \theta$$

In the last step, we applied the rule from algebra that $\sqrt{a^2x^2} = \sqrt{a^2}\sqrt{x^2} = ax$. Substitute the above expressions for dy and $\sqrt{L^2 + y^2}$ into the previous integral.

$$V = k\lambda \int_{y=0}^{L} \frac{dy}{\sqrt{L^2 + y^2}} = k\lambda \int_{\theta=0°}^{45°} \frac{L \sec^2 \theta \, d\theta}{L \sec \theta} = k\lambda \int_{\theta=0°}^{45°} \sec \theta \, d\theta$$

Find the anti-derivative on page 42.

$$V = k\lambda[\ln|\sec \theta + \tan \theta|]_{\theta=0°}^{45°}$$

Evaluate the anti-derivative over the limits.

$$V = k\lambda(\ln|\sec 45° + \tan 45°| - \ln|\sec 0° + \tan 0°|)$$
$$V = k\lambda\left(\ln\left|\sqrt{2} + 1\right| - \ln|1 + 0|\right) = k\lambda\left(\ln\left|\sqrt{2} + 1\right| - \ln|1|\right) = k\lambda\left(\ln\left|\sqrt{2} + 1\right|\right)$$

Note that $\ln(1) = 0$. We're not finished yet. We must eliminate the constant λ from our answer. The way to do this is to integrate over dq to find the total charge of the rod, Q. This integral is convenient since we use the same substitutions from before.

$$Q = \int dq = \int \lambda \, ds = \lambda \int ds = \lambda \int_{y=0}^{L} dy = \lambda L$$

Solve for λ in terms of Q: Divide both sides of the equation by L.

$$\lambda = \frac{Q}{L}$$

Substitute this expression for λ into our previous expression for V.

$$V = k\lambda\left(\ln\left|\sqrt{2} + 1\right|\right) = \frac{kQ}{L}\left(\ln\left|\sqrt{2} + 1\right|\right)$$

The electric potential at the point $(L, 0)$ is $V = \frac{kQ}{L}\left(\ln\left|\sqrt{2} + 1\right|\right)$.

Solid Disc: Another Prelude to Examples 52-53. A uniformly charged solid disc lies in the xy plane, centered about the origin as illustrated below. The radius of the disc is denoted by the symbol a and the positively charged disc has total charge Q. Derive an equation for the electric potential at the point $(0, 0, p)$ in terms of k, Q, a, and p.

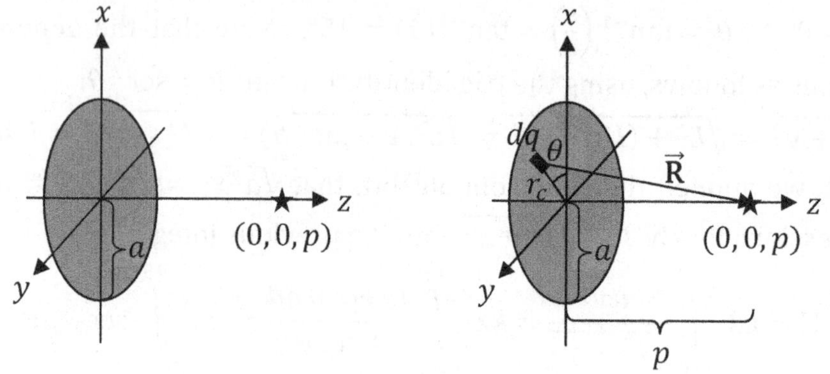

Solution. Begin with a labeled diagram. See the diagram above on the right. Draw a representative dq: Since this is a solid disc, most of the dq's lie within the area of the disc, so we drew dq inside the disc (not on its circumference). Draw $\vec{\mathbf{R}}$ from the source, dq, to the field point $(0, 0, p)$. When we perform the integration, we effectively integrate over every dq that makes up the disc. Start the math with the electric potential integral.

$$V = k \int \frac{dq}{R}$$

Examine the right diagram above. The vector $\vec{\mathbf{R}}$ has a different length for each dq that makes up the solid disc. Therefore, its magnitude, R, can't come out of the integral. What we can do is express R in terms of the variable r_c and the constant p using the Pythagorean theorem: $R = \sqrt{r_c^2 + p^2}$. Substitute the expression for R into the electric potential integral.

$$V = k \int \frac{dq}{\sqrt{r_c^2 + p^2}}$$

For a solid disc, we write $dq = \sigma dA$ (see page 126).

$$V = k \int \frac{\sigma dA}{\sqrt{r_c^2 + p^2}}$$

Since the solid disc has uniform charge density, we may pull σ out of the integral.

$$V = k\sigma \int \frac{dA}{\sqrt{r_c^2 + p^2}}$$

For a solid disc, we write the differential area element as $dA = r_c dr_c d\theta$ (see page 126), and since the field point at $(0, 0, p)$ does not lie in the xy plane, we work with cylindrical coordinates (which simply adds the z-coordinate to the 2D polar coordinates). The limits of integration are from $r_c = 0$ to $r_c = a$ and $\theta = 0$ to $\theta = 2\pi$ <u>**radians**</u>.

$$V = k\sigma \int_{r_c=0}^{a} \int_{\theta=0}^{2\pi} \frac{r_c dr_c d\theta}{\sqrt{r_c^2 + p^2}}$$

When we integrate over θ, we treat the independent variable r_c as a constant. In fact, nothing in the integrand depends on the variable θ, so we can factor everything out of the θ-integration.

$$V = k\sigma \int_{r_c=0}^{a} \frac{r_c dr_c}{\sqrt{r_c^2 + p^2}} \int_{\theta=0}^{2\pi} d\theta = k\sigma \int_{r_c=0}^{a} \frac{r_c dr_c}{\sqrt{r_c^2 + p^2}} [\theta]_{\theta=0}^{2\pi} = 2\pi k\sigma \int_{r_c=0}^{a} \frac{r_c dr_c}{\sqrt{r_c^2 + p^2}}$$

This integral can be performed via the following trigonometric substitution:

$$r_c = p \tan\psi$$
$$dr_c = p \sec^2\psi \, d\psi$$

Solving for ψ, we get $\theta = \tan^{-1}\left(\frac{r_c}{p}\right)$, which shows that the new limits of integration are from $\psi = \tan^{-1}(0) = 0°$ to $\psi = \tan^{-1}\left(\frac{a}{p}\right)$, which for now we will simply call ψ_{max}. The denominator of the integral simplifies through the trig identity $\tan^2\psi + 1 = \sec^2\psi$:

$$\sqrt{r_c^2 + p^2} = \sqrt{(p\tan\psi)^2 + p^2} = \sqrt{p^2(\tan^2\psi + 1)} = \sqrt{p^2 \sec^2\psi} = p\sec\psi$$

In the last step, we applied the rule from algebra that $\sqrt{a^2x^2} = \sqrt{a^2}\sqrt{x^2} = ax$. Substitute the above expressions for r_c, dr_c, and $\sqrt{r_c^2 + p^2}$ into the previous integral.

$$V = 2\pi k\sigma \int_{r_c=0}^{a} \frac{r_c dr_c}{\sqrt{r_c^2 + p^2}} = 2\pi k\sigma \int_{\psi=0°}^{\psi_{max}} \frac{(p\tan\psi)(p\sec^2\psi \, d\psi)}{p\sec\psi}$$

Note that $\frac{p^2}{p} = p$ and $\frac{\sec^2\psi}{\sec\psi} = \sec\psi$.

$$V = 2\pi k\sigma \int_{\psi=0°}^{\psi_{max}} p\tan\psi\sec\psi \, d\psi = 2\pi k\sigma p \int_{\psi=0°}^{\psi_{max}} \tan\psi\sec\psi \, d\psi$$

Recall from trig that $\sec\psi = \frac{1}{\cos\psi}$ and $\tan\psi = \frac{\sin\psi}{\cos\psi}$. Therefore, it follows that:

$$\tan\psi\sec\psi = \frac{\sin\psi}{\cos\psi}\frac{1}{\cos\psi} = \frac{\sin\psi}{\cos^2\psi}$$

The electric potential integral becomes:

$$V = 2\pi k\sigma p \int_{\psi=0°}^{\psi_{max}} \frac{\sin\psi}{\cos^2\psi} d\psi$$

One way to perform this integral is through the following substitution:

$$u = \cos\psi$$
$$du = -\sin\psi \, d\psi$$

Since $u = \cos\psi$, the new limits of integration are from $u(0) = \cos 0° = 1$ to $u(\psi_{max}) = \cos\psi_{max}$. For now, we will just call the upper limit u_{new} and worry about it later.

$$V = 2\pi k\sigma p \int_{u=1}^{u_{new}} \frac{-du}{u^2} = -2\pi k\sigma p \int_{u=1}^{u_{new}} \frac{du}{u^2}$$

Apply the rule from algebra that $\frac{1}{u^2} = u^{-2}$.

$$V = -2\pi k\sigma p \int_{u=1}^{u_{new}} u^{-2}\, du = -2\pi k\sigma p \left[\frac{u^{-1}}{-1}\right]_{u=1}^{u_{new}} = 2\pi k\sigma p [u^{-1}]_{u=1}^{u_{new}} = 2\pi k\sigma p \left(\frac{1}{u_{u_{new}}} - 1\right)$$

Recall that what we called u_{new} is equal to $u_{new} = \cos\psi_{max}$ and that ψ_{max} is given by $\psi_{max} = \tan^{-1}\left(\frac{a}{p}\right)$, such that u_{new} is the complicated looking expression $u_{new} = \cos\left(\tan^{-1}\left(\frac{a}{p}\right)\right)$. When you find yourself taking the cosine of an inverse tangent, you can find a simpler way to write it by drawing a right triangle and applying the Pythagorean theorem. Since $\psi_{max} = \tan^{-1}\left(\frac{a}{p}\right)$, it follows that $\tan\psi_{max} = \frac{a}{p}$. We can make a right triangle from this: Since the tangent of ψ_{max} equals the opposite over the adjacent, we draw a right triangle with a opposite and p adjacent to ψ_{max}.

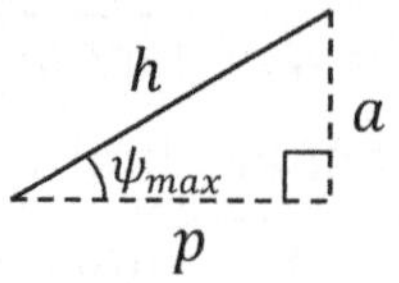

Find the hypotenuse, h, of the right triangle from the Pythagorean theorem.

$$h = \sqrt{p^2 + a^2}$$

Now we can write an expression for the cosine of ψ_{max}. It equals the adjacent (p) over the hypotenuse ($h = \sqrt{p^2 + a^2}$).

$$u_{new} = \cos\psi_{max} = \frac{p}{h} = \frac{p}{\sqrt{p^2 + a^2}}$$

The reciprocal of u_{new} is:

$$\frac{1}{u_{new}} = \frac{\sqrt{p^2 + a^2}}{p}$$

Substitute this expression into the previous equation for electric potential.

$$V = 2\pi k\sigma p \left(\frac{1}{u_{new}} - 1\right) = 2\pi k\sigma p \left(\frac{\sqrt{p^2 + a^2}}{p} - 1\right)$$

To add fractions, make a **common denominator**.

$$V = 2\pi k\sigma p \left(\frac{\sqrt{p^2 + a^2}}{p} - \frac{p}{p}\right) = 2\pi k\sigma p \left(\frac{\sqrt{p^2 + a^2} - p}{p}\right)$$

Note that the p from out front cancels with the $\frac{1}{p}$.

$$V = 2\pi k\sigma \left(\sqrt{p^2 + a^2} - p\right)$$

We're not finished yet because we need to eliminate the charge density σ from our answer. The way to do this is to integrate over dq to find the total charge of the disc, Q. This integral involves the same substitutions from before.

$$Q = \int dq = \int \sigma\, dA = \sigma \int dA = \sigma \int_{r_c=0}^{a} \int_{\theta=0}^{2\pi} r_c dr_c d\theta = \sigma \int_{r_c=0}^{a} r_c dr_c \int_{\theta=0}^{2\pi} d\theta$$

$$Q = \sigma \left[\frac{r_c^2}{2}\right]_{r_c=0}^{a} [\theta]_{\theta=0}^{2\pi} = \sigma\left(\frac{a^2}{2} - \frac{0^2}{2}\right)(2\pi - 0) = \sigma\left(\frac{a^2}{2}\right)(2\pi) = \pi\sigma a^2$$

Solve for σ in terms of Q: Divide both sides of the equation by πa^2.

$$\sigma = \frac{Q}{\pi a^2}$$

Substitute this expression for σ into our previous expression for electric potential.

$$V = 2\pi k\sigma\left(\sqrt{p^2 + a^2} - p\right) = 2\pi k\left(\frac{Q}{\pi a^2}\right)\left(\sqrt{p^2 + a^2} - p\right)$$

$$V = \frac{2kQ}{a^2}\left(\sqrt{p^2 + a^2} - p\right)$$

Example 52. A thin wire is bent into the shape of the semicircle illustrated below. The radius of the semicircle is denoted by the symbol a and the positively charged wire has total charge Q and uniform charge density. Derive an equation for the electric potential at the origin, in terms of k, Q, and a.

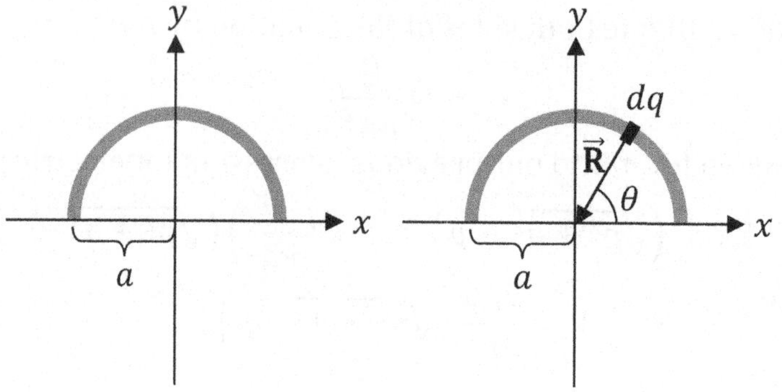

Solution. Begin with a labeled diagram. See the diagram above on the right. Draw a representative dq: Since this is a semicircular wire (and **not** a solid semicircle – that is, it's **not** half of a solid disc), dq must lie on the circumference and not inside the semicircle. Draw $\vec{\mathbf{R}}$ from the source, dq, to the field point (at the origin). When we perform the integration, we effectively integrate over every dq that makes up the semicircle. Start the math with the electric potential integral.

$$V = k\int \frac{dq}{R}$$

Examine the right diagram above. The vector $\vec{\mathbf{R}}$ extends from the source (each dq that makes up the semicircle) to the field point $(0, 0)$. The vector $\vec{\mathbf{R}}$ has the same length for each dq that makes up the semicircle. Its magnitude, R, equals the radius of the semicircle: $R = a$. (Note that this would **not** be the case for a thick ring or a solid semicircle.)

$$V = k\int \frac{dq}{a}$$

We may pull $\frac{1}{a}$ out of the integral, since the radius (a) is constant.

$$V = \frac{k}{a}\int dq$$

This turns out to be a trivial solution: $\int dq = Q$ is the total charge of the semicircle.

$$V = \frac{kQ}{a}$$

The electric potential at the origin is $V = \frac{kQ}{a}$.

Example 53. A very thin uniformly charged ring lies in the xy plane, centered about the origin as illustrated below. The radius of the ring is denoted by the symbol a and the positively charged ring has total charge Q. Derive an equation for the electric potential at the point $(0, 0, p)$ in terms of k, Q, a, and p.

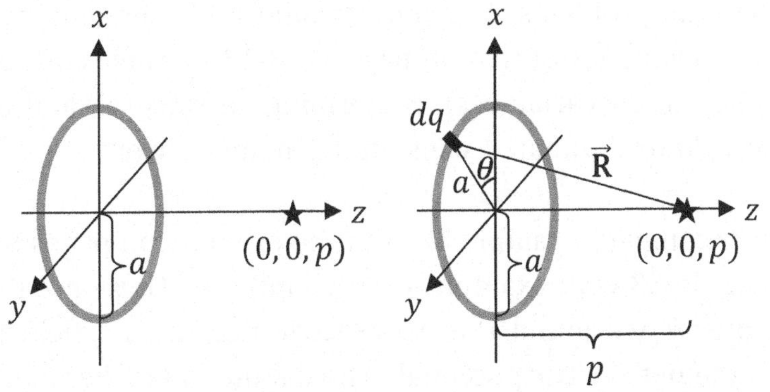

Solution. Begin with a labeled diagram. See the diagram above on the right. Draw a representative dq on the thin ring. Draw $\vec{\mathbf{R}}$ from the source, dq, to the field point $(0, 0, p)$. When we perform the integration, we effectively integrate over every dq that makes up the thin ring. Start the math with the electric potential integral.

$$V = k \int \frac{dq}{R}$$

Examine the right diagram above. The vector $\vec{\mathbf{R}}$ extends from the source (each dq that makes up the ring) to the field point $(0, 0, p)$. Express the magnitude of $\vec{\mathbf{R}}$, denoted by R, in terms of the radius a and the constant p using the Pythagorean theorem.

$$R = \sqrt{a^2 + p^2}$$

Substitute this expression for R into the electric potential integral.

$$V = k \int \frac{dq}{\sqrt{a^2 + p^2}}$$

Note that $\sqrt{a^2 + p^2}$ is constant (since a and p are both constants as defined in this problem) and may come out of the integral.

$$V = \frac{k}{\sqrt{a^2 + p^2}} \int dq$$

Just as in the previous example, this turns out to be a trivial solution: $\int dq = Q$ is the total charge of the thin ring.

$$V = \frac{kQ}{\sqrt{a^2 + p^2}}$$

The electric potential at the origin is $V = \frac{kQ}{\sqrt{a^2+p^2}}$.

It's instructive to compare examples 52-53 with the examples on pages 128-133. In examples 52-53, **every** charge on the thin circle happens to be equidistant from the field point. In that case, R is constant and comes out of the integral, leaving $\int dq = Q$ as a trivial integral to perform.

In most electric potential problems, R is **not** constant and thus may not come out of the integral. We saw two examples of this on pages 128-133. This required expressing R in terms of suitable integration variable(s), performing an integral, and then performing a second integral $Q = \int dq$ to eliminate a constant from the answer.

It's also instructive to compare example 52 with the example on pages 65-67 in Chapter 7, and to compare example 53 with example 41 in Chapter 7. These are the same problems, except that in Chapter 7 we found the net electric **field** at the field point, whereas in Chapter 9 we found the net electric **potential**. The distinction makes a big difference in the solution. If you look back to the similar problems in Chapter 7, we had to do quite a bit of work in order to perform the integrals for those similar problems because the **direction** of $\vec{\mathbf{R}}$ was **not** constant. Since electric potential is a scalar, whereas electric field is a vector, we didn't need to worry about the direction of $\vec{\mathbf{R}}$ in Chapter 9, which made these two integrals much simpler in Chapter 9.

10 MOTION OF A CHARGED PARTICLE IN A UNIFORM ELECTRIC FIELD

Electric Force	Weight (Gravitational Force)
$\vec{\mathbf{F}}_e = q\vec{\mathbf{E}}$	$\vec{\mathbf{F}}_g = m\vec{\mathbf{g}}$
Newton's Second Law	**Potential Difference**
$\sum \vec{\mathbf{F}} = m\vec{\mathbf{a}}$	$\Delta V = V_f - V_i$
Electric Field (if Uniform)	**Electrical Work**
$E = \dfrac{\Delta V}{d}$	$W_e = q\Delta V = -\Delta PE_e$

1D Uniform Acceleration

$$\Delta y = v_{y0}t + \frac{1}{2}a_y t^2 \quad , \quad v_y = v_{y0} + a_y t \quad , \quad v_y^2 = v_{y0}^2 + 2a_y\Delta y$$

Conservation of Energy

$$PE_0 + KE_0 + W_{nc} = PE + KE$$
$$q\Delta V = -\Delta PE_e$$
$$PE_{g0} = mgh_0 \quad , \quad PE_g = mgh$$
$$KE_0 = \frac{1}{2}mv_0^2 \quad , \quad KE = \frac{1}{2}mv^2$$

Acceleration through a Potential Difference (no Other Fields)

$$|q|\Delta V = \frac{1}{2}m(v^2 - v_0^2) \quad \text{(gaining speed)}$$
$$|q|\Delta V = \frac{1}{2}m(v_0^2 - v^2) \quad \text{(losing speed)}$$

Symbol	Name	SI Units
E	electric field	N/C or V/m
F_e	electric force	N
m	mass	kg
g	gravitational acceleration	m/s^2
a_y	y-component of acceleration	m/s^2
v_{y0}	y-component of initial velocity	m/s
v_y	y-component of final velocity	m/s
Δy	y-component of net displacement	m
t	time	s
d	separation between parallel plates	m
V	electric potential	V
ΔV	potential difference	V
PE_e	electric potential energy	J
W_e	electrical work	J
q	charge	C

Example 54. As illustrated below, two large parallel charged plates are separated by a distance of 20 cm. The potential difference between the plates is 120 V. A tiny object with a charge of +1500 μC and mass of 18 g begins from rest at the positive plate.

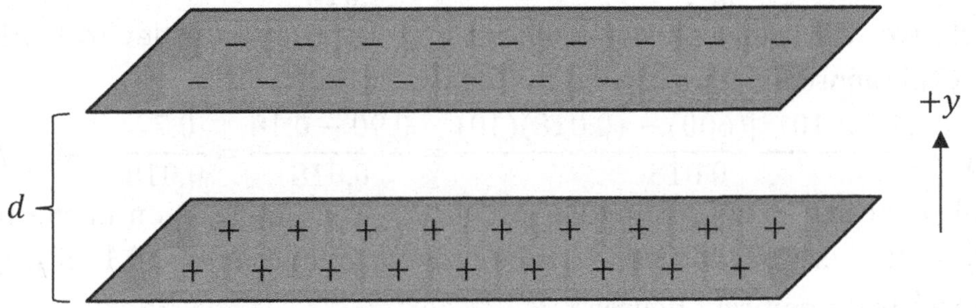

(A) Determine the acceleration of the tiny charged object.

Solution. We will eventually need to know the magnitude of the electric field (E). We can get that from the potential difference (ΔV) and the separation between the plates (d). Convert d from cm to m: $d = 20$ cm $= 0.20$ m.

$$E = \frac{\Delta V}{d} = \frac{120}{0.2}$$

Note that $\frac{1}{0.20} = 5$.

$$E = (120)(5) = 600 \text{ N/C}$$

Draw a free-body diagram (FBD) for the tiny charged object.

- The electric force ($q\vec{\mathbf{E}}$) pushes the tiny charged object straight up. Since the charged object is positive, the electric force ($q\vec{\mathbf{E}}$) is parallel to the electric field ($\vec{\mathbf{E}}$), and since the electric field lines travel from the positive plate to the negative plate, $\vec{\mathbf{E}}$ points upward.
- The weight ($m\vec{\mathbf{g}}$) of the object pulls straight down.
- We choose $+y$ to point upward.

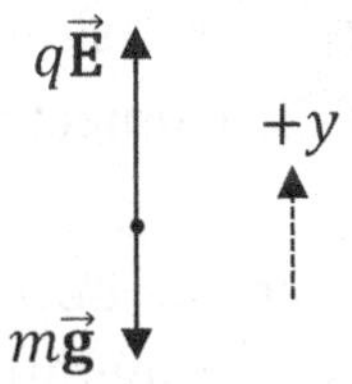

Apply Newton's second law to the tiny charged object. Since we chose $+y$ to point upward, $|q|E$ is positive while mg is negative in the sum of the y-components of the forces.

$$\sum F_y = ma_y$$

$$|q|E - mg = ma_y$$

Divide both sides of the equation by mass.

$$a_y = \frac{|q|E - mg}{m}$$

Convert the charge from μC to C and the mass from g to kg: $q = 1500\ \mu\text{C} = 1.500 \times 10^{-3}$ C and $m = 18$ g $= 0.018$ kg.

$$a_y = \frac{|q|E - mg}{m} = \frac{-|1.5 \times 10^{-3}|(600) - (0.018)(9.81)}{0.018}$$

In this book, we will round gravity from 9.81 to 10 m/s^2 in order to work without a calculator. (This approximation is good to 19 parts in 1000.)

$$a_y \approx \frac{|1.5 \times 10^{-3}|(600) - (0.018)(10)}{0.018} = \frac{0.90 - 0.18}{0.018} = \frac{0.72}{0.018} = 40\ \text{m/s}^2$$

The symbol ≈ means "is approximately equal to." The acceleration of the tiny charged object is $a_y \approx 40$ m/s^2. (If you don't round gravity, you get $a_y = 40.19$ m/s^2, which still equals 40 m/s^2 to 2 significant figures.)

(B) How fast is the charged object moving just before it reaches the negative plate?
Solution. Now that we know the acceleration, we can use the equations of one-dimensional uniform acceleration. Begin by listing the knowns (out of Δy, v_{y0}, v_y, a_y, and t).

- $a_y \approx 40$ m/s^2. We know this from part (A). It's positive because the object accelerates upward (and because we chose $+y$ to point upward).
- $\Delta y = 0.20$ m. The net displacement of the tiny charged object equals the separation between the plates. It's positive because it finishes above where it started (and because we chose $+y$ to point upward).
- $v_{y0} = 0$. The initial velocity is zero because the object starts from rest.
- We're solving for the final velocity (v_y).

We know a_y, Δy, and v_{y0}. We're looking for v_y. Choose the equation with these symbols.

$$v_y^2 = v_{y0}^2 + 2a_y\Delta y$$
$$v_y^2 = 0^2 + 2(40)(0.2) = 16$$
$$v_y = \sqrt{16} = \pm 4 = 4.0\ \text{m/s}$$

We chose the positive root because the object is heading upward in the final position (and because we chose $+y$ to point upward). The object moves 4.0 m/s just before impact.

Note: We couldn't use the equations from the bottom row on page 137 in part (B) of this example because there is a significant gravitational field in the problem. That is, $mg = 0.18$ N compared to $|q|E = 0.90$ N. If you encounter a problem where a charged particle accelerates through a potential difference for which the gravitational force is negligible compared to the electric force, then you could apply the equations from the bottom row on page 137.

11 EQUIVALENT CAPACITANCE

Capacitance

$$C = \frac{Q}{\Delta V} \quad , \quad Q = C\Delta V \quad , \quad \Delta V = \frac{Q}{C}$$

Capacitors in Series

$$\frac{1}{C_s} = \frac{1}{C_1} + \frac{1}{C_2} + \cdots + \frac{1}{C_N}$$

Capacitors in Parallel

$$C_p = C_1 + C_2 + \cdots + C_N$$

Energy Stored in a Capacitor

$$U = \frac{1}{2} Q\Delta V \quad , \quad U = \frac{1}{2} C\Delta V^2 \quad , \quad U = \frac{Q^2}{2C}$$

Schematic Representation	Symbol	Name
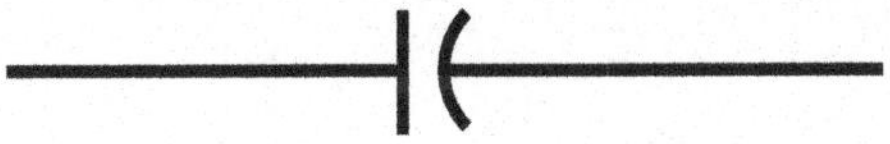	C	capacitor
	ΔV	battery or DC power supply

Symbol	Name	SI Units
C	capacitance	F
Q	the charge stored on the positive plate of a capacitor	C
ΔV	the potential difference between two points in a circuit	V
U	the energy stored by a capacitor	J

Prefix	Name	Power of 10
m	milli	10^{-3}
μ	micro	10^{-6}
n	nano	10^{-9}
p	pico	10^{-12}

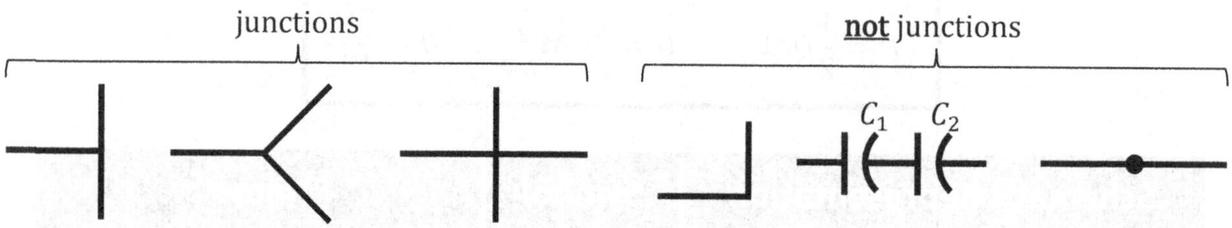

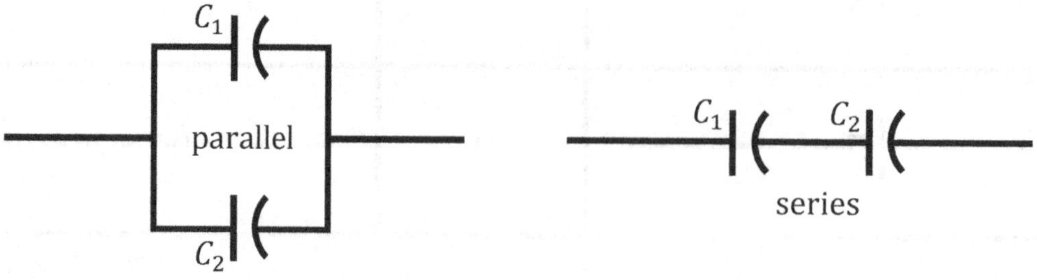

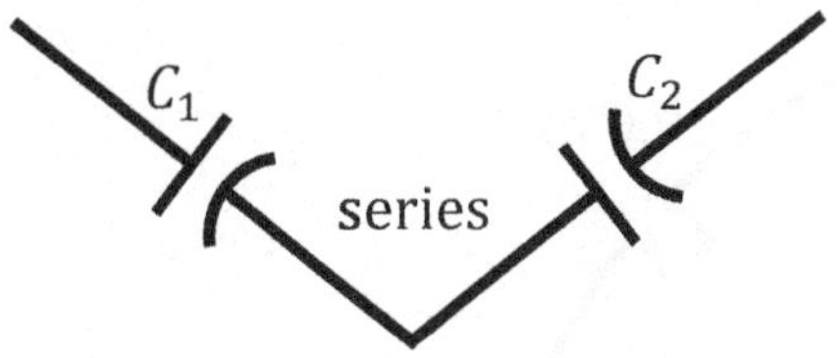
C_1
C_2
series

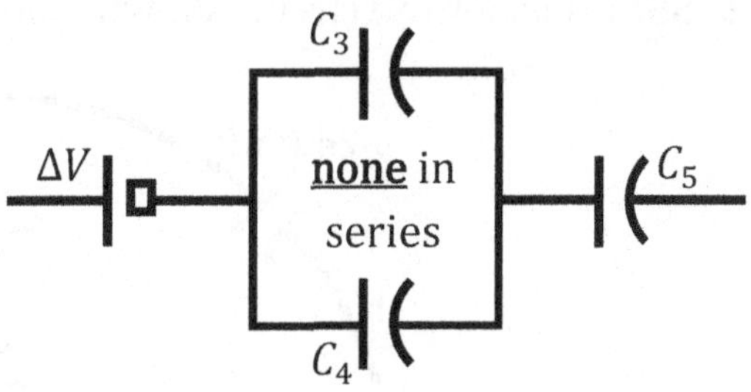
C_3
ΔV
none in series
C_5
C_4

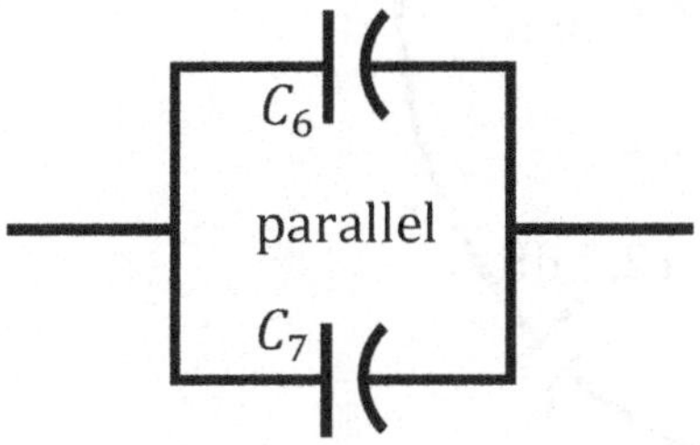
C_6
parallel
C_7

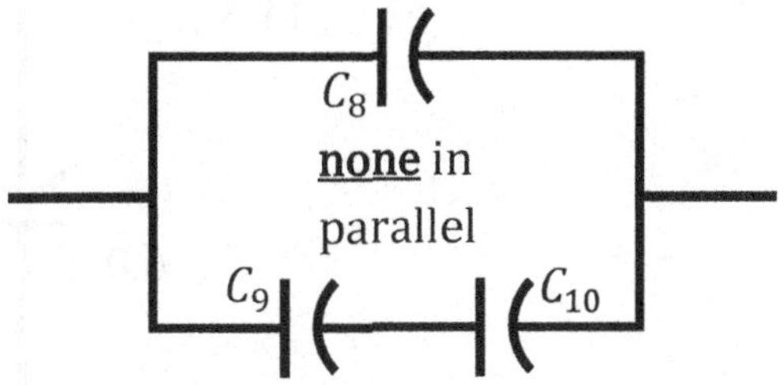
C_8
none in parallel
C_9
C_{10}

series
C_1
C_2
replace with
C_{s1}

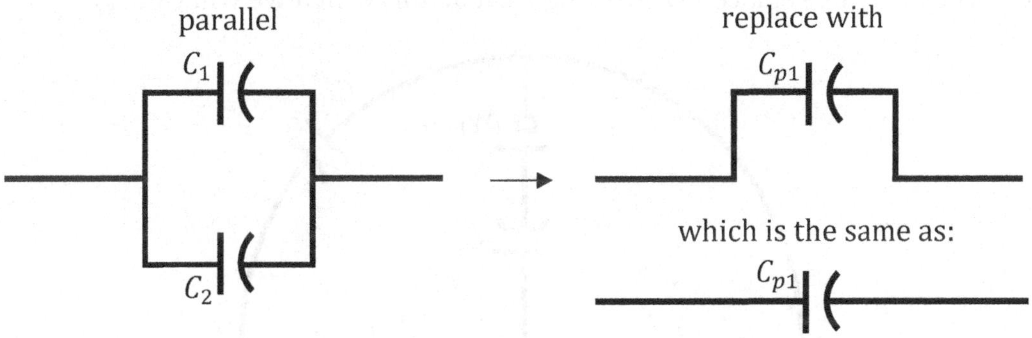
parallel
C_1
C_2
replace with
C_{p1}
which is the same as:
C_{p1}

Example 55. Consider the circuit shown below.

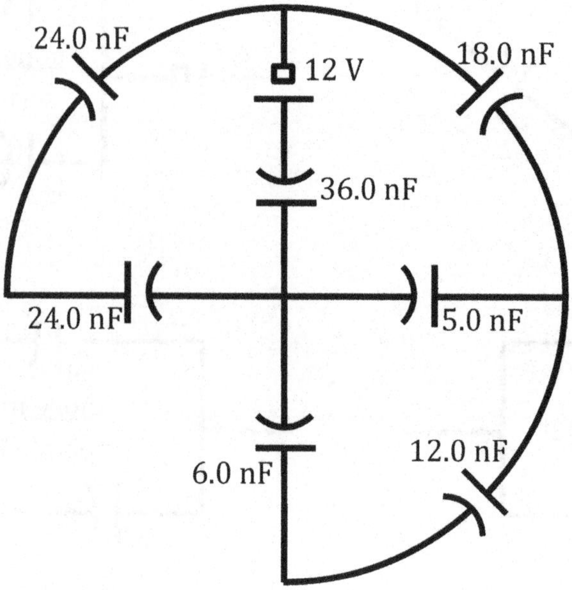

(A) Redraw the circuit, simplifying it one step at a time by identifying series and parallel combinations.

Solution. The two 24-nF capacitors are in **series** because an electron could travel from one capacitor to the other without crossing a junction. They may be replaced with a single capacitor, which we will call C_{s1}. Similarly, the 6-nF capacitor is in **series** with the 12-nF capacitor. They may be replaced with a single capacitor, which we will call C_{s2}.

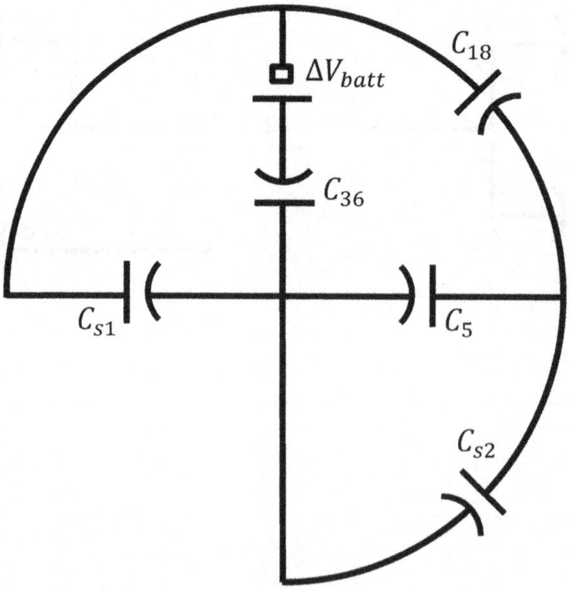

In the diagram above, C_5 and C_{s2} are in **parallel**. One way to see this is to place both of your forefingers across C_5: Since <u>**both**</u> fingers can reach C_{s2} without crossing another capacitor or the battery, they are in parallel. (In parallel, unlike series, it's okay to cross a junction.) The C_5 and C_{s2} capacitors may be replaced with a single capacitor, which we will call C_{p1}. We will erase one path and rename the remaining capacitor when we make this replacement. We may change the shape of the circuit, provided that we preserve the overall structure.

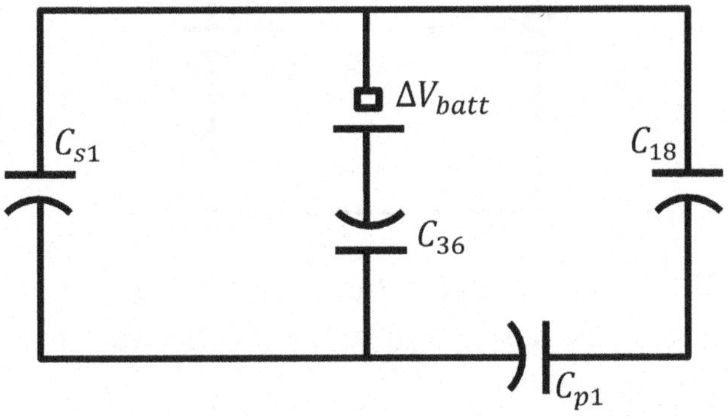

Capacitors C_{p1} and C_{18} are in **series** because an electron could travel from C_{p1} to C_{18} without crossing a junction. They may be replaced with a single capacitor, which we will call C_{s3}.

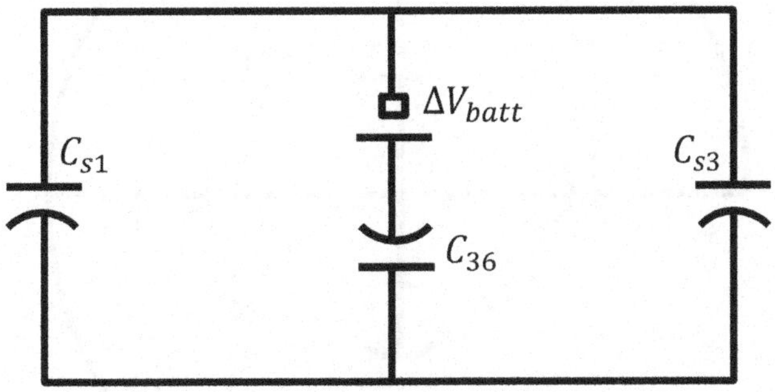

Capacitors C_{s1} and C_{s3} are in **parallel**. Place both of your forefingers across the C_{s1} capacitor: Since **both** fingers can reach C_{s3} without crossing another capacitor or the battery, they are in parallel. They may be replaced with a single capacitor, which we will call C_{p2}.

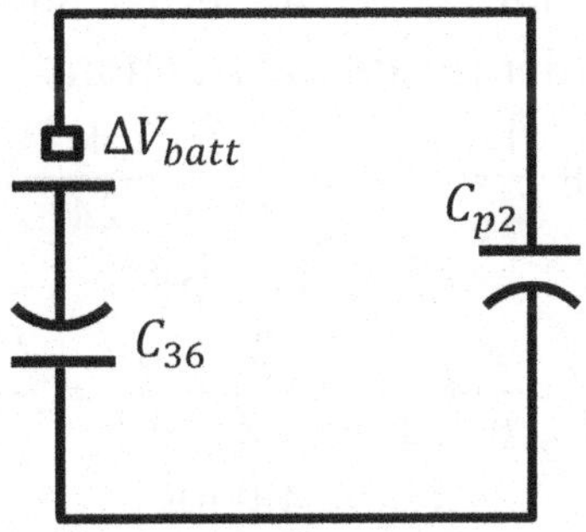

Capacitors C_{36} and C_{p2} are in **series** because an electron could travel from C_{36} to C_{p2}

without crossing a junction. They may be replaced with a single capacitor, which we will call C_{eq}. Since this is the last capacitor remaining, it is the **equivalent capacitance.**

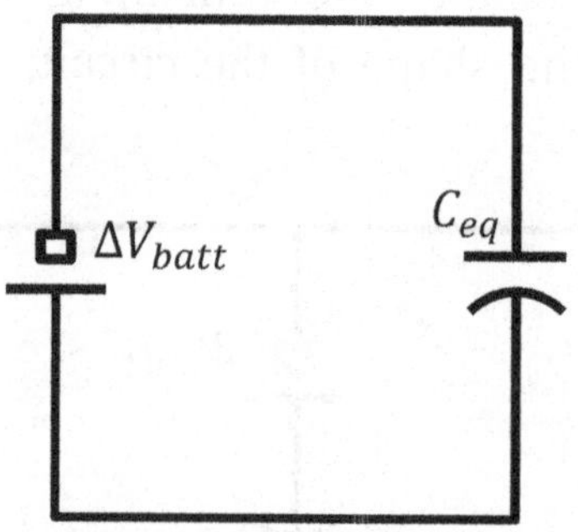

(B) Determine the equivalent capacitance of the circuit.

Solution. Apply the formulas for series and parallel capacitors to each reduction that we made in part (A).

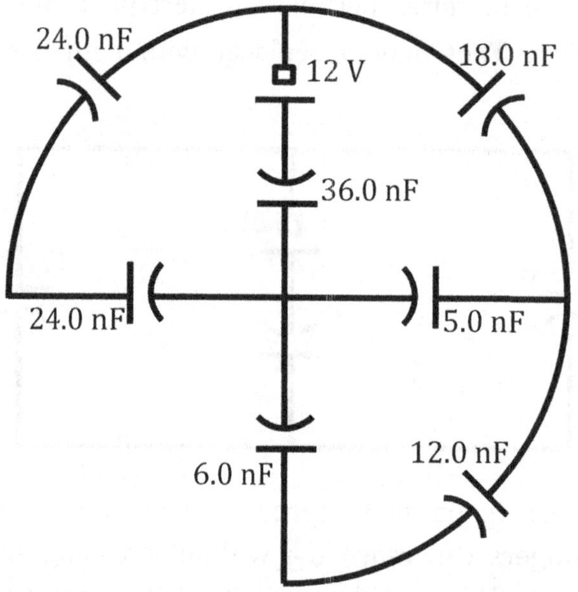

Begin by combining the two 24-nF capacitors in **series** to form C_{s1} and combining the 6-nF and 12-nF capacitors in **series** to form C_{s2}. In series, add the capacitances in **reciprocal.*** Find a **common denominator** in order to add the fractions.

$$\frac{1}{C_{s1}} = \frac{1}{C_{24}} + \frac{1}{C_{24}} = \frac{1}{24} + \frac{1}{24} = \frac{1+1}{24} = \frac{2}{24} = \frac{1}{12}$$

$$C_{s1} = 12.0 \text{ nF}$$

$$\frac{1}{C_{s2}} = \frac{1}{C_6} + \frac{1}{C_{12}} = \frac{1}{6} + \frac{1}{12} = \frac{2}{12} + \frac{1}{12} = \frac{2+1}{12} = \frac{3}{12} = \frac{1}{4}$$

$$C_{s2} = 4.0 \text{ nF}$$

* Note that this is backwards compared to resistors, as we will see in Chapter 13.

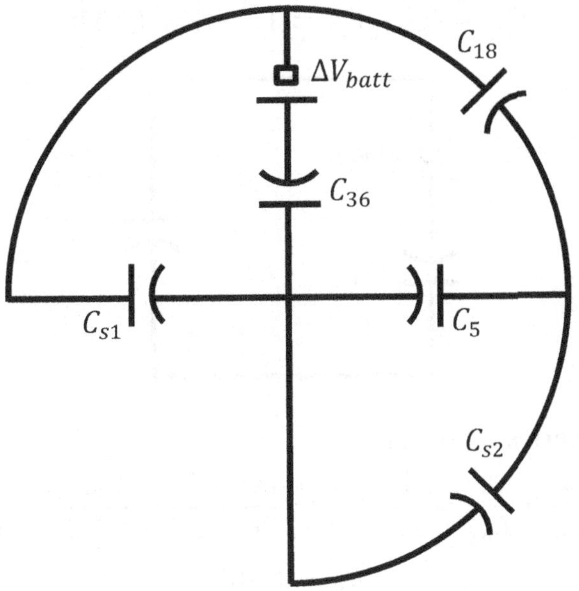

Combine C_5 and C_{s2} in **parallel** to form C_{p1}. In parallel, add the capacitances.

$$C_{p1} = C_5 + C_{s2} = 5 + 4 = 9.0 \text{ nF}$$

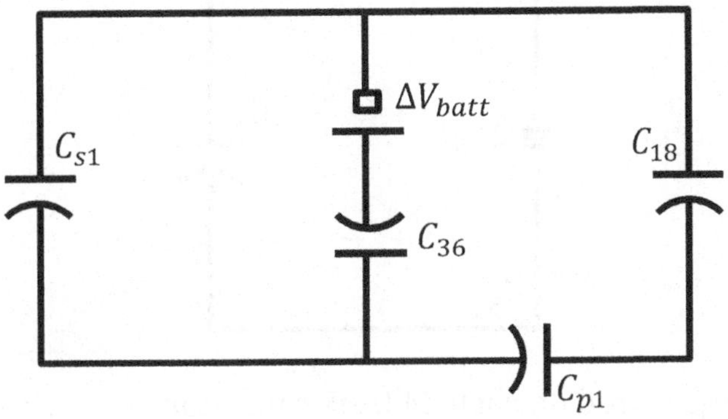

Combine C_{p1} and C_{18} in **series** to form C_{s3}.

$$\frac{1}{C_{s3}} = \frac{1}{C_{p1}} + \frac{1}{C_{18}} = \frac{1}{9} + \frac{1}{18} = \frac{2}{18} + \frac{1}{18} = \frac{2+1}{18} = \frac{3}{18} = \frac{1}{6}$$

$$C_{s3} = 6.0 \text{ nF}$$

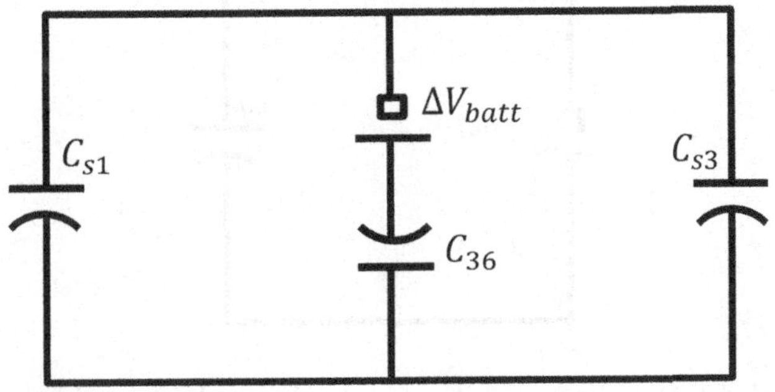

Combine C_{s1} and C_{s3} in **parallel** to form C_{p2}.

$$C_{p2} = C_{s1} + C_{s3} = 12 + 6 = 18.0 \text{ nF}$$

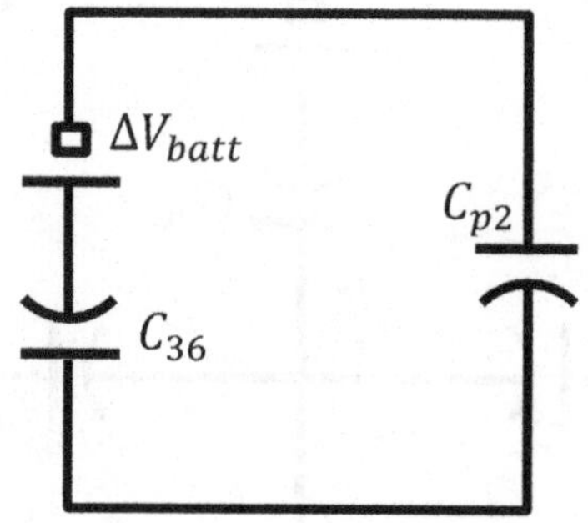

Combine C_{36} and C_{p2} in **series** to form C_{eq}.

$$\frac{1}{C_{eq}} = \frac{1}{C_{36}} + \frac{1}{C_{p2}} = \frac{1}{36} + \frac{1}{18} = \frac{1}{36} + \frac{2}{36} = \frac{1+2}{36} = \frac{3}{36} = \frac{1}{12}$$

$$C_{eq} = 12 \text{ nF}$$

The equivalent capacitance for the circuit (that is, from one terminal of the battery to the other) is $C_{eq} = 12$ nF.

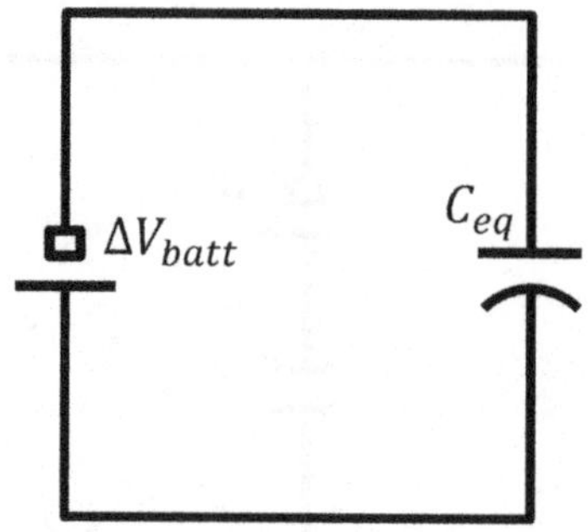

(C) Determine the charge stored on each 24.0-nF capacitor.

Solution. We must work "backwards" through our simplified circuits, beginning with the circuit that just has C_{eq}, in order to solve for charge (or potential difference or energy stored).

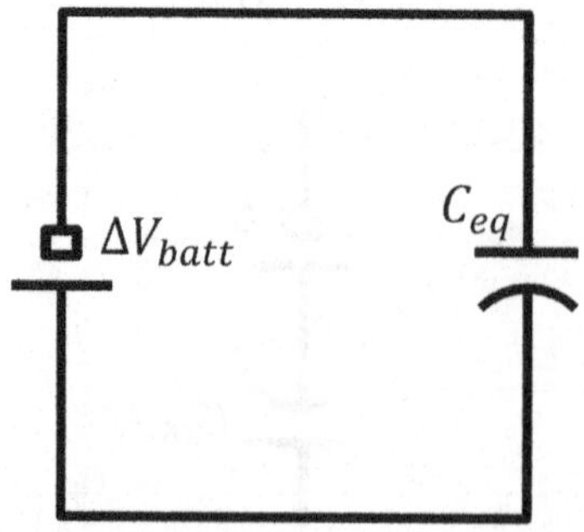

The math begins with the following equation, which applies to the last circuit. Recall from the original diagram that the battery provides a potential difference across its terminals equal to $\Delta V_{batt} = 12$ V. We also found that the equivalent capacitance is $C_{eq} = 12$ nF in part (B).

$$Q_{eq} = C_{eq}\Delta V_{batt} = (12)(12) = 144 \text{ nC}$$

Note that 1 nF $\times$ V = 1 nC since 1 F $\times$ V = 1 C.

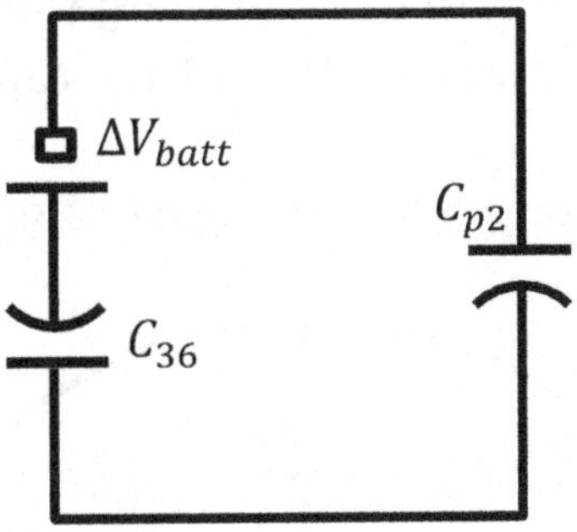

Now we will go one step "backwards" from the simplest circuit (with just C_{eq}) to the second-to-last circuit (which has C_{36} and C_{p2}). Recall that C_{36} and C_{p2} are in **series. Charge** is the same in **series.** Therefore, we set the charges of C_{36} and C_{p2} equal to one another and also set them equal to the charge of the capacitor that replaced them (C_{eq}). This is expressed in the following equation.

$$Q_{36} = Q_{p2} = Q_{eq} = 144 \text{ nC}$$

When we set charges equal to one another, we must calculate potential difference. Based on the question for part (C), which is to find the charge stored by each 24-nF capacitor, we need ΔV_{p2} in order to work our way to the 24-nF capacitors. You can see this by studying the sequence of diagrams in part (A). Recall that $C_{p2} = 18$ nF from part (B).

$$\Delta V_{p2} = \frac{Q_{p2}}{C_{p2}} = \frac{144 \text{ nC}}{18 \text{ nF}} = 8.0 \text{ V}$$

Note that all of the subscripts match in the above equation. Note also that the n's cancel: $\frac{1 \text{ nC}}{1 \text{ nF}} = \frac{10^{-9} \text{ C}}{10^{-9} \text{ F}} = 1$ V. It should make sense that the potential difference across C_{p2}, which equals $\Delta V_{p2} = 8.0$ V, is less than the potential difference across the battery, $\Delta V_{batt} = 12$ V.

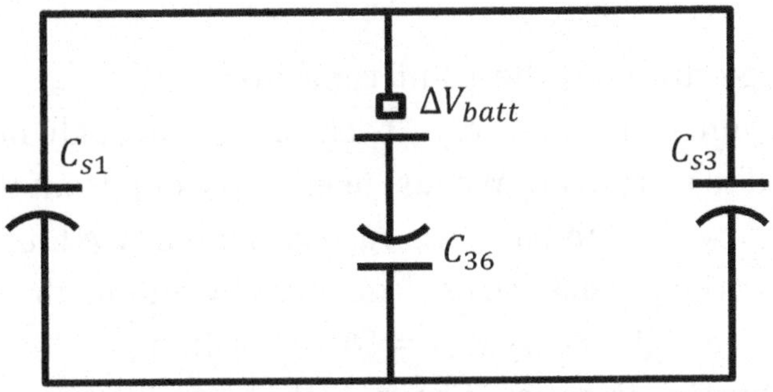

Going one more step "backwards" through our simplified circuits, C_{s1} and C_{s3} are in **parallel. Potential difference** is the same in parallel. Therefore, we set the potential differences across C_{s1} and C_{s3} equal to one another and also set them equal to the potential difference across the capacitor that replaced them (C_{p2}).

$$\Delta V_{s1} = \Delta V_{s3} = \Delta V_{p2} = 8.0 \text{ V}$$

When we set potential differences equal to one another, we must calculate charge. Based on the question for part (C), which is to find the charge stored by each 24-nF capacitor, we need Q_{s1} in order to work our way to the 24-nF capacitors. Recall that we found $C_{s1} = 12$ nF in part (B).

$$Q_{s1} = C_{s1}\Delta V_{s1} = (12)(8) = 96 \text{ nC}$$

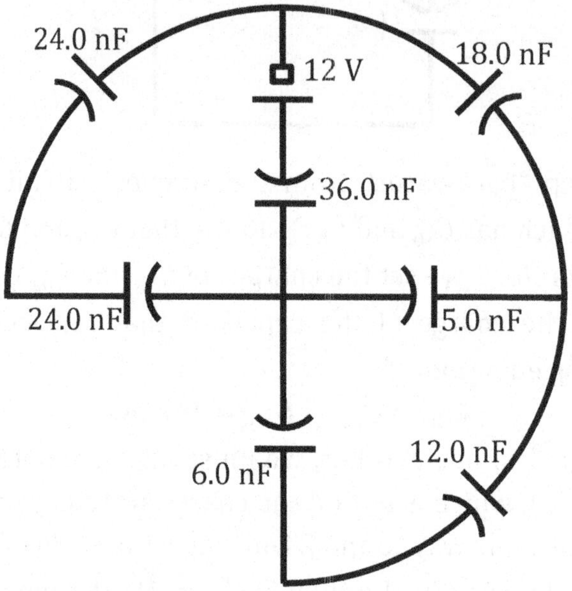

We must go one more step "backwards" in order to reach the 24-nF capacitors. The two 24-nF capacitors are in **series**. **Charge** is the same in **series**. Therefore, we set the charges of the 24-nF capacitors equal to one another and also set them equal to the charge of the capacitor that replaced them (C_{s1}). This is expressed in the following equation.

$$Q_{24} = Q_{24} = Q_{s1} = 96 \text{ nC}$$

The charge stored on each 24-nF capacitor is $Q_{24} = 96$ nC.

(D) Determine the energy stored by the 18-nF capacitor.

Solution. We must again work our way "backwards" through our simplified circuits. However, we don't need to start over. We just need to pick up from the relevant place from part (C). We found $\Delta V_{s3} = 8.0$ V in part (C). Also recall that we found $C_{s3} = 6.0$ nF in part (B). We will continue working "backwards" from here by finding the charge stored on C_{s3}.

$$Q_{s3} = C_{s3}\Delta V_{s3} = (6)(8) = 48 \text{ nC}$$

The C_{p1} and C_{18} capacitors are in **series. Charge** is the same in **series**. Set the charges of C_{p1} and C_{18} equal to one another and to the charge of the capacitor that replaced them (C_{s3}).

$$Q_{p1} = Q_{18} = Q_{s3} = 48 \text{ nC}$$

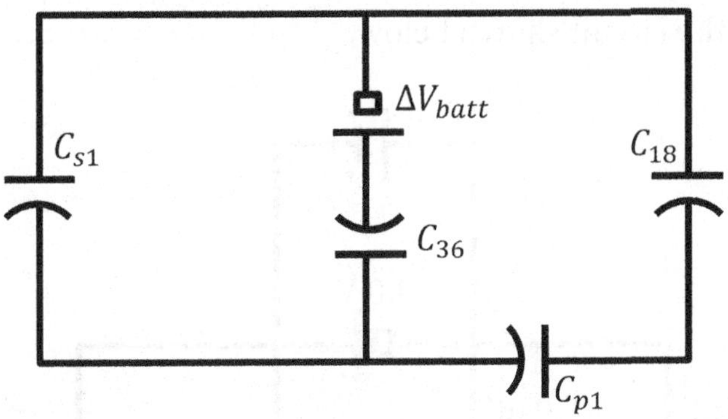

We now know that $Q_{18} = 48$ nC. We also know that $C_{18} = 18$ nF (that's why we named this capacitor C_{18} in our simplified circuits.) Use the appropriate energy equation.

$$U_{18} = \frac{Q_{18}^2}{2C_{18}} = \frac{(48)^2}{2(18)} = \frac{(48)(48)}{(2)(3)(6)} = \frac{(48)(48)}{(6)(6)} = (8)(8) = 64 \text{ nJ}$$

The energy stored by the 18-nF capacitor is $U_{18} = 64$ nJ. Note that factoring 18 as (3)(6) in the denominator made the arithmetic simple enough to do without the aid of a calculator. (If you instead square 48 to get 2304 and then divide by 36, you still get 64.)

Example 56. Consider the circuit shown below.

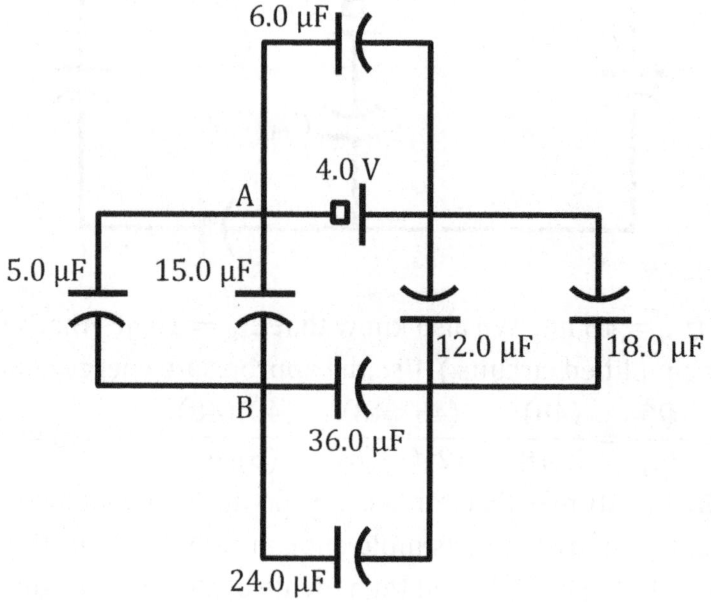

(A) Redraw the circuit, simplifying it one step at a time by identifying series and parallel combinations.

Solution. In the diagram above, the 5-μF and 15-μF capacitors are in **parallel**. Place both of your forefingers across the 5-μF capacitor: Since <u>**both**</u> fingers can reach the 15-μF capacitor without crossing another capacitor or the battery, they are in parallel. The 5-μF and 15-μF capacitors may be replaced with a single capacitor, which we will call C_{p1}. Similarly, the 36-μF and 24-μF capacitors are in **parallel**: We will replace them with C_{p2}. Furthermore, the 12-μF and 18-μF capacitors are in **parallel**: We will replace them with C_{p3}. For each parallel combination, we will erase one path and rename the remaining capacitor when we make the replacement.

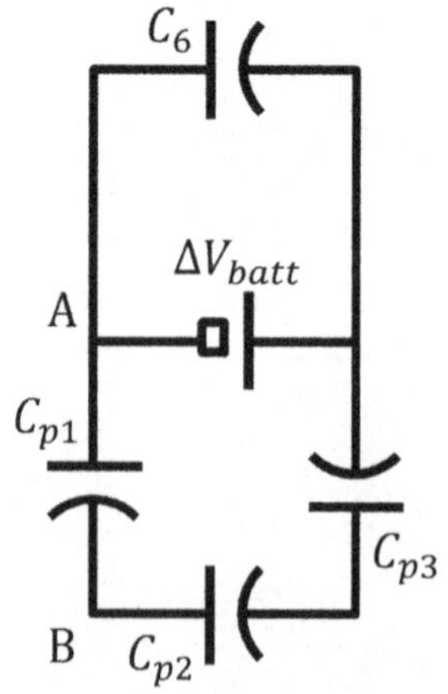

Capacitors C_{p1}, C_{p2}, and C_{p3} are all in **series** because an electron could travel from C_{p1} to C_{p2} to C_{p3} without crossing a junction. These three capacitors may be replaced with a single capacitor, which we will call C_{s1}.

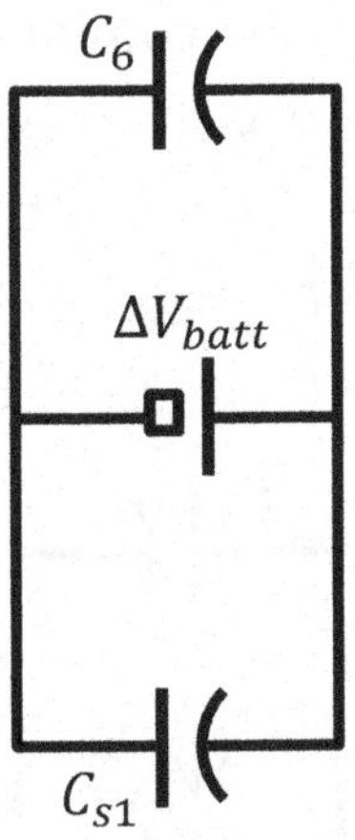

Capacitors C_{s1} and C_6 are in **parallel**. Place both of your forefingers across C_{s1}: Since **<u>both</u>** fingers can reach C_6 without crossing another capacitor or the battery, they are in parallel. They may be replaced with a single capacitor, which we will call C_{eq}. Since this is the last capacitor remaining, it is the **equivalent capacitance.**

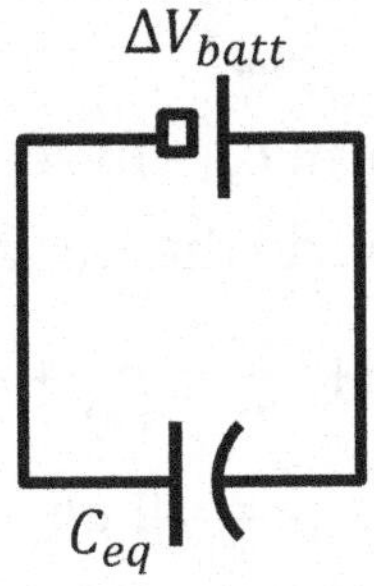

(B) Determine the equivalent capacitance of the circuit.

Solution. Apply the formulas for series and parallel capacitors to each reduction.

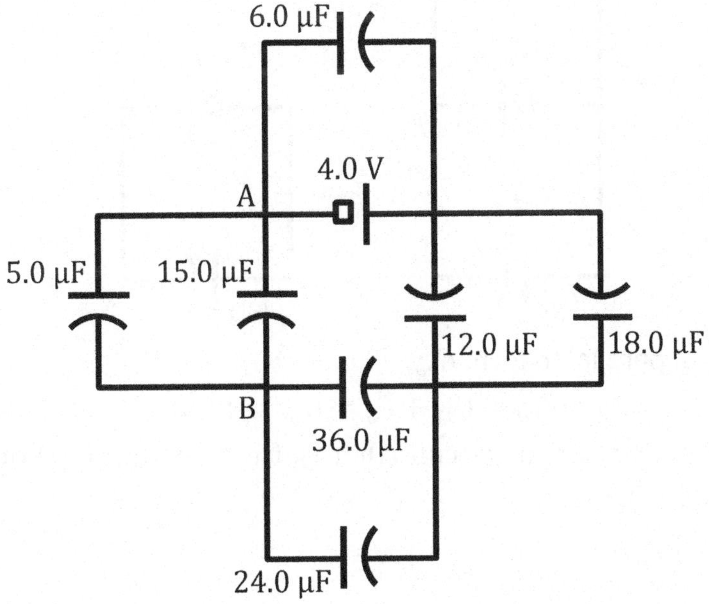

Combine the 5-μF and 15-μF capacitors in **parallel** to form C_{p1}, combine the 24-μF and 36-μF capacitors in **parallel** to form C_{p2}, and combine the 12-μF and 18-μF capacitors in **parallel** to form C_{p3}. In parallel, add the capacitances.

$$C_{p1} = C_5 + C_{15} = 5 + 15 = 20\ \mu\text{F}$$
$$C_{p2} = C_{24} + C_{36} = 24 + 36 = 60\ \mu\text{F}$$
$$C_{p3} = C_{12} + C_{18} = 12 + 18 = 30\ \mu\text{F}$$

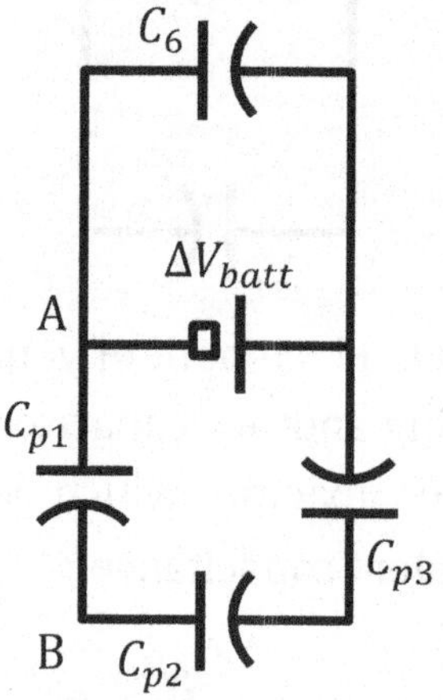

Combine C_{p1}, C_{p2}, and C_{p3} in **series** to form C_{s1}. In series, add the capacitances in **reciprocal**. Find a **common denominator** in order to add the fractions.

$$\frac{1}{C_{s1}} = \frac{1}{C_{p1}} + \frac{1}{C_{p2}} + \frac{1}{C_{p3}} = \frac{1}{20} + \frac{1}{60} + \frac{1}{30} = \frac{3}{60} + \frac{1}{60} + \frac{2}{60} = \frac{3+1+2}{60} = \frac{6}{60} = \frac{1}{10}$$
$$C_{s1} = 10\ \mu\text{F}$$

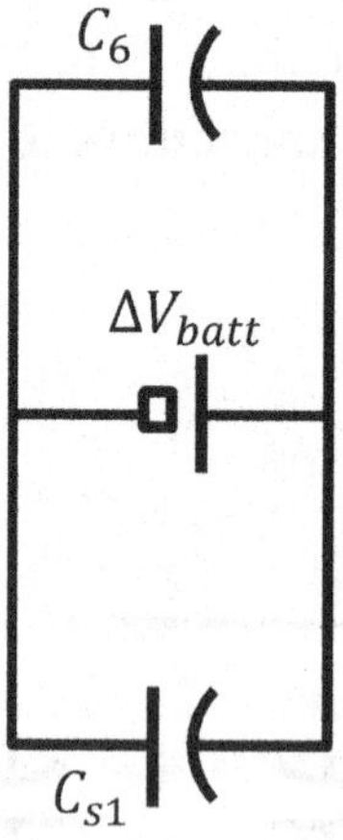

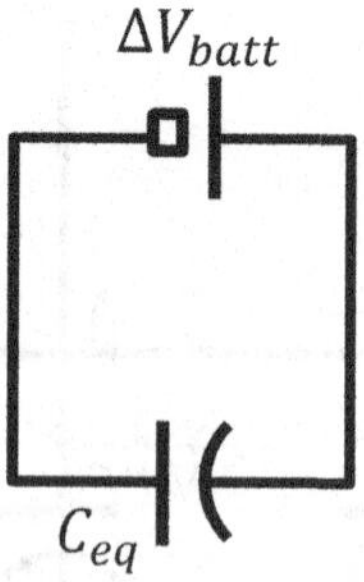

Combine C_{s1} and C_6 in **parallel** to form C_{eq}.

$$C_{eq} = C_{s1} + C_6 = C_{eq} = 16\ \mu\text{F}$$

The equivalent capacitance for the circuit (that is, from one terminal of the battery to the other) is $C_{eq} = 16\ \mu\text{F}$.

(C) Determine the potential difference between points A and B.

Solution. We must work "backwards" through our simplified circuits, beginning with the circuit that just has C_{eq}, in order to solve for the potential difference between two points.

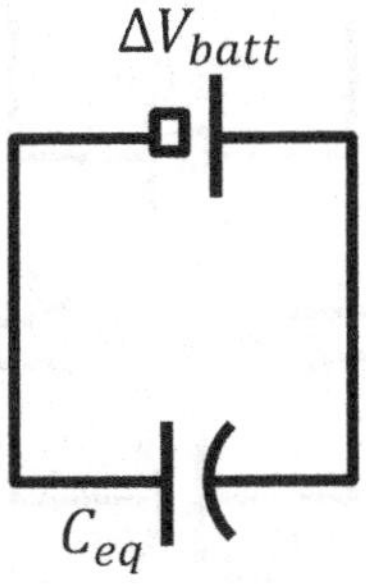

Begin with the following equation, which applies to the last circuit. Recall from the original diagram that the battery provides a potential difference across its terminals equal to $\Delta V_{batt} = 4.0$ V. We also found that the equivalent capacitance is $C_{eq} = 16$ µF in part (B).

$$Q_{eq} = C_{eq}\Delta V_{batt} = (16)(4) = 64 \text{ µC}$$

(Technically, when the second-to-last circuit is a parallel combination, which you can see is the case in the diagram below, we could skip the previous step.)

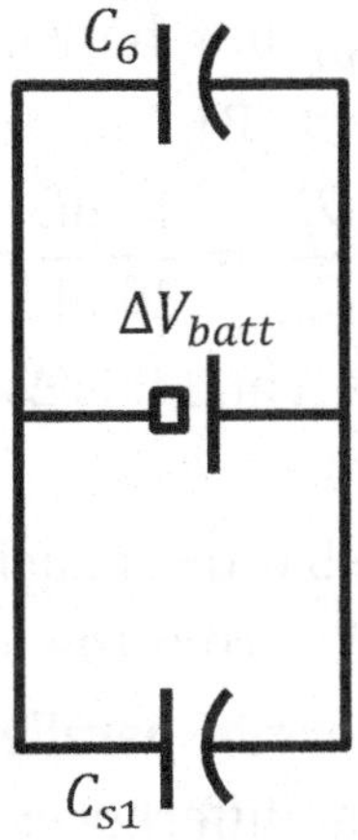

Now we will go one step "backwards" from the simplest circuit (with just C_{eq}) to the second-to-last circuit (which has C_6 and C_{s1}). In the second-to-last circuit, C_6 and C_{s1} are in **parallel**. **Potential difference** is the same in parallel. Therefore, we set the potential differences across C_6 and C_{s1} equal to one another and also set them equal to the potential difference across the battery (since they happen to be in parallel with the battery).

$$\Delta V_{s1} = \Delta V_6 = \Delta V_{batt} = 4.0 \text{ V}$$

When we set potential differences equal to one another, we must calculate charge. Based on the question for part (C), which is to find the potential difference between points A and B, we need Q_{s1}. See the following diagram. Recall that we found $C_{s1} = 10$ µF in part (B).

$$Q_{s1} = C_{s1}\Delta V_{s1} = (10)(4) = 40 \text{ µC}$$

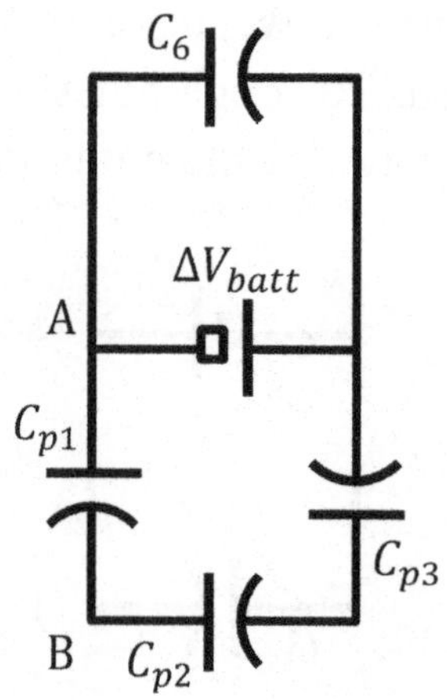

Going back one more step "backwards," C_{p1}, C_{p2}, and C_{p3} are in **series**. **Charge** is the same in **series**. Therefore, we set the charges of C_{p1}, C_{p2}, and C_{p3} equal to one another and also set them equal to the charge of the capacitor that replaced them (C_{s1}).

$$Q_{p1} = Q_{p2} = Q_{p3} = Q_{s1} = 40\ \mu\text{C}$$

When we set charges equal to one another, we must calculate potential difference. Based on the question for part (C), which is to find the potential difference between points A and B, we need ΔV_{p1} because capacitor C_{p1} lies between points A and B (see the diagram above). Recall that $C_{p1} = 20\ \mu\text{F}$ from part (B).

$$\Delta V_{p1} = \frac{Q_{p1}}{C_{p1}} = \frac{40\ \mu\text{C}}{20\ \mu\text{F}} = 2.0\ \text{V}$$

The potential difference between points A and B is $\Delta V_{AB} = \Delta V_{p1} = 2.0$ V.

(D) Determine the energy stored on the 5.0-μF capacitor.

Solution. Continue working "backwards" from the value of $\Delta V_{p1} = 2.0$ V that we found in part (C). The 5-μF and 15-μF capacitors in **parallel**. **Potential difference** is the same in parallel. Therefore, we set the potential differences across the 5-μF and 15-μF capacitors equal to one another and also set them equal to the potential difference across the capacitor that replaced them (C_{p1}).

$$\Delta V_5 = \Delta V_{15} = \Delta V_{p1} = 2.0\ \text{V}$$

We now know that $\Delta V_5 = 2.0$ V. We also know that $C_5 = 5.0\ \mu\text{F}$ (that's why we named this capacitor C_5 in our simplified circuits.) Use the appropriate energy equation.

$$U_5 = \frac{1}{2}C_5\Delta V_5^2 = \frac{1}{2}(5)(2)^2 = \frac{1}{2}(5)(4) = 10\ \mu\text{J}$$

The energy stored by the 5.0-μF capacitor is $U_5 = 10\ \mu\text{J}$.

12 PARALLEL-PLATE AND OTHER CAPACITORS

Parallel-plate Capacitor

$$C = \frac{\kappa \epsilon_0 A}{d}$$

Capacitance

$$C = \frac{Q}{\Delta V} \quad , \quad Q = C\Delta V \quad , \quad \Delta V = \frac{Q}{C}$$

Parallel and Series Capacitors

$$C_p = C_1 + C_2 + \cdots + C_N \quad , \quad \frac{1}{C_s} = \frac{1}{C_1} + \frac{1}{C_2} + \cdots + \frac{1}{C_N}$$

Energy Stored in a Capacitor

$$U = \frac{1}{2} Q\Delta V \quad , \quad U = \frac{1}{2} C\Delta V^2 \quad , \quad U = \frac{Q^2}{2C}$$

Dielectric Strength

$$E_{max} = \frac{\Delta V_{max}}{d}$$

Coulomb's Constant

$$k = 8.99 \times 10^9 \ \frac{\text{N}\cdot\text{m}^2}{\text{C}^2} \approx 9.0 \times 10^9 \ \frac{\text{N}\cdot\text{m}^2}{\text{C}^2}$$

Permittivity of Free Space

$$\epsilon_0 = \frac{1}{4\pi k} = 8.8 \times 10^{-12} \ \frac{\text{C}^2}{\text{N}\cdot\text{m}^2} \approx \frac{10^{-9}}{36\pi} \ \frac{\text{C}^2}{\text{N}\cdot\text{m}^2}$$

Symbol	Name	SI Units
C	capacitance	F
ϵ_0	permittivity of free space	$\frac{C^2}{N \cdot m^2}$ or $\frac{C^2 \cdot s^2}{kg \cdot m^3}$
κ	dielectric constant	unitless
A	area of one plate	m^2
d	separation between the plates	m
E_{max}	dielectric strength	N/C or V/m
ΔV_{max}	maximum potential difference	V
Q	the charge stored on the positive plate of a capacitor	C
ΔV	the potential difference between two points in a circuit	V
U	the energy stored by a capacitor	J

Note: The symbol κ is the lowercase Greek letter kappa and ϵ is epsilon.

Prefix	Name	Power of 10
m	milli	10^{-3}
μ	micro	10^{-6}
n	nano	10^{-9}
p	pico	10^{-12}

Example 57. A parallel-plate capacitor has circular plates with a radius of 30 mm. The separation between the plates is of 2.0 mm. A dielectric is inserted between the plates. The dielectric constant is 8.0 and the dielectric strength is $6.0 \times 10^6 \frac{\text{V}}{\text{m}}$.

(A) Determine the capacitance.

Solution. Make a list of the known quantities and identify the desired unknown symbol:

- The capacitor plates have a radius of $a = 30$ mm.
- The separation between the plates is $d = 2.0$ mm.
- The dielectric constant is $\kappa = 8.0$ and the dielectric strength is $E_{max} = 6.0 \times 10^6 \frac{\text{V}}{\text{m}}$.
- We also know $\epsilon_0 = 8.8 \times 10^{-12} \frac{\text{C}^2}{\text{N}\cdot\text{m}^2}$, which we will approximate as $\epsilon_0 \approx \frac{10^{-9}}{36\pi} \frac{\text{C}^2}{\text{N}\cdot\text{m}^2}$ so that we may solve the problem without the aid of a calculator.
- The unknown we are looking for is capacitance (C).

Convert the radius and plate separation to SI units.

$$a = 30 \text{ mm} = 0.030 \text{ m} \quad , \quad d = 2.0 \text{ mm} = 0.0020 \text{ m}$$

The equation for the capacitance of a parallel-plate capacitor involves area, so we need to find area first. The area of a circular plate is:

$$A = \pi a^2 = \pi(0.03)^2 = 0.0009\pi \text{ m}^2 = 9\pi \times 10^{-4} \text{ m}^2$$

If you use a calculator, the area comes out to $A = 0.00283 \text{ m}^2$. Now we are ready to use the equation for capacitance.

$$C = \frac{\kappa\epsilon_0 A}{d} = \frac{(8)\left(\frac{10^{-9}}{36\pi}\right)(9\pi \times 10^{-4})}{(0.002)} = \frac{(8)\left(\frac{9\pi}{36\pi}\right)(10^{-9} \times 10^{-4})}{(0.002)} = \frac{(8)\left(\frac{1}{4}\right)(10^{-13})}{(0.002)}$$

$$C = \frac{2 \times 10^{-13}}{2 \times 10^{-3}} = 1.0 \times 10^{-10} \text{ F} = 0.10 \text{ nF}$$

Note that $10^{-9} \times 10^{-4} = 10^{-13}$ and $\frac{10^{-13}}{10^{-3}} = 10^{-13-(-3)} = 10^{-13+3} = 10^{-10}$ according to the rules $x^m x^n = x^{m+n}$ and $\frac{x^m}{x^n} = x^{m-n}$. Also note that $-13 - (-3) = -13 + 3 = -10$ and that $0.002 = 2 \times 10^{-3}$. The capacitance is $C = 0.10$ nF, which is the same as $C = 1.0 \times 10^{-10}$ F.

(B) What is the maximum charge that this capacitor can store on its plates?

Solution. The key to this solution is to note the word "maximum." First find the maximum potential difference (ΔV_{max}) across the plates. The maximum potential difference across the plates is related to the dielectric strength (E_{max}).

$$E_{max} = \frac{\Delta V_{max}}{d}$$

Multiply both sides of the equation by d.

$$\Delta V_{max} = E_{max} d = (6 \times 10^6)(0.002) = (6 \times 10^6)(2 \times 10^{-3}) = 12 \times 10^3 = 12{,}000 \text{ V}$$

Use the equation $Q = C\Delta V$ to find the maximum charge that can be stored on the plates.

$$Q_{max} = C\Delta V_{max} = (1 \times 10^{-10})(12{,}000) = (1 \times 10^{-10})(1.2 \times 10^4) = 1.2 \times 10^{-6} = 1.2 \text{ μC}$$

The maximum charge is $Q_{max} = 1.2$ μC, which can also be expressed as $Q_{max} = 1.2 \times 10^{-6}$ C.

Deriving an Equation for Capacitance

$C = \frac{Q}{\lvert \Delta V \rvert}$	$\Delta V = -\int \vec{\mathbf{E}} \cdot d\vec{\mathbf{s}}$	$Q = \int dq$

Gauss's Law

$$\oint_S \vec{\mathbf{E}} \cdot d\vec{\mathbf{A}} = \frac{q_{enc}}{\epsilon_0}$$

Differential Charge Element

$dq = \lambda ds$ (line or thin arc)	$dq = \sigma dA$ (surface area)	$dq = \rho dV$ (volume)

Relation Among Coordinate Systems and Unit Vectors

$x = r\cos\theta$ $y = r\sin\theta$ $\hat{\mathbf{r}} = \hat{\mathbf{x}}\cos\theta + \hat{\mathbf{y}}\sin\theta$ (2D polar)	$x = r_c\cos\theta$ $y = r_c\sin\theta$ $\hat{\mathbf{r}}_c = \hat{\mathbf{x}}\cos\theta + \hat{\mathbf{y}}\sin\theta$ (cylindrical)	$x = r\sin\theta\cos\varphi$ $y = r\sin\theta\sin\varphi$ $z = r\cos\theta$ $\hat{\mathbf{r}} = \hat{\mathbf{x}}\cos\varphi\sin\theta$ $+\hat{\mathbf{y}}\sin\varphi\sin\theta + \hat{\mathbf{z}}\cos\theta$ (spherical)

Differential Arc Length

$ds = dx$ (along x)	$ds = dy$ or dz (along y or z)	$ds = ad\theta$ (circular arc of radius a)

Differential Area Element

$dA = dxdy$ (polygon in xy plane)	$dA = rdrd\theta$ (pie slice, disc, thick ring)	$dA = a^2\sin\theta\, d\theta d\varphi$ (sphere of radius a)

Differential Volume Element

$dV = dxdydz$ (bounded by flat sides)	$dV = r_c dr_c d\theta dz$ (cylinder or cone)	$dV = r^2\sin\theta\, drd\theta d\varphi$ (spherical)

Symbol	Name	SI Units
dq	differential charge element	C
Q	total charge of the object	C
q_{enc}	the charge enclosed by the Gaussian surface	C
V	electric potential	V
$\vec{\mathbf{E}}$	electric field	N/C or V/m
ϵ_0	permittivity of free space	$\frac{C^2}{N \cdot m^2}$ or $\frac{C^2 \cdot s^2}{kg \cdot m^3}$
$\vec{\mathbf{R}}$	a vector from each dq to the field point	m
R	the distance from each dq to the field point	m
$d\vec{\mathbf{s}}$	differential displacement vector (see Chapter 21 of Volume 1)	m
x, y, z	Cartesian coordinates of dq	m, m, m
$\hat{\mathbf{x}}, \hat{\mathbf{y}}, \hat{\mathbf{z}}$	unit vectors along the $+x$-, $+y$-, $+z$-axes	unitless
r, θ	2D polar coordinates of dq	m, rad
r_c, θ, z	cylindrical coordinates of dq	m, rad, m
r, θ, φ	spherical coordinates of dq	m, rad, rad
$\hat{\mathbf{r}}, \hat{\boldsymbol{\theta}}, \hat{\boldsymbol{\varphi}}$	unit vectors along spherical coordinate axes	unitless
$\hat{\mathbf{r}}_c$	a unit vector pointing away from the $+z$-axis	unitless
λ	linear charge density (for an arc)	C/m
σ	surface charge density (for an area)	C/m^2
ρ	volume charge density	C/m^3
ds	differential arc length	m
dA	differential area element	m^2
dV	differential volume element	m^3

Note: The symbols λ, σ, and ρ are the lowercase Greek letters lambda, sigma, and rho.

Parallel-plate Example: A Prelude to Example 58. Derive an equation for the capacitance of a parallel-plate capacitor with vacuum* or air between its plates.

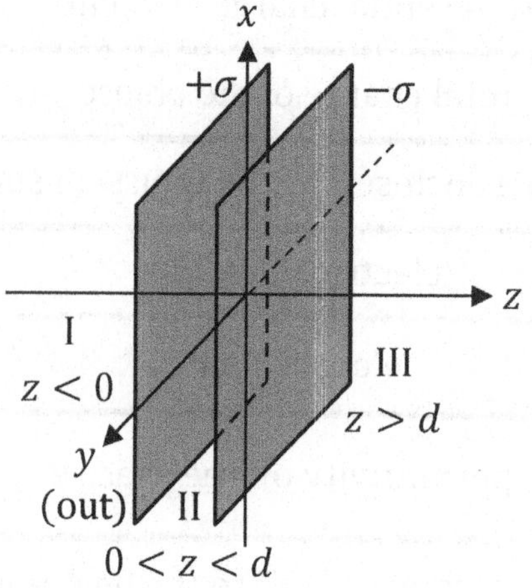

Solution. It's common for the distance between the plates (d) to be small compared to the length and width of the plates, in which case we may approximate the parallel-plate capacitor as consisting of two infinite charged planes. We found the electric field between two infinitely large, parallel, oppositely charged plates in Example 47 of Chapter 8. Applying Gauss's law, the electric field in the region between the plates is $\vec{\mathbf{E}} = \frac{\sigma}{\epsilon_0}\hat{\mathbf{z}}$. (See Chapter 8, Example 47.) We'll perform the potential difference integral from the positive plate (i) to the negative plate (f), such that $d\vec{\mathbf{s}}$ points to the right (along $\hat{\mathbf{z}}$). Therefore, $\vec{\mathbf{E}} \cdot d\vec{\mathbf{s}} = E\cos 0°\, ds = Eds$. Since $d\vec{\mathbf{s}}$ is along the z-axis, $ds = dz$. The limits of integration are from $z = 0$ to $z = d$, where d is the separation between the plates.

$$\Delta V = V_f - V_i = -\int_i^f \vec{\mathbf{E}} \cdot d\vec{\mathbf{s}} = -\int_i^f E\, ds = -\int_{z=0}^{d} \frac{\sigma}{\epsilon_0} dz = -\frac{\sigma}{\epsilon_0}\int_{z=0}^{d} dz = -\frac{\sigma}{\epsilon_0}[z]_{z=0}^{d} = -\frac{\sigma d}{\epsilon_0}$$

For uniformly charged plates, σ is constant, and σ and ϵ_0 may both come out of the integral. Substitute the above expression into the general equation for capacitance.

$$C = \frac{Q}{|\Delta V|} = \frac{Q}{\left|-\frac{\sigma d}{\epsilon_0}\right|} = \frac{Q}{\frac{\sigma d}{\epsilon_0}} = Q\frac{\epsilon_0}{\sigma d} = \frac{\epsilon_0 Q}{\sigma d}$$

Perform the following integral to eliminate the charge density (σ) from the answer. For a plane of charge, $dq = \sigma dA$ (see page 160). For uniform plates, σ is constant.

$$Q = \int dq = \int \sigma\, dA = \sigma \int dA = \sigma A$$

Substitute this expression into the previous equation for capacitance.

$$C = \frac{\epsilon_0 Q}{\sigma d} = \frac{\epsilon_0(\sigma A)}{\sigma d} = \frac{\epsilon_0 A}{d}$$

* For vacuum $\kappa = 1$, and for air $\kappa \approx 1$, such that we won't need to worry about the dielectric constant. (If there is a dielectric, it simply introduces a factor of κ into the final expression.)

Coaxial Cylinders Example: Another Prelude to Example 58. A cylindrical capacitor consists of two very long coaxial cylindrical conductors: One is a solid cylinder with radius a, while the other is a thin cylindrical shell of radius b. There is vacuum between the two cylinders. Derive an equation for the capacitance of this cylindrical capacitor.

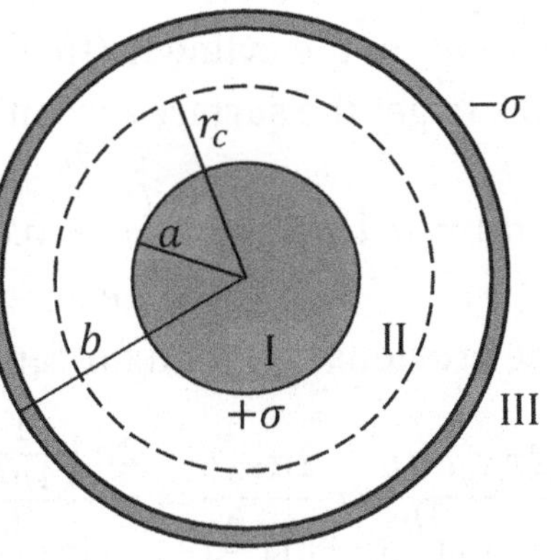

Solution. In the diagram above, we are looking at a cross section of the coaxial cylinders. The first step is to apply Gauss's law to region II ($a < r_c < b$). As discussed in Chapter 8, the charge resides on the surface of a conductor in electrostatic equilibrium. Thus, we will work with σ (and not ρ) for the charge density. In Chapter 8, we learned that the conducting shell doesn't matter in region II (since the Gaussian cylinder drawn for region II won't enclose any charge from the outer cylinder). Hence, we just need to find the electric field in region II for a solid conducting cylinder. We did that in an example in Chapter 8 (see page 100).

$$\vec{\mathbf{E}} = \frac{\sigma a}{\epsilon_0 r_c} \hat{\mathbf{r}}_c$$

We will integrate from the positive cylinder (i) to the negative shell (f), such that $d\vec{\mathbf{s}}$ points outward (along $\hat{\mathbf{r}}_c$) like $\vec{\mathbf{E}}$ does. Therefore, $\vec{\mathbf{E}} \cdot d\vec{\mathbf{s}} = E \cos 0° \, ds = E ds$. For a conductor in electrostatic equilibrium, the charge density (σ) is uniform (see Chapter 8), which means that we may pull σ out of the integral.

$$\Delta V = V_f - V_i = -\int_i^f \vec{\mathbf{E}} \cdot d\vec{\mathbf{s}} = -\int_i^f E \, ds = -\int_{r_c=a}^{b} \frac{\sigma a}{\epsilon_0 r_c} dr_c = -\frac{\sigma a}{\epsilon_0} \int_{r_c=a}^{b} \frac{dr_c}{r_c}$$

The anti-derivative for $\int \frac{dx}{x}$ is a natural logarithm (see Chapter 17).

$$\Delta V = -\frac{\sigma a}{\epsilon_0} [\ln(r_c)]_{r_c=a}^{b} = -\frac{\sigma a}{\epsilon_0} [\ln(b) - \ln(a)] = -\frac{\sigma a}{\epsilon_0} \ln\left(\frac{b}{a}\right)$$

We applied the rule $\ln\left(\frac{b}{a}\right) = \ln(b) - \ln(a)$. Plug this into the capacitance formula.

$$C = \frac{Q}{|\Delta V|} = \frac{Q}{\left|-\frac{\sigma a}{\epsilon_0} \ln\left(\frac{b}{a}\right)\right|} = \frac{Q}{\frac{\sigma a}{\epsilon_0} \ln\left(\frac{b}{a}\right)} = \frac{\epsilon_0 Q}{\sigma a \ln\left(\frac{b}{a}\right)}$$

Perform the following integral to eliminate the charge density (σ) from the answer. Treat the charge density (σ) as a constant. To integrate over the surface area of the cylinder, we integrate over z (along the length of the cylinder) and θ (around the body). We write $dA = ad\theta dz$, which is like $r_c d\theta$ times dz with $r_c = a$ (since the charge that we are integrating over resides on the surface of the inner conductor which has radius a). Note that $r_c = a$ is constant over the surface of the cylinder (that is, we're not integrating over r_c – we instead integrate over θ and z to get the surface area of a cylinder).

$$Q = \int dq = \int \sigma \, dA = \sigma \int dA = \sigma \int_{z=0}^{L} \int_{\theta=0}^{2\pi} a \, d\theta \, dz = 2\pi a \sigma L$$

Substitute this expression into the previous equation for capacitance. Recall that $\epsilon_0 = \frac{1}{4\pi k}$.

$$C = \frac{\epsilon_0 Q}{\sigma a \ln\left(\frac{b}{a}\right)} = \frac{2\pi\epsilon_0 a\sigma L}{\sigma a \ln\left(\frac{b}{a}\right)} = \frac{2\pi\epsilon_0 L}{\ln\left(\frac{b}{a}\right)} = \frac{2\pi\left(\frac{1}{4\pi k}\right)L}{\ln\left(\frac{b}{a}\right)} = \frac{L}{2k\ln\left(\frac{b}{a}\right)}$$

For a cylindrical capacitor, it's customary to divide both sides of the equation by L to get **capacitance per unit length** (since capacitance per unit length is finite).

$$\frac{C}{L} = \frac{1}{2k\ln\left(\frac{b}{a}\right)}$$

Example 58. A spherical capacitor consists of two spherical conductors: One is a solid sphere with radius a, while the other is a thin spherical shell of radius b. There is vacuum between the two spheres. Derive an equation for the capacitance of this spherical capacitor.

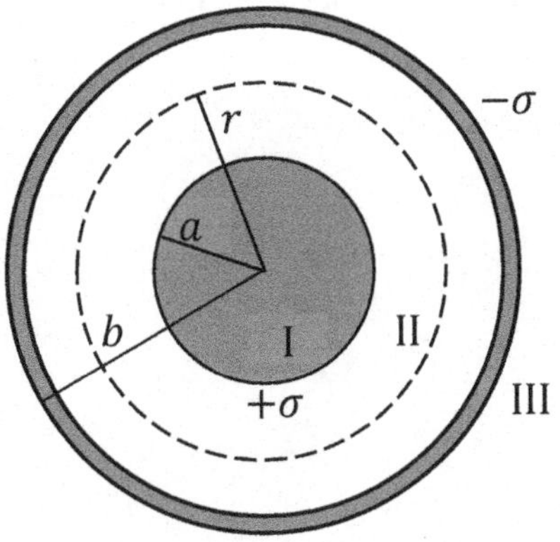

Solution. The first step is to apply Gauss's law to region II ($a < r < b$). As discussed in Chapter 8, the charge resides on the surface of a conductor in electrostatic equilibrium. Thus, we will work with σ (and not ρ) for the charge density. In Chapter 8, we learned that the conducting shell doesn't matter in region II (since the Gaussian sphere drawn for region II won't enclose any charge from the outer sphere). Hence, we just need to find the electric field in region II for a solid conducting sphere. As we learned in Chapter 8, the electric field outside of a uniformly charged sphere is no different from the electric field due to a pointlike charge (Chapter 2). For example, see page 86 in Chapter 8.

$$\vec{\mathbf{E}} = \frac{kQ}{r^2}\hat{\mathbf{r}}$$

We will integrate from the positive sphere (i) to the negative shell (f), such that $d\vec{\mathbf{s}}$ points outward (along $\hat{\mathbf{r}}$) like $\vec{\mathbf{E}}$ does. Therefore, $\vec{\mathbf{E}} \cdot d\vec{\mathbf{s}} = E \cos 0° \, ds = E ds$. For a conductor in electrostatic equilibrium, the charge density (σ) is uniform (see Chapter 8), which means that we may pull σ out of the integral.

$$\Delta V = V_f - V_i = -\int_i^f \vec{\mathbf{E}} \cdot d\vec{\mathbf{s}} = -\int_i^f E\, ds = -\int_{r=a}^{b} \frac{kQ}{r^2} dr = -kQ \int_{r=a}^{b} \frac{dr}{r^2} = -kQ \int_{r=a}^{b} r^{-2}\, dr$$

$$\Delta V = -kQ[-r^{-1}]_{r=a}^{b} = -kQ\left[-\frac{1}{r}\right]_{r=a}^{b} = -kQ\left(-\frac{1}{b} + \frac{1}{a}\right)$$

To subtract fractions, find a **common denominator**.

$$\Delta V = -kQ\left(-\frac{a}{ab} + \frac{b}{ab}\right) = -kQ\left(\frac{-a+b}{ab}\right) = -kQ\left(\frac{b-a}{ab}\right)$$

Note that $-a + b = b - a$. Plug the above expression into the capacitance formula.

$$C = \frac{Q}{|\Delta V|} = \frac{Q}{\left|-kQ\left(\frac{b-a}{ab}\right)\right|} = \frac{Q}{kQ\left(\frac{b-a}{ab}\right)} = \frac{1}{k\left(\frac{b-a}{ab}\right)}$$

To divide by a fraction, multiply by its **reciprocal**. The reciprocal of $\frac{b-a}{ab}$ is $\frac{ab}{b-a}$.

$$C = \frac{ab}{k(b-a)}$$

13 EQUIVALENT RESISTANCE

Ohm's Law

$$\Delta V = IR \quad , \quad I = \frac{\Delta V}{R} \quad , \quad R = \frac{\Delta V}{I}$$

Resistors in Series

$$R_s = R_1 + R_2 + \cdots + R_N$$

Resistors in Parallel

$$\frac{1}{R_p} = \frac{1}{R_1} + \frac{1}{R_2} + \cdots + \frac{1}{R_N}$$

Power

$$P = I\Delta V \quad , \quad P = I^2 R \quad , \quad P = \frac{\Delta V^2}{R}$$

Schematic Representation	Symbol	Name
	R	resistor
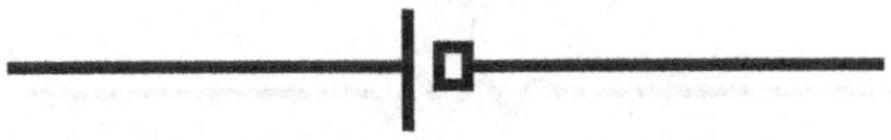	ΔV	battery or DC power supply
	measures ΔV	voltmeter
	measures I	ammeter

Symbol	Name	SI Units
R	resistance	Ω
I	electric current	A
ΔV	the potential difference between two points in a circuit	V
P	electric power	W

Prefix	Name	Power of 10
k	kilo	10^3

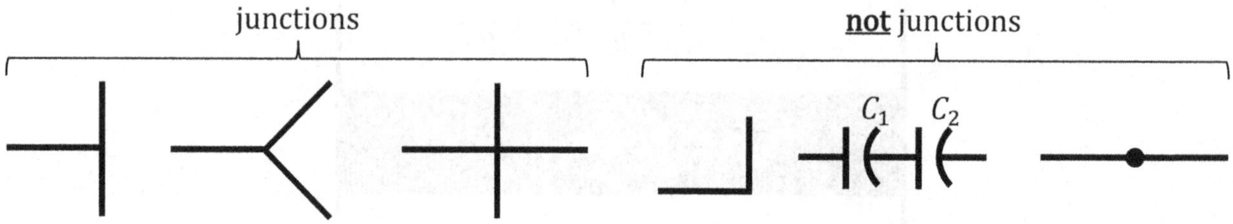

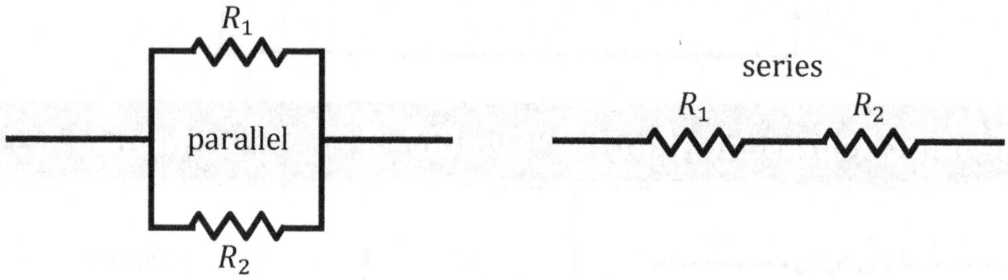

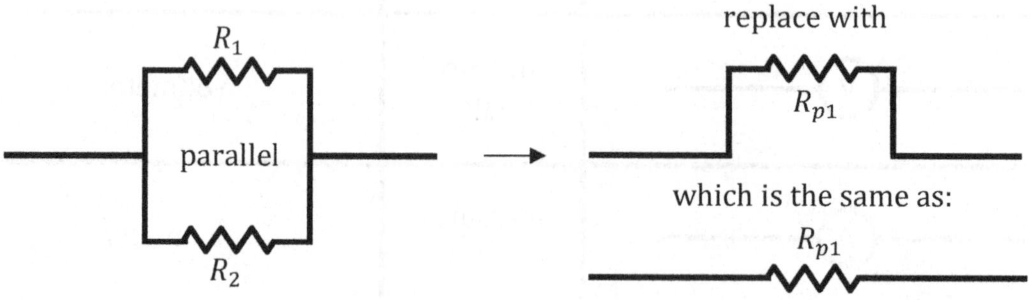

Example 59. Consider the circuit shown below.

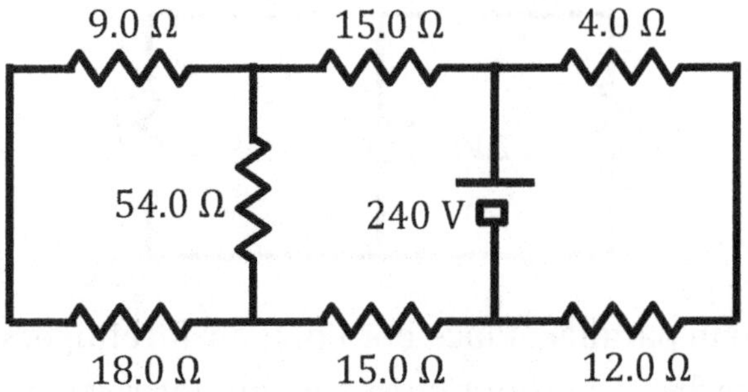

(A) Redraw the circuit, simplifying it one step at a time by identifying series and parallel combinations.

Solution. The 9-Ω and 18-Ω resistors are in **series** because an electron could travel from one resistor to the other without crossing a junction. They may be replaced with a single resistor, which we will call R_{s1}. Similarly, the 4-Ω and 12-Ω resistors are in **series**. They may also be replaced with a single resistor, which we will call R_{s2}.

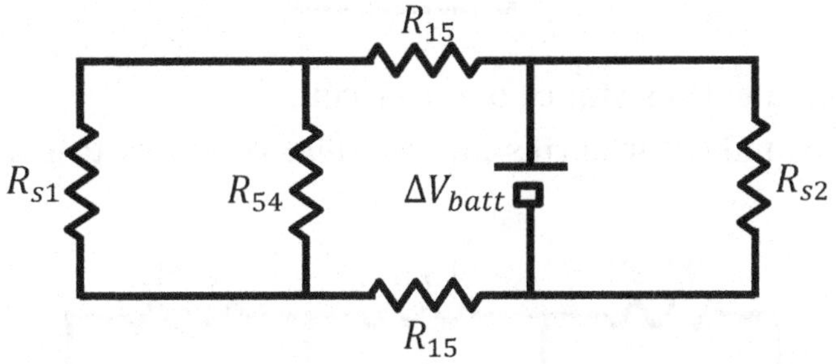

In the diagram above, R_{s1} and R_{54} are in **parallel**. One way to see this is to place both of your forefingers across R_{s1}: Since <u>**both**</u> fingers can reach R_{54} without crossing another resistor or the battery, they are in parallel. (In parallel, unlike series, it's okay to cross a junction.) The R_{s1} and R_{54} resistors may be replaced with a single resistor, which we will call R_{p1}. We will erase one path and rename the remaining resistor when we make this replacement. Compare the diagrams above and below.

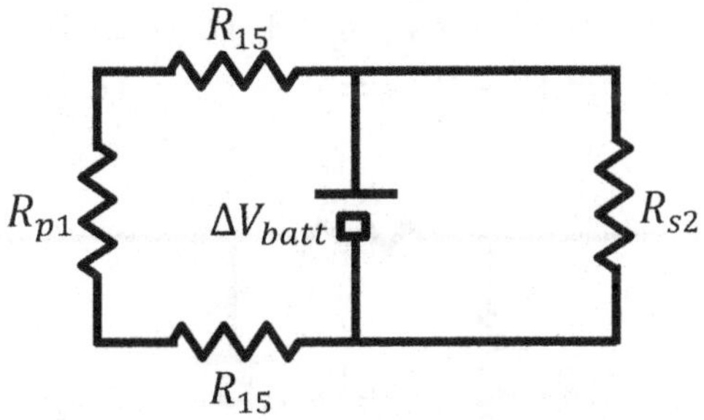

Resistors R_{15}, R_{p1}, and R_{15} are all in **series** because an electron could travel from R_{15} to R_{p1} to R_{15} without crossing a junction. All three resistors may be replaced with a single resistor, which we will call R_{s3}.

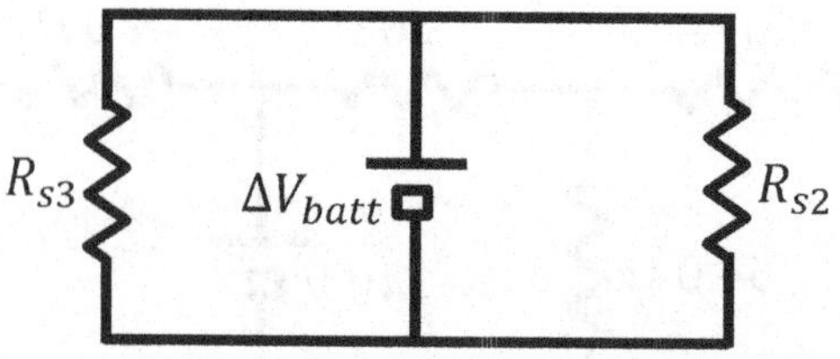

Resistors R_{s3} and R_{s2} are in **parallel**. Place both of your forefingers across the R_{s3} resistor: Since **both** fingers can reach R_{s2} without crossing another resistor or the battery, they are in parallel. They may be replaced with a single resistor, which we will call R_{eq}. Since this is the last resistor remaining, it is the **equivalent resistance**.

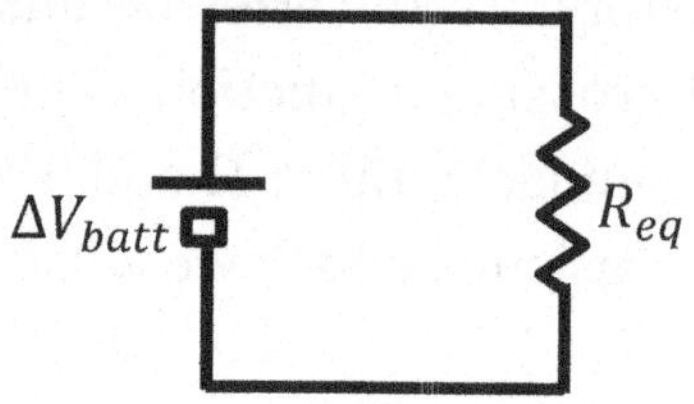

(B) Determine the equivalent resistance of the circuit.

Solution. Apply the formulas for series and parallel resistors to each reduction that we made in part (A).

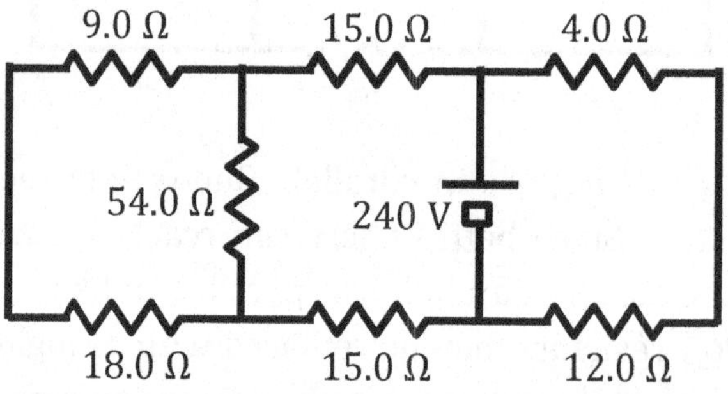

Begin by combining the 9-Ω and 18-Ω resistors in **series** to form R_{s1} and combining the 4-Ω and 12-Ω resistors in **series** to form R_{s2}. In series, add the resistances.

$$R_{s1} = R_9 + R_{18} = 9 + 18 = 27\ \Omega$$
$$R_{s2} = R_4 + R_{12} = 4 + 12 = 16\ \Omega$$

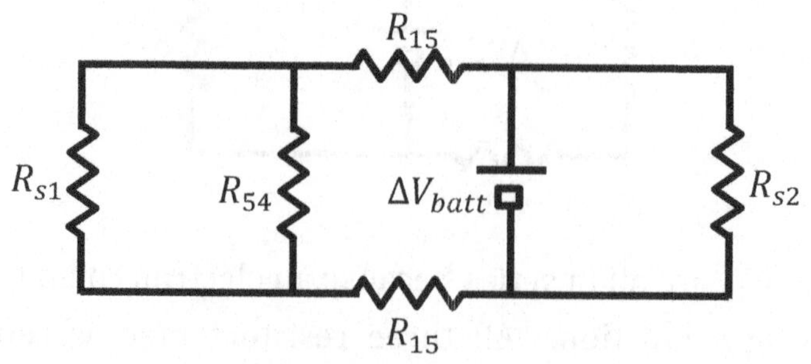

Combine R_{s1} and R_{54} in **parallel** to form R_{p1}. In parallel, add the resistances in reciprocal. Find a **common denominator** in order to add the fractions.

$$\frac{1}{R_{p1}} = \frac{1}{R_{s1}} + \frac{1}{R_{54}} = \frac{1}{27} + \frac{1}{54} = \frac{2}{54} + \frac{1}{54} = \frac{2+1}{54} = \frac{3}{54} = \frac{1}{18}$$

$$R_{p1} = 18\ \Omega$$

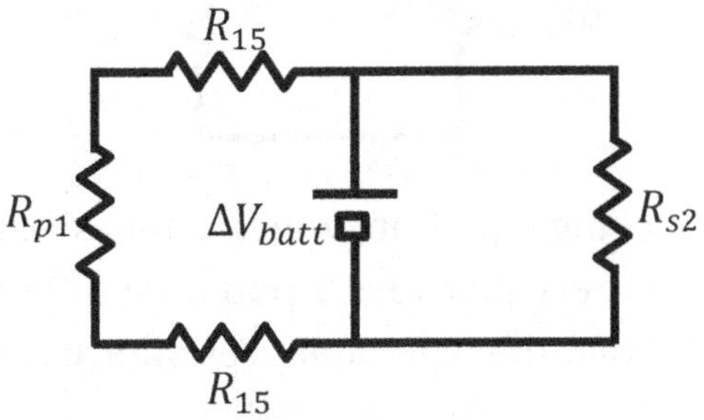

Combine R_{15}, R_{p1}, and R_{15} in **series** to form R_{s3}.

$$R_{s3} = R_{15} + R_{p1} + R_{15} = 15 + 18 + 15 = 48\ \Omega$$

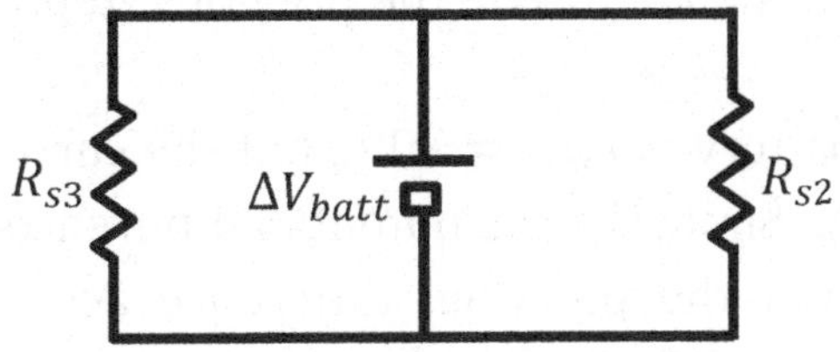

Combine R_{s3} and R_{s2} in **parallel** to form R_{eq}.

$$\frac{1}{R_{eq}} = \frac{1}{R_{s3}} + \frac{1}{R_{s2}} = \frac{1}{48} + \frac{1}{16} = \frac{1}{48} + \frac{3}{48} = \frac{1+3}{48} = \frac{4}{48} = \frac{1}{12}$$

$$R_{eq} = 12\ \Omega$$

The equivalent resistance for the circuit (that is, from one terminal of the battery to the other) is $R_{eq} = 12\ \Omega$.

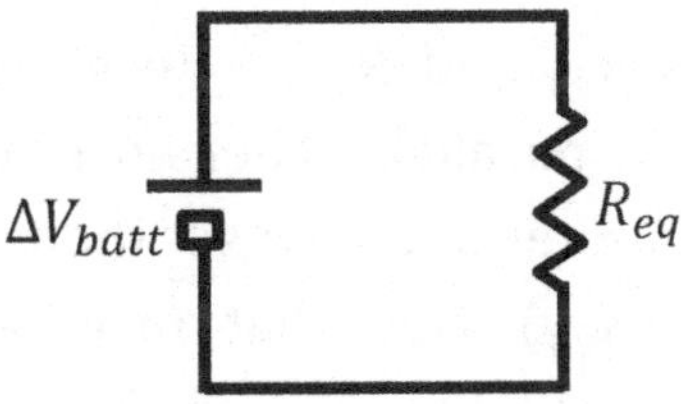

(C) Determine the power dissipated in either 15-Ω resistor.

Solution. We must work "backwards" through our simplified circuits, beginning with the circuit that just has R_{eq}, in order to solve for power (or current or potential difference).

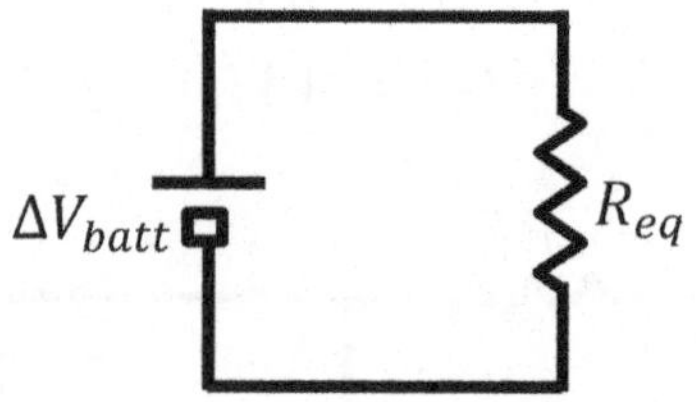

The math begins with the following equation, which applies to the last circuit. Recall from the original diagram that the battery provides a potential difference across its terminals of $\Delta V_{batt} = 240$ V. We also found that the equivalent resistance is $R_{eq} = 12\ \Omega$ in part (B).

$$I_{batt} = \frac{\Delta V_{batt}}{R_{eq}} = \frac{240}{12} = 20 \text{ A}$$

(Technically, when the second-to-last circuit is a parallel combination, which you can see is the case in the diagram below, we could skip the previous step.)

Notes: The current in the battery is $I_{batt} = 20$ A, but the current in other branches in the circuit will be less than 20 A. Similarly, the potential difference across the terminals of the battery is $\Delta V_{batt} = 240$ V, but the potential difference across most (if not all) of the resistors in the original circuit will be less than 240 V.

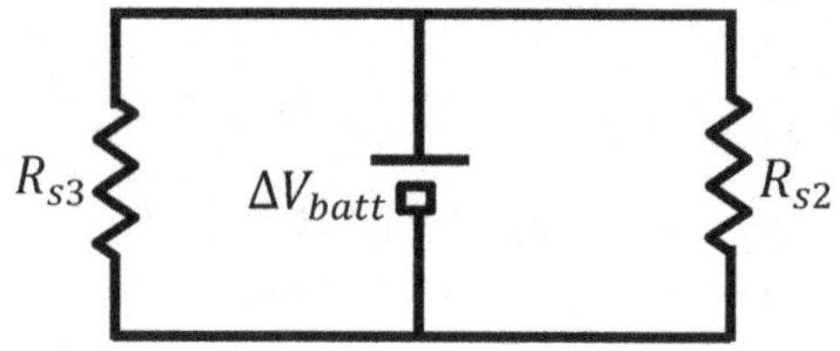

Now we will go one step "backwards" from the simplest circuit (with just R_{eq}) to the second-to-last circuit (which has R_{s3} and R_{s2}). Recall that R_{s3} and R_{s2} are in **parallel**. **Potential difference** is the same in **parallel**. Therefore, we set the potential differences across R_{s3} and R_{s2} equal to one another and also set them equal to the potential difference of the resistor that replaced them (R_{eq}). Note that the potential difference across R_{eq} is the same as the potential difference supplied by the battery. This is expressed in the following equation.

$$\Delta V_{s3} = \Delta V_{s2} = \Delta V_{batt} = 240 \text{ V}$$

When we set potential differences equal to one another, we must calculate current. Based on the question for part (C), which is to find the power dissipated in either 15-Ω resistor, we need I_{s3} in order to work our way to the 15-Ω resistors. You can see this by studying the sequence of diagrams in part (A). Recall that $R_{s3} = 48\ \Omega$ from part (B).

$$I_{s3} = \frac{\Delta V_{s3}}{R_{s3}} = \frac{240 \text{ V}}{48 \ \Omega} = 5.0 \text{ A}$$

Note that all of the subscripts match in the above equation. It should make sense that the current through R_{s3}, which equals $I_{s3} = 5.0$ A, is less than the current through the battery, $I_{batt} = 20$ A.

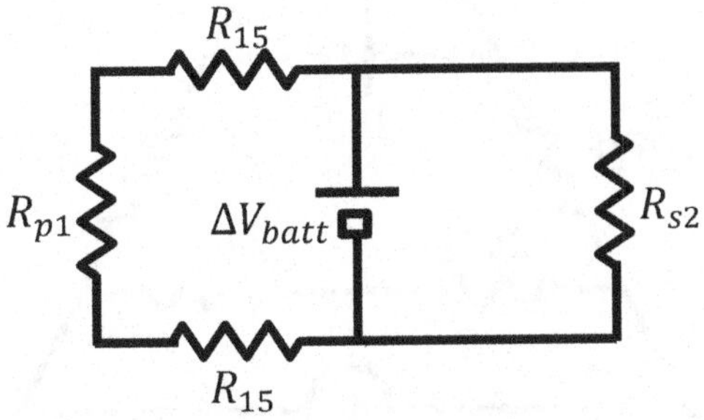

Going one more step "backwards" through our simplified circuits, R_{15}, R_{p1}, and R_{15} are in **series**. **Current** is the same in series. Therefore, we set the currents through R_{15}, R_{p1}, and R_{15} equal to one another and also set them equal to the current through the resistor that replaced them (R_{s3}).

$$I_{15} = I_{p1} = I_{15} = I_{s3} = 5.0 \text{ A}$$

We now know that $I_{15} = 5.0$ A. We also know that $R_{15} = 15\ \Omega$ (that's why we named this resistor R_{15} in our simplified circuits.) Use the appropriate power equation.

$$P_{15} = I_{15}^2 R_{15} = (5)^2(15) = (25)(15) = 375 \text{ W}$$

The power dissipated in either 15-Ω resistor is $P_{15} = 375$ W.

Example 60. Consider the circuit shown below.

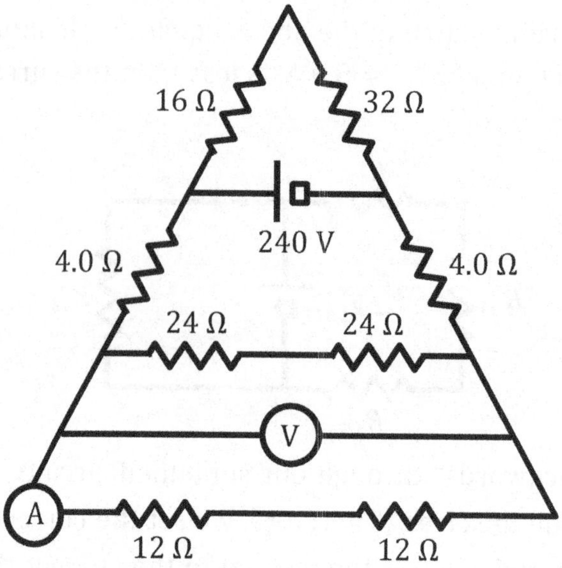

(A) Redraw the circuit, simplifying it one step at a time by identifying series and parallel combinations.

Solution. The first step is to treat the meters as follows:

- Remove the voltmeter and also remove its connecting wires.
- Remove the ammeter, patching it up with a line.

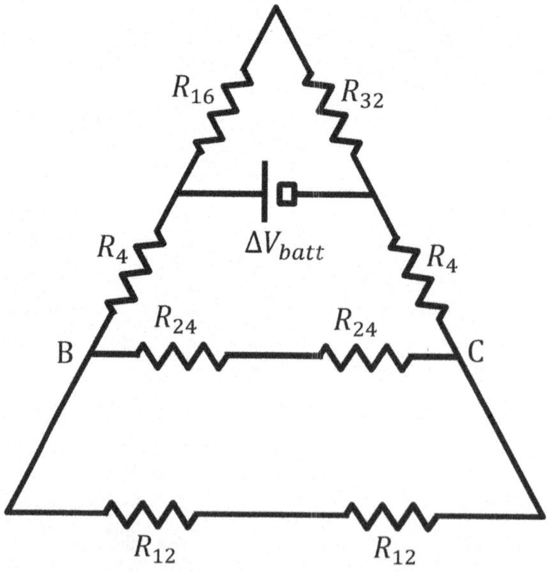

The R_{16} and R_{32} resistors are in **series** because an electron could travel from one resistor to the other without crossing a junction. They may be replaced with a single resistor, which we will call R_{s1}. Similarly, the two R_{24}'s and the two R_{12}'s are each in **series**. Each pair may be replaced with a single resistor, which we will call R_{s2} and R_{s3}.

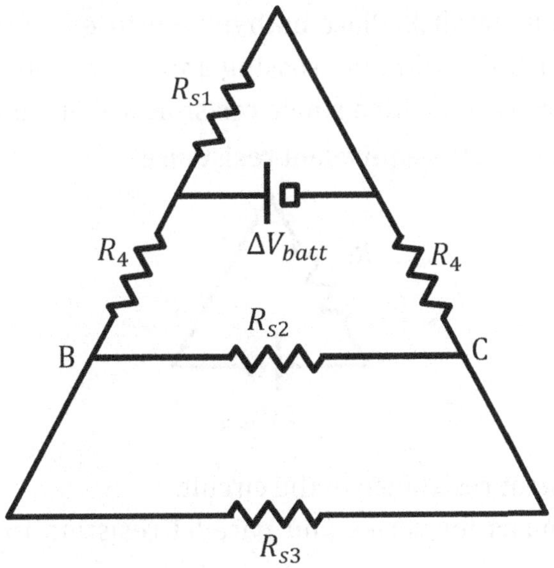

In the diagram above, R_{s2} and R_{s3} are in **parallel**. Place both of your forefingers across R_{s2}: Since **both** fingers can reach R_{s3} without crossing another resistor or the battery, they are in parallel. The R_{s2} and R_{53} resistors may be replaced with a single resistor, which we will call R_{p1}.

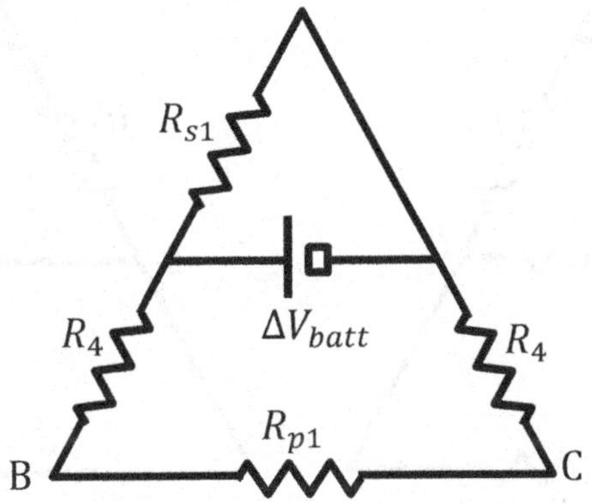

Resistors R_4, R_{p1}, and R_4 are all in **series** because an electron could travel from R_4 to R_{p1} to R_4 without crossing a junction. All three resistors may be replaced with a single resistor, which we will call R_{s4}.

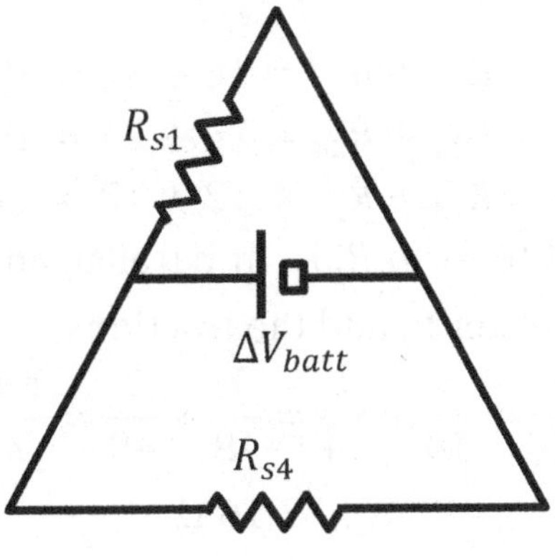

Resistors R_{s1} and R_{s4} are in **parallel**. Place both of your forefingers across the R_{s1} resistor: Since <u>**both**</u> fingers can reach R_{s4} without crossing another resistor or the battery, they are in parallel. They may be replaced with a single resistor, which we will call R_{eq}. Since this is the last resistor remaining, it is the **equivalent resistance.**

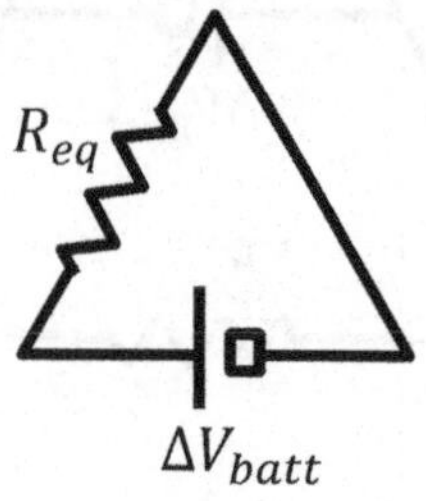

(B) Determine the equivalent resistance of the circuit.

Solution. Apply the formulas for series and parallel resistors to each reduction that we made in part (A).

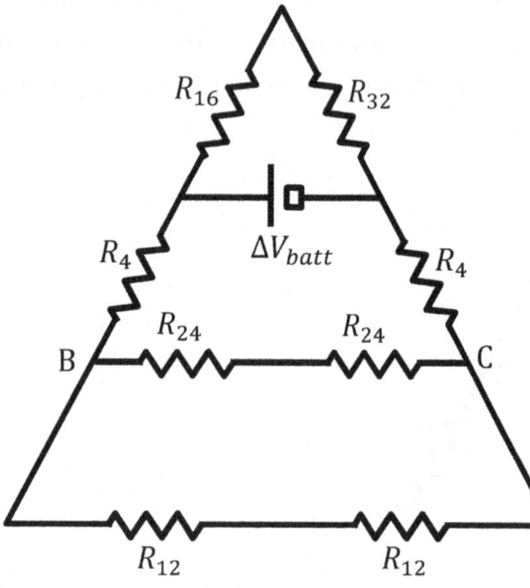

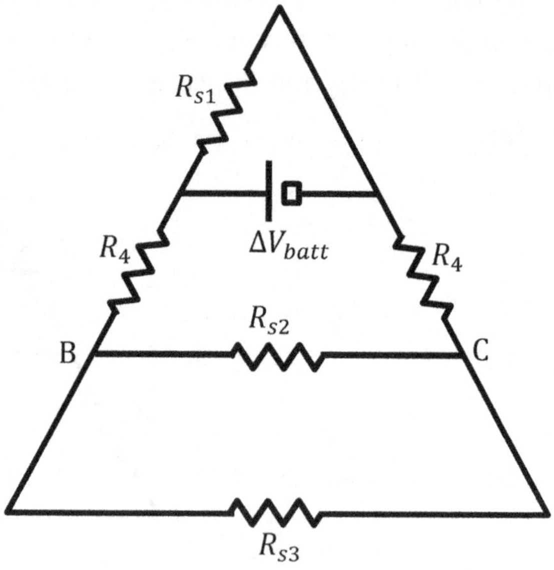

Begin by combining the 16-Ω and 32-Ω resistors in **series** to form R_{s1}, combining the two 24-Ω resistors in **series** to form R_{s2}, and combining two 12-Ω resistors in **series** to form R_{s3}. In series, add the resistances.

$$R_{s1} = R_{16} + R_{32} = 16 + 32 = 48\ \Omega$$
$$R_{s2} = R_{24} + R_{24} = 24 + 24 = 48\ \Omega$$
$$R_{s3} = R_{12} + R_{12} = 12 + 12 = 24\ \Omega$$

Combine R_{s2} and R_{s3} in **parallel** to form R_{p1}. In parallel, add the resistances in reciprocal. Find a **common denominator** in order to add the fractions.

$$\frac{1}{R_{p1}} = \frac{1}{R_{s2}} + \frac{1}{R_{s3}} = \frac{1}{48} + \frac{1}{24} = \frac{1}{48} + \frac{2}{48} = \frac{1+2}{48} = \frac{3}{48} = \frac{1}{16}$$
$$R_{p1} = 16\ \Omega$$

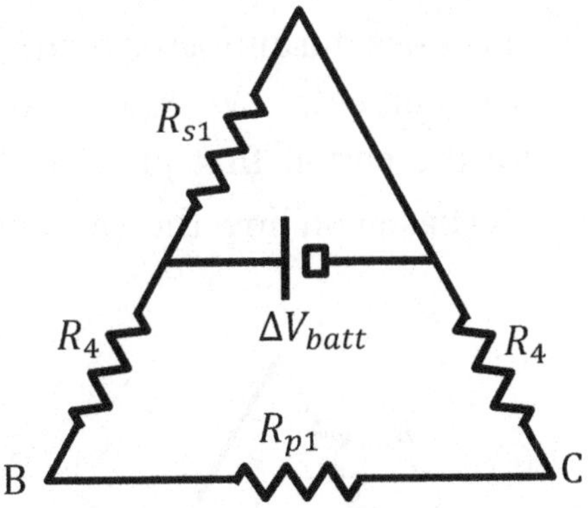

Combine R_4, R_{p1}, and R_4 in **series** to form R_{s4}.

$$R_{s4} = R_4 + R_{p1} + R_4 = 4 + 16 + 4 = 24\ \Omega$$

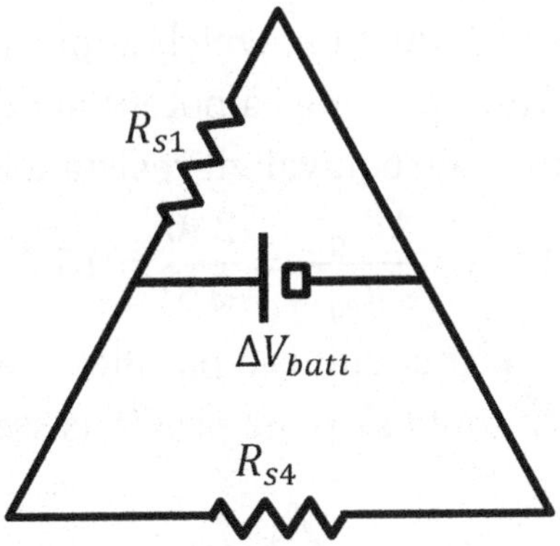

Combine R_{s1} and R_{s4} in **parallel** to form R_{eq}.

$$\frac{1}{R_{eq}} = \frac{1}{R_{s1}} + \frac{1}{R_{s4}} = \frac{1}{48} + \frac{1}{24} = \frac{1}{48} + \frac{2}{48} = \frac{1+2}{48} = \frac{3}{48} = \frac{1}{16}$$

$$R_{eq} = 16\ \Omega$$

The equivalent resistance for the circuit (that is, from one terminal of the battery to the other) is $R_{eq} = 16\ \Omega$.

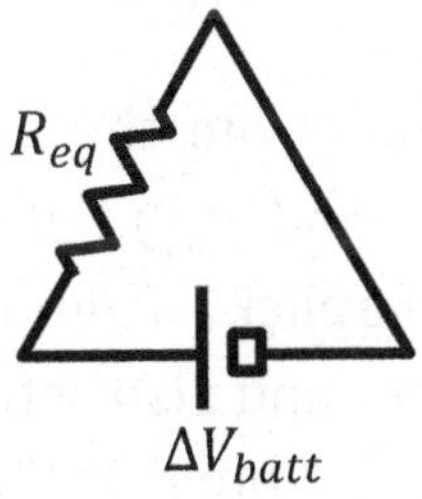

(C) What numerical value with units does the ammeter read?

Solution. The ammeter measures current. We must work "backwards" through our simplified circuits, beginning with the circuit that just has R_{eq}, in order to solve for the current through the 12-Ω resistors (based on how the ammeter is connected in the original diagram).

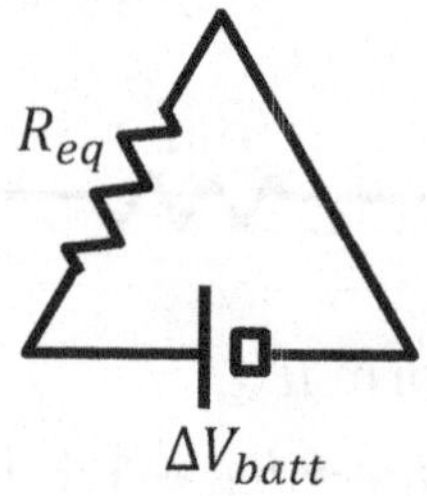

The math begins with the following equation, which applies to the last circuit. Recall from the original diagram that the battery provides a potential difference across its terminals of $\Delta V_{batt} = 240$ V. We also found that the equivalent resistance is $R_{eq} = 16\ \Omega$ in part (B).

$$I_{batt} = \frac{\Delta V_{batt}}{R_{eq}} = \frac{240}{16} = 15 \text{ A}$$

(Technically, when the second-to-last circuit is a parallel combination, which you can see is the case in the diagram below, we could skip the previous step.)

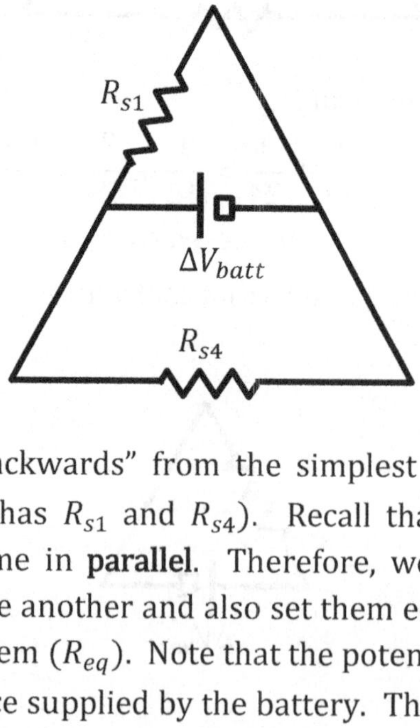

Now we will go one step "backwards" from the simplest circuit (with just R_{eq}) to the second-to-last circuit (which has R_{s1} and R_{s4}). Recall that R_{s1} and R_{s4} are in **parallel. Potential difference** is the same in **parallel**. Therefore, we set the potential differences across R_{s1} and R_{s4} equal to one another and also set them equal to the potential difference of the resistor that replaced them (R_{eq}). Note that the potential difference across R_{eq} is the same as the potential difference supplied by the battery. This is expressed in the following equation.

$$\Delta V_{s1} = \Delta V_{s4} = \Delta V_{batt} = 240 \text{ V}$$

When we set potential differences equal to one another, we must calculate current. Based on the question for part (C), which is to determine what the ammeter reads, we need I_{s4} in order to work our way to the ammeter. You can see this by studying the sequence of diagrams in part (A). Recall that $R_{s4} = 24\ \Omega$ from part (B).

$$I_{s4} = \frac{\Delta V_{s4}}{R_{s4}} = \frac{240\text{ V}}{24\ \Omega} = 10\text{ A}$$

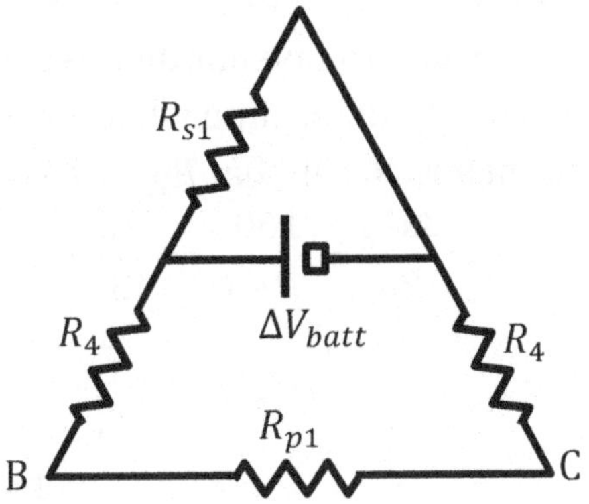

Going one more step "backwards" through our simplified circuits, R_4, R_{p1}, and R_4 are in **series. Current** is the same in series. Therefore, we set the currents through R_4, R_{p1}, and R_4 equal to one another and also set them equal to the current through the resistor that replaced them (R_{s4}).

$$I_4 = I_{p1} = I_4 = I_{s4} = 10\text{ A}$$

When we set currents equal to one another, we must calculate potential difference. Based on the question for part (C), which is to determine what the ammeter reads, we need ΔV_{p1} in order to work our way to the ammeter. Recall that $R_{p1} = 16\ \Omega$ from part (B).

$$\Delta V_{p1} = I_{p1} R_{p1} = (10)(16) = 160\text{ V}$$

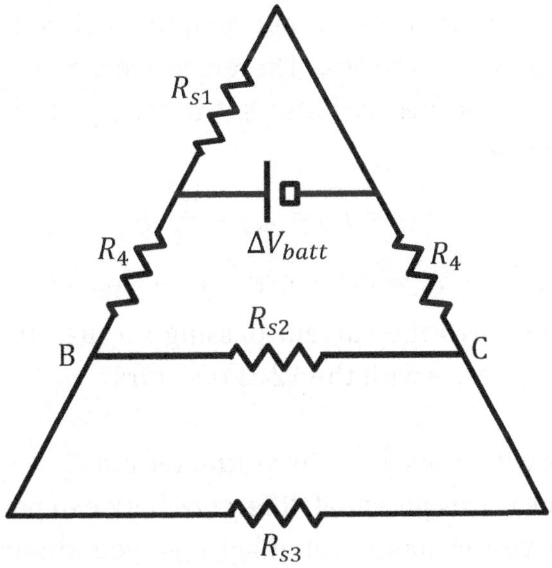

Going one more step "backwards" through our simplified circuits, R_{s2} and R_{s3} are in **parallel**. **Potential difference** is the same in parallel. Therefore, we set the potential differences across R_{s2} and R_{s3} equal to one another and also set them equal to the potential difference across the resistor that replaced them (R_{p1}).

$$\Delta V_{s2} = \Delta V_{s3} = \Delta V_{p1} = 160 \text{ V}$$

When we set potential differences equal to one another, we must calculate current. Based on the question for part (C), which is to determine what the ammeter reads, we need I_{s3} in order to work our way to the ammeter. Recall that $R_{s3} = 24\ \Omega$ from part (B).

$$I_{s3} = \frac{\Delta V_{s3}}{R_{s3}} = \frac{160 \text{ V}}{24\ \Omega} = \frac{20}{3} \text{ A}$$

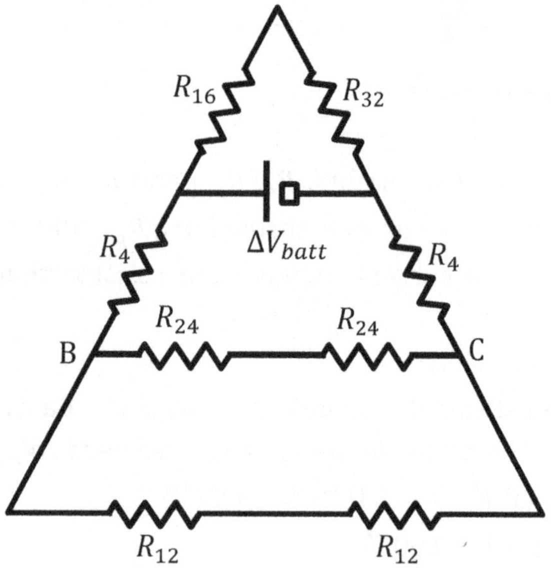

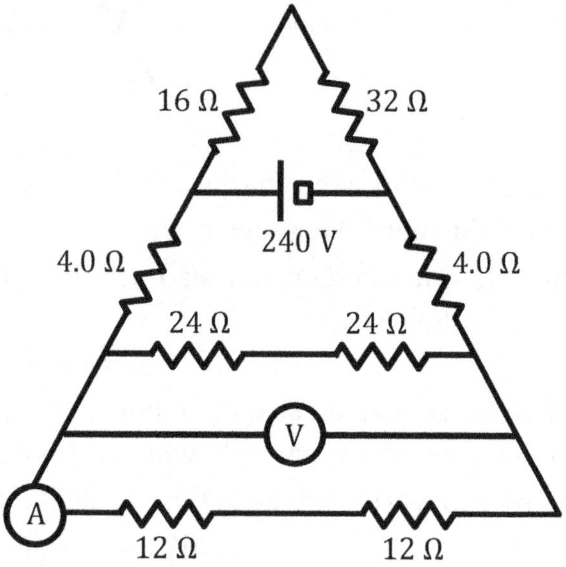

Going one more step "backwards" through our simplified circuits, the two 12-Ω resistors are in **series**. **Current** is the same in series. Therefore, we set the currents through the two 12-Ω resistors equal to one another and also set them equal to the current through the resistor that replaced them (R_{s3}).

$$I_{12} = I_{12} = I_{s3} = \frac{20}{3} \text{ A}$$

The ammeter measures $I_{12} = \frac{20}{3}$ A, which works out to $I_{12} = 6.7$ A if you use a calculator. Note that the ammeter measures the current passing through the 12-Ω resistors because the ammeter is connected in series with the 12-Ω resistors.

(D) What numerical value with units does the voltmeter read?

Solution. The **voltmeter** measures **potential difference** between points B and C (since that is how it is connected). If you examine our diagrams, you should see that the potential difference between points B and C is the same as ΔV_{p1}, which we determined to equal $\Delta V_{p1} = 160$ V in part (C). Therefore, the voltmeter reads $\Delta V_{p1} = 160$ V.

Example 61. Determine the equivalent resistance of the circuit shown below.

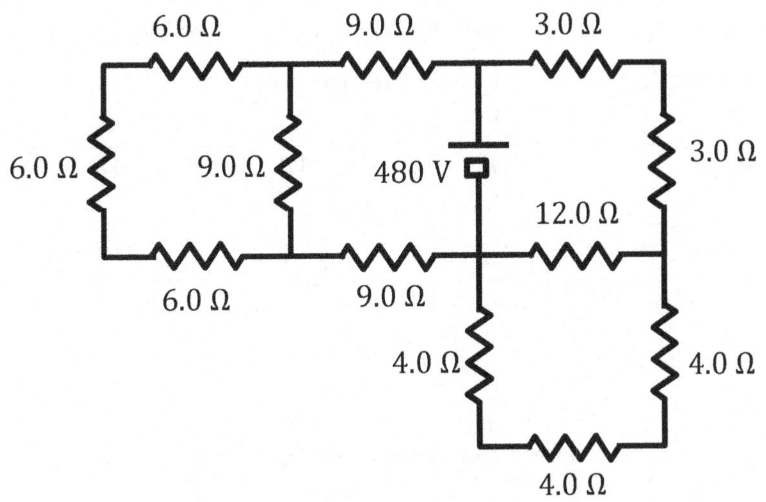

Solution. Redraw the circuit, simplifying it one step at a time by identifying series and parallel combinations. Apply the formulas for series and parallel resistors to each reduced circuit. The three 6-Ω are in **series** because an electron could travel from one resistor to the other without crossing a junction. They may be replaced with a single resistor, which we will call R_{s1}. Similarly, the three 4-Ω resistors are in **series**. They may be replaced with a single resistor, which we will call R_{s2}. (Although the two 3-Ω resistors are in series, it will be convenient to save these for a later step.) In series, add the resistances.

$$R_{s1} = R_6 + R_6 + R_6 = 6 + 6 + 6 = 18\ \Omega$$
$$R_{s2} = R_4 + R_4 + R_4 = 4 + 4 + 4 = 12\ \Omega$$

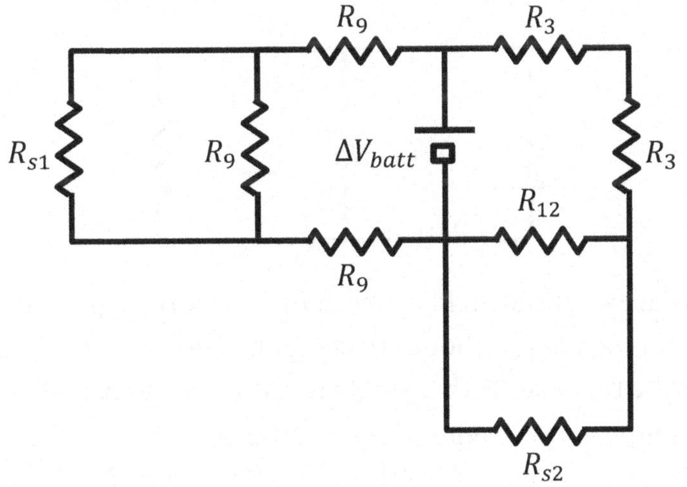

In the diagram above, R_{s1} and the left R_9 are in **parallel**. Place both of your forefingers across R_{s1}: Since <u>**both**</u> fingers can reach R_9 without crossing another resistor or the battery, they are in parallel. The R_{s1} and left R_9 resistors may be replaced with a single resistor, which we will call R_{p1}. Similarly, R_{12} and R_{s2} are in **parallel**. They may be replaced with a single resistor, which we will call R_{p2}. In parallel, add the resistances in reciprocal. Find a **common denominator** in order to add the fractions.

$$\frac{1}{R_{p1}} = \frac{1}{R_{s1}} + \frac{1}{R_9} = \frac{1}{18} + \frac{1}{9} = \frac{1}{18} + \frac{2}{18} = \frac{1+2}{18} = \frac{3}{18} = \frac{1}{6}$$

$$R_{p1} = 6\ \Omega$$

$$\frac{1}{R_{p2}} = \frac{1}{R_{12}} + \frac{1}{R_{s2}} = \frac{1}{12} + \frac{1}{12} = \frac{1+1}{12} = \frac{2}{12} = \frac{1}{6}$$

$$R_{p2} = 6\ \Omega$$

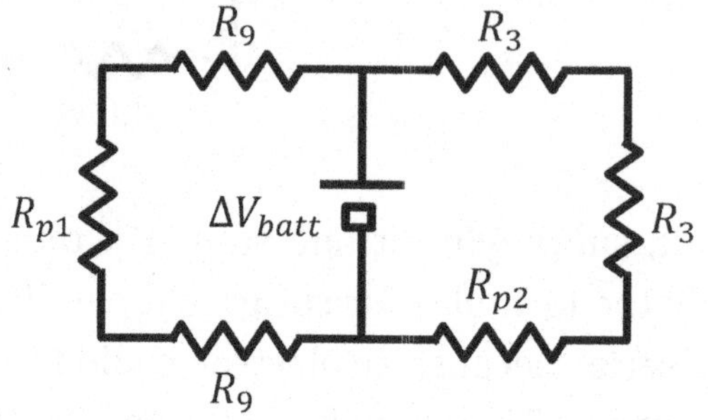

Resistors R_9, R_{p1}, and R_9 are all in **series** because an electron could travel from R_9 to R_{p1} to R_9 without crossing a junction. All three resistors may be replaced with a single resistor, which we will call R_{s3}. Similarly, R_3, R_3, and R_{p2} are all in **series**. These three resistors may also be replaced with a single resistor, which we will call R_{s4}.

$$R_{s3} = R_9 + R_{p1} + R_9 = 9 + 6 + 9 = 24\ \Omega$$

$$R_{s4} = R_3 + R_3 + R_{p2} = 3 + 3 + 6 = 12\ \Omega$$

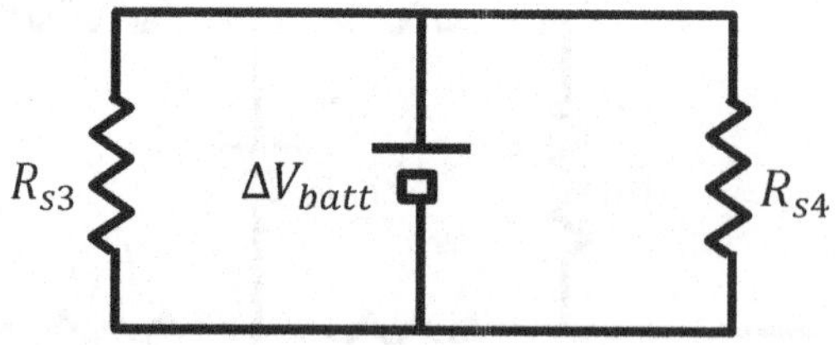

Resistors R_{s3} and R_{s4} are in **parallel**. Place both of your forefingers across the R_{s3} resistor: Since **both** fingers can reach R_{s4} without crossing another resistor or the battery, they are in parallel. They may be replaced with a single resistor, which we will call R_{eq}. Since this is the last resistor remaining, it is the **equivalent resistance**.

$$\frac{1}{R_{eq}} = \frac{1}{R_{s3}} + \frac{1}{R_{s4}} = \frac{1}{24} + \frac{1}{12} = \frac{1}{24} + \frac{2}{24} = \frac{1+2}{24} = \frac{3}{24} = \frac{1}{8}$$

$$R_{eq} = 8.0\ \Omega$$

The equivalent resistance for the circuit is $R_{eq} = 8.0\ \Omega$.

14 CIRCUITS WITH SYMMETRY

Ohm's Law

$$\Delta V = IR \quad , \quad I = \frac{\Delta V}{R} \quad , \quad R = \frac{\Delta V}{I}$$

Resistors in Series

$$R_s = R_1 + R_2 + \cdots + R_N$$

Resistors in Parallel

$$\frac{1}{R_p} = \frac{1}{R_1} + \frac{1}{R_2} + \cdots + \frac{1}{R_N}$$

Power

$$P = I\Delta V \quad , \quad P = I^2 R \quad , \quad P = \frac{\Delta V^2}{R}$$

Schematic Representation	Symbol	Name
	R	resistor
	ΔV	battery or DC power supply

Symbol	Name	SI Units
R	resistance	Ω
I	electric current	A
ΔV	the potential difference between two points in a circuit	V
P	electric power	W

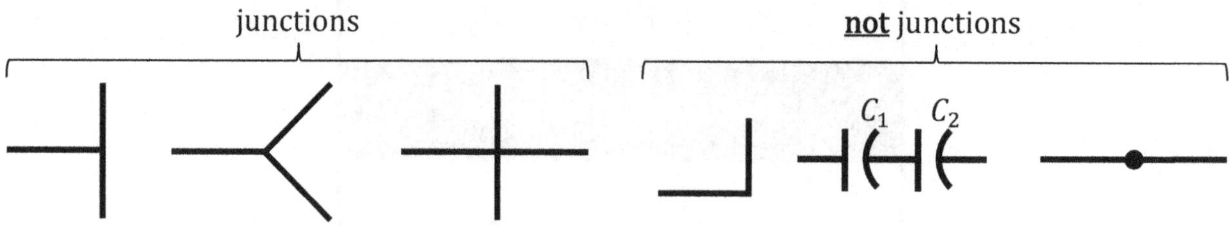

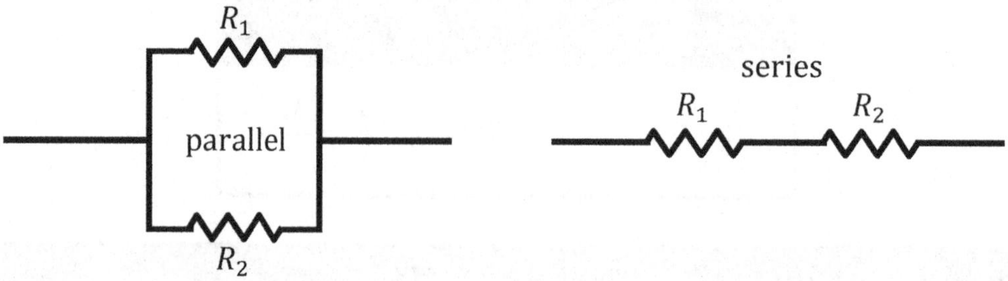

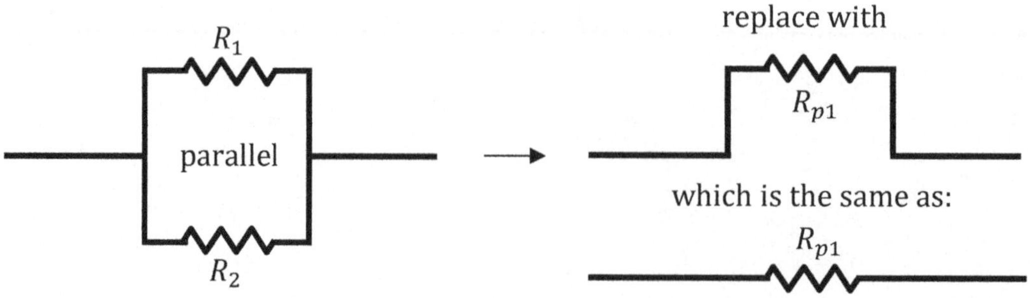

Conceptual Symmetry Example: A Prelude to Examples 62-63. A wire only impacts the equivalent resistance of a circuit if current flows through it.

- If there is **no current** flowing through a wire, the wire may be removed from the circuit without affecting the equivalent resistance.
- If a wire is added to a circuit in such a way that **no current** flows through the wire, the wire's presence won't affect the equivalent resistance.

Let's apply Ohm's law to such a wire. The potential difference across the length of the wire equals the current through the wire times the resistance of the wire. (Although the wire's resistance may be small compared to other resistances in the circuit, every wire does have some resistance.)

$$\Delta V_{wire} = I_{wire} R_{wire}$$

If the **potential difference** across the wire is **zero**, Ohm's law tells us that **there won't be any current in the wire.**

With a symmetric circuit, we consider the electric potential (V) at points between resistors, and try to find two such points that definitely have the **same electric potential**. If two points have the same electric potential, then the potential difference between those two points will be zero. We may then add or remove a wire between those two points without disturbing the equivalent resistance of the circuit.

For a circuit with a single power supply, electric potential is highest at the positive terminal and lowest at the negative terminal. Our goal is to find two points between resistors that are equal percentages – in terms of **electric potential, <u>not</u>** in terms of distance – between the two terminals of the battery.

For example, in the circuit on the left below, points B and C are each exactly halfway (in terms of electric potential) from the negative terminal to the positive terminal. In the circuit in the middle, points F and G are each one-third of the way from the negative to the positive, since 4 Ω is one-third of 12 Ω (note that 4 Ω + 8 Ω = 12 Ω). In the circuit on the right, J and K are also each one-third of the way from the negative to the positive, since 2 Ω is one-third of 6 Ω (note that 2 Ω + 4 Ω = 6 Ω) and 4 Ω is one-third of 12 Ω.

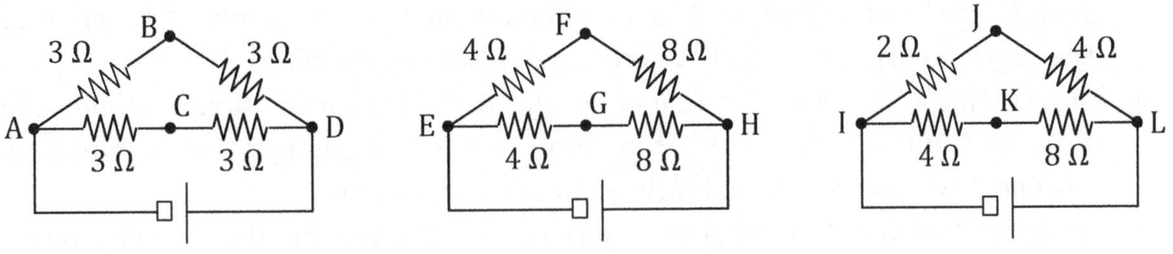

Body Diagonal Example: A Prelude to Example 62. Twelve identical 12-Ω resistors are joined together to form a cube, as illustrated below. If a battery is connected by joining its negative terminal to point F and its positive terminal to point C, what will be the equivalent resistance of the cube? (This connection is across a **body diagonal** between opposite corners of the cube.)

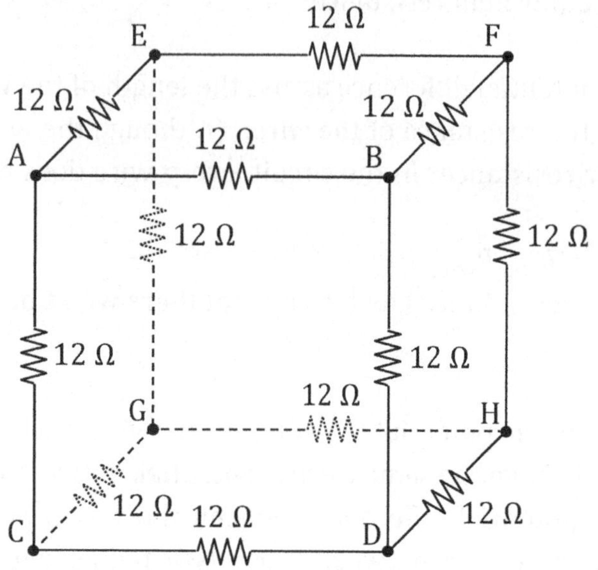

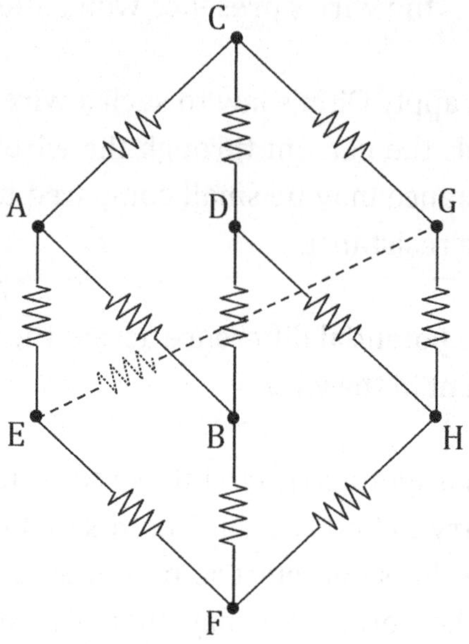

Solution. Note that no two resistors presently appear to be in series or parallel.

- Since there is a junction between any two resistors, none are in series.
- If you try to find a parallel combination, you should find that another resistor always gets in the way. (It may help to study the second diagram from the top on page 143.) No two resistors currently appear to be in parallel.

Fortunately, there is enough symmetry in the circuit to find points with the same potential.

We will "unfold" the circuit with point F at the bottom (call it the "ground") and point C at the top (call it the "roof"). This is based on how the battery is connected.

- Points E, B, and H are each one step from point F (the "ground") and two steps from point C (the "roof"). Neither of these points is closer to the "ground" or the "roof." Therefore, points E, B, and H have the same electric potential.
- Points A, D, and G are each two steps from point F (the "ground") and one step from point C (the "roof"). Neither of these points is closer to the "ground" or the "roof." Therefore, points A, D, and G have the same electric potential.
- Draw points E, B, and H at the same "height" in the "unfolded" circuit. Draw points A, D, and G at the same "height," too. Draw points E, B, and H closer to point F (the "ground") and points A, D, and G closer to point C (the "roof").
- Study the "unfolded" circuit at the top right and compare it to the original circuit at the top left. Try to understand the reasoning behind how it was drawn.

Consider points E, B, and H, which have the same electric potential:

- There are presently no wires connecting these three points.
- Therefore, we will add wires to connect points E, B, and H.
- Make these new wires so short that you have to move points E, B, and H toward one another. Make it so extreme that points E, B, and H merge into a single point, which we will call EBH.

Do the same thing with points A, D, and G, merging them into point ADG. With these changes, the diagram from the top right of the previous page turns into the diagram at the left below. Study the two diagrams to try to understand the reasoning behind it.

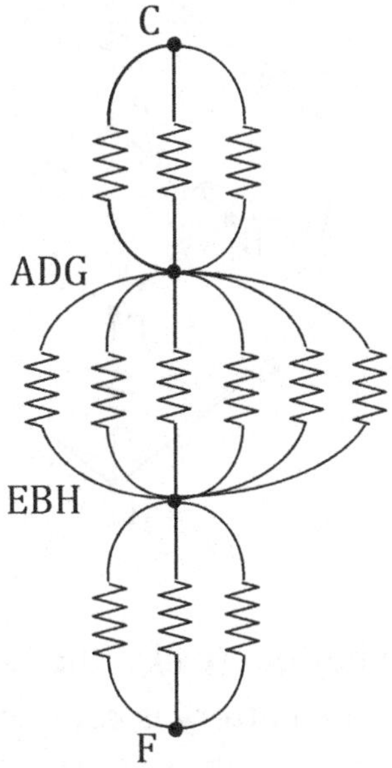

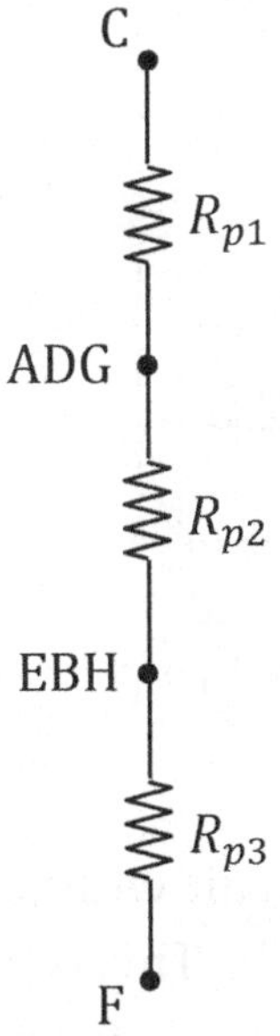

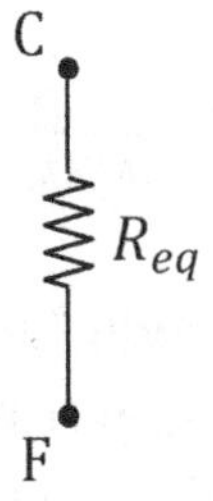

It's a good idea to count corners and resistors to make sure you don't forget one:

- The cube has 8 corners: A, B, C, D, E, F, G, and H.
- The cube has 12 resistors: one along each edge.

If you count, you should find 12 resistors in the left diagram above. The top 3 are in parallel (R_{p1}), the middle 6 are in parallel (R_{p2}), and the bottom 3 are in parallel (R_{p3}).

$$\frac{1}{R_{p1}} = \frac{1}{12} + \frac{1}{12} + \frac{1}{12} = \frac{3}{12} = \frac{1}{4} \quad \Rightarrow \quad R_{p1} = 4.0\ \Omega$$

$$\frac{1}{R_{p2}} = \frac{1}{12} + \frac{1}{12} + \frac{1}{12} + \frac{1}{12} + \frac{1}{12} + \frac{1}{12} = \frac{6}{12} = \frac{1}{2} \quad \Rightarrow \quad R_{p2} = 2.0\ \Omega$$

$$\frac{1}{R_{p3}} = \frac{1}{12} + \frac{1}{12} + \frac{1}{12} = \frac{3}{12} = \frac{1}{4} \quad \Rightarrow \quad R_{p3} = 4.0\ \Omega$$

After drawing the reduced circuit, R_{p1}, R_{p2}, and R_{p3} are in series, forming R_{eq}.

$$R_{eq} = R_{p1} + R_{p2} + R_{p3} = 4 + 2 + 4 = 10.0\ \Omega$$

The equivalent resistance of the cube along a **body diagonal** is $R_{eq} = 10.0\ \Omega$.

Face Diagonal Example: Another Prelude to Example 62. Twelve identical 12-Ω resistors are joined together to form a cube, as illustrated below. If a battery is connected by joining its negative terminal to point H and its positive terminal to point C, what will be the equivalent resistance of the cube? (This connection is across a **face diagonal** between opposite corners of one square face of the cube. Contrast this with the previous example. It will make a huge difference in the solution.)

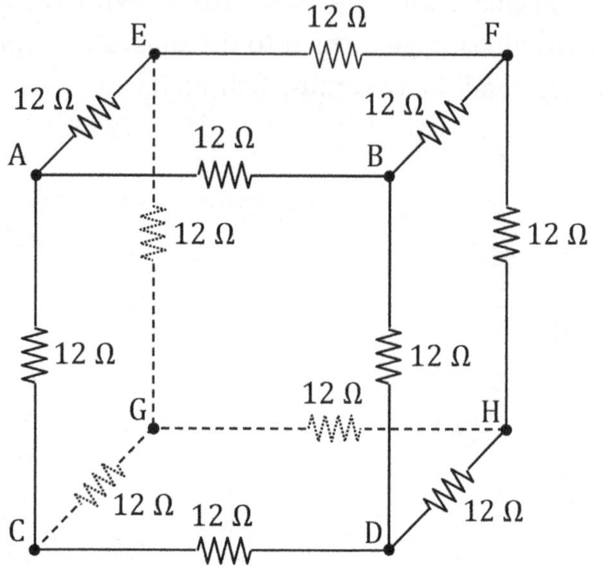

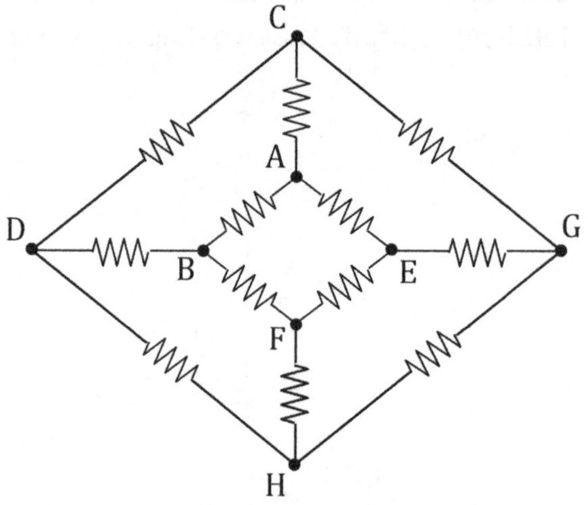

Solution. We will "unfold" the circuit with point H at the bottom (call it the "ground") and point C at the top (call it the "roof"). This is based on how the battery is connected.

- Points G and D are each one step from point H (the "ground") and also one step from point C (the "roof"). Each of these points is exactly halfway between the "ground" and the "roof." Therefore, points D and G have the same electric potential.
- Points B and E are each two steps from point H (the "ground") and also two steps from point C (the "roof"). Each of these points is exactly halfway between the "ground" and the "roof." Therefore, points B and E have the same electric potential.
- Furthermore, points G, D, B, and E all have the same electric potential, since we have reasoned that all 4 points are exactly halfway between the "ground" and the "roof."
- Draw points G, D, B, and E at the same height in the "unfolded" circuit.
- Draw point F closer to point H (the "ground") and point A closer to point C (the "roof").

Study the "unfolded" circuit at the top right and compare it to the original circuit at the top left. Try to understand the reasoning behind how it was drawn.

Count corners and resistors to make sure you don't forget one:

- See if you can find all 8 corners (A, B, C, D, E, F, G, and H) in the "unfolded" circuit.
- See if you can find all 12 resistors in the "unfolded" circuit.

Consider points G, D, B, and E, which have the same electric potential:

- There are presently resistors between points D and B and also between E and G.
- Therefore, we will remove these two wires.

With these changes, the diagram from the top right of the previous page turns into the diagram at the left below.

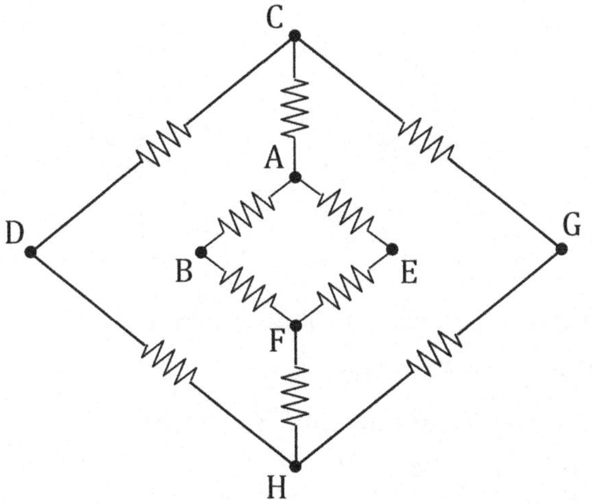

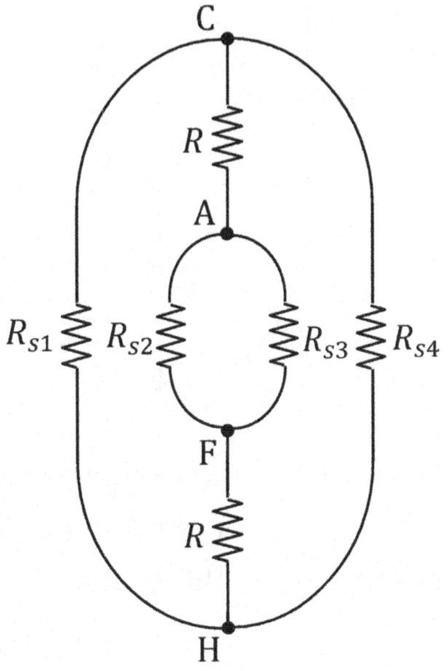

In the diagram above on the left:

- The 2 resistors from H to D and D to C are in series. They form R_{s1}.
- The 2 resistors from F to B and B to A are in series. They form R_{s2}.
- The 2 resistors from F to E and E to A are in series. They form R_{s3}.
- The 2 resistors from H to G and G to C are in series. They form R_{s4}.
- Note that points D, B, E and G are **not** junctions now that the wires connecting D to B and E to G have been removed.

$$R_{s1} = 12 + 12 = 24.0\ \Omega$$
$$R_{s2} = 12 + 12 = 24.0\ \Omega$$
$$R_{s3} = 12 + 12 = 24.0\ \Omega$$
$$R_{s4} = 12 + 12 = 24.0\ \Omega$$

In the diagram above on the right:

- R_{s1} and R_{s4} are in parallel. They form R_{p1}.
- R_{s2} and R_{s3} are in parallel. They form R_{p2}.

$$\frac{1}{R_{p1}} = \frac{1}{R_{s1}} + \frac{1}{R_{s4}} = \frac{1}{24} + \frac{1}{24} = \frac{2}{24} = \frac{1}{12} \quad \Rightarrow \quad R_{p1} = 12.0\ \Omega$$
$$\frac{1}{R_{p2}} = \frac{1}{R_{s2}} + \frac{1}{R_{s3}} = \frac{1}{24} + \frac{1}{24} = \frac{2}{24} = \frac{1}{12} \quad \Rightarrow \quad R_{p2} = 12.0\ \Omega$$

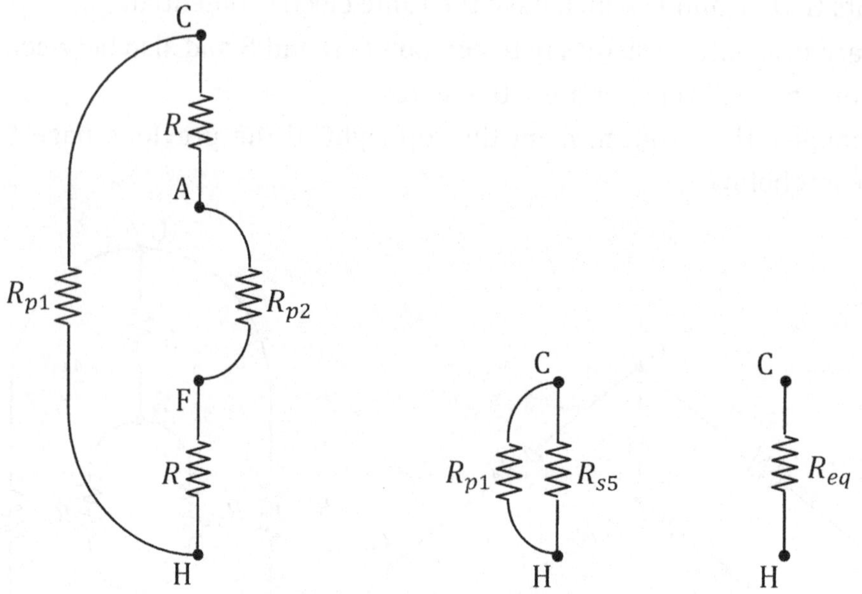

In the diagram above on the left, R, R_{p2}, and R are in series. They form R_{s5}.

$$R_{s5} = R + R_{p2} + R = 12 + 12 + 12 = 36.0\ \Omega$$

In the diagram above in the middle, R_{p1} and R_{s5} are in parallel. They form R_{eq}.

$$\frac{1}{R_{eq}} = \frac{1}{R_{p1}} + \frac{1}{R_{s5}} = \frac{1}{12} + \frac{1}{36} = \frac{3}{36} + \frac{1}{36} = \frac{4}{36} = \frac{1}{9}$$

$$R_{eq} = 9.0\ \Omega$$

The equivalent resistance of the cube along a **face diagonal** is $R_{eq} = 9.0\ \Omega$.

Example 62. Twelve identical 12-Ω resistors are joined together to form a cube, as illustrated below. If a battery is connected by joining its negative terminal to point D and its positive terminal to point C, what will be the equivalent resistance of the cube? (This connection is across an **edge**. Contrast this with the previous examples. It will make a significant difference in the solution.)

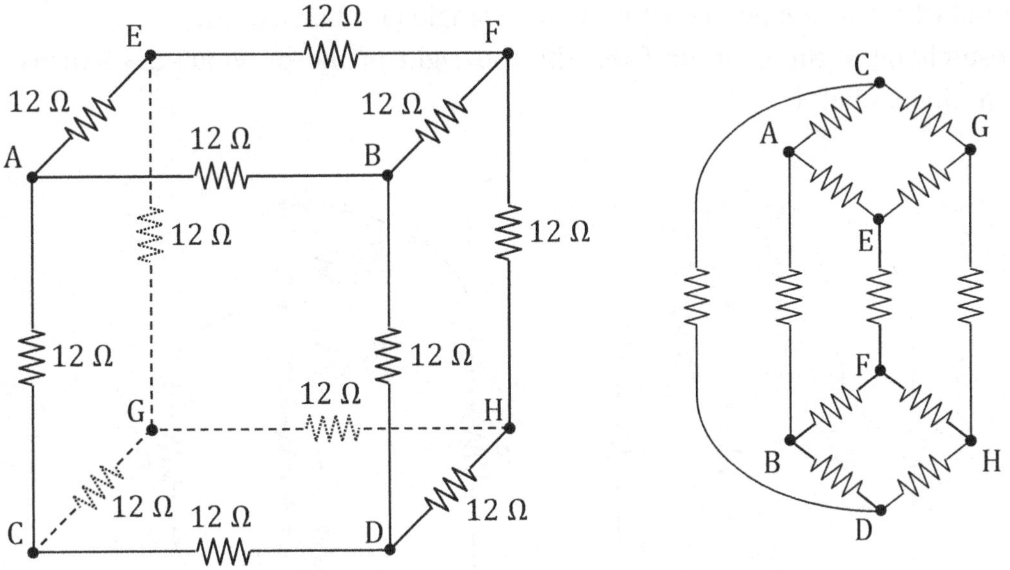

Solution. We will "unfold" the circuit with point D at the bottom (call it the "ground") and point C at the top (call it the "roof"). This is based on how the battery is connected.

- Points B and H are each one step from point D (the "ground") and two steps from point C (the "roof"). Points B and H have the same electric potential.
- Points A and G are each two steps from point D (the "ground") and one step from point C (the "roof"). Points A and G have the same electric potential.
- Draw points B and H at the same height in the "unfolded" circuit.
- Draw points A and G at the same height in the "unfolded" circuit.
- Draw points B and H closer to point D (the "ground") and points A and G closer to point C (the "roof").
- Between those two pairs (B and H, and A and G), draw point F closer to point D (the "ground") and point E closer to point C (the "roof").

Study the "unfolded" circuit at the top right and compare it to the original circuit at the top left. Try to understand the reasoning behind how it was drawn.

Count corners and resistors to make sure you don't forget one:

- See if you can find all 8 corners (A, B, C, D, E, F, G, and H) in the "unfolded" circuit.
- See if you can find all 12 resistors in the "unfolded" circuit.

Consider points B and H, which have the same electric potential, and also consider points A and G, which have the same electric potential:

- Points B and H have the same electric potential. There are presently no wires connecting points B and H. Add wires to connect points B and H. Make these new wires so short that you have to move points B and H toward one another. Make it so extreme that points B and H merge into a single point, which we will call BH.
- Similarly, collapse points A and G into a single point called AG.

With these changes, the diagram from the top right of the previous page turns into the diagram at the left below.

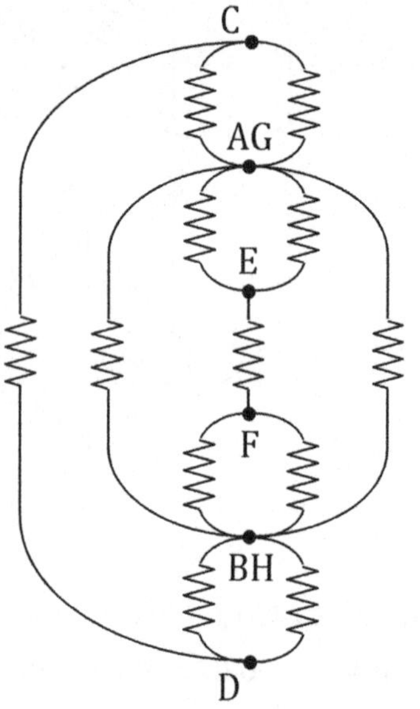

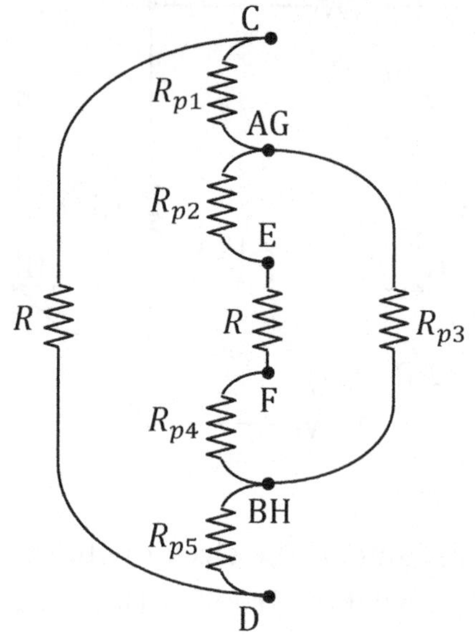

There are 5 pairs of parallel resistors in the diagram above on the left:

- The two resistors between C and AG form R_{p1}.
- The two resistors between AG and E form R_{p2}.
- The two resistors between AG and BH form R_{p3}.
- The two resistors between F and BH form R_{p4}.
- The two resistors between BH and D form R_{p5}.

$$\frac{1}{R_{p1}} = \frac{1}{12} + \frac{1}{12} = \frac{2}{12} = \frac{1}{6} \quad \Rightarrow \quad R_{p1} = 6.0\ \Omega$$

$$R_{p1} = R_{p2} = R_{p3} = R_{p4} = R_{p5} = 6.0\ \Omega$$

In the diagram above on the right, R_{p2}, R, and R_{p4} are in series. They form R_{s1}.

$$R_{s1} = R_{p2} + R + R_{p4} = 6 + 12 + 6 = 24.0\ \Omega$$

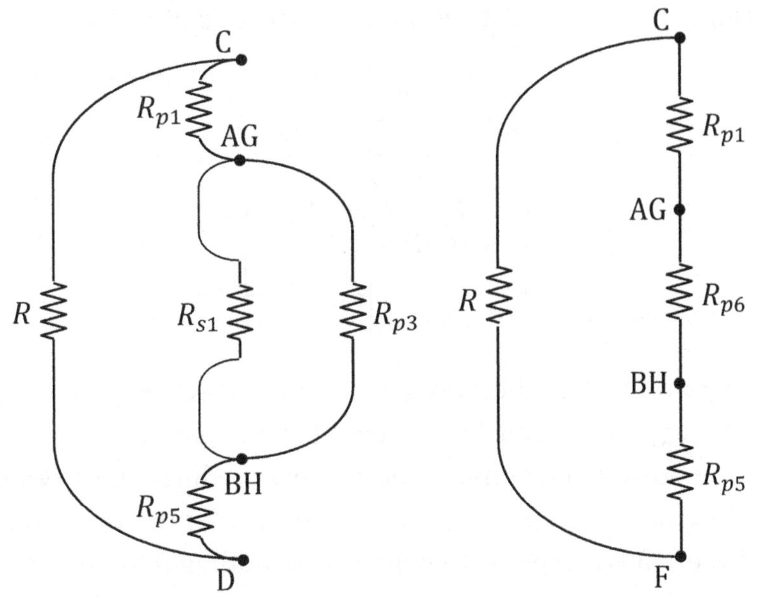

In the diagram above on the left, R_{s1} and R_{p3} are in parallel. They form R_{p6}.

$$\frac{1}{R_{p6}} = \frac{1}{R_{s1}} + \frac{1}{R_{p3}} = \frac{1}{24} + \frac{1}{6} = \frac{1}{24} + \frac{4}{24} = \frac{5}{24} \quad \Rightarrow \quad R_{p1} = \frac{24}{5}\ \Omega$$

In the diagram above on the right, R_{p1}, R_{p6}, and R_{p5} are in series. They form R_{s2}.

$$R_{s2} = R_{p1} + R_{p6} + R_{p5} = 6 + \frac{24}{5} + 6 = \frac{30}{5} + \frac{24}{5} + \frac{30}{5} = \frac{84}{5}\ \Omega$$

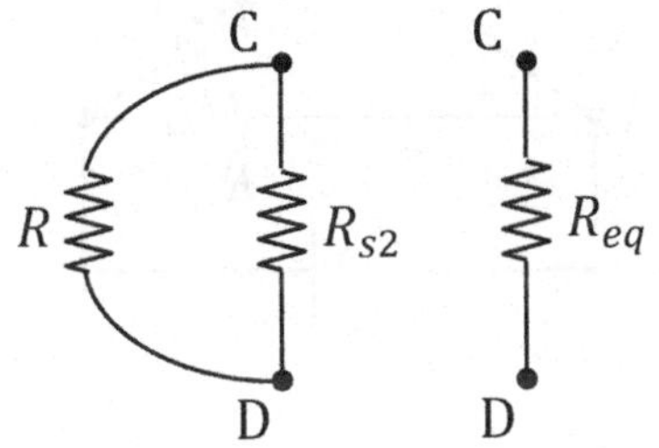

In the diagram above on the left, R and R_{s2} are in parallel. They form R_{eq}.

$$\frac{1}{R_{eq}} = \frac{1}{R} + \frac{1}{R_{s2}} = \frac{1}{84/5} + \frac{1}{12} = \frac{5}{84} + \frac{1}{12} = \frac{5}{84} + \frac{7}{84} = \frac{12}{84} = \frac{1}{7}$$

$$R_{eq} = 7.0\ \Omega$$

Note that $\frac{1}{84/5} = 5/84$ (to divide by a fraction, multiply by its **reciprocal**). The equivalent resistance of the cube along an **edge** is $R_{eq} = 7.0\ \Omega$.

Example 63. Determine the equivalent resistance of the circuit shown below.

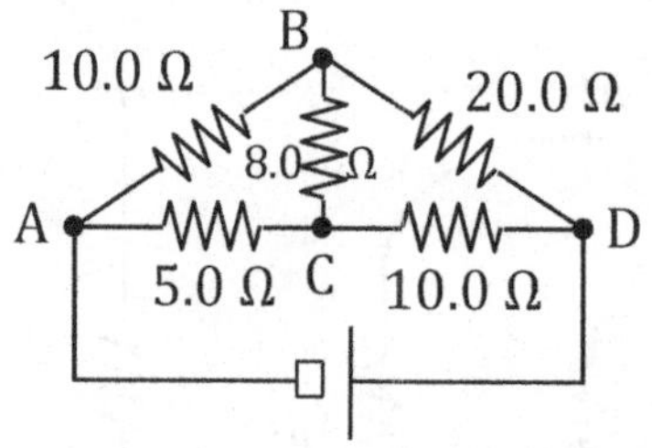

Solution. Point B is one-third of the way (in terms of electric potential, **not** in terms of distance) from point A (the "ground," since it's at the negative terminal) to point D (the "roof," since it's at the positive terminal), since 10 is one third of 30 (where the 30 comes from adding 10 to 20). Similarly, point C is also one-third of the way from point A to point D, since 5 is one third of 15 (where the 15 comes from adding 5 to 10).

Therefore, points B and C have the same electric potential: $V_B = V_C$. The potential difference between points B and C is zero: $\Delta V_{BC} = V_C - V_B = 0$. From Ohm's law, $\Delta V_{BC} = I_{BC} R_{BC}$. Since $\Delta V_{BC} = 0$, the current from B to C must be zero: $I_{BC} = 0$. Since there is no current in the 8.0-Ω resistor, we may remove this wire without affecting the equivalent resistance of the circuit. See the diagram below.

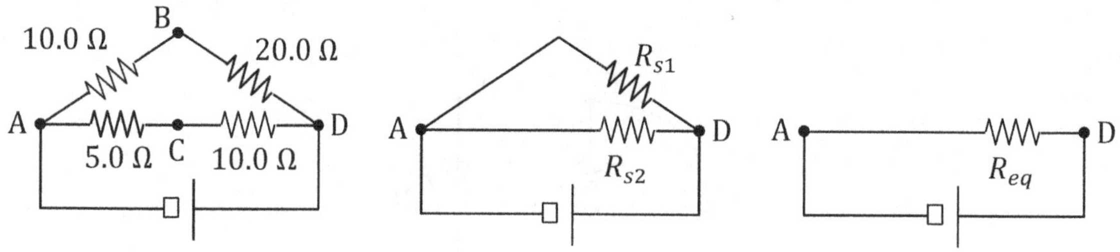

In the diagram above on the left:

- The 10.0-Ω and 20.0-Ω resistors are in series. They form R_{s1}.
- The 5.0-Ω and 10.0-Ω resistors are in series. They form R_{s2}.

$$R_{s1} = 10 + 20 = 30.0\ \Omega$$
$$R_{s2} = 5 + 10 = 15.0\ \Omega$$

In the center diagram above, R_{s1} and R_{s2} are in parallel.

$$\frac{1}{R_{eq}} = \frac{1}{R_{s1}} + \frac{1}{R_{s2}} = \frac{1}{30} + \frac{1}{15} = \frac{1}{30} + \frac{2}{30} = \frac{3}{30} = \frac{1}{10}$$
$$R_{eq} = 10.0\ \Omega$$

The equivalent resistance for the circuit (that is, from one terminal of the battery to the other) is $R_{eq} = 10.0\ \Omega$.

15 KIRCHHOFF'S RULES

Kirchhoff's Junction Rule	Kirchhoff's Loop Rule
$\sum_{entering} I_i = \sum_{exiting} I_j$	$\sum_{loop} \Delta V_i = 0$

Symbol	Name	SI Units
R	resistance	Ω
I	electric current	A
ΔV	the potential difference between two points in a circuit	V
P	electric power	W

Schematic Representation	Symbol	Name
	R	resistor
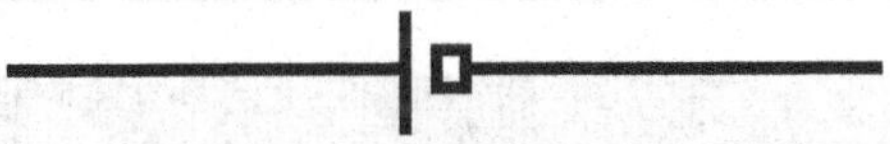	ΔV	battery or DC power supply

Note: The long line of the battery symbol represents the **positive** terminal.

Schematic Representation	Symbol	Name
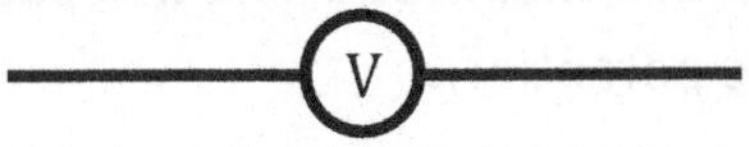	measures ΔV	voltmeter
	measures I	ammeter

Junction Examples

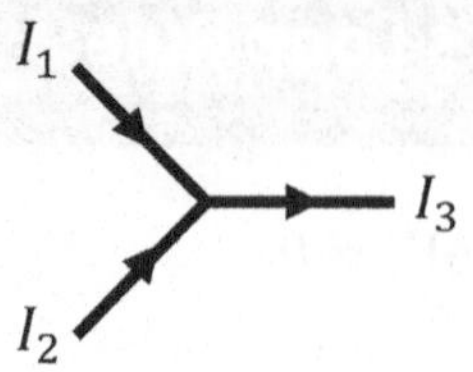

$I_1 + I_2 = I_3$

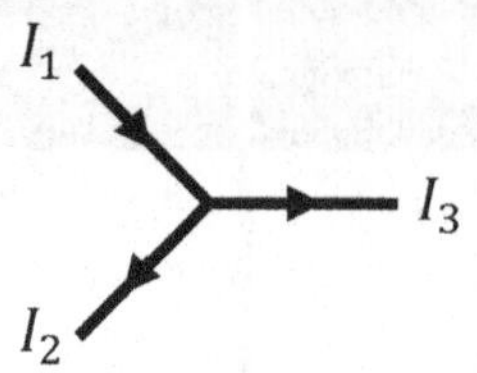

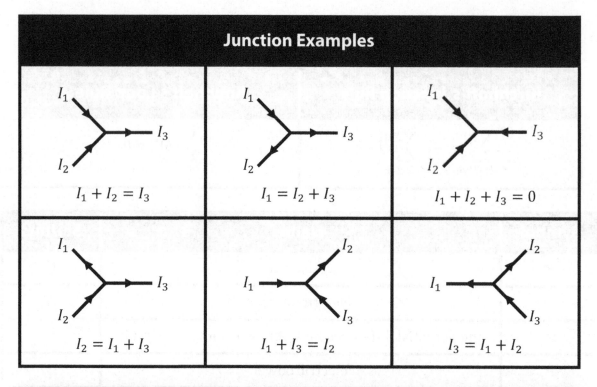

$I_1 = I_2 + I_3$

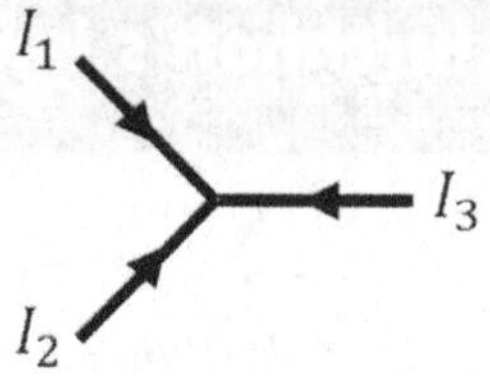

$I_1 + I_2 + I_3 = 0$

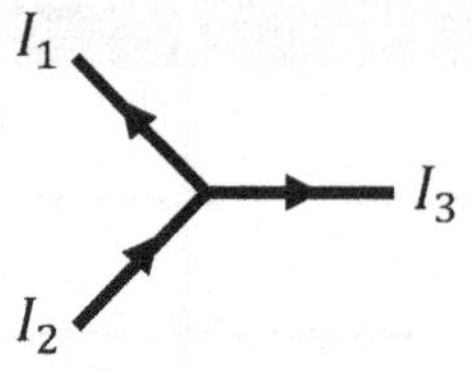

$I_2 = I_1 + I_3$

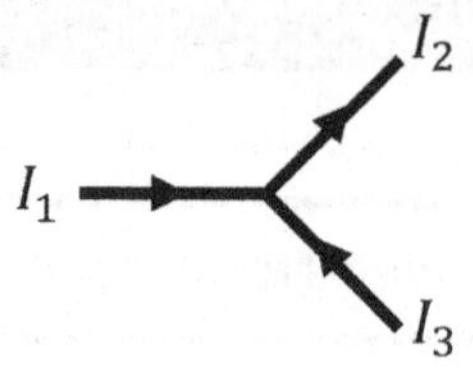

$I_1 + I_3 = I_2$

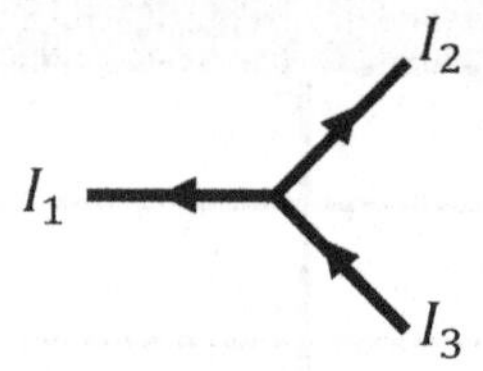

$I_3 = I_1 + I_2$

Current	Sense of Traversal
 solid arrow	 dashed arrow

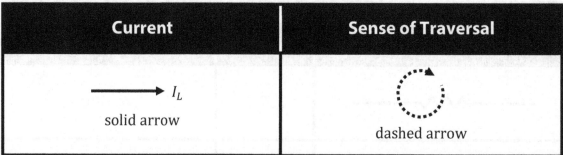

Sign Conventions for Loops

When your "test" charge comes to a battery:

- If the "test" charge comes to the **negative** terminal first, write a **positive** potential difference (since this is a rise in electric potential).
- If the "test" charge comes to the **positive** terminal first, write a **negative** potential difference (since this is a drop in electric potential).

When your "test" charge comes to a resistor:

- If the "test" charge is "swimming" opposite to the current, write $+IR$ (since it rises in potential when it swims "upstream" against the current).
- If the "test" charge is "swimming" in the same direction as the current, write $-IR$ (since it drops in potential when it swims "downstream" with the current).

Example 64. Consider the circuit illustrated below.

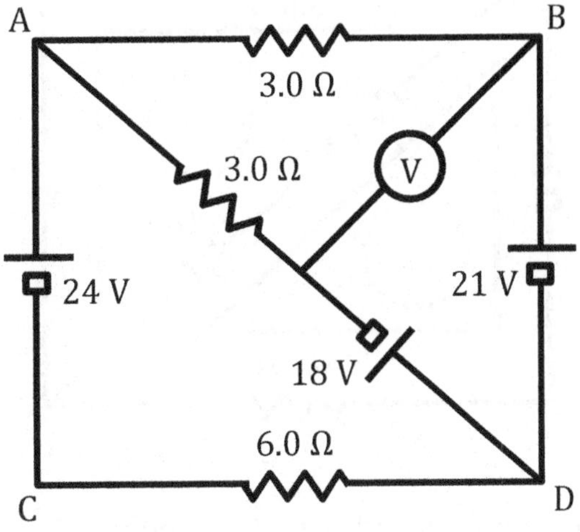

(A) Find each of the currents.

Solution. First remove the voltmeter and its connecting wires. (Since a voltmeter has a very large internal resistance, virtually no current passes through it, which is why it may be removed from the circuit.) See the diagram below.

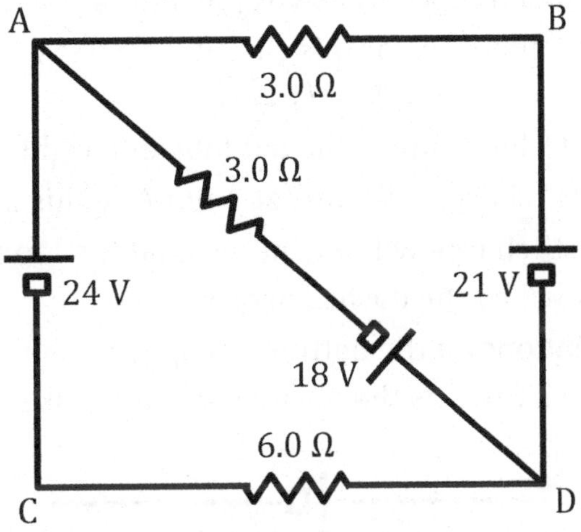

There are three distinct **currents**. See the solid arrows (→) in the following diagram:

- I_L exits junction D and enters junction A. (Note that neither B nor C is a junction.)
- I_M exits junction A and enters junction D.
- I_R exits junction D and enters junction A.

Draw the sense of **traversal** in each loop. The sense of traversal is different from current. The sense of traversal shows how your "test" charge will "swim" around each loop (which will sometimes be opposite to an actual current). See the dashed arrows (⤏) in the diagram above: We chose our sense of traversal to be **clockwise** in each loop.

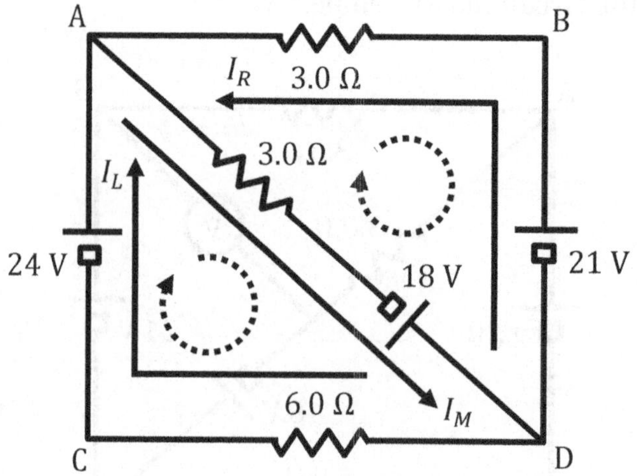

Count the number of distinct currents and "smallest" loops:

- There are $N_C = 3$ distinct currents: I_L, I_M, and I_R. These are the unknowns.
- There are $N_L = 2$ "smallest" loops: the left loop (ADCA) and the right loop (ABDA).

Therefore, we need $N_J = N_C - N_L = 3 - 2 = 1$ junction equation. We choose **junction A**. (We would obtain an equivalent answer for junction D in this example.)

- Which currents enter junction A? I_L and I_R go into junction A.
- Which currents exit junction A? I_M leaves junction A.

According to Kirchhoff's junction rule, for this circuit we get:

$$I_L + I_R = I_M$$

Next we will apply Kirchhoff's loop rule to the left loop and right loop.

- In each case, our "test" charge will start at point A. (This is our choice.)
- In each case, our "test" charge will "swim" around the loop with a **clockwise** sense of **traversal**. This is shown by the dashed arrows (⋯→).
- Study the **sign conventions** on the bottom of page 196.

Note that the long line of the battery is the positive terminal, as indicated below.

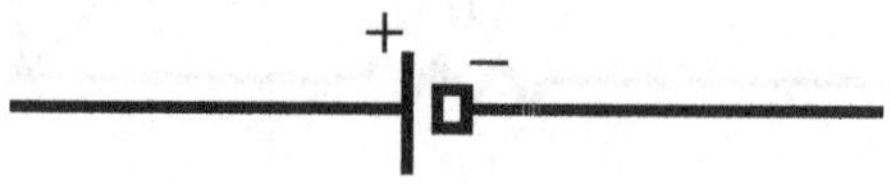

Let's apply Kirchhoff's loop rule to the **left loop**, starting at point A and heading clockwise:

- Our "test" charge first comes to a 3-Ω resistor. The "test" charge is heading in the **same** direction as the current I_M. According to page 196, we write $-3\,I_M$ (it's **negative** for a resistor when the "test" charge "swims" with the current), multiplying current (I_M) times resistance (3 Ω) according to Ohm's law, $\Delta V = IR$.
- Our "test" charge next comes to an 18-V battery. The "test" charge comes to the **negative terminal** of the battery **first**. According to page 196, we write +18 V (since the "test" charge rises in electric potential, going from negative to positive). **Note:** The direction of the current does **not** matter for a **battery**.

- Our "test" charge next comes to a 6-Ω resistor. The "test" charge is heading left, which is the **same** direction as the current I_L. Since the "test" charge "swims" with the current here, we write $-6\ I_L$.
- Finally, our "test" charge comes to a 24-V battery. The "test" charge comes to the **negative terminal** of the battery **first**. Therefore, we write +24 V.

Add these four terms together and set the sum equal to zero. As usual, we'll suppress the units (V for Volts and Ω for Ohms) to avoid clutter until the calculation is complete. Also as usual, we'll not worry about significant figures (like 3.0) until the end of the calculation. (When using a calculator, it's proper technique to keep extra digits and not to round until the end of the calculation. This reduces round-off error. In our case, however, we're not losing anything to rounding by writing 3.0 as 3.)

$$-3\ I_M + 18 - 6\ I_L + 24 = 0$$

Now apply Kirchhoff's loop rule to the **right loop**, starting at point A and heading clockwise. Study the diagram and try to follow along.

- Our "test" charge first comes to a 3-Ω resistor. The "test" charge is heading to the right, whereas the current I_R is heading to the left (the **opposite** direction). According to page 196, we write $+3\ I_R$ (it's **positive** for a resistor when the "test" charge "swims" against the current).
- Our "test" charge next comes to a 21-V battery. The "test" charge comes to the **positive terminal** of the battery **first**. According to page 196, we write −21 V (since the "test" charge drops in electric potential, going from positive to negative). **Reminder**: The direction of the current does **not** matter for a **battery**.
- Our "test" charge next comes to an 18-V battery. The "test" charge comes to the **positive terminal** of the battery **first**. Therefore, we write −18 V.
- Finally, our "test" charge comes to a 3-Ω resistor. The "test" charge is heading **opposite** to the current I_M. Since the "test" charge "swims" against the current here, we write $+3\ I_M$.

Add these four terms together and set the sum equal to zero.

$$+3\ I_R - 21 - 18 + 3\ I_M = 0$$

We now have three equations in three unknowns.

$$I_L + I_R = I_M$$

$$-3\ I_M + 18 - 6\ I_L + 24 = 0$$

$$3\ I_R - 21 - 18 + 3\ I_M = 0$$

First, we will substitute the junction equation into the loop equations. Since $I_L + I_R = I_M$, we could substitute the expression $(I_L + I_R)$ in place of I_M in the two loop equations, but we would need to do this twice. If instead we solve for I_L (or I_R), we only have to do this once, which is less work. So let's subtract I_R from both sides of the junction equation in order to solve for I_L:

$$I_L = I_M - I_R$$

Now we'll substitute $(I_M - I_R)$ in place of I_L in the equation for the left loop:

$$-3\ I_M + 18 - 6\ I_L + 24 = 0$$

$$-3\ I_M + 18 - 6\ (I_M - I_R) + 24 = 0$$

Distribute the -6 to both terms. Note that the two minus signs make a plus sign.

$$-3\ I_M + 18 - 6\ I_M + 6\ I_R + 24 = 0$$

Combine **like terms**. The 18 and 24 are like terms: They make $18 + 24 = 42$. The $-3\ I_M$ and $-6\ I_M$ are like terms: They make $-3\ I_M - 6\ I_M = -9\ I_M$.

$$-9\ I_M + 6\ I_R + 42 = 0$$

Subtract 42 from both sides of the equation.

$$-9\ I_M + 6\ I_R = -42$$

We'll return to this equation in a moment. Let's work with the other loop equation now.

$$3\ I_R - 21 - 18 + 3\ I_M = 0$$

Combine **like terms**. The -21 and -18 are like terms: They make $-21 - 18 = -39$.

$$3\ I_R - 39 + 3\ I_M = 0$$

Add 39 to both sides of the equation.

$$3\ I_R + 3\ I_M = 39$$

Let's put our two simplified equations together.

$$-9\ I_M + 6\ I_R = -42$$

$$3\ I_R + 3\ I_M = 39$$

It helps to write the terms in the same order. Note that $-9\ I_M + 6\ I_R = 6\ I_R - 9\ I_M$.

$$6\ I_R - 9\ I_M = -42$$

$$3\ I_R + 3\ I_M = 39$$

The "trick" is to make equal and opposite coefficients for one of the currents. If we multiply the bottom equation by 3, we will have $-9\ I_M$ in the top equation and $+9\ I_M$ in the bottom.

$$6\ I_R - 9\ I_M = -42$$

$$9\ I_R + 9\ I_M = 117$$

Now I_M cancels out if we add the two equations together. The sum of the left-hand sides equals the sum of the right-hand sides.

$$6\ I_R - 9\ I_M + 9\ I_R + 9\ I_M = -42 + 117$$

$$6\ I_R + 9\ I_R = 75$$

$$15\ I_R = 75$$

Divide both sides of the equation by 15.

$$I_R = \frac{75}{15} = 5.0\text{ A}$$

Once you get a numerical value for one of your unknowns, you may plug this value into any of the previous equations. Look for one that will make the algebra simple. We choose:

$$3\ I_R + 3\ I_M = 39$$

$$3(5) + 3\ I_M = 39$$

$$15 + 3\ I_M = 39$$

$$3\ I_M = 39 - 15$$

$$3\,I_M = 24$$
$$I_M = \frac{24}{3} = 8.0 \text{ A}$$

Once you have two currents, plug them into the junction equation.

$$I_L + I_R = I_M$$
$$I_L + 5 = 8$$
$$I_L = 8 - 5 = 3.0 \text{ A}$$

The currents are $I_L = 3.0$ A, $I_M = 8.0$ A, and $I_R = 5.0$ A.

Tip: If you get a minus sign when you solve for a current:

- Keep the minus sign.
- Don't go back and alter your diagram.
- Don't rework the solution. Don't change any equations.
- If you need to plug the current into an equation, keep the minus sign.
- The minus sign simply means that the current's actual direction is opposite to the arrow that you drew in the beginning of the problem. It's not a big deal.

Check the answers. Plug the currents ($I_L = 3.0$ A, $I_M = 8.0$ A, and $I_R = 5.0$ A) into the three original equations (one junction and two loops) and check that both sides of each equation are equal.

$$I_L + I_R = I_M$$
$$3 + 5 = 8 \checkmark$$

$$-3\,I_M + 18 - 6\,I_L + 24 = 0$$
$$-3(8) + 18 - 6(3) + 24 = -24 + 18 - 18 + 24 = 0 \checkmark$$

$$3\,I_R - 21 - 18 + 3\,I_M = 0$$
$$3(5) - 21 - 18 + 3(8) = 15 - 21 - 18 + 24 = 39 - 39 = 0 \checkmark$$

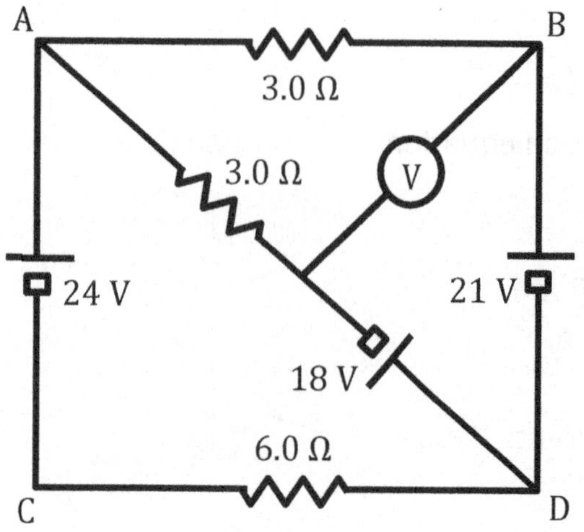

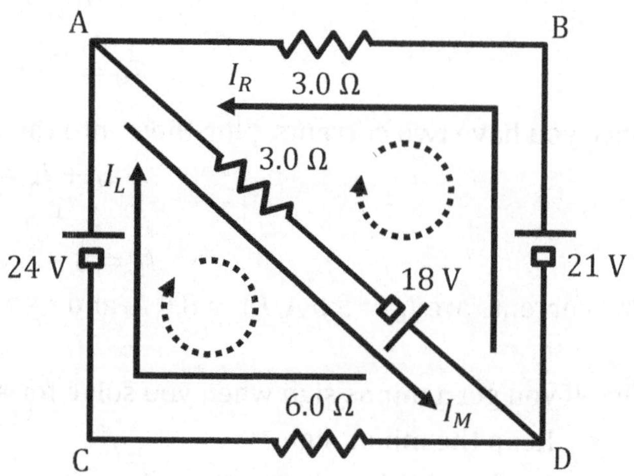

(B) What numerical value, with units, does the voltmeter read?

Solution. Apply Kirchhoff's loop rule, beginning at one probe of the voltmeter and ending at the other probe of the voltmeter. We won't set the sum to zero, since we're not going around a complete loop. Instead, we will simply plug in the currents and add up the values. It doesn't matter which path we take (we'll get the same answer either way), but we'll choose the route along the two batteries (since that makes the math simpler – we won't even need to plug in any currents). We choose to start at the probe in the center of the original circuit and work our way to point B going diagonally down to the right and then going up.

$$\Delta V = 18 + 21 = 39 \text{ V}$$

Both terms are positive according to the sign conventions on page 196 because we come to the positive terminal first. The voltmeter reads $\Delta V = 39$ V.

Check for consistency. Do the same thing for a different path between the two probes and check that you get the same answer. We must start at the same point: We chose the center probe as our starting point earlier. This time we will go diagonally up to the left and then go right. First, we come to a 3-Ω resistor and our "test" charge is swimming opposite to I_M, so we write $+3\ I_M$. (Review the sign conventions on page 196, if necessary. Note that the original sense of traversal is irrelevant now: Our current sense of traversal is from A to B.) Next, we come to the other 3-Ω resistor and our "test" charge is swimming opposite to I_R, so we write $+3\ I_R$.

$$\Delta V = 3I_M + 3I_R$$

Plug in the values of the currents from part (A).

$$\Delta V = 3(8) + 3(5) = 24 + 15 = 39 \text{ V}$$

Since we got the same answer as before, our answer checks out.

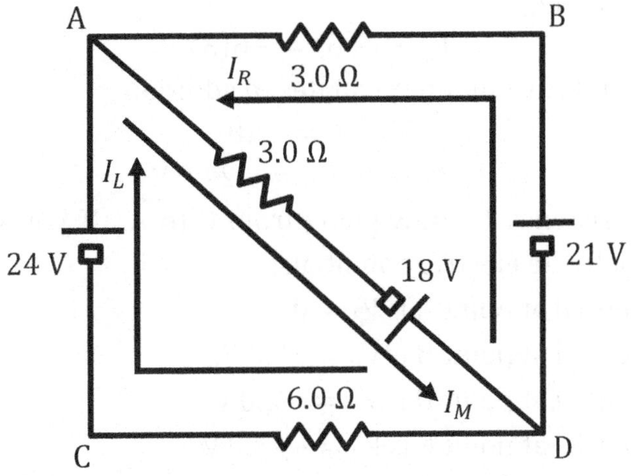

(C) Rank the electric potential at points A, B, C, and D.

Solution. To **rank** electric potential at two or more points, set the electric potential at one point equal to zero and then apply Kirchhoff's loop rule the same way that we did in part (B). It won't matter which point we choose to have zero electric potential, as the relative values will still come out in the same order. We choose point A to have zero electric potential.

$$V_A = 0$$

Traverse from A to B to find the potential difference $V_B - V_A$ (final minus initial). There is just a 3-Ω resistor between A and B. Going from A to B, we are traversing **opposite** to the current I_R, so we write $+3\ I_R$. (Review the sign conventions on page 196, if necessary. Note that the original sense of traversal is irrelevant now: Our current sense of traversal is from A to B.)

$$V_B - V_A = 3\ I_R$$

Plug in the values for V_A and I_R.

$$V_B - 0 = 3(5)$$
$$V_B = 15.0 \text{ V}$$

Now traverse from B to D to find the potential difference $V_D - V_B$. There is just a 21.0-V battery between B and D. Going from B to D, we come to the **positive terminal** first, so we write −21.0 V.

$$V_D - V_B = -21$$

Plug in the value for V_B.

$$V_D - 15 = -21$$
$$V_D = -21 + 15 = -6.0 \text{ V}$$

Now traverse from D to C to find the potential difference $V_C - V_D$. There is just a 6.0-Ω resistor between D and C. Going from D to C, we are traversing in the **same** direction as the current I_L, so we write $-6\ I_L$.

$$V_C - V_D = -6\ I_L$$

Plug in the values for V_D and I_L.

$$V_C - (-6) = -6(3)$$

Note that subtracting a negative number equates to addition.

$$V_C + 6 = -18$$

$$V_C = -18 - 6 = -24.0 \text{ V}$$

(You can check for consistency by now going from C to A. If you get $V_A = 0$, everything checks out). Let's tabulate the electric potentials:

- The electric potential at point A is $V_A = 0$.
- The electric potential at point B is $V_B = 15.0$ V.
- The electric potential at point D is $V_D = -6.0$ V.
- The electric potential at point C is $V_C = -24.0$ V.

Now it should be easy to rank them: B is highest, then A, then D, and C is lowest.

$$V_B > V_A > V_D > V_C$$

Check for consistency. As we mentioned earlier, we can check for consistency by going from C to A. Traversing from C to A, we will find the potential difference $V_A - V_C$. (Remember, it's final minus initial.) There is just a 24.0-V battery between C and A. Going from C to A, we come to the **negative terminal** first, so we write +24.0 V.

$$V_A - V_C = +24$$

Plug in the value for V_C.

$$V_A - (-24) = 24$$

Note that subtracting a negative number equates to addition.

$$V_A + 24 = 24$$

$$V_A = 24 - 24 = 0$$

Since we obtained the same value as our starting point, $V_A = 0$, our answers check out.

Note: Not every physics textbook solves Kirchhoff's rules problems the same way. If you're taking a physics course, it's possible that your instructor or textbook will apply a different (but equivalent) method.

16 MORE RESISTANCE EQUATIONS

Ohm's Law	Emf and Internal Resistance
$\Delta V = IR$	$\varepsilon = \Delta V + Ir = I(R + r)$
Resistance of a Cylindrical Wire	**Temperature Dependence**
$R = \frac{\rho L}{A}$, $\sigma = \frac{1}{\rho}$	$\rho = \rho_0(1 + \alpha\Delta T)$ $R \approx R_0(1 + \alpha\Delta T)$
Parallel Resistors	**Series Resistors**
$\frac{1}{R_p} = \frac{1}{R_1} + \frac{1}{R_2} + \cdots + \frac{1}{R_N}$	$R_s = R_1 + R_2 + \cdots + R_N$
Power	**Current Density**
$P = I\Delta V = I^2R = \frac{\Delta V^2}{R}$	$J = \frac{dI}{dA}$, $I = \int \vec{\mathbf{J}} \cdot d\vec{\mathbf{A}}$, $J = \sigma E$

Schematic Representation	Symbol	Name
	R	resistor
	ΔV	battery or DC power supply

Symbol	Name	SI Units
R	resistance	Ω
R_0	resistance at a reference temperature	Ω
r	internal resistance of a battery or power supply	Ω
ρ	resistivity	$\Omega \cdot \text{m}$
ρ_0	resistivity at a reference temperature	$\Omega \cdot \text{m}$
σ	conductivity	$\frac{1}{\Omega \cdot \text{m}}$
L	the length of the wire	m
A	cross-sectional area	m^2
α	temperature coefficient of resistivity	$\frac{1}{\text{K}}$ or $\frac{1}{°\text{C}}$*
T	temperature	K
T_0	reference temperature	K
ΔT	change in temperature $(T - T_0)$	K or °C*
I	electric current	A
J	current density	A/m^2
ΔV	the potential difference between two points in a circuit	V
ε	emf	V
P	electric power	W
$\vec{\mathbf{E}}$	electric field	N/C or V/m
$d\vec{\mathbf{s}}$	differential displacement vector (see Chapter 21 of Volume 1)	m
$d\vec{\mathbf{A}}$	differential area element	m^2

Note: The symbols ρ, σ, and α are the lowercase Greek letters rho, sigma, and alpha, while ε is a variation of the Greek letter epsilon.

* Note that α has the same numerical value in both $\frac{1}{\text{K}}$ and $\frac{1}{°\text{C}}$* since the change in temperature, $\Delta T = T - T_0$, works out the same in both Kelvin and Celsius (because $T_K = T_c + 273.15$).

Example 65. A long, straight wire has the shape of a right-circular cylinder. The length of the wire is π m and the thickness of the wire is 0.80 mm. The resistance of the wire is 5.0 Ω. What is the resistivity of the wire?

Solution. Make a list of the known quantities and identify the desired unknown symbol:

- The length of the wire is $L = \pi$ m.
- The thickness of the wire is $T = 0.80$ mm.
- The resistance of the wire is $R = 5.0\ \Omega$.
- The unknown we are looking for is resistivity (ρ).

The equation relating resistance to resistivity involves area, so we need to find area first. Before we do that, let's convert the thickness to SI units. Recall that the prefix milli (m) stands for 10^{-3}.

$$T = 0.80 \text{ mm} = 0.00080 \text{ m} = 8.0 \times 10^{-4} \text{ m}$$

The radius of the circular cross section is one-half the diameter, and the diameter is the same as the thickness.

$$a = \frac{D}{2} = \frac{T}{2} = \frac{8 \times 10^{-4}}{2} = 4 \times 10^{-4} \text{ m}$$

The cross-sectional area of the wire is the area of a circle. Note that $a^2 = (4 \times 10^{-4})^2 = (4)^2(10^{-4})^2 = 16 \times 10^{-8}$ m^2.

$$A = \pi a^2 = \pi(4 \times 10^{-4})^2 = \pi(16 \times 10^{-8}) \text{ m}^2 = 16\pi \times 10^{-8} \text{ m}^2$$

If you use a calculator, this works out to $A = 5.03 \times 10^{-7}$ m^2. Now we are ready to use the equation for resistance.

$$R = \frac{\rho L}{A}$$

Multiply both sides of the equation by area and divide by length in order to solve for the resistivity.

$$\rho = \frac{RA}{L} = \frac{(5)(16\pi \times 10^{-8})}{(\pi)} = 80 \times 10^{-8}\ \Omega\text{·m} = 8.0 \times 10^{-7}\ \Omega\text{·m}$$

Note that the π's cancel. The resistivity is $\rho = 8.0 \times 10^{-7}$ Ω·m, which is equivalent to $\rho = 80 \times 10^{-8}$ Ω·m.

Example 66. A resistor is made from a material that has a temperature coefficient of resistivity of 2.0×10^{-3} /°C. Its resistance is 30 Ω at 20 °C. At what temperature is its resistance equal to 33 Ω?

Solution. Make a list of the known quantities and identify the desired unknown symbol:

- The reference temperature is $T_0 = 20$ °C.
- The reference resistance is $R_0 = 30$ Ω.
- The resistance is $R = 33$ Ω at the desired temperature.
- The temperature coefficient of resistivity is $\alpha = 2.0 \times 10^{-3}$ /°C.
- The unknown we are looking for is the temperature T for which $R = 33$ Ω.

Apply the equation for resistance that depends on temperature.

$$R \approx R_0(1 + \alpha\Delta T) = R_0[1 + \alpha(T - T_0)]$$

Plug in the known values.

$$33 = (30)[1 + (2 \times 10^{-3})(T - 20)]$$

Divide both sides of the equation by 30.

$$\frac{33}{30} = 1 + (2 \times 10^{-3})(T - 20)$$

Note that $\frac{33}{30} = 1.1$.

$$1.1 = 1 + (2 \times 10^{-3})(T - 20)$$

Subtract 1 from both sides of the equation.

$$1.1 - 1 = (2 \times 10^{-3})(T - 20)$$

Note that and $1.1 - 1 = 0.1$. Also note that $2 \times 10^{-3} = 0.002$.

$$0.1 = 0.002(T - 20)$$

Divide both sides of the equation by 0.002.

$$\frac{0.1}{0.002} = T - 20$$

Note that $\frac{0.1}{0.002} = 50$.

$$50 = T - 20$$

Add 20 to both sides of the equation.

$$70 \text{ °C} = T$$

The temperature for which $R \approx 33$ Ω is $T = 70$ °C.†

† The similar equation for resistivity involves an equal sign, whereas with resistance it is approximately equal. The reason this is approximate is that we're neglecting the effect that thermal expansion has on the length and thickness. The effect that temperature has on resistivity is generally more significant to the change in resistance than the effect that temperature has on the length and thickness of the wire, so the approximation is usually very good.

Example 67. A battery has an emf of 36 V. When the battery is connected across an 8.0-Ω resistor, a voltmeter measures the potential difference across the resistor to be 32 V.

(A) What is the internal resistance of the battery?

Solution. Make a list of the known quantities and identify the desired unknown symbol:

- The emf of the battery is $\varepsilon = 36$ V.
- The potential difference across the resistor is $\Delta V = 32$ V.
- The load resistance (or external resistance) is $R = 8.0\ \Omega$.
- The unknown we are looking for is the internal resistance (r) of the battery.

Since all of the relevant equations involve current (I), and since we don't already know the current, let's begin by solving for the current. We can find it from Ohm's law.

$$\Delta V = IR$$

Divide both sides of the equation by the load resistance.

$$I = \frac{\Delta V}{R} = \frac{32}{8} = 4.0 \text{ A}$$

Now we are ready to apply one of the emf equations.

$$\varepsilon = \Delta V + Ir$$

Plug in the known values.

$$36 = 32 + 4r$$

Subtract 32 from both sides of the equation.

$$4 = 4r$$

Divide both sides of the equation by 4.

$$r = 1.0\ \Omega$$

The internal resistance of the battery is $r = 1.0\ \Omega$.

(B) How much power is dissipated in the 8.0-Ω resistor?

Solution. Use one of the power equations.

$$P = I\Delta V = (4)(32) = 128 \text{ W}$$

The power dissipated in the load resistor is $P = 128$ W.

Deriving an Equation for Resistance		
$R = \frac{\lvert \Delta V \rvert}{I}$	$\Delta V = -\int \vec{\mathbf{E}} \cdot d\vec{\mathbf{s}}$	$I = \int \vec{\mathbf{J}} \cdot d\vec{\mathbf{A}}$
Gauss's Law		
$\oint_S \vec{\mathbf{E}} \cdot d\vec{\mathbf{A}} = \frac{q_{enc}}{\epsilon_0}$		
Differential Charge Element		
$dq = \lambda ds$ (line or thin arc)	$dq = \sigma dA$ (surface area)	$dq = \rho dV$ (volume)
Relation Among Coordinate Systems and Unit Vectors		
$x = r\cos\theta$ $y = r\sin\theta$ $\hat{\mathbf{r}} = \hat{\mathbf{x}}\cos\theta + \hat{\mathbf{y}}\sin\theta$ (2D polar)	$x = r_c\cos\theta$ $y = r_c\sin\theta$ $\hat{\mathbf{r}}_c = \hat{\mathbf{x}}\cos\theta + \hat{\mathbf{y}}\sin\theta$ (cylindrical)	$x = r\sin\theta\cos\varphi$ $y = r\sin\theta\sin\varphi$ $z = r\cos\theta$ $\hat{\mathbf{r}} = \hat{\mathbf{x}}\cos\varphi\sin\theta$ $+\hat{\mathbf{y}}\sin\varphi\sin\theta + \hat{\mathbf{z}}\cos\theta$ (spherical)
Differential Arc Length		
$ds = dx$ (along x)	$ds = dy$ or dz (along y or z)	$ds = a d\theta$ (circular arc of radius a)
Differential Area Element		
$dA = dxdy$ (polygon in xy plane)	$dA = rdrd\theta$ (pie slice, disc, thick ring)	$dA = a^2\sin\theta\, d\theta d\varphi$ (sphere of radius a)
Differential Volume Element		
$dV = dxdydz$ (bounded by flat sides)	$dV = r_c dr_c d\theta dz$ (cylinder or cone)	$dV = r^2\sin\theta\, drd\theta d\varphi$ (spherical)

Long Cylinder Example: A Prelude to Example 68. Derive an equation for the resistance of a right-circular cylinder of length L, radius a, uniform cross-section, and uniform resistivity, where the terminals of a battery are connected across the circular ends of the cylinder to create an approximately uniform electric field along the length of the cylinder.

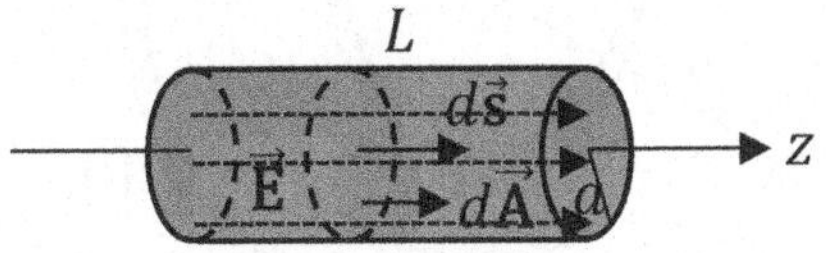

Solution. Apply Ohm's law, and then integrate over electric field to find potential difference and integrate over current density to find current.

$$R = \frac{|\Delta V|}{I} = \frac{\left|\int \vec{\mathbf{E}} \cdot d\vec{\mathbf{s}}\right|}{\int \vec{\mathbf{J}} \cdot d\vec{\mathbf{A}}}$$

The calculus of this example will be trivial because the electric field ($\vec{\mathbf{E}}$) running along the length of the cylinder is uniform. Since $\vec{\mathbf{E}}$ is uniform (constant), we may pull it out of the integrals. (In this example, unlike the following example and most problems that involve deriving an equation for resistance, we won't need to integrate or use Gauss's law to find $\vec{\mathbf{E}}$. The only reason is that in this problem, $\vec{\mathbf{E}}$ happens to be constant.) In this example, we may simply write $\vec{\mathbf{E}} = E\hat{\mathbf{z}}$ since the electric field is along the $+z$-axis. Similarly, $\vec{\mathbf{J}}$ is along the $+z$-axis since the current density is $\vec{\mathbf{J}} = \sigma\vec{\mathbf{E}}$. Note that $d\vec{\mathbf{s}}$ and $d\vec{\mathbf{A}}$ are also along the $+z$-axis, such that $\vec{\mathbf{E}} \cdot d\vec{\mathbf{s}} = E \cos 0° \, ds = Eds$ and $\vec{\mathbf{J}} \cdot d\vec{\mathbf{A}} = J \cos 0° \, dA = JdA$. (In most problems, you can't pull E out of the integrals, but since the electric field is uniform in this example, we may now.)

$$R = \frac{\int E\, ds}{\int J\, dA} = \frac{\int E\, ds}{\int \sigma E\, dA} = \frac{E \int ds}{\sigma E \int dA} = \frac{EL}{\sigma EA} = \frac{L}{\sigma A} = \frac{\rho L}{A}$$

The integral over ds corresponds to potential difference: Since the battery is connected to the ends of the cylinder, the integral over ds is along the length (L) of the cylinder. In the last step, we used the equation $\sigma = \frac{1}{\rho}$, which can also be written $\rho = \frac{1}{\sigma}$ (the conductivity and resistivity have a reciprocal relationship, as each is the reciprocal of the other).

Coaxial Cylinders Example: Another Prelude to Example 68. A cylindrical resistor consists of two very long thin coaxial cylindrical conducting shells: The inner cylindrical shell has radius a, while the outer shell has radius b. The region between the two shells is filled with a semiconductor (such as silicon). Derive an equation for the resistance of this cylindrical resistor when one terminal of a battery is connected to the inner conductor while the other terminal is connected to the outer conductor, creating electric field lines that radiate outward from the axis of the cylinders.

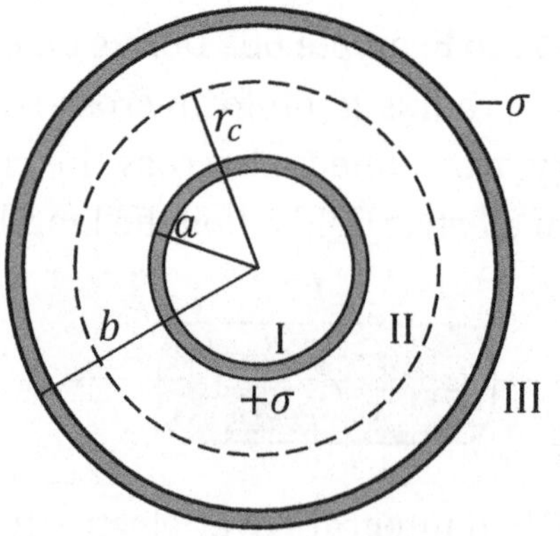

Solution. In the diagram above, we are looking at a cross section of the coaxial cylinders. The first step is to apply Gauss's law to region II ($a < r_c < b$). In Chapter 8, we learned that the outer conducting shell doesn't matter in region II (since the Gaussian cylinder drawn for region II won't enclose any charge from the outer cylinder). Hence, we just need to find the electric field in region II for the inner thin conducting cylindrical shell. We will get the same result as in an example from Chapter 8 that featured a long conducting cylinder (see page 100).

$$\vec{\mathbf{E}} = \frac{\sigma_{cd} a}{\epsilon_0 r_c} \hat{\mathbf{r}}_c$$

Note the symbol σ_{cd}, where the cd stands for "charge density." Unfortunately, we use the same symbol (σ) for conductivity and for surface charge density. So to help avoid confusion, we will add subscripts to the symbol for surface charge density, σ_{cd}, to distinguish it from the symbol for conductivity, σ.

To derive an equation for resistance, apply Ohm's law, and then integrate over electric field to find potential difference and integrate over current density to find current. Note that $\vec{\mathbf{E}}$, $\vec{\mathbf{J}}$, $d\vec{\mathbf{s}}$, and $d\vec{\mathbf{A}}$ all radiate outward along $\hat{\mathbf{r}}_c$. (It's common for these four vectors to be parallel in problems where you're deriving an equation for resistance.) When we integrate over $\vec{\mathbf{E}} \cdot d\vec{\mathbf{s}}$ to get ΔV, we integrate outward along dr_c from $r_c = a$ to $r_c = b$. When we integrate over $\vec{\mathbf{J}} \cdot d\vec{\mathbf{A}}$ to get I, the area is over the surface area of a cylinder of radius r_c, where $dA = r_c d\theta dz$. Note that θ and z vary with r_c fixed to make a cylinder (see page 164).

$$R = \frac{|\Delta V|}{I} = \frac{\left|\int \vec{\boldsymbol{E}} \cdot d\vec{\boldsymbol{s}}\right|}{\int \vec{\mathbf{J}} \cdot d\vec{\mathbf{A}}} = \frac{\int E\, ds}{\int J\, dA} = \frac{\int E\, ds}{\int \sigma E\, dA} = \frac{\int E\, ds}{\sigma \int E\, dA}$$

Note that the scalar products simplify to $\int \vec{\boldsymbol{E}} \cdot d\vec{\boldsymbol{s}} = \int E\, ds$ and $\int \vec{\mathbf{J}} \cdot d\vec{\mathbf{A}} = \int J\, dA$ because $\vec{\mathbf{E}}$, $\vec{\mathbf{J}}$, $d\vec{\mathbf{s}}$, and $d\vec{\mathbf{A}}$ are all parallel. Substitute the equation $E = \frac{\sigma_{cd} a}{\epsilon_0 r_c}$, as we discussed earlier.

$$R = \frac{\int_{r_c=a}^{r_c=b} \frac{\sigma_{cd} a}{\epsilon_0 r_c} dr_c}{\sigma \int_{z=0}^{L} \int_{\theta=0}^{2\pi} \frac{\sigma_{cd} a}{\epsilon_0 r_c} r_c\, d\theta dz} = \frac{\frac{\sigma_{cd} a}{\epsilon_0} \int_{r_c=a}^{r_c=b} \frac{dr_c}{r_c}}{\sigma \frac{\sigma_{cd} a}{\epsilon_0} \int_{z=0}^{L} \int_{\theta=0}^{2\pi} d\theta\, dz}$$

In the denominator, note that the r_c from $dA = r_c d\theta dz$ cancels the r_c from $E = \frac{\sigma_{cd}a}{\epsilon_0 r_c}$. Also, the constants $\frac{\sigma_{cd}a}{\epsilon_0}$ from the numerator and denominator all cancel.

$$R = \frac{\int_{r_c=a}^{r_c=b} \frac{dr_c}{r_c}}{\sigma \int_{z=0}^{L} \int_{\theta=0}^{2\pi} d\theta\, dz} = \frac{[\ln(r_c)]_{r_c=a}^{b}}{2\pi\sigma L}$$

The anti-derivative for $\int \frac{dx}{x}$ is a natural logarithm (see Chapter 17).

$$R = \frac{\ln(b) - \ln(a)}{2\pi\sigma L}$$

The conductivity can be written in terms of the resistivity according to $\sigma = \frac{1}{\rho}$. Substitute this into the previous equation. Note that $\frac{1}{1/\rho} = \rho$. Apply the rule for logarithms that $\ln\left(\frac{b}{a}\right) = \ln(b) - \ln(a)$. (See Chapter 17.)

$$R = \frac{\rho}{2\pi L}[\ln(b) - \ln(a)] = \frac{\rho}{2\pi L}\ln\left(\frac{b}{a}\right)$$

Example 68. A spherical resistor consists of two thin spherical conducting shells: The inner shell has radius a while the outer shell has radius b. There is a semiconductor between the two spheres. Derive an equation for the resistance of this spherical resistor.

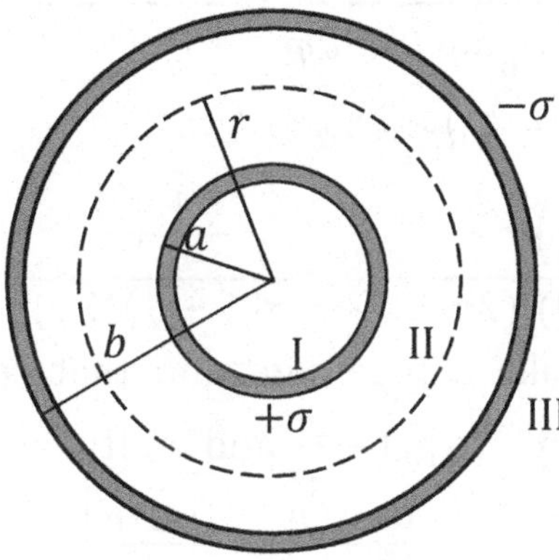

Solution. This is similar to the previous example, except for the shape being a sphere rather than a cylinder. In the diagram above, we are looking at a cross section of the spheres. The first step is to apply Gauss's law to region II ($a < r < b$). In Chapter 8, we learned that the outer conducting shell doesn't matter in region II (since the Gaussian sphere drawn for region II won't enclose any charge from the outer sphere). Hence, we just need to find the electric field in region II for the inner thin conducting spherical shell. We will get the same result as in an example from Chapter 8 that featured a uniform sphere (see the bottom of page 86).

$$\vec{\mathbf{E}} = \frac{kQ}{r^2}\hat{\mathbf{r}}$$

To derive an equation for resistance, apply Ohm's law, and then integrate over electric field to find potential difference and integrate over current density to find current. Note that $\vec{\mathbf{E}}$, $\vec{\mathbf{J}}$, $d\vec{\mathbf{s}}$, and $d\vec{\mathbf{A}}$ all radiate outward along $\hat{\mathbf{r}}$. (It's common for these four vectors to be parallel in problems where you're deriving an equation for resistance.) When we integrate over $\vec{\mathbf{E}} \cdot d\vec{\mathbf{s}}$ to get ΔV, we integrate outward along dr from $r = a$ to $r = b$. When we integrate over $\vec{\mathbf{J}} \cdot d\vec{\mathbf{A}}$ to get I, the area is over the surface area of a sphere of radius r, where $dA = r^2 \sin\theta \, d\theta d\varphi$ (see page 210). Note that θ and φ vary with r fixed to make a sphere.

$$R = \frac{|\Delta V|}{I} = \frac{\left|\int \vec{\boldsymbol{E}} \cdot d\vec{\boldsymbol{s}}\right|}{\int \vec{\mathbf{J}} \cdot d\vec{\mathbf{A}}} = \frac{\int E\, ds}{\int J\, dA} = \frac{\int E\, ds}{\int \sigma E\, dA} = \frac{\int E\, ds}{\sigma \int E\, dA}$$

Note that the scalar products simplify to $\int \vec{\boldsymbol{E}} \cdot d\vec{\boldsymbol{s}} = \int E\, ds$ and $\int \vec{\mathbf{J}} \cdot d\vec{\mathbf{A}} = \int J\, dA$ because $\vec{\mathbf{E}}$, $\vec{\mathbf{J}}$, $d\vec{\mathbf{s}}$, and $d\vec{\mathbf{A}}$ are all parallel. Substitute the equation $E = \frac{kQ}{r^2}$, as we discussed earlier.

$$R = \frac{\int_{r=a}^{r=b} \left(\frac{kQ}{r^2}\right) dr}{\sigma \int_{\theta=0}^{\pi} \int_{\varphi=0}^{2\pi} \left(\frac{kQ}{r^2}\right) r^2 \sin\theta \, d\theta d\varphi} = \frac{kQ \int_{r=a}^{r=b} \frac{dr}{r^2}}{\sigma kQ \int_{\theta=0}^{\pi} \int_{\varphi=0}^{2\pi} \sin\theta \, d\theta d\varphi}$$

In the denominator, note that the r^2 from $dA = r^2 \sin\theta \, d\theta d\varphi$ cancels the r^2 from $E = \frac{kQ}{r^2}$. Also, the constants kQ from the numerator and denominator all cancel.

$$R = \frac{\int_{r=a}^{r=b} \frac{dr}{r^2}}{\sigma \int_{\theta=0}^{\pi} \int_{\varphi=0}^{2\pi} \sin\theta \, d\theta d\varphi} = \frac{\left[-\frac{1}{r}\right]_{r=a}^{b}}{\sigma[-\cos\theta]_{\theta=0}^{\pi} [\varphi]_{\varphi=0}^{2\pi}}$$

Recall from calculus that $\int \frac{dr}{r^2} = \int r^{-2}\, dr = -r^{-1} = -\frac{1}{r}$.

$$R = \frac{-\frac{1}{b} - \left(-\frac{1}{a}\right)}{\sigma[-\cos\pi - (-\cos 0)](2\pi - 0)} = \frac{-\frac{1}{b} + \frac{1}{a}}{\sigma(1+1)(2\pi)} = \frac{-\frac{1}{b} + \frac{1}{a}}{\sigma(2)(2\pi)} = \frac{-\frac{1}{b} + \frac{1}{a}}{4\pi\sigma}$$

Recall that two minus signs make a plus sign and that $\cos\pi = -1$. Make a **common denominator** by multiplying $-\frac{1}{b}$ by $\frac{a}{a}$ to get $-\frac{a}{ab}$ and multiplying $\frac{1}{a}$ by $\frac{b}{b}$ to get $\frac{b}{ab}$.

$$R = \frac{-\frac{a}{ab} + \frac{b}{ab}}{4\pi\sigma} = \frac{\frac{-a+b}{ab}}{4\pi\sigma}$$

To divide with fractions, multiply by the **reciprocal**. Note that the reciprocal of $4\pi\sigma$ is $\frac{1}{4\pi\sigma}$. Also note that $-a + b = b - a$.

$$R = \frac{\frac{-a+b}{ab}}{4\pi\sigma} = \frac{-a+b}{ab} \div 4\pi\sigma = \frac{-a+b}{ab} \times \frac{1}{4\pi\sigma} = \frac{-a+b}{4\pi\sigma ab} = \frac{b-a}{4\pi\sigma ab}$$

The conductivity can be written in terms of the resistivity according to $\sigma = \frac{1}{\rho}$. Substitute this into the previous equation. Note that $\frac{1}{1/\rho} = \rho$.

$$R = \frac{\rho(b-a)}{4\pi ab}$$

17 LOGARITHMS AND EXPONENTIALS

Logarithm Identities

$$\ln(1) = 0 \quad , \quad \ln(y^a) = a\ln(y)$$

$$\ln(xy) = \ln(x) + \ln(y) \quad , \quad \ln\left(\frac{x}{y}\right) = \ln(x) - \ln(y) \quad , \quad \ln\left(\frac{1}{y}\right) = \ln(y^{-1}) = -\ln(y)$$

Change of Base Formula

$$\log_b y = \frac{\log_{10} y}{\log_{10} b}$$

Exponential Identities

$$e = 2.718281828 \ldots \quad , \quad e^0 = 1 \quad , \quad (e^y)^a = e^{ay}$$

$$e^{x+y} = e^x e^y \quad , \quad e^{x-y} = e^x e^{-y} \quad , \quad e^{-y} = \frac{1}{e^y}$$

Identities with Logarithms and Exponentials

$$\ln(e) = 1 \quad , \quad \ln(e^y) = y \quad , \quad e^{\ln(y)} = y$$

Calculus with Simple Logarithms and Exponentials

$$\frac{de^x}{dx} = e^x \quad , \quad \frac{de^{ax}}{dx} = ae^{ax} \quad , \quad \int e^x \, dx = e^x \quad , \quad \int e^{ax} \, dx = \frac{e^{ax}}{a}$$

$$\frac{d}{dx}\ln(x) = \frac{1}{x} \quad , \quad \int \frac{dx}{x} = \int x^{-1} \, dx = \ln(x)$$

Graphs of Exponentials and Logarithms

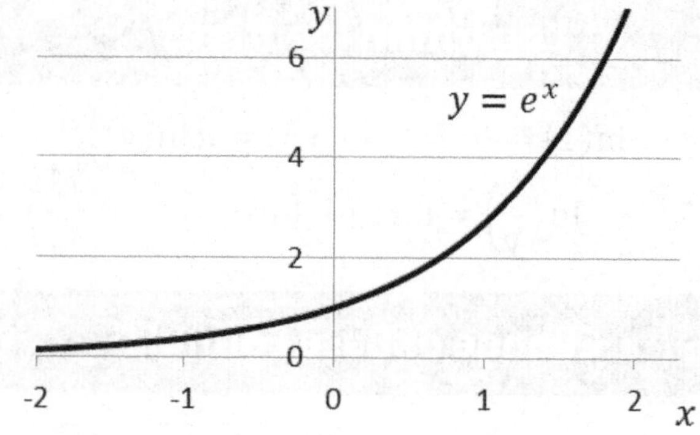

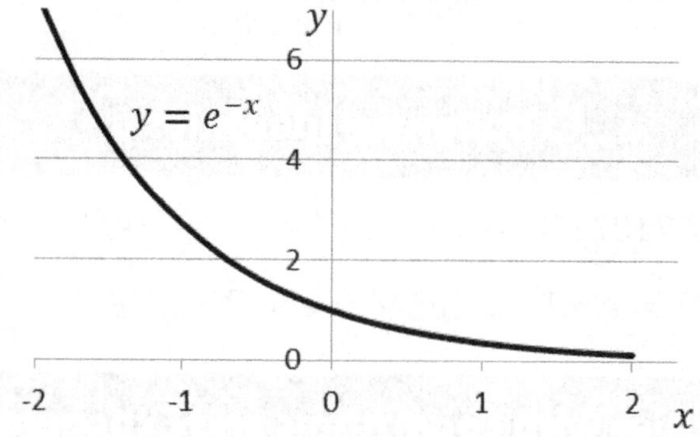

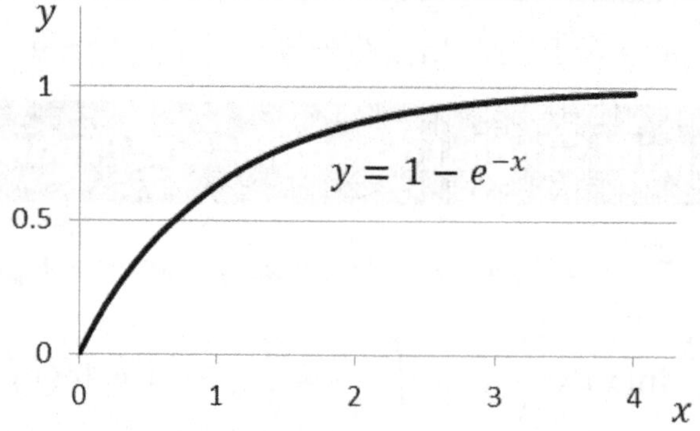

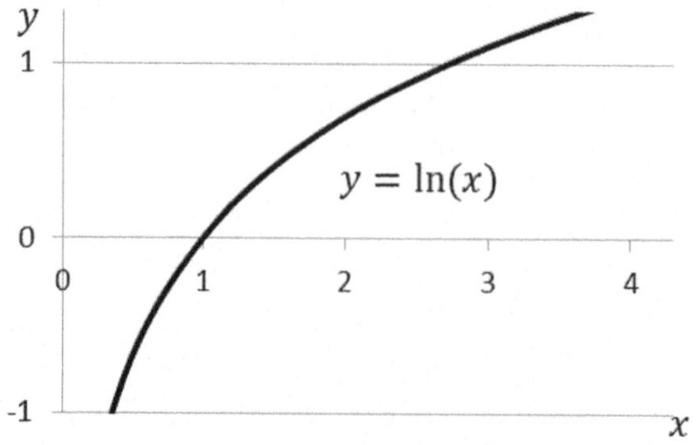

Example 69. Evaluate the following expression.

$$\log_5 625$$

Solution. The expression $\log_5 625$ means: "5 raised to what power equals 625?" The answer is 4 because $5^4 = 625$.

Example 70. Solve for x in the following equation.

$$6e^{-x/2} = 2$$

Solution. First isolate the exponential function. Divide both sides of the equation by 6.

$$e^{-x/2} = \frac{2}{6}$$

Note that $\frac{2}{6}$ **reduces** to $\frac{1}{3}$ if you divide the numerator and denominator both by 2.

$$e^{-x/2} = \frac{1}{3}$$

Take the natural logarithm of both sides of the equation.

$$\ln\left(e^{-x/2}\right) = \ln\left(\frac{1}{3}\right)$$

Apply the rule $\ln(e^y) = y$. In this case, $y = -\frac{x}{2}$ such that $\ln\left(e^{-x/2}\right) = -\frac{x}{2}$.

$$-\frac{x}{2} = \ln\left(\frac{1}{3}\right)$$

Multiply both sides of the equation by -2.

$$x = -2\ln\left(\frac{1}{3}\right)$$

Apply the rule $\ln\left(\frac{1}{y}\right) = -\ln(y)$. Let $y = 3$ to see that $\ln\left(\frac{1}{3}\right) = -\ln(3)$.

$$x = 2\ln(3)$$

The answer is $x = 2\ln(3)$, which is approximately $x = 2.197$.

Example 71. Perform the following definite integral.

$$\int_{x=0}^{2} \left(1 - e^{-x/2}\right) dx$$

Solution. Separate the integral into two terms.

$$\int_{x=0}^{2} \left(1 - e^{-x/2}\right) dx = \int_{x=0}^{2} dx - \int_{x=0}^{2} e^{-x/2}\, dx$$

Note that $\int e^{ax}\, dx = \frac{e^{ax}}{a}$. Compare $e^{-x/2}$ to e^{ax} to see that $a = -\frac{1}{2}$.

$$\int_{x=0}^{2} \left(1 - e^{-x/2}\right) dx = [x]_{x=0}^{2} - \left[\frac{e^{-x/2}}{-1/2}\right]_{x=0}^{2}$$

Note that $\frac{1}{1/2} = 2$. Also note that the two minus signs make a plus sign.

$$\int_{x=0}^{2} \left(1 - e^{-x/2}\right) dx = [x]_{x=0}^{2} - \left[\frac{e^{-x/2}}{-1/2}\right]_{x=0}^{2} = [x]_{x=0}^{2} + \left[2e^{-x/2}\right]_{x=0}^{2}$$

$$= (2 - 0) + \left(2e^{-2/2} - 2e^{-0/2}\right) = 2 + 2e^{-1} - 2e^{0}$$

Note that $e^0 = 1$. Also note that $e^{-1} = \frac{1}{e}$.

$$\int_{x=0}^{2} \left(1 - e^{-x/2}\right) dx = 2 + \frac{2}{e} - 2 = \frac{2}{e}$$

The final answer to the definite integral is $\frac{2}{e}$, which can also be expressed as $2e^{-1}$. If you use a calculator, this comes out to approximately 0.7358.

Example 72. Perform the following definite integral.

$$\int_{x=1}^{2} \frac{dx}{x}$$

Solution. The anti-derivative for $\int \frac{dx}{x}$ is $\ln(x)$. Evaluate $\ln(x)$ over the limits.

$$\int_{x=1}^{2} \frac{dx}{x} = [\ln(x)]_{x=1}^{2} = \ln(2) - \ln(1)$$

Note that $\ln(1) = 0$.

$$\int_{x=1}^{2} \frac{dx}{x} = \ln(2)$$

The answer to the definite integral is $\ln(2)$, which is approximately 0.6931.

18 RC CIRCUITS

Discharging a Capacitor in an RC Circuit

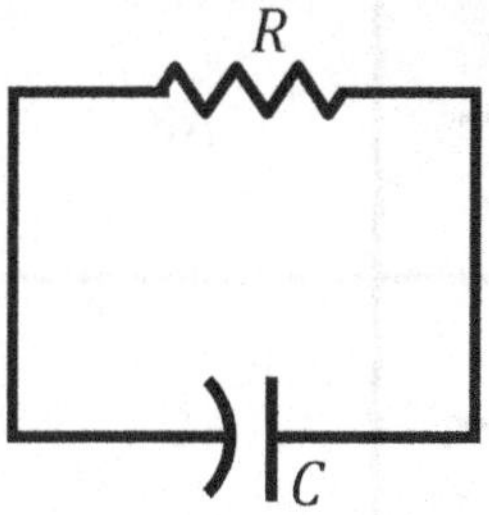

$$Q = Q_m e^{-t/\tau} \quad , \quad I = I_m e^{-t/\tau}$$
$$Q = C\Delta V \quad , \quad \Delta V = IR \quad , \quad Q_m = C\Delta V_m \quad , \quad \Delta V_m = I_m R$$

Charging a Capacitor in an RC Circuit

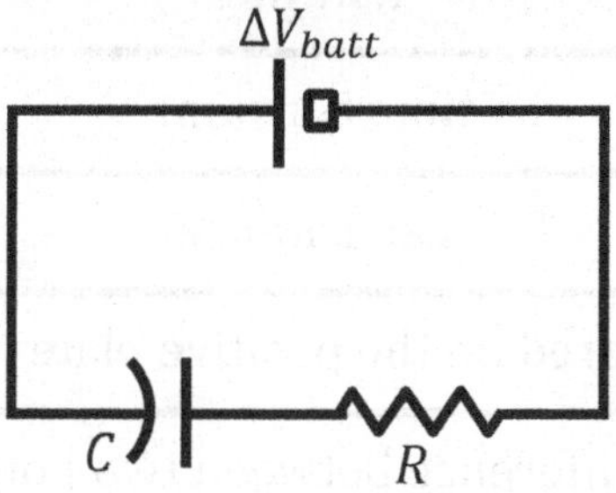

$$Q = Q_m(1 - e^{-t/\tau}) \quad , \quad I = I_m e^{-t/\tau}$$
$$Q = C\Delta V_C \quad , \quad \Delta V_R = IR \quad , \quad Q_m = C\Delta V_{batt} \quad , \quad \Delta V_{batt} = I_m R$$

Time Constant and Half-life

$$\tau = RC \quad , \quad t_{1/2} = \tau \ln(2)$$

Schematic Representation	Symbol	Name
	R	resistor
	ΔV	battery or DC power supply
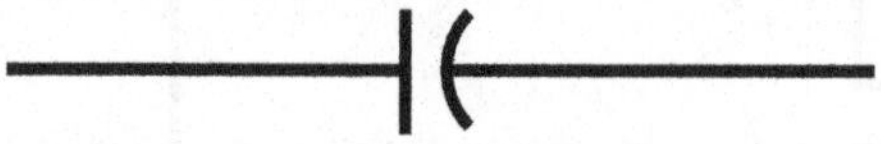	C	capacitor

Symbol	Name	SI Units
t	time	s
$t_{1/2}$	half-life	s
τ	time constant	s
C	capacitance	F
Q	the charge stored on the positive plate of a capacitor	C
ΔV	the potential difference between two points in a circuit	V
R	resistance	Ω
I	electric current	A

Note: The symbol τ is the lowercase Greek letter tau.

Prefix	Name	Power of 10
k	kilo	10^3
m	milli	10^{-3}
μ	micro	10^{-6}
n	nano	10^{-9}
p	pico	10^{-12}

Example 73. A 5.0-μF capacitor with an initial charge of 60 μC discharges while connected in series with a 20-kΩ resistor.

(A) What is the initial potential difference across the capacitor?

Solution. Make a list of the known symbols.

- The capacitance is $C = 5.0$ μF.
- The initial charge is the maximum charge: $Q_m = 60$ μC.
- The resistance is $R = 20$ kΩ.

Convert these values to SI units using $\mu = 10^{-6}$ and $\mathrm{k} = 10^3$.

$C = 5.0\ \mu\text{F} = 5.0 \times 10^{-6}$ F , $Q_m = 60\ \mu\text{C} = 60 \times 10^{-6}$ C , $R = 20\ \text{k}\Omega = 2.0 \times 10^4\ \Omega$

Apply the capacitance equation to find the initial potential difference across the capacitor.

$$Q_m = C\Delta V_m$$

Divide both sides of the equation by capacitance.

$$\Delta V_m = \frac{Q_m}{C} = \frac{60 \times 10^{-6}}{5.0 \times 10^{-6}} = 12.0 \text{ V}$$

The 10^{-6} cancels out. The initial potential difference across the capacitor is $\Delta V_m = 12.0$ V.

(B) What is the initial current?

Solution. Apply Ohm's law to find the initial current.

$$\Delta V_m = I_m R$$

Divide both sides of the equation by resistance.

$$I_m = \frac{\Delta V_m}{R} = \frac{12}{2.0 \times 10^4} = 6.0 \times 10^{-4} \text{ A} = 0.00060 \text{ A} = 0.60 \text{ mA}$$

The initial current is $I_m = 0.60$ mA, which is the same as 0.00060 A or 6.0×10^{-4} A.

(C) What is the half-life?

Solution. First determine the time constant (τ) for this circuit.

$$\tau = RC = (2.0 \times 10^4)(5.0 \times 10^{-6}) = 10 \times 10^{-2} \text{ s} = 0.10 \text{ s} = \frac{1}{10} \text{ s}$$

Now apply the half-life equation.

$$t_{1/2} = \tau \ln(2) = \frac{\ln(2)}{10} \text{ s}$$

The half-life is $t_{1/2} = \frac{\ln(2)}{10}$ s. If you use a calculator, it works out to $t_{1/2} = 0.069$ s.

(D) How much charge is stored on the capacitor after 0.20 s?

Solution. Use the equation for exponential decay. Plug in $t = 0.20$ s.

$$Q = Q_m e^{-t/\tau} = (60 \times 10^{-6}) e^{-0.2/0.1} = (60 \times 10^{-6}) e^{-2} = \frac{60}{e^2} \ \mu\text{C}$$

Note that $-\frac{0.20}{0.10} = -2$. Also note that $e^{-2} = \frac{1}{e^2}$. The charge stored on the capacitor at $t = 0.20$ s is $Q = \frac{60}{e^2}$ μC. If you use a calculator, this works out to $Q = 8.1$ μC.

Example 74. A 4.0-μF capacitor discharges while connected in series with a resistor. The current drops from its initial value of 6.0 A down to 3.0 A after 200 ms.

(A) What is the time constant?

Solution. The "trick" to this problem is to realize that 200 ms is the half-life because the current has dropped to one-half of its initial value. Make a list of the known symbols.

- The capacitance is $C = 4.0\ \mu\text{F}$.
- The initial current is $I_m = 6.0$ A.
- The current is $I = 3.0$ A after one half-life.
- The half-life is $t_{½} = 200$ ms.

Convert the half-life and capacitance to SI units using $\mu = 10^{-6}$ and $\text{m} = 10^{-3}$.

$$C = 4.0\ \mu\text{F} = 4.0 \times 10^{-6}\ \text{F} \quad , \quad t_{½} = 200\ \text{ms} = 0.200\ \text{s}$$

Find the time constant (τ) from the half-life ($t_{½}$).

$$\tau \ln(2) = t_{½}$$

Divide both sides of the equation by $\ln(2)$.

$$\tau = \frac{t_{½}}{\ln(2)} = \frac{0.2}{\ln(2)}\ \text{s} = \frac{1}{5\ln(2)}\ \text{s}$$

Note that $0.2 = \frac{1}{5}$. The time constant is $\tau = \frac{1}{5\ln(2)}$ s. If you use a calculator, this works out to $\tau = 0.29$ s.

(B) What is the resistance of the resistor?

Solution. Use the other equation for time constant.

$$\tau = RC$$

Divide both sides of the equation by C.

$$R = \frac{\tau}{C} = \left[\frac{1}{5\ln(2)}\right]\left(\frac{1}{4.0 \times 10^{-6}}\right)$$

Note that $\frac{1}{10^{-6}} = 10^6$.

$$R = \frac{10^6}{20\ln(2)}$$

It's convenient to write $10^6 = 10^2 10^4 = 100 \times 10^4$.

$$R = \frac{100 \times 10^4}{20\ln(2)} = \frac{5 \times 10^4}{\ln(2)}\ \Omega$$

Note that $5 \times 10^4 = 50 \times 10^3$ and that $1\ \text{k}\Omega = 1{,}000\ \Omega$.

$$R = \frac{50}{\ln(2)}\ \text{k}\Omega$$

The resistance is $R = \frac{5\times10^4}{\ln(2)}\ \Omega = \frac{50}{\ln(2)}\ \text{k}\Omega$. If you use a calculator, this works out to approximately $R = 72\ \text{k}\Omega = 72{,}000\ \Omega$.

19 SCALAR AND VECTOR PRODUCTS

Vector Equations

$$\vec{\mathbf{A}} = A_x\hat{\mathbf{x}} + A_y\hat{\mathbf{y}} + A_z\hat{\mathbf{z}} \quad , \quad A = \sqrt{A_x^2 + A_y^2 + A_z^2}$$

Scalar Product

$$\vec{\mathbf{A}} \cdot \vec{\mathbf{B}} = AB\cos\theta = A_xB_x + A_yB_y + A_zB_z$$

Vector Product

$$\vec{\mathbf{A}} \times \vec{\mathbf{B}} = \begin{vmatrix} \hat{\mathbf{x}} & \hat{\mathbf{y}} & \hat{\mathbf{z}} \\ A_x & A_y & A_z \\ B_x & B_y & B_z \end{vmatrix} = \hat{\mathbf{x}}\begin{vmatrix} A_y & A_z \\ B_y & B_z \end{vmatrix} - \hat{\mathbf{y}}\begin{vmatrix} A_x & A_z \\ B_x & B_z \end{vmatrix} + \hat{\mathbf{z}}\begin{vmatrix} A_x & A_y \\ B_x & B_y \end{vmatrix}$$

$$\vec{\mathbf{A}} \times \vec{\mathbf{B}} = (A_yB_z - A_zB_y)\hat{\mathbf{x}} - (A_xB_z - A_zB_x)\hat{\mathbf{y}} + (A_xB_y - A_yB_x)\hat{\mathbf{z}}$$

$$\vec{\mathbf{A}} \times \vec{\mathbf{B}} = A_yB_z\hat{\mathbf{x}} - A_zB_y\hat{\mathbf{x}} + A_zB_x\hat{\mathbf{y}} - A_xB_z\hat{\mathbf{y}} + A_xB_y\hat{\mathbf{z}} - A_yB_x\hat{\mathbf{z}}$$

Magnitude of the Vector Product

$$\|\vec{\mathbf{C}}\| = \|\vec{\mathbf{A}} \times \vec{\mathbf{B}}\| = AB\sin\theta = \sqrt{C_x^2 + C_y^2 + C_z^2}$$

Scalar Products among Cartesian Unit Vectors

$$\hat{\mathbf{x}} \cdot \hat{\mathbf{x}} = 1 \quad , \quad \hat{\mathbf{y}} \cdot \hat{\mathbf{y}} = 1 \quad , \quad \hat{\mathbf{z}} \cdot \hat{\mathbf{z}} = 1$$
$$\hat{\mathbf{x}} \cdot \hat{\mathbf{y}} = 0 \quad , \quad \hat{\mathbf{y}} \cdot \hat{\mathbf{z}} = 0 \quad , \quad \hat{\mathbf{z}} \cdot \hat{\mathbf{x}} = 0$$
$$\hat{\mathbf{y}} \cdot \hat{\mathbf{x}} = 0 \quad , \quad \hat{\mathbf{z}} \cdot \hat{\mathbf{y}} = 0 \quad , \quad \hat{\mathbf{x}} \cdot \hat{\mathbf{z}} = 0$$

Vector Products among Cartesian Unit Vectors

$$\hat{\mathbf{x}} \times \hat{\mathbf{x}} = 0 \quad , \quad \hat{\mathbf{y}} \times \hat{\mathbf{y}} = 0 \quad , \quad \hat{\mathbf{z}} \times \hat{\mathbf{z}} = 0$$
$$\hat{\mathbf{x}} \times \hat{\mathbf{y}} = \hat{\mathbf{z}} \quad , \quad \hat{\mathbf{y}} \times \hat{\mathbf{z}} = \hat{\mathbf{x}} \quad , \quad \hat{\mathbf{z}} \times \hat{\mathbf{x}} = \hat{\mathbf{y}}$$
$$\hat{\mathbf{y}} \times \hat{\mathbf{x}} = -\hat{\mathbf{z}} \quad , \quad \hat{\mathbf{z}} \times \hat{\mathbf{y}} = -\hat{\mathbf{x}} \quad , \quad \hat{\mathbf{x}} \times \hat{\mathbf{z}} = -\hat{\mathbf{y}}$$

Example 75. Consider the following vectors.

$$\vec{\mathbf{A}} = 5\,\hat{\mathbf{x}} + 2\,\hat{\mathbf{y}} + 3\,\hat{\mathbf{z}}$$
$$\vec{\mathbf{B}} = 4\,\hat{\mathbf{x}} - 6\,\hat{\mathbf{y}} - 2\,\hat{\mathbf{z}}$$

(A) Find the scalar product $\vec{\mathbf{A}} \cdot \vec{\mathbf{B}}$.

Solution. Compare $\vec{\mathbf{A}} = 5\,\hat{\mathbf{x}} + 2\,\hat{\mathbf{y}} + 3\,\hat{\mathbf{z}}$ with $\vec{\mathbf{A}} = A_x\hat{\mathbf{x}} + A_y\hat{\mathbf{y}} + A_z\hat{\mathbf{z}}$ to see that $A_x = 5$, $A_y = 2$, and $A_z = 3$. Similarly, compare $\vec{\mathbf{B}} = 4\,\hat{\mathbf{x}} - 6\,\hat{\mathbf{y}} - 2\,\hat{\mathbf{z}}$ with $\vec{\mathbf{B}} = B_x\hat{\mathbf{x}} + B_y\hat{\mathbf{y}} + B_z\hat{\mathbf{z}}$ to see that $B_x = 4$, $B_y = -6$, and $B_z = -2$. Since we know the components of the given vectors, use the component form of the scalar product.

$$\vec{\mathbf{A}} \cdot \vec{\mathbf{B}} = A_x B_x + A_y B_y + A_z B_z = (5)(4) + (2)(-6) + (3)(-2) = 20 - 12 - 6 = 2$$

The scalar product between these two vectors is $\vec{\mathbf{A}} \cdot \vec{\mathbf{B}} = 2$.

(B) Find the vector product $\vec{\mathbf{A}} \times \vec{\mathbf{B}}$.

Solution. Use the determinant form of the vector product.

$$\vec{\mathbf{A}} \times \vec{\mathbf{B}} = \begin{vmatrix} \hat{\mathbf{x}} & \hat{\mathbf{y}} & \hat{\mathbf{z}} \\ A_x & A_y & A_z \\ B_x & B_y & B_z \end{vmatrix} = \hat{\mathbf{x}} \begin{vmatrix} A_y & A_z \\ B_y & B_z \end{vmatrix} - \hat{\mathbf{y}} \begin{vmatrix} A_x & A_z \\ B_x & B_z \end{vmatrix} + \hat{\mathbf{z}} \begin{vmatrix} A_x & A_y \\ B_x & B_y \end{vmatrix}$$

Plug in the numbers for the components that we found in part (A).

$$\vec{\mathbf{A}} \times \vec{\mathbf{B}} = \begin{vmatrix} \hat{\mathbf{x}} & \hat{\mathbf{y}} & \hat{\mathbf{z}} \\ 5 & 2 & 3 \\ 4 & -6 & -2 \end{vmatrix} = \hat{\mathbf{x}} \begin{vmatrix} 2 & 3 \\ -6 & -2 \end{vmatrix} - \hat{\mathbf{y}} \begin{vmatrix} 5 & 3 \\ 4 & -2 \end{vmatrix} + \hat{\mathbf{z}} \begin{vmatrix} 5 & 2 \\ 4 & -6 \end{vmatrix}$$

$$\vec{\mathbf{A}} \times \vec{\mathbf{B}} = \hat{\mathbf{x}}[(2)(-2) - (3)(-6)] - \hat{\mathbf{y}}[(5)(-2) - (3)(4)] + \hat{\mathbf{z}}[(5)(-6) - (2)(4)]$$

$$\vec{\mathbf{A}} \times \vec{\mathbf{B}} = \hat{\mathbf{x}}(-4 + 18) - \hat{\mathbf{y}}(-10 - 12) + \hat{\mathbf{z}}(-30 - 8)$$

$$\vec{\mathbf{A}} \times \vec{\mathbf{B}} = 14\,\hat{\mathbf{x}} - (-22)\hat{\mathbf{y}} - 38\,\hat{\mathbf{z}}$$

$$\vec{\mathbf{A}} \times \vec{\mathbf{B}} = 14\,\hat{\mathbf{x}} + 22\,\hat{\mathbf{y}} - 38\,\hat{\mathbf{z}}$$

The vector product between these two vectors is $\vec{\mathbf{A}} \times \vec{\mathbf{B}} = 14\,\hat{\mathbf{x}} + 22\,\hat{\mathbf{y}} - 38\,\hat{\mathbf{z}}$.

Example 76. Consider the following vectors.

$$\vec{\mathbf{A}} = 3\,\hat{\mathbf{x}} - \hat{\mathbf{y}} - 4\,\hat{\mathbf{z}}$$
$$\vec{\mathbf{B}} = 2\,\hat{\mathbf{x}} - \hat{\mathbf{z}}$$

(A) Find the scalar product $\vec{\mathbf{A}} \cdot \vec{\mathbf{B}}$.

Solution. Compare $\vec{\mathbf{A}} = 3\,\hat{\mathbf{x}} - \hat{\mathbf{y}} - 4\,\hat{\mathbf{z}}$ with $\vec{\mathbf{A}} = A_x\hat{\mathbf{x}} + A_y\hat{\mathbf{y}} + A_z\hat{\mathbf{z}}$ to see that $A_x = 3$, $A_y = -1$, and $A_z = -4$. Similarly, compare $\vec{\mathbf{B}} = 2\,\hat{\mathbf{x}} - \hat{\mathbf{z}}$ with $\vec{\mathbf{B}} = B_x\hat{\mathbf{x}} + B_y\hat{\mathbf{y}} + B_z\hat{\mathbf{z}}$ to see that $B_x = 2$, $B_y = 0$, and $B_z = -1$. (Observe that $\vec{\mathbf{B}}$ doesn't have a $\hat{\mathbf{y}}$: That's why $B_y = 0$.) Since we know the components of the given vectors, use the component form of the scalar product.

$$\vec{\mathbf{A}} \cdot \vec{\mathbf{B}} = A_x B_x + A_y B_y + A_z B_z = (3)(2) + (-1)(0) + (-4)(-1) = 6 - 0 + 4 = 10$$

The scalar product between these two vectors is $\vec{\mathbf{A}} \cdot \vec{\mathbf{B}} = 10$.

(B) Find the vector product $\vec{\mathbf{A}} \times \vec{\mathbf{B}}$.

Solution. Use the determinant form of the vector product.

$$\vec{\mathbf{A}} \times \vec{\mathbf{B}} = \begin{vmatrix} \hat{\mathbf{x}} & \hat{\mathbf{y}} & \hat{\mathbf{z}} \\ A_x & A_y & A_z \\ B_x & B_y & B_z \end{vmatrix} = \hat{\mathbf{x}} \begin{vmatrix} A_y & A_z \\ B_y & B_z \end{vmatrix} - \hat{\mathbf{y}} \begin{vmatrix} A_x & A_z \\ B_x & B_z \end{vmatrix} + \hat{\mathbf{z}} \begin{vmatrix} A_x & A_y \\ B_x & B_y \end{vmatrix}$$

Plug in the numbers for the components that we found in part (A).

$$\vec{\mathbf{A}} \times \vec{\mathbf{B}} = \begin{vmatrix} \hat{\mathbf{x}} & \hat{\mathbf{y}} & \hat{\mathbf{z}} \\ 3 & -1 & -4 \\ 2 & 0 & -1 \end{vmatrix} = \hat{\mathbf{x}} \begin{vmatrix} -1 & -4 \\ 0 & -1 \end{vmatrix} - \hat{\mathbf{y}} \begin{vmatrix} 3 & -4 \\ 2 & -1 \end{vmatrix} + \hat{\mathbf{z}} \begin{vmatrix} 3 & -1 \\ 2 & 0 \end{vmatrix}$$

$$\vec{\mathbf{A}} \times \vec{\mathbf{B}} = \hat{\mathbf{x}}[(-1)(-1) - (-4)(0)] - \hat{\mathbf{y}}[(3)(-1) - (-4)(2)] + \hat{\mathbf{z}}[(3)(0) - (-1)(2)]$$

$$\vec{\mathbf{A}} \times \vec{\mathbf{B}} = \hat{\mathbf{x}}(1 + 0) - \hat{\mathbf{y}}(-3 + 8) + \hat{\mathbf{z}}(0 + 2)$$

$$\vec{\mathbf{A}} \times \vec{\mathbf{B}} = \hat{\mathbf{x}} - (5)\hat{\mathbf{y}} + 2\,\hat{\mathbf{z}}$$

$$\vec{\mathbf{A}} \times \vec{\mathbf{B}} = \hat{\mathbf{x}} - 5\,\hat{\mathbf{y}} + 2\,\hat{\mathbf{z}}$$

The vector product between these two vectors is $\vec{\mathbf{A}} \times \vec{\mathbf{B}} = \hat{\mathbf{x}} - 5\,\hat{\mathbf{y}} + 2\,\hat{\mathbf{z}}$.

Example 77. The magnitude of vector $\vec{\mathbf{A}}$ is 8.0 and the magnitude of vector $\vec{\mathbf{B}}$ is 5.0. The angle between vectors $\vec{\mathbf{A}}$ and $\vec{\mathbf{B}}$ is 60°.

(A) Find the scalar product between $\vec{\mathbf{A}}$ and $\vec{\mathbf{B}}$.
Solution. Identify the given values:

- Vector $\vec{\mathbf{A}}$ has a magnitude of $A = 8$.
- Vector $\vec{\mathbf{B}}$ has a magnitude of $B = 5$.
- The angle between vectors $\vec{\mathbf{A}}$ and $\vec{\mathbf{B}}$ is $\theta = 60°$.

Apply the trigonometric form of the scalar product.

$$\vec{\mathbf{A}} \cdot \vec{\mathbf{B}} = AB\cos\theta = (8)(5)\cos 60° = (40)\left(\frac{1}{2}\right) = 20$$

The scalar product between these two vectors is $\vec{\mathbf{A}} \cdot \vec{\mathbf{B}} = 20$.

(B) Find the magnitude of the vector product between $\vec{\mathbf{A}}$ and $\vec{\mathbf{B}}$.
Solution. Apply the trigonometric form of the vector product.

$$\left\|\vec{\mathbf{A}} \times \vec{\mathbf{B}}\right\| = AB\sin\theta = (8)(5)\sin 60° = (40)\left(\frac{\sqrt{3}}{2}\right) = 20\sqrt{3}$$

The magnitude of the vector product between these two vectors is $\left\|\vec{\mathbf{A}} \times \vec{\mathbf{B}}\right\| = 20\sqrt{3}$.

20 BAR MAGNETS

Symbol	Name	SI Units
$\vec{\mathbf{B}}$	magnetic field	T

Magnetic Field Lines of a Bar Magnet

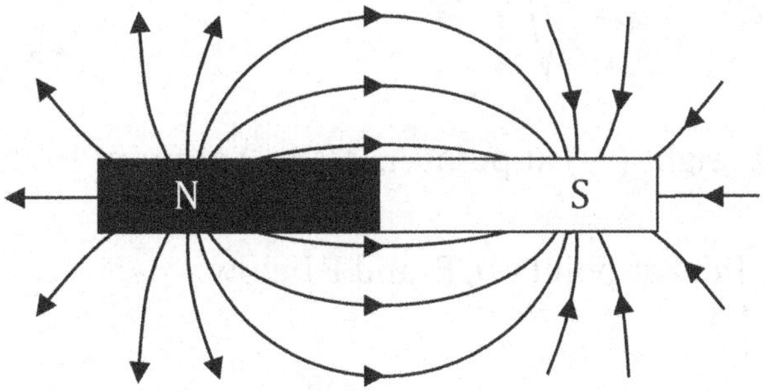

Earth's Magnetic Field

Image of earth from NASA.

Example 78. (A) Sketch the magnetic field at points A, B, and C below.

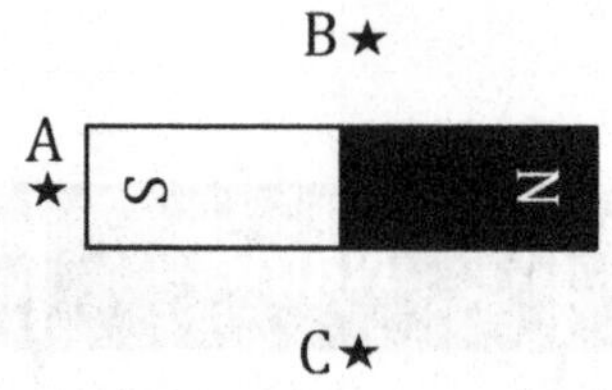

Flip the diagram from the previous page (since these poles are backwards).

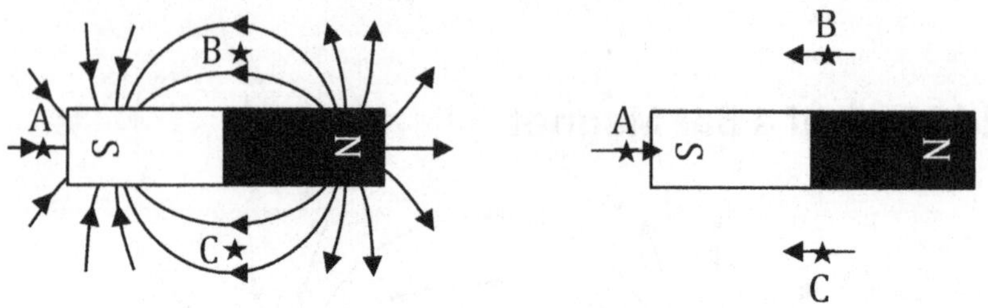

The magnetic field points right (→) at point A, left (←) at point B, and left (←) at point C.

(B) Sketch the magnetic field at points D, E, and F below.

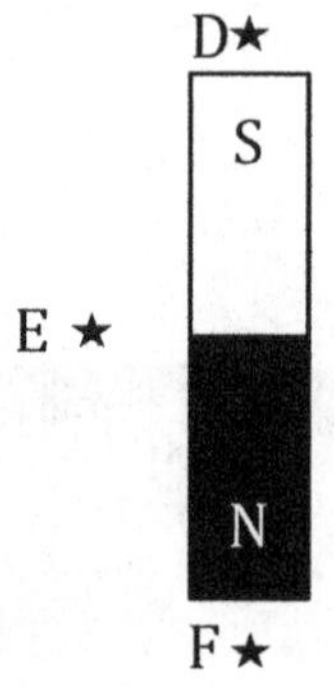

Rotate the diagram from the previous page.

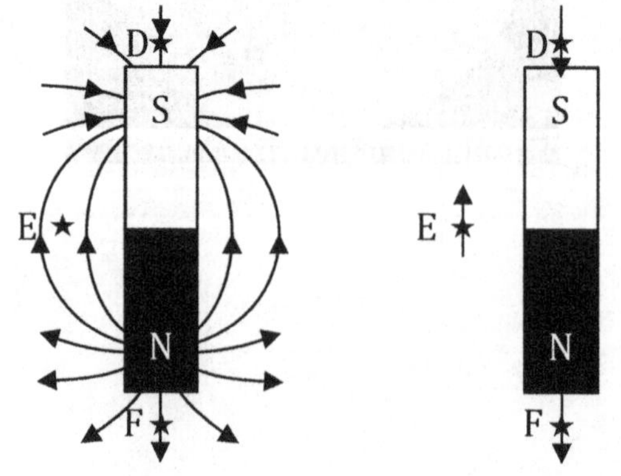

The magnetic field points down (↓) at point D, up (↑) at point E, and down (↓) at point F.

(C) Sketch the magnetic field at points G, H, and I below.

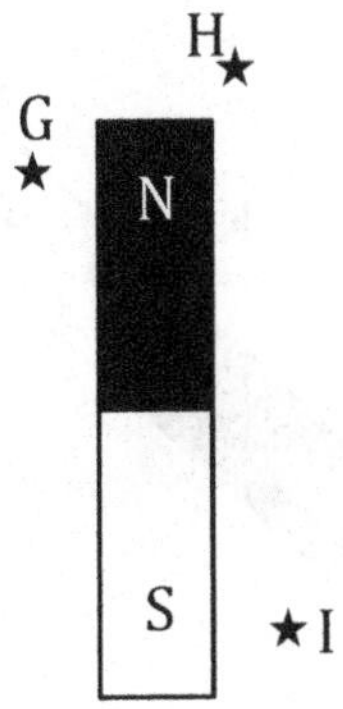

Rotate the diagram from page 227.

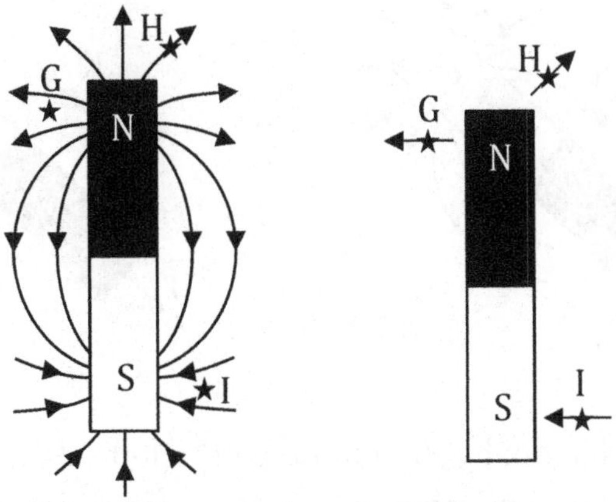

The magnetic field points (approximately) left (←) at point G, diagonally up and to the right (↗) at point H, and (approximately) left (←) at point I.

(D) Sketch the magnetic field at points J, K, and L below.

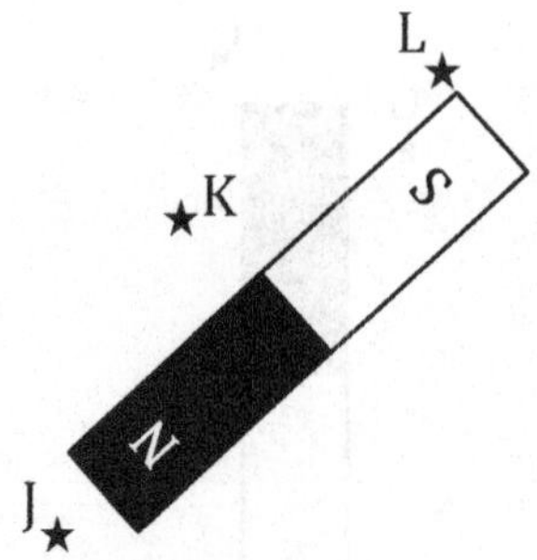

Rotate the diagram from page 227.

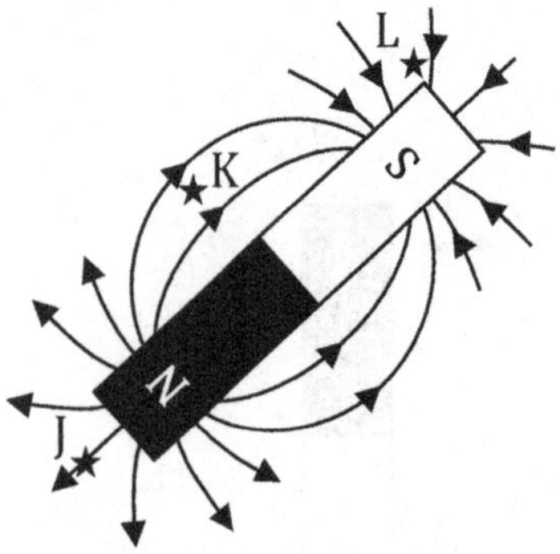

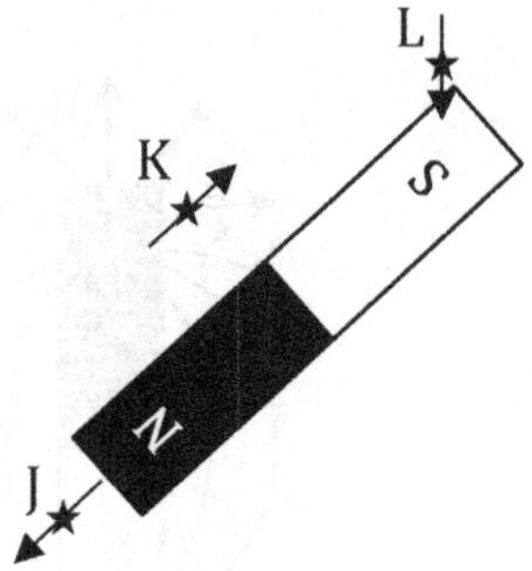

The magnetic field points diagonally down and to the left (↙) at point J, diagonally up and to the right (↗) at point K, and (approximately) down (↓) at point L.

21 RIGHT-HAND RULE FOR MAGNETIC FORCE

Right-hand Rule for Magnetic Force

To find the direction of the magnetic force ($\vec{\mathbf{F}}_m$) exerted on a moving charge (or current-carrying wire) in the presence of an external magnetic field ($\vec{\mathbf{B}}$):

- Point the extended fingers of your right hand along the velocity ($\vec{v}$) or current (I).
- Rotate your forearm until your palm faces the magnetic field ($\vec{\mathbf{B}}$), meaning that your palm will be perpendicular to the magnetic field.
- When your right-hand is simultaneously doing both of the first two steps, your extended thumb will point along the magnetic force ($\vec{\mathbf{F}}_m$).
- As a check, if your fingers are pointing toward the velocity ($\vec{v}$) or current (I) while your thumb points along the magnetic force ($\vec{\mathbf{F}}_m$), if you then bend your fingers as shown below, your fingers should now point along the magnetic field ($\vec{\mathbf{B}}$).

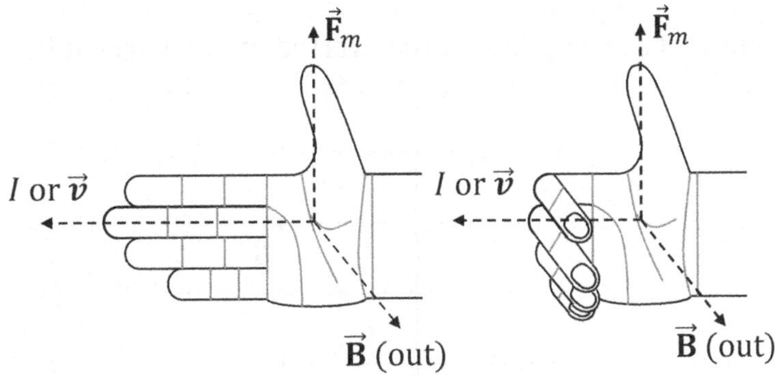

Important Exception

If the velocity ($\vec{v}$) or current (I) is **parallel** or **anti-parallel** to the magnetic field ($\vec{\mathbf{B}}$), the magnetic force ($\vec{\mathbf{F}}_m$) is **zero**. In Chapter 24, we'll learn that in these two extreme cases, $\theta = 0°$ or $180°$ such that $\sin\theta = 0$.

Symbol	Name	SI Units
$\vec{\mathbf{B}}$	magnetic field	T
$\vec{\mathbf{F}}_m$	magnetic force	N
$\vec{v}$	velocity	m/s
I	current	A

Symbol	Meaning
$\otimes$	into the page
$\odot$	out of the page
p	proton
n	neutron
e^-	electron
N	north pole
S	south pole

Example 79. Apply the right-hand rule for magnetic force to answer each question.

(A) What is the direction of the magnetic force exerted on the current below?

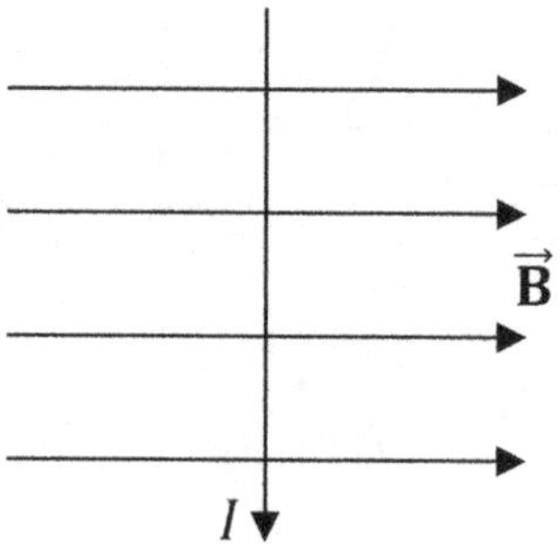

Solution. Apply the right-hand rule for magnetic force:

- Point your fingers down (↓), along the current (I).
- At the same time, face your palm to the right (→), along the magnetic field ($\vec{\mathbf{B}}$).
- If your fingers point down (↓) at the same time as your palm faces right (→), your thumb will be pointing out of the page ($\odot$).
- Your thumb is the answer: The magnetic force ($\vec{\mathbf{F}}_m$) is out of the page ($\odot$).

Note: The symbol $\odot$ represents an arrow pointing out of the page.

Tip: Make sure you're using your **right** hand.*

Tip: If you find it difficult to physically get your right hand into the correct position, try turning your book around until you find an angle that makes it more comfortable.

* This should seem obvious, right? But guess what: If you're right-handed, when you're taking a test, your right hand is busy writing, so it's instinctive to want to use your free hand, which is the wrong one.

(B) What is the direction of the magnetic force exerted on the proton below?

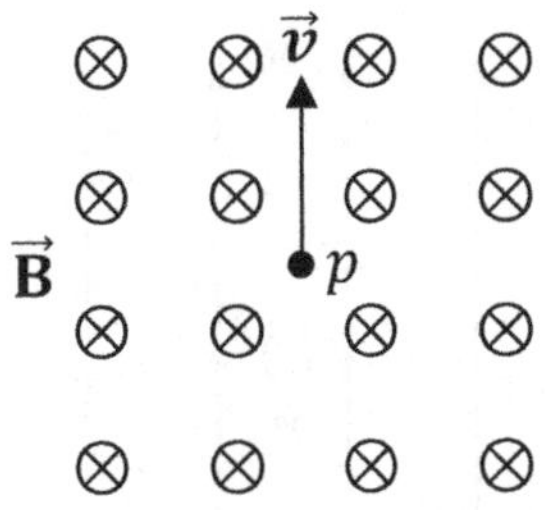

Solution. Apply the right-hand rule for magnetic force:

- Point your fingers up (↑), along the velocity ($\vec{v}$).
- At the same time, face your palm into the page (⊗), along the magnetic field ($\vec{\mathbf{B}}$).
- Your thumb points to the left: The magnetic force ($\vec{\mathbf{F}}_m$) is to the left (←).

Note: The symbol ⊗ represents an arrow pointing into the page.

(C) What is the direction of the magnetic force exerted on the current below?

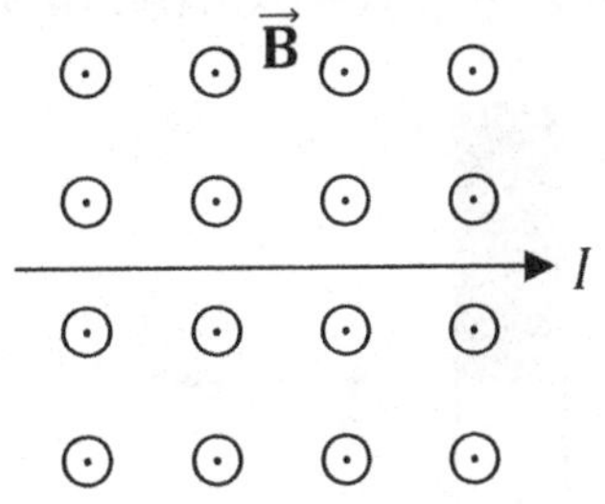

Tip: Turn the book to make it easier to get your right hand into the correct position.

Solution. Apply the right-hand rule for magnetic force:

- Point your fingers to the right (→), along the current (I).
- At the same time, face your palm out of the page (⊙), along the magnetic field ($\vec{\mathbf{B}}$).
- Your thumb points down: The magnetic force ($\vec{\mathbf{F}}_m$) is down (↓).

(D) What is the direction of the magnetic force exerted on the electron below?

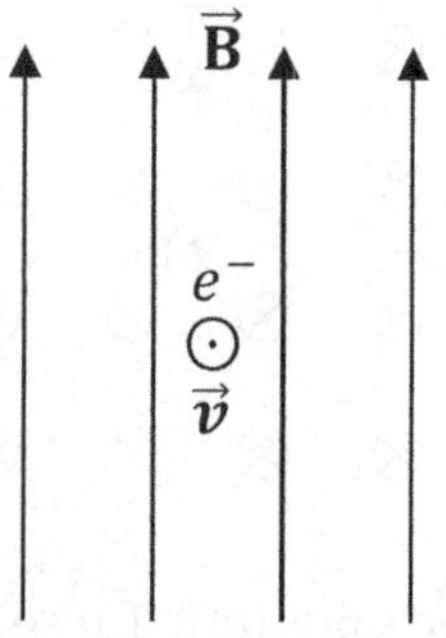

Solution. Apply the right-hand rule for magnetic force:
- Point your fingers out of the page (⊙), along the velocity ($\vec{v}$).
- At the same time, face your palm up (↑), along the magnetic field ($\vec{\mathbf{B}}$).
- Your thumb points to the left, but that's **not** the answer: The electron has **negative** charge, so the answer is **backwards**: The magnetic force ($\vec{\mathbf{F}}_m$) is to the right (→).

(E) What is the direction of the magnetic force exerted on the proton below?

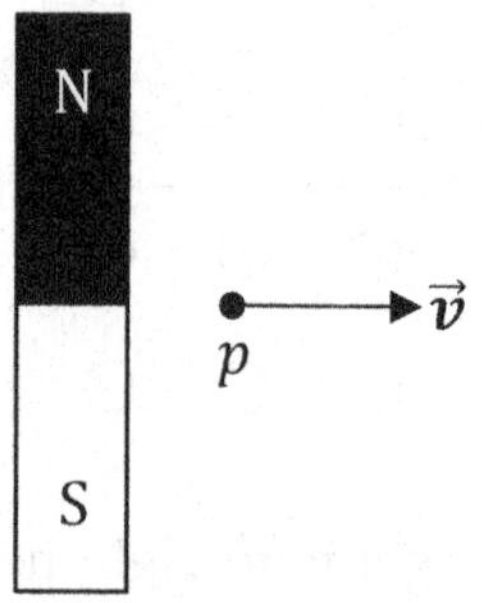

Solution. First sketch the magnetic field lines for the bar magnet (see Chapter 20). What is the direction of the magnetic field lines where the proton is? See point A below: $\vec{\mathbf{B}}$ is down. However, this is **not** the answer. We still need to find magnetic **force**.

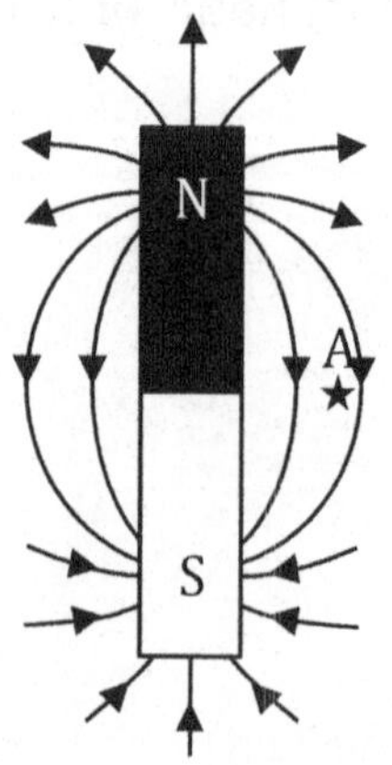

Apply the right-hand rule for magnetic force:

- Point your fingers to the right (→), along the velocity ($\vec{v}$).
- At the same time, face your palm down (↓), along the magnetic field ($\vec{\mathbf{B}}$). This is because the magnetic field lines in the previous diagram point down at point A, where the proton is located.
- Your thumb points into the page: The magnetic force ($\vec{\mathbf{F}}_m$) is into the page (⊗).

(F) What is the direction of the magnetic field, assuming that the magnetic field is perpendicular to the current below?

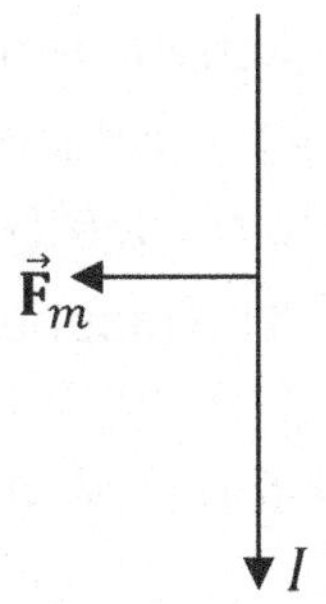

Solution. Invert the right-hand rule for magnetic force. This time, we're solving for magnetic field ($\vec{\mathbf{B}}$), <u>not</u> magnetic force ($\vec{\mathbf{F}}_m$).

- Point your fingers down (↓), along the current (I).
- At the same time, point your <u>thumb</u> (it's <u>not</u> your palm in this example) to the left (←), along the magnetic force ($\vec{\mathbf{F}}_m$).
- Your <u>palm</u> faces out of the page: The magnetic field ($\vec{\mathbf{B}}$) is out of the page (⊙).

Note: The magnetic force is always perpendicular to both the current and the magnetic field, but the current and magnetic field need not be perpendicular. We will work with the angle between I and $\vec{\mathbf{B}}$ in Chapter 24.

Example 80. Apply the right-hand rule for magnetic force to answer each question.

(A) What is the direction of the magnetic force exerted on the proton below?

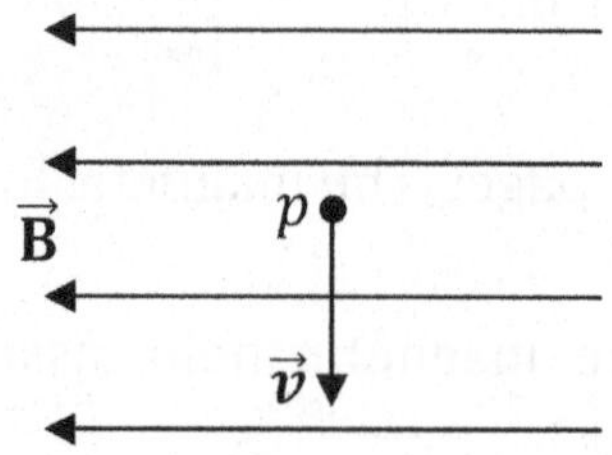

Solution. Apply the right-hand rule for magnetic force:

- Point your fingers down (↓), along the velocity ($\vec{v}$).
- At the same time, face your palm to the left (←), along the magnetic field ($\vec{\mathbf{B}}$).
- Your thumb points into the page: The magnetic force ($\vec{\mathbf{F}}_m$) is into the page (⊗).

(B) What is the direction of the magnetic force exerted on the current below?

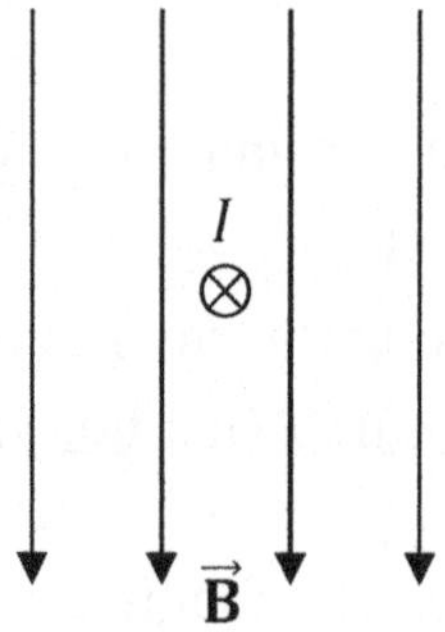

Solution. Apply the right-hand rule for magnetic force:

- Point your fingers into the page (⊗), along the current (I).
- At the same time, face your palm down (↓), along the magnetic field ($\vec{\mathbf{B}}$).
- Your thumb points to the left: The magnetic force ($\vec{\mathbf{F}}_m$) is to the left (←).

(C) What is the direction of the magnetic force exerted on the current below?

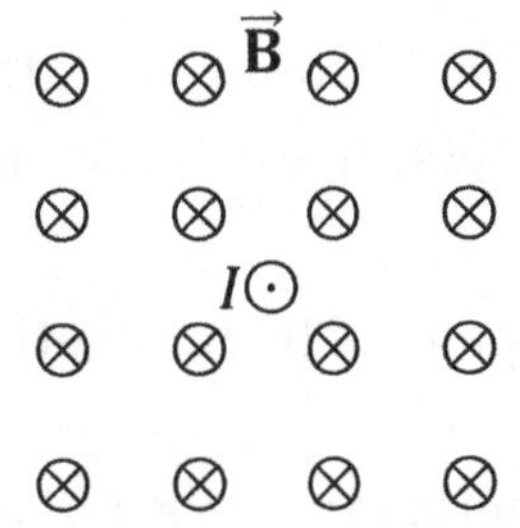

Solution. This is a "trick" question. In this example, the current (I) is anti-parallel to the magnetic field ($\vec{\mathbf{B}}$). According to the Important Exception on page 231, the magnetic force ($\vec{\mathbf{F}}_m$) is therefore **zero** (and thus has no direction). In Chapter 24, we'll see that in this case, $\theta = 180°$ such that $F_m = ILB \sin 180° = 0$.

(D) What is the direction of the magnetic force exerted on the proton below?

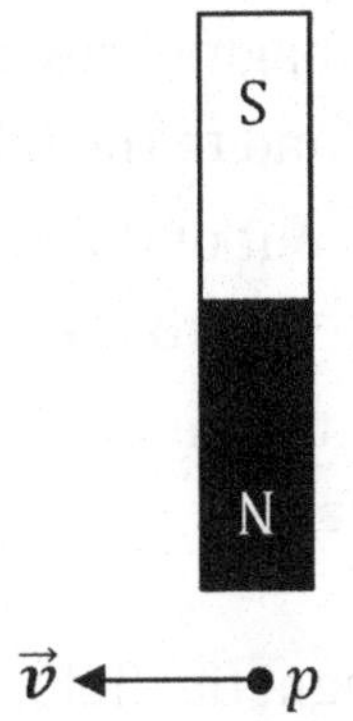

Solution. First sketch the magnetic field lines for the bar magnet (see Chapter 20). What is the direction of the magnetic field lines where the proton is? See point A below: $\vec{\mathbf{B}}$ is down. However, this is **not** the answer. We still need to find magnetic **force**. (Note that the north pole is at the bottom of the figure. The magnet is upside down.)

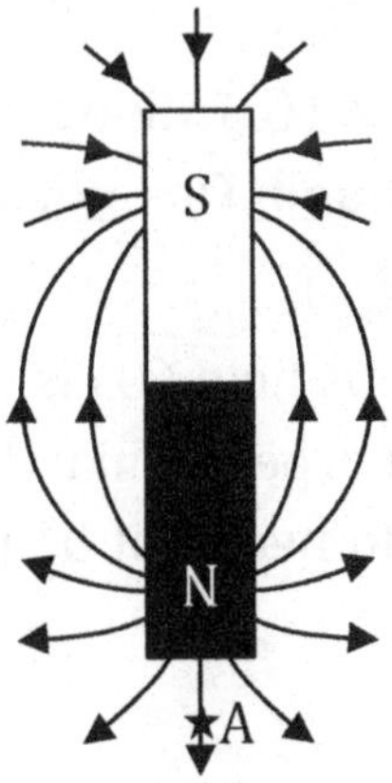

Apply the right-hand rule for magnetic force:

- Point your fingers to the left (←), along the velocity ($\vec{v}$).
- At the same time, face your palm down (↓), along the magnetic field ($\vec{\mathbf{B}}$). This is because the magnetic field lines in the previous diagram point down at point A, where the proton is located.
- Your thumb points out of the page: The magnetic force ($\vec{\mathbf{F}}_m$) is out of the page (⊙).

(E) What is the direction of the magnetic force exerted on the electron below?

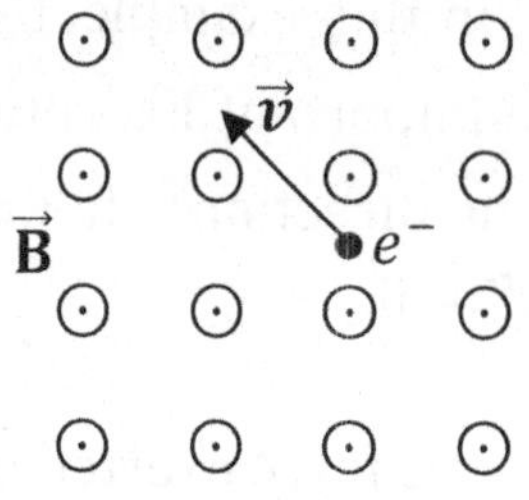

Solution. Apply the right-hand rule for magnetic force:

- Point your fingers diagonally up and to the left (↖), along the velocity ($\vec{v}$).
- At the same time, face your palm out of the page (⊙), along the magnetic field ($\vec{\mathbf{B}}$).
- Your thumb points diagonally up and to the right (↗), but that's **not** the answer: The electron has **negative** charge, so the answer is **backwards**: The magnetic force ($\vec{\mathbf{F}}_m$) is diagonally down and to the left (↙).

(F) What is the direction of the magnetic field, assuming that the magnetic field is perpendicular to the current below?

Solution. Invert the right-hand rule for magnetic force. This time, we're solving for magnetic field ($\vec{\mathbf{B}}$), **not** magnetic force ($\vec{\mathbf{F}}_m$).

- Point your fingers out of the page (⊙), along the current (I).
- At the same time, point your **thumb** (it's **not** your palm in this example) to the left (←), along the magnetic force ($\vec{\mathbf{F}}_m$).
- Your **palm** faces up: The magnetic field ($\vec{\mathbf{B}}$) is up (↑).

Note: The magnetic force is always perpendicular to both the current and the magnetic field, but the current and magnetic field need not be perpendicular. We will work with the angle between I and $\vec{\mathbf{B}}$ in Chapter 24.

Example 81. Apply the right-hand rule for magnetic force to answer each question.

(A) What is the direction of the magnetic force exerted on the current below?

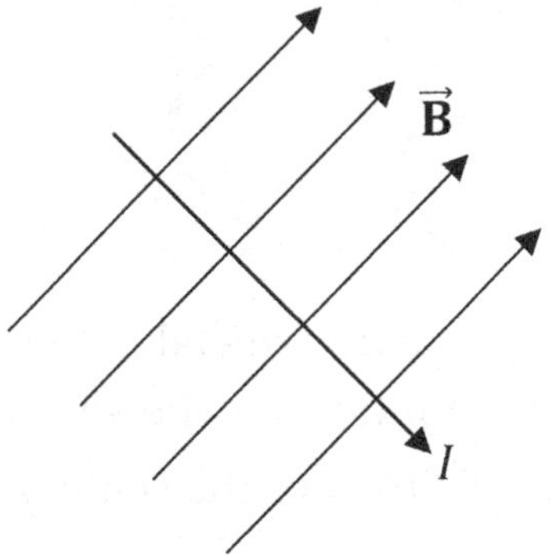

Solution. Apply the right-hand rule for magnetic force:

- Point your fingers diagonally down and to the right (↘), along the current (I).
- At the same time, face your palm diagonally up and to the right (↗), along the magnetic field ($\vec{\mathbf{B}}$).
- Your thumb points out of the page: The magnetic force ($\vec{\mathbf{F}}_m$) is out of the page (⊙).

(B) What is the direction of the magnetic force exerted on the electron below?

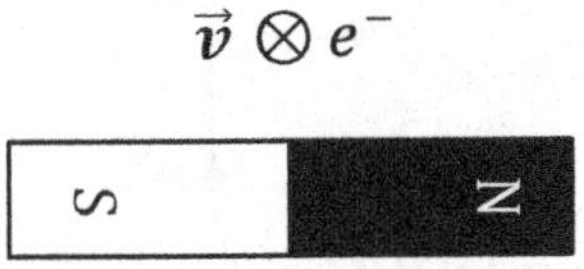

Solution. First sketch the magnetic field lines for the bar magnet (see Chapter 20). What is the direction of the magnetic field lines where the electron is? See point A below: $\vec{\mathbf{B}}$ is left.

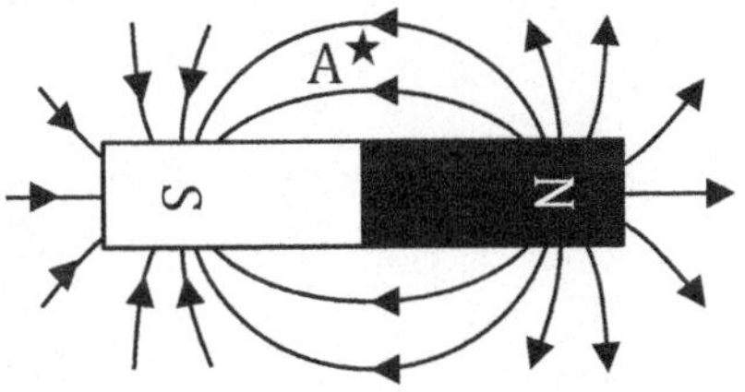

Apply the right-hand rule for magnetic force:

- Point your fingers into the page (⊗), along the velocity ($\vec{v}$).
- At the same time, face your palm to the left (←), along the magnetic field ($\vec{\mathbf{B}}$).
- Your thumb points up (↑), but that's **not** the answer: The electron has **negative** charge, so the answer is **backwards**: The magnetic force ($\vec{\mathbf{F}}_m$) is down (↓).

(C) What must be the direction of the magnetic field in order for the proton to travel in the circle shown below?

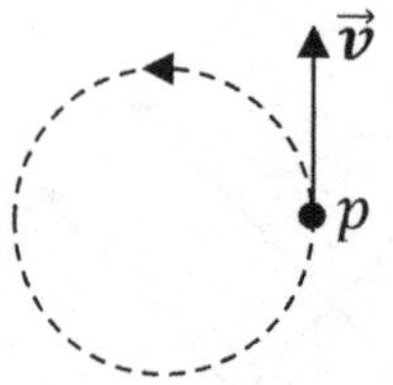

Solution. The magnetic force ($\vec{\mathbf{F}}_m$) is a **centripetal** force: It pushes the proton towards the center of the circle. For the position indicated, the velocity ($\vec{v}$) is up (along a tangent) and the magnetic force ($\vec{\mathbf{F}}_m$) is to the left (toward the center). Invert the right-hand rule to find the magnetic field.

- Point your fingers up (↑), along the instantaneous velocity ($\vec{v}$).
- At the same time, point your **thumb** (it's **not** your palm in this example) to the left (←), along the magnetic force ($\vec{\mathbf{F}}_m$).
- Your **palm** faces into the page: The magnetic field ($\vec{\mathbf{B}}$) is into the page (⊗).

(D) What must be the direction of the magnetic field in order for the electron to travel in the circle shown below?

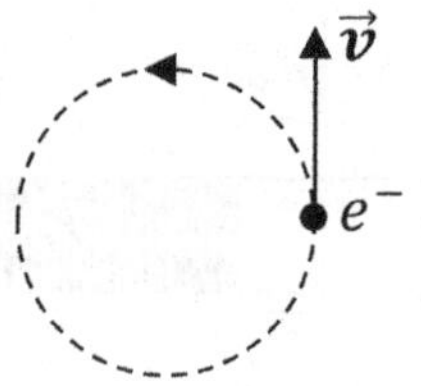

Solution. It's the same as part (C), except that the electron has **negative** charge, so the answer is **backwards**: The magnetic field ($\vec{\mathbf{B}}$) is out of the page (⊙).

(E) Would the rectangular loop below tend to rotate, expand, or contract?

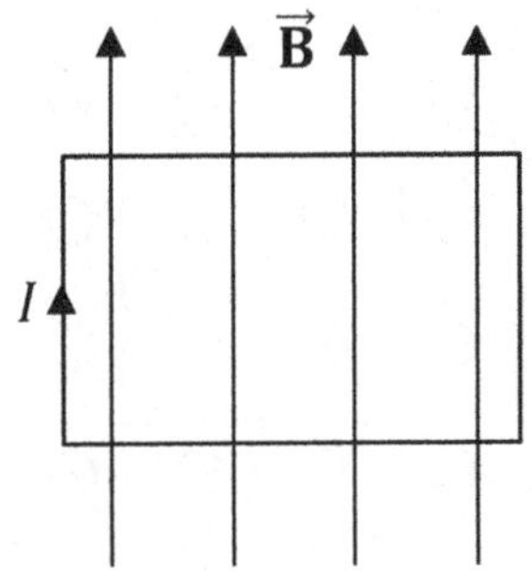

Solution. First, apply the right-hand rule for magnetic force to each side of the rectangular loop:

- Left side: The current (I) points up (↑) and the magnetic field ($\vec{\mathbf{B}}$) also points up (↑). Since I and $\vec{\mathbf{B}}$ are parallel, the magnetic force ($\vec{\mathbf{F}}_{left}$) is zero.
- Top side: Point your fingers to the right (→) along the current (I) and your palm up (↑) along the magnetic field ($\vec{\mathbf{B}}$). The magnetic force ($\vec{\mathbf{F}}_{top}$) is out of the page (⊙).
- Right side: The current (I) points down (↓) and the magnetic field ($\vec{\mathbf{B}}$) points up (↑). Since I and $\vec{\mathbf{B}}$ are anti-parallel, the magnetic force ($\vec{\mathbf{F}}_{right}$) is zero.
- Bottom side: Point your fingers to the left (←) along the current (I) and your palm up (↑) along the magnetic field ($\vec{\mathbf{B}}$). The magnetic force ($\vec{\mathbf{F}}_{bot}$) is into the page (⊗).

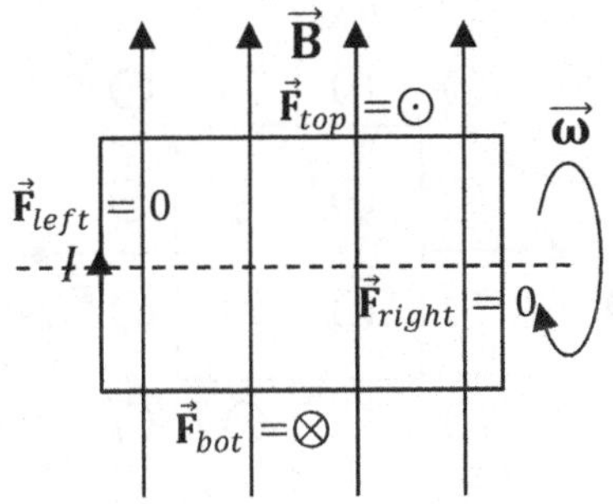

The top side of the loop is pulled out of the page, while the bottom side of the loop is pushed into the page. What will happen? The loop will rotate about the dashed axis.

(F) Would the rectangular loop below tend to rotate, expand, or contract?

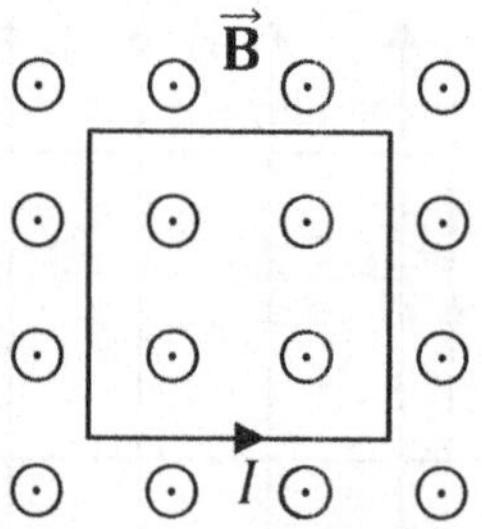

Solution. First, apply the right-hand rule for magnetic force to each side of the rectangular loop:

- **Bottom side**: Point your fingers to the right (→) along the current (I) and your palm out of the page (⊙) along the magnetic field ($\vec{\mathbf{B}}$). The magnetic force ($\vec{\mathbf{F}}_{bot}$) is down (↓).
- **Right side**: Point your fingers up (↑) along the current (I) and your palm out of the page (⊙) along the magnetic field ($\vec{\mathbf{B}}$). The magnetic force ($\vec{\mathbf{F}}_{right}$) is to the right (→).
- **Top side**: Point your fingers to the left (←) along the current (I) and your palm out of the page (⊙) along the magnetic field ($\vec{\mathbf{B}}$). The magnetic force ($\vec{\mathbf{F}}_{top}$) is up (↑).
- **Left side**: Point your fingers down (↓) along the current (I) and your palm out of the page (⊙) along the magnetic field ($\vec{\mathbf{B}}$). The magnetic force ($\vec{\mathbf{F}}_{left}$) is to the left (←).

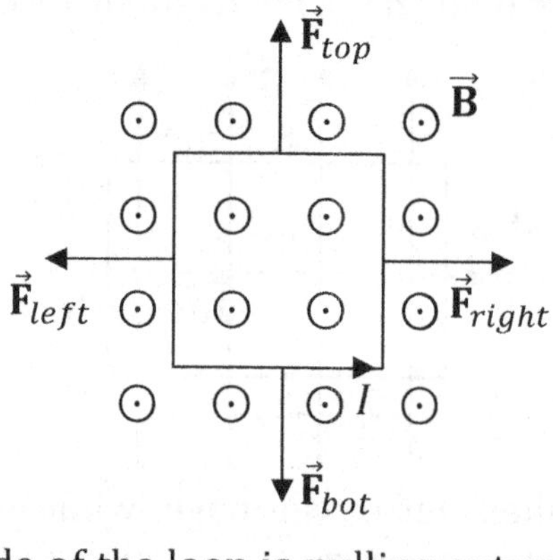

The magnetic force on each side of the loop is pulling outward, which would tend to make the loop try to expand. (That's the tendency: Whether or not it actually will expand depends on such factors as how strong the forces are and the rigidity of the materials.)

22 RIGHT-HAND RULE FOR MAGNETIC FIELD

Right-hand Rule for Magnetic Field

A long straight wire creates magnetic field lines that circulate around the current, as shown below.

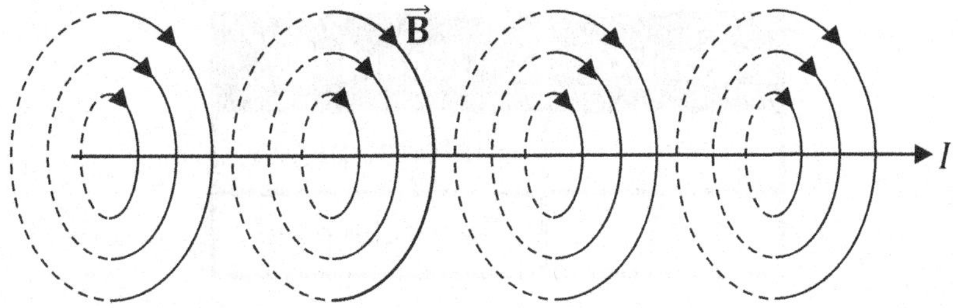

We use a right-hand rule (different from the right-hand rule that we learned in Chapter 21) to determine which way the magnetic field lines circulate. This right-hand rule gives you the direction of the magnetic field ($\vec{\mathbf{B}}$) created by a current (I) or moving charge:

- Imagine grabbing the wire with your right hand. **Tip**: You can use a pencil to represent the wire and actually grab the pencil.
- Grab the wire with your thumb pointing along the current (I).
- Your fingers represent **circular** magnetic field lines traveling around the wire **toward your fingertips**. At a given point, the direction of the magnetic field is **tangent** to these circles (your fingers).

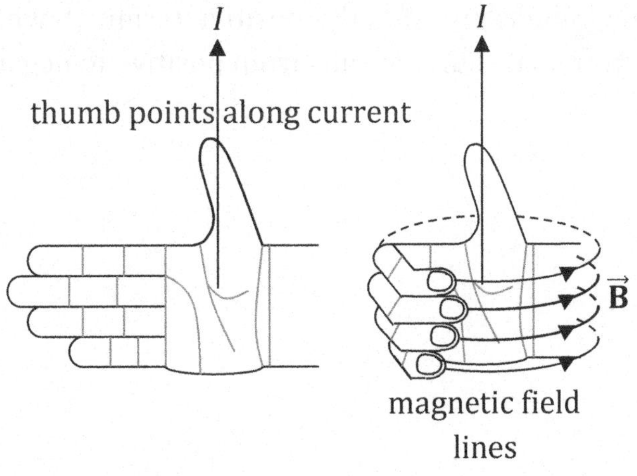

Symbol	Name	SI Units
$\vec{\mathbf{B}}$	magnetic field	T
I	current	A

Symbol	Meaning
⊗	into the page
⊙	out of the page

Schematic Representation	Symbol	Name
	R	resistor
	ΔV	battery or DC power supply

Note: Recall that the long line represents the positive terminal, while the small rectangle represents the negative terminal. Current runs from positive to negative.

Example 82. Apply the right-hand rule for magnetic field to answer each question.

(A) Determine the direction of the magnetic field at each point indicated below.

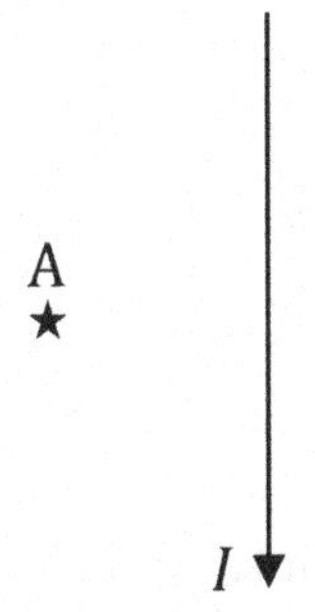

Solution. Apply the right-hand rule for magnetic field:

- Grab the current with your thumb pointing down (↓), along the current (I).
- Your fingers make circles around the wire (toward your fingertips), as shown in the diagram below on the left.
- The magnetic field ($\vec{\mathbf{B}}$) at a specified point is tangent to these circles, as shown in the diagram below on the right. Try to visualize the circles that your fingers make: Left of the wire, your fingers are going into the page, while to the right of the wire, your fingers are coming back out of the page. (Note that in order to truly "grab" the wire, your fingers would actually go "through" the page, with part of your fingers on each side of the page. Your fingers would intersect the paper to the left of the wire, where they are headed into the page, and also intersect the paper to the right of the wire, where they are headed back out of the page.) At point A the magnetic field ($\vec{\mathbf{B}}_A$) points into the page ($\otimes$), while at point C the magnetic field ($\vec{\mathbf{B}}_C$) points out of the page ($\odot$).
- It may help to study the diagram on the top of page 243 and compare it with the diagrams shown below. If you rotate the diagram from the top of page 243 to the right, so that the current points down, it actually makes the picture shown below on the left.

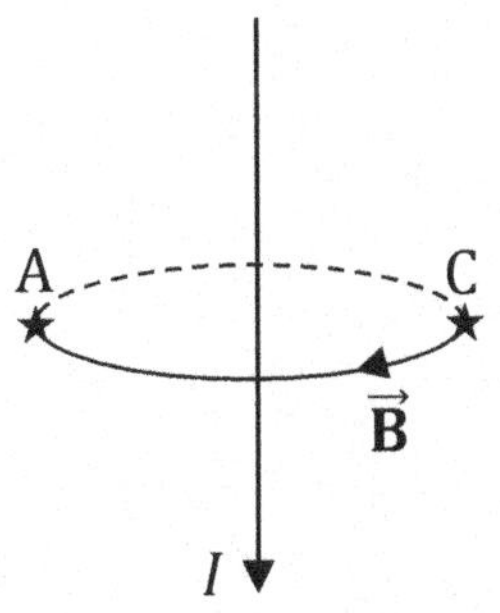

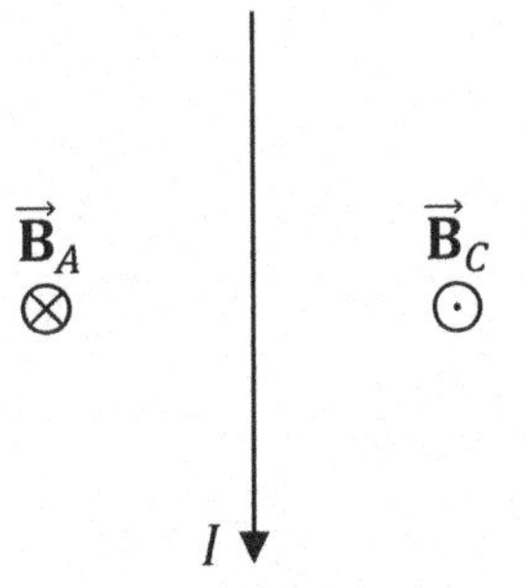

(B) Determine the direction of the magnetic field at each point indicated below.

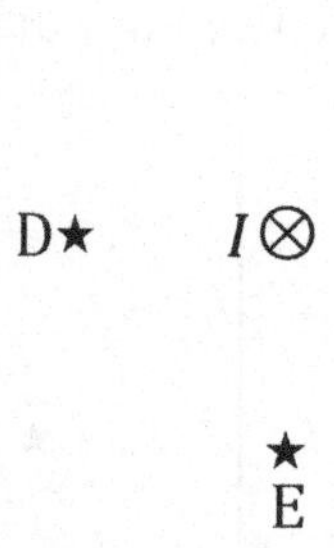

Solution. Apply the right-hand rule for magnetic field:

- Grab the current with your thumb pointing into the page (⊗), along the current (I).
- Your fingers make **clockwise** (use the right-hand rule to see this) circles around the wire (toward your fingertips), as shown in the diagram below on the left.
- The magnetic field ($\vec{\mathbf{B}}$) at a specified point is **tangent** to these circles, as shown in the diagram below on the right. Draw tangent lines at points D, E, and F with the arrows headed **clockwise**. See the diagram below on the right. At point D the magnetic field ($\vec{\mathbf{B}}_D$) points up (↑), at point E the magnetic field ($\vec{\mathbf{B}}_E$) points left (←), and at point F the magnetic field ($\vec{\mathbf{B}}_F$) points diagonally down and to the right (↘).
- (Note: If the current were heading out of the page, rather than into the page, then the magnetic field would be counterclockwise instead of clockwise.)

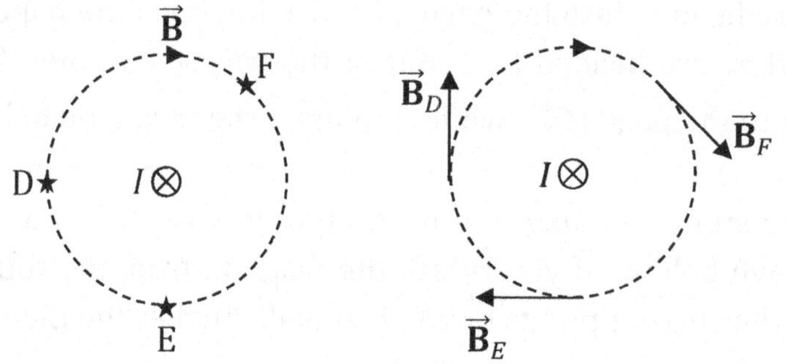

(C) Determine the direction of the magnetic field at each point indicated below.

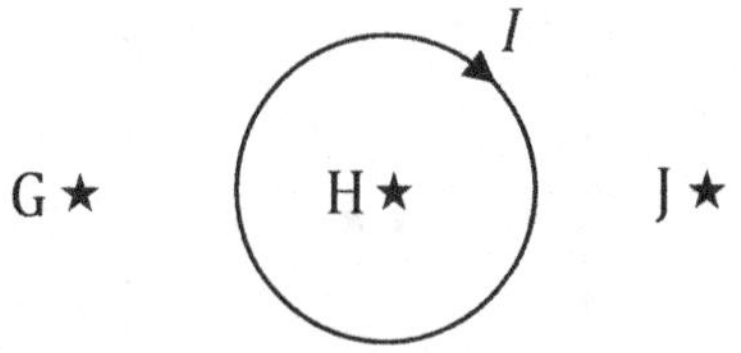

Solution. Apply the right-hand rule for magnetic field:

- Imagine grabbing the steering wheel of a car with your right hand, such that your thumb points **clockwise** (since that's how the current is drawn above). No matter where you grab the steering wheel, your fingers are going into the page (⊗) at point H. The magnetic field ($\vec{\mathbf{B}}_H$) points into the page (⊗) at point H.
- For point G, grab the loop at the leftmost point (that point is nearest to point G, so it will have the dominant effect). For point J, grab the loop at the rightmost point (the point nearest to point J). Your fingers are coming out of the page (⊙) at points G and J. The magnetic field ($\vec{\mathbf{B}}_G$ and $\vec{\mathbf{B}}_J$) points out of the page (⊙) at points G and J.

Tip: The magnetic field **outside** of the loop is **opposite** to its direction **inside** the loop.

(D) Determine the direction of the magnetic field at each point indicated below.

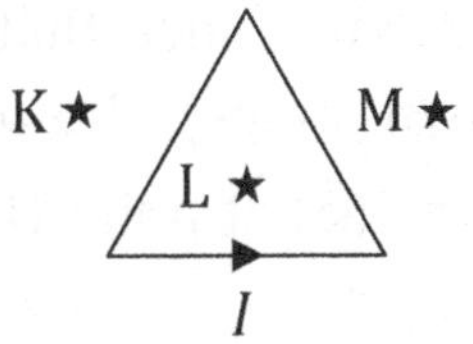

Solution. Apply the right-hand rule for magnetic field:

- Grab the loop with your right hand, such that your thumb points **counterclockwise** (since that's how the current is drawn in the problem). No matter where you grab the loop, your fingers are coming out of the page (⊙) at point L. The magnetic field ($\vec{\mathbf{B}}_L$) points out of the page (⊙) at point L.
- For point K, grab the loop at the left side (that side is nearest to point K, so it will have the dominant effect). For point M, grab the loop at the right side (the point nearest to point M). Your fingers are going into the page (⊗) at points K and M. The magnetic field ($\vec{\mathbf{B}}_K$ and $\vec{\mathbf{B}}_M$) points into the page (⊗)at points K and M.

Note: These answers are the opposite inside and outside of the loop compared to the previous question because the current is counterclockwise in this question, whereas the current was clockwise in the previous question.

(E) Determine the direction of the magnetic field at point N below.

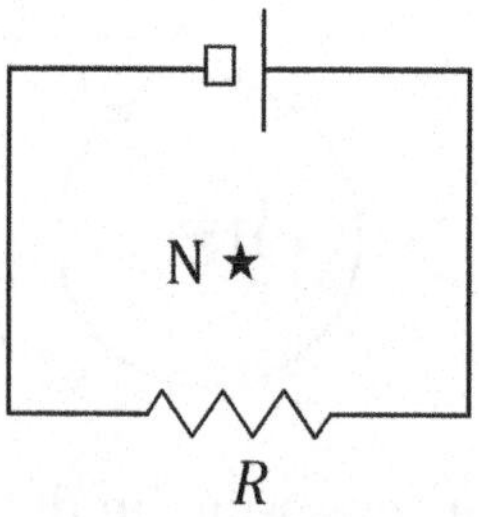

Solution. First label the positive (+) and negative (−) terminals of the battery (the long line is the positive terminal) and draw the current from the positive terminal to the negative terminal. See the diagram below on the left.

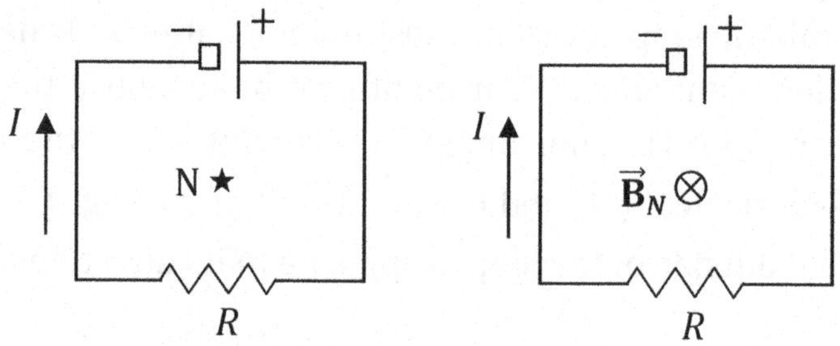

Now apply the right-hand rule for magnetic field:

- Grab the loop with your right hand, such that your thumb points **clockwise** (since that's how the current is drawn above). No matter where you grab the loop, your fingers are going into the page ($\otimes$) at point N. The magnetic field ($\vec{\mathbf{B}}_N$) points into the page ($\otimes$) at point N.

Note: This problem is the same as point H from part (C), since the current is traveling through the loop in a **clockwise** path. (This problem is also opposite compared to point L from part (D), where the current is traveling through the loop in a counterclockwise path.)

Note: In physics courses and textbooks, we draw the **conventional** current from the positive terminal to the negative terminal. This is because all of the sign conventions in physics are based on what a **positive** charge would do.* However, if you consider what is actually happening in the wire, the actual current – called the **electronic** current – involves the flow of electrons, which are negatively charged. Since we draw the conventional current based on what positive charges would do, the electrons actually flow in the opposite direction to the arrows that we draw for the conventional current: Electrons flow from negative to positive, whereas we draw the conventional current from positive to negative. (This note doesn't change our solution to the above problem in any way.)

* We have seen this in right-hand rule problems, for example: When a question involves an electron, the answer is backwards. See Example 79, part (D).

(F) Determine the direction of the magnetic field at point O below.

Solution. Note that this loop (unlike the three previous problems) does **not** lie in the plane of the paper. This loop is a **horizontal** circle with the solid (—) semicircle in front of the paper and the dashed (---) semicircle behind the paper. It's like the rim of a basketball hoop. Apply the right-hand rule for magnetic field:

- Imagine grabbing the front of the rim of a basketball hoop with your right hand, such that your thumb points to your left (since in the diagram, the current is heading to the left in the front of the loop).
- Your fingers are going down (↓) at point O inside of the loop. The magnetic field ($\vec{\mathbf{B}}_O$) points down (↓) at point O.
- (Note: If the current were heading to the right in the front of the loop, rather than to the left in the front of the loop, then the magnetic field would point up inside of the loop instead of down inside of the loop.)

(G) Determine the direction of the magnetic field at point P below.

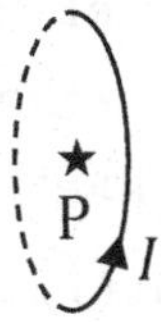

Solution. This is similar to part (F), except that now the loop is vertical instead of horizontal. Apply the right-hand rule for magnetic field:

- Imagine grabbing the front of the loop with your right hand, such that your thumb points up (since in the diagram, the current is heading up in the front of the loop).
- Your fingers are going to the left (←) at point P inside of the loop. The magnetic field ($\vec{\mathbf{B}}_P$) points to the left (←) at point P.

(H) Determine the direction of the magnetic field at point Q below.

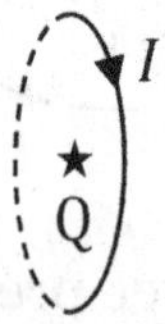

Solution. This is the same as part (G), except that the current is heading in the opposite direction. The answer is simply the opposite of part (G)'s answer. The magnetic field ($\vec{\mathbf{B}}_Q$) points to the right (→) at point Q.

(I) Determine the direction of the magnetic field at point R below.

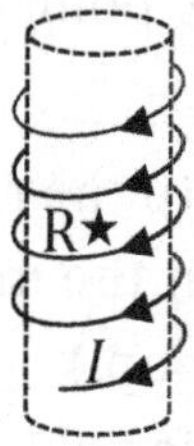

Solution. This **solenoid** essentially consists of several (approximately) horizontal loops. Each horizontal loop is just like part (F). Note that the current (I) is heading the same way (it is pointing to the left in the front of each loop) in parts (I) and (F). Therefore, just as in part (F), the magnetic field ($\vec{\mathbf{B}}_R$) points down (↓) at point R inside of the solenoid.

23 COMBINING THE TWO RIGHT-HAND RULES

Right-hand Rule for Magnetic Force

Recall the right-hand rule for magnetic **force** from Chapter 21: To find the direction of the magnetic force ($\vec{\mathbf{F}}_m$) exerted on a moving charge (or current-carrying wire) in the presence of an external magnetic field ($\vec{\mathbf{B}}$):

- Point the extended fingers of your right hand along the velocity ($\vec{v}$) or current (I).
- Rotate your forearm until your palm faces the magnetic field ($\vec{\mathbf{B}}$), meaning that your palm will be perpendicular to the magnetic field.
- When your right-hand is simultaneously doing both of the first two steps, your extended thumb will point along the magnetic force ($\vec{\mathbf{F}}_m$).
- As a check, if your fingers are pointing toward the velocity ($\vec{v}$) or current (I) while your thumb points along the magnetic force ($\vec{\mathbf{F}}_m$), if you then bend your fingers as shown below, your fingers should now point along the magnetic field ($\vec{\mathbf{B}}$).

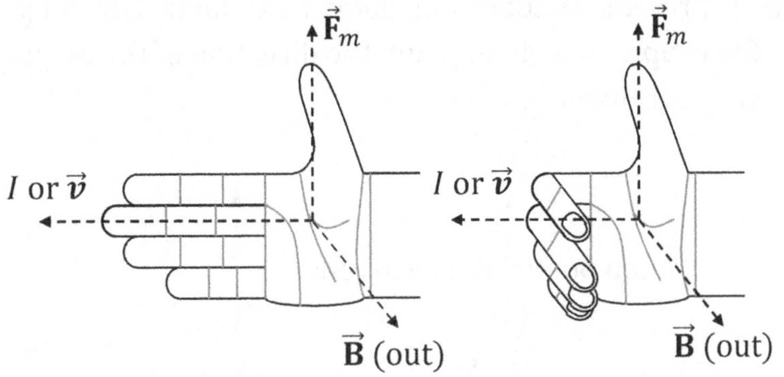

Important Exception

If the velocity ($\vec{v}$) or current (I) is <u>**parallel**</u> or <u>**anti-parallel**</u> to the magnetic field ($\vec{\mathbf{B}}$), the magnetic force ($\vec{\mathbf{F}}_m$) is <u>**zero**</u>. In Chapter 24, we'll learn that in these two extreme cases, $\theta = 0°$ or $180°$ such that $\sin\theta = 0$.

Symbol	Name	SI Units
$\vec{\mathbf{B}}$	magnetic field	T
$\vec{\mathbf{F}}_m$	magnetic force	N
$\vec{\mathbf{v}}$	velocity	m/s
I	current	A

Right-hand Rule for Magnetic Field

A long straight wire creates magnetic field lines that circulate around the current, as shown below.

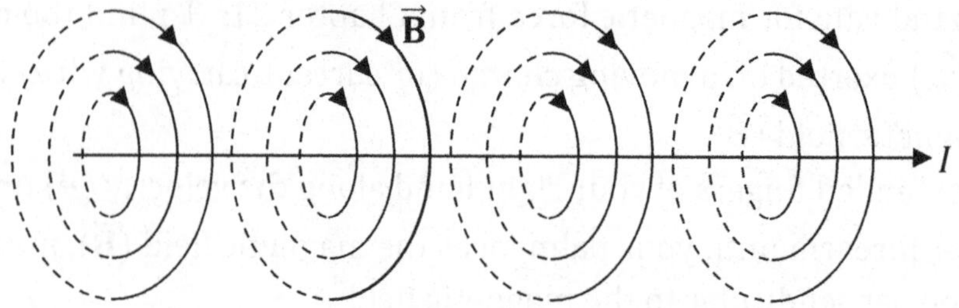

Recall the right-hand rule for magnetic **field** from Chapter 22: This right-hand rule gives you the direction of the magnetic field ($\vec{\mathbf{B}}$) created by a current (I) or moving charge:

- Imagine grabbing the wire with your right hand. **Tip**: You can use a pencil to represent the wire and actually grab the pencil.
- Grab the wire with your thumb pointing along the current (I).
- Your fingers represent **circular** magnetic field lines traveling around the wire **toward your fingertips**. At a given point, the direction of the magnetic field is **tangent** to these circles (your fingers).

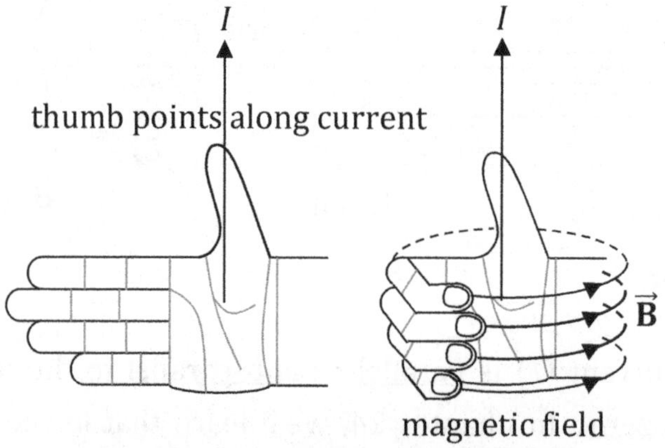

Symbol	Meaning
⊗	into the page
⊙	out of the page
p	proton
n	neutron
e^-	electron

The Force that One Current Exerts on Another

To find the magnetic force that one current-carrying wire (call it I_a) exerts on another current-carrying wire (call it I_b), apply both right-hand rules in combination. In this example, we're thinking of I_a as exerting the force and I_b as being pushed or pulled by the force. (Of course, it's mutual: I_a exerts a force on I_b, and I_b also exerts a force on I_a. However, we will calculate just one force at a time. So for the purposes of the calculation, let's consider the force that I_a exerts on I_b.)

- First find the direction of the magnetic field ($\vec{\mathbf{B}}_a$) created by the current (I_a), which we're thinking of as exerting the force, **at the location of the second current** (I_b). Put the field point (★) on I_b and ask yourself, "What is the direction of $\vec{\mathbf{B}}_a$ at the ★?" Apply the right-hand rule for magnetic **field** (Chapter 22) to find $\vec{\mathbf{B}}_a$ from I_a.
- Now find the direction of the magnetic force ($\vec{\mathbf{F}}_a$) that the first current (I_a) exerts on the second current (I_b). Apply the right-hand rule for magnetic **force** (Chapter 21) to find $\vec{\mathbf{F}}_a$ from I_b and $\vec{\mathbf{B}}_a$. In this step, we use the second current (I_b) because that is the current we're thinking of as being pushed or pulled by the force.
- This technique is illustrated in the following example.
- Note that both currents get used: The first current (I_a), which we're thinking of as exerting the force, is used in the first step to find the magnetic **field** ($\vec{\mathbf{B}}_a$), and the second current (I_b), which we're thinking of as being pushed or pulled by the force, is used in the second step to find the magnetic **force** ($\vec{\mathbf{F}}_a$).

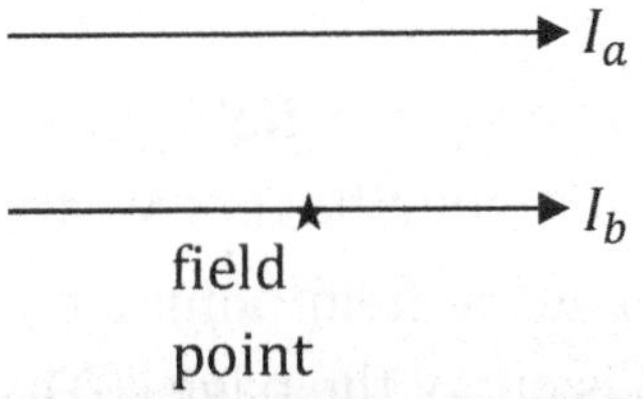

Schematic Representation	Symbol	Name
	R	resistor
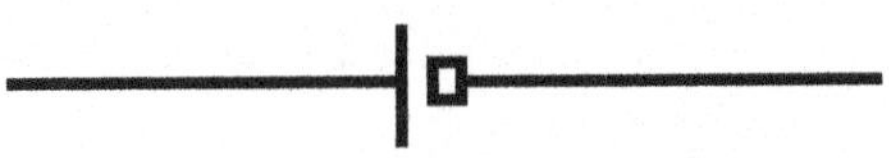	ΔV	battery or DC power supply

Recall that the long line represents the positive terminal.

Example 83. Combine the two right-hand rules together to answer each question.

(A) What is the direction of the magnetic force that the left current (I_1) exerts on the right current (I_2) in the diagram below?

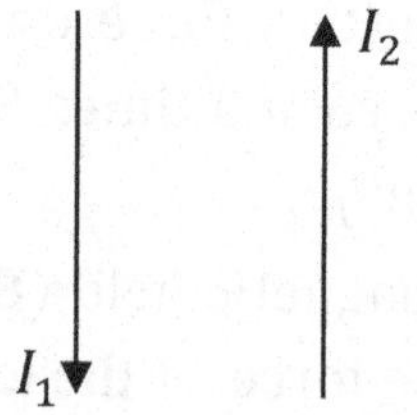

Solution. This question is thinking of the magnetic force that is exerted by I_1 and which is pushing or pulling I_2. We therefore draw a field point on I_2, as shown below.

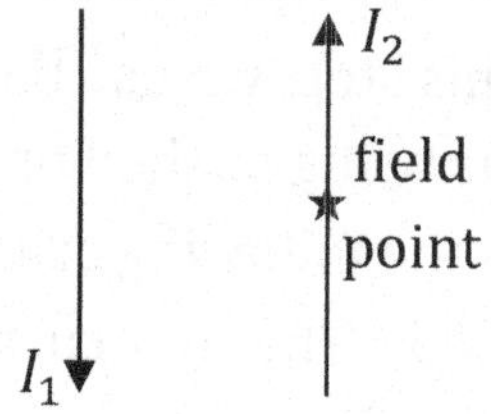

First apply the right-hand rule from Chapter 22 to find the magnetic **field** that I_1 creates at the field point, and then apply the right-hand rule from Chapter 21 to find the magnetic **force** exerted on I_2.

- Apply the right-hand rule for magnetic **field** (Chapter 22) to I_1. Grab I_1 with your thumb down (↓), along I_1, and your fingers wrapped around I_1. Your fingers are coming out of the page (⊙) at the field point (★). The magnetic field ($\vec{\mathbf{B}}_1$) that I_1 makes at the field point (★) is out of the page (⊙).
- Now apply the right-hand rule for magnetic force (Chapter 21) to I_2. Point your fingers up (↑), along I_2. At the same time, face your palm out of the page (⊙), along $\vec{\mathbf{B}}_1$. Your thumb points to the right (→), along the magnetic force ($\vec{\mathbf{F}}_1$).
- The left current (I_1) pushes the right current (I_2) to the **right** (→). These anti-parallel currents repel.

(B) What is the direction of the magnetic force that the left current (I_1) exerts on the right current (I_2) in the diagram below?

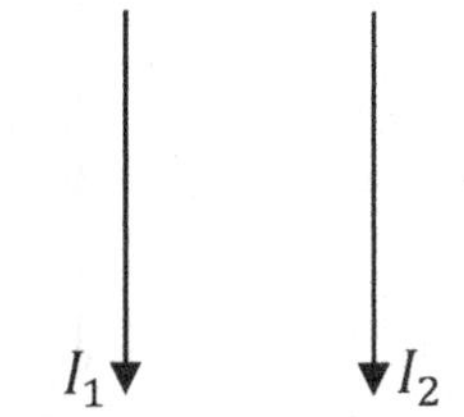

Solution. This question is thinking of the magnetic force that is exerted by I_1 and which is pushing or pulling I_2. We therefore draw a field point on I_2, as shown below.

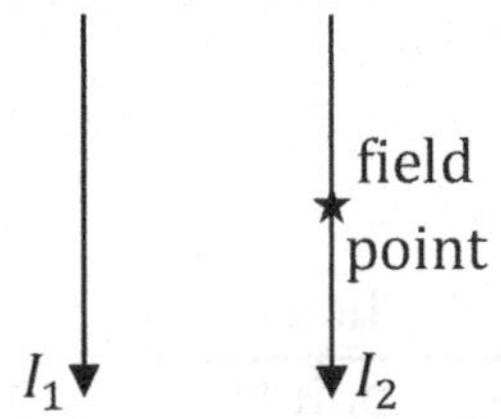

First apply the right-hand rule from Chapter 22 to find the magnetic **field** that I_1 creates at the field point, and then apply the right-hand rule from Chapter 21 to find the magnetic **force** exerted on I_2.

- Apply the right-hand rule for magnetic **field** (Chapter 22) to I_1. Grab I_1 with your thumb down (↓), along I_1, and your fingers wrapped around I_1. Your fingers are coming out of the page (⊙) at the field point (★). The magnetic field ($\vec{\mathbf{B}}_1$) that I_1 makes at the field point (★) is out of the page (⊙).
- Now apply the right-hand rule for magnetic force (Chapter 21) to I_2. Point your fingers down (↓), along I_2. At the same time, face your palm out of the page (⊙), along $\vec{\mathbf{B}}_1$. Your thumb points to the left (←), along the magnetic force ($\vec{\mathbf{F}}_1$).
- The left current (I_1) pulls the right current (I_2) to the <u>**left**</u> (←). These parallel currents attract.

(C) What is the direction of the magnetic force that the right current (I_2) exerts on the left current (I_1) in the diagram below?

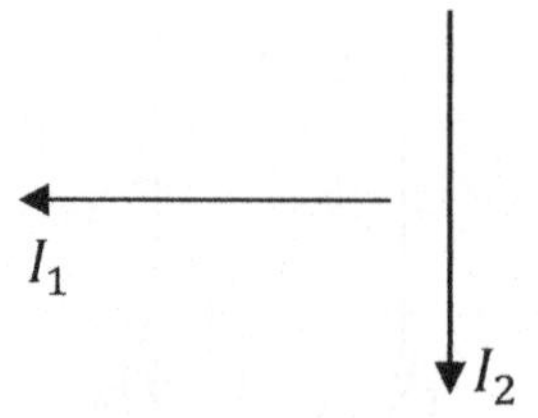

Solution. Note that the **wording** of this problem is **different** compared to the previous questions: Note the difference in the **subscripts**. This question is thinking of the magnetic force that is exerted by I_2 and which is pushing or pulling I_1. We therefore draw a field point on I_1, as shown below.

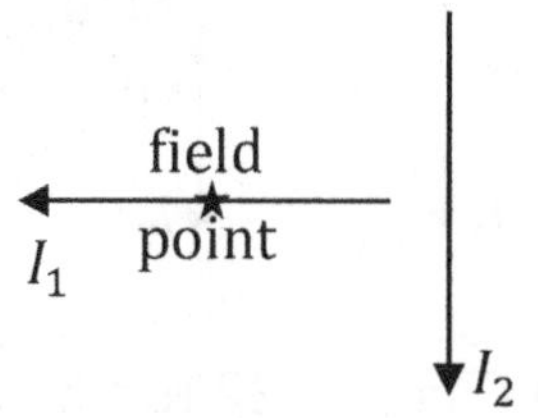

First apply the right-hand rule from Chapter 22 to find the magnetic **field** that I_2 creates at the field point, and then apply the right-hand rule from Chapter 21 to find the magnetic **force** exerted on I_1.

- Apply the right-hand rule for magnetic **field** (Chapter 22) to I_2 (**not** I_1). Grab I_2 with your thumb down (↓), along I_2, and your fingers wrapped around I_2. Your fingers are going into the page (⊗) at the field point (★). The magnetic field ($\vec{\mathbf{B}}_2$) that I_2 (it's **not** I_1 in this problem) makes at the field point (★) is into the page (⊗).
- Now apply the right-hand rule for magnetic force (Chapter 21) to I_1 (**not** I_2). Point your fingers to the left (←), along I_1 (it's **not** I_2 in this problem). At the same time, face your palm into the page (⊗), along $\vec{\mathbf{B}}_2$. Your thumb points down (↓), along the magnetic force ($\vec{\mathbf{F}}_2$).
- The right current (I_2) pushes the left current (I_1) **downward** (↓).

Note: We found $\vec{\mathbf{B}}_2$ and $\vec{\mathbf{F}}_2$ in this problem (**not** $\vec{\mathbf{B}}_1$ and $\vec{\mathbf{F}}_1$) because this problem asked us to find the magnetic force exerted by I_2 on I_1 (**not** by I_1 on I_2).

(D) What is the direction of the magnetic force that the outer current (I_1) exerts on the inner current (I_2) at point A in the diagram below?

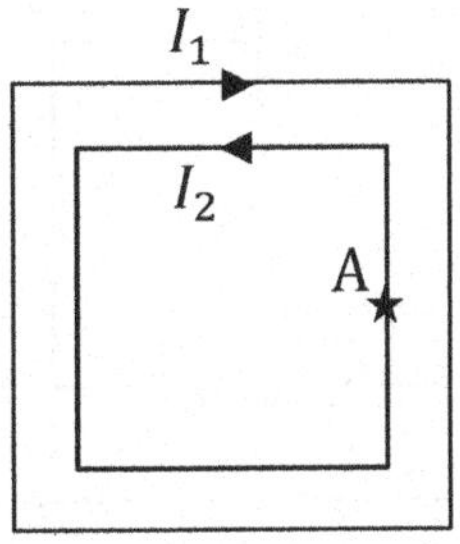

Solution. This question is thinking of the magnetic force that is exerted by I_1 and which is pushing or pulling I_2 at point A. Therefore, point A is the field point.

First apply the right-hand rule from Chapter 22 to find the magnetic **field** that I_1 creates at the field point, and then apply the right-hand rule from Chapter 21 to find the magnetic **force** exerted on I_2 at point A.

- Apply the right-hand rule for magnetic **field** (Chapter 22) to I_1. Grab I_1 with your thumb along I_1 and your fingers wrapped around I_1. Your fingers are going into the page ($\otimes$) at the field point (point A). The magnetic field ($\vec{\mathbf{B}}_1$) that I_1 makes at the field point (point A) is into the page ($\otimes$).
- Now apply the right-hand rule for magnetic force (Chapter 21) to I_2 at point A. Point your fingers up ($\uparrow$), since I_2 runs upward at point A. At the same time, face your palm into the page ($\otimes$), along $\vec{\mathbf{B}}_1$. Your thumb points to the left ($\leftarrow$), along the magnetic force ($\vec{\mathbf{F}}_1$).
- The outer current (I_1) pushes the inner current (I_2) to the **left** ($\leftarrow$) at point A. More generally, the outer current pushes the inner current inward. These anti-parallel currents repel.

(E) What is the direction of the magnetic force that the loop exerts on the bottom current (I_2) in the diagram below?

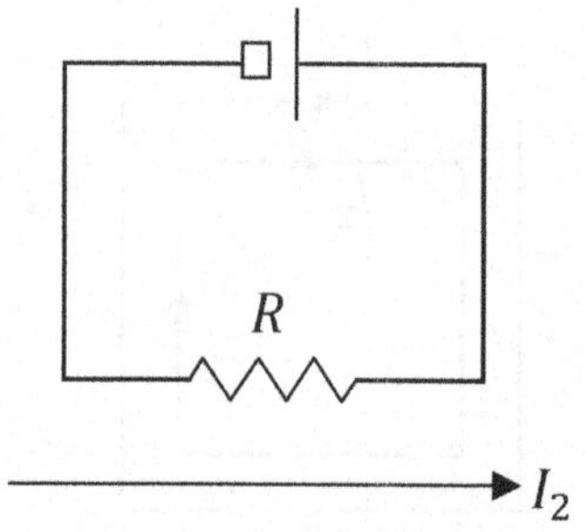

Solution. This question is thinking of the magnetic force that is exerted by the current (I_1) in the loop and which is pushing or pulling I_2. We therefore draw a field point on I_2, as shown below.

Label the positive (+) and negative (−) terminals of the battery: The long line is the positive (+) terminal. We draw the current from the positive terminal to the negative terminal: In this problem, the current (I_1) runs clockwise, as shown below.

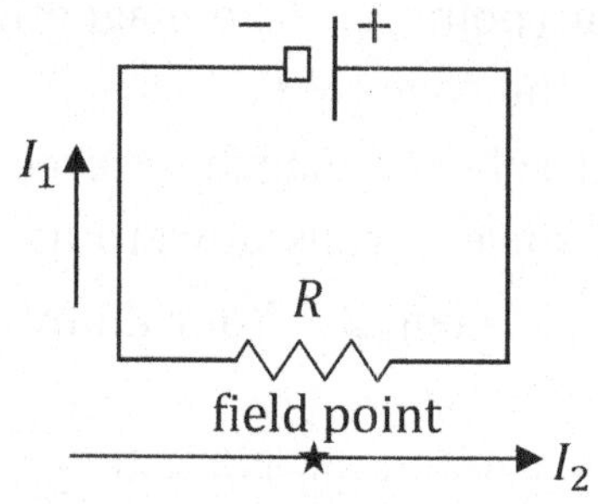

First apply the right-hand rule from Chapter 22 to find the magnetic **field** that I_1 creates at the field point, and then apply the right-hand rule from Chapter 21 to find the magnetic **force** exerted on I_2.

- Apply the right-hand rule for magnetic **field** (Chapter 22) to I_1. Grab I_1 at the **bottom** of the loop, with your thumb left (←), since I_1 runs to the left at the **bottom** of the loop. Your fingers are coming out of the page (⊙) at the field point (★). The magnetic field ($\vec{\mathbf{B}}_1$) that I_1 makes at the field point (★) is out of the page (⊙).
- Now apply the right-hand rule for magnetic force (Chapter 21) to I_2. Point your fingers to the right (→), along I_2. At the same time, face your palm out of the page (⊙), along $\vec{\mathbf{B}}_1$. **Tip**: Turn the book to make it more comfortable to position your hand as needed. Your thumb points down (↓), along the magnetic force ($\vec{\mathbf{F}}_1$).
- The top current (I_1) pushes the bottom current (I_2) <u>**downward**</u> (↓). These anti-parallel currents repel.

(F) What is the direction of the magnetic force that the outer current (I_1) exerts on the inner current (I_2) in the diagram below?

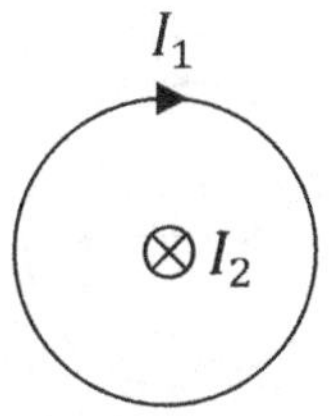

Solution. This question is thinking of the magnetic force that is exerted by I_1 and which is pushing or pulling I_2. We therefore draw a field point on I_2, as shown below.

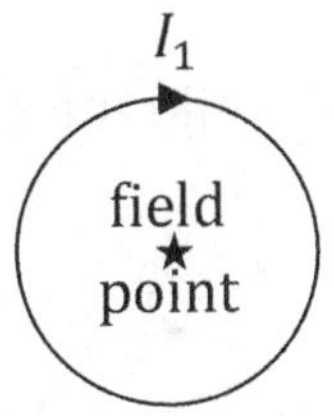

First apply the right-hand rule from Chapter 22 to find the magnetic **field** that I_1 creates at the field point, and then apply the right-hand rule from Chapter 21 to find the magnetic **force** exerted on I_2.

- Apply the right-hand rule for magnetic **field** (Chapter 22) to I_1. Grab I_1 with your thumb along I_1 and your fingers wrapped around I_1. Your fingers are going into the page ($\otimes$) at the field point (★). The magnetic field ($\vec{\mathbf{B}}_1$) that I_1 makes at the field point (★) is into the page ($\otimes$).
- Now apply the right-hand rule for magnetic force (Chapter 21) to I_2. The current (I_2) runs into the page ($\otimes$), and the magnetic field ($\vec{\mathbf{B}}_1$) is also into the page ($\otimes$). Recall from Chapter 21 that the magnetic force ($\vec{\mathbf{F}}_1$) is **zero** when the current (I_2) and magnetic field ($\vec{\mathbf{B}}_1$) are **parallel**.
- The outer current (I_1) exerts **no force** on the inner current (I_2): $\vec{\mathbf{F}}_1 = 0$.

(G) What is the direction of the magnetic force that the current (I_1) exerts on the proton (p) in the diagram below?

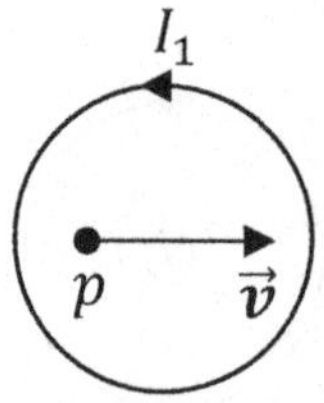

Solution. This question is thinking of the magnetic force that is exerted by I_1 and which is pushing or pulling the proton (p). We therefore draw a field point where the proton is, as shown below.

Note: The symbol p represents a proton, while the symbol $\vec{v}$ represents velocity.

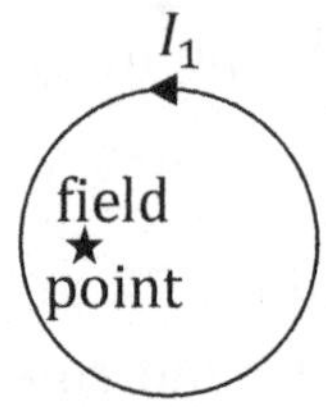

First apply the right-hand rule from Chapter 22 to find the magnetic **field** that I_1 creates at the field point, and then apply the right-hand rule from Chapter 21 to find the magnetic **force** exerted on the proton.

- Apply the right-hand rule for magnetic **field** (Chapter 22) to I_1. Grab I_1 with your thumb along I_1 and your fingers wrapped around I_1. Your fingers are coming out of the page (⊙) at the field point (★). The magnetic field ($\vec{\mathbf{B}}_1$) that I_1 makes at the field point (★) is out of the page (⊙).
- Now apply the right-hand rule for magnetic force (Chapter 21) to the proton (p). Point your fingers to the right (→), along $\vec{v}$. At the same time, face your palm out of the page (⊙), along $\vec{\mathbf{B}}_1$. **Tip**: Turn the book to make it more comfortable to position your hand as needed. Your thumb points down (↓), along the magnetic force ($\vec{\mathbf{F}}_1$).
- The current (I_1) pushes the proton <u>**downward**</u> (↓).

(H) What is the direction of the magnetic force that the left current (I_1) exerts on the right current (I_2) in the diagram below?

Solution. This question is thinking of the magnetic force that is exerted by I_1 and which is pushing or pulling I_2. We therefore draw a field point on I_2, as shown below on the left.

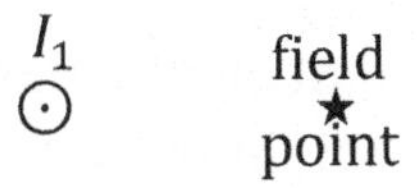

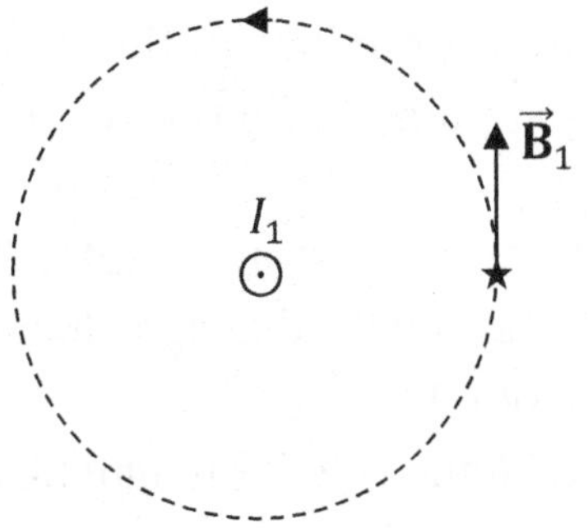

First apply the right-hand rule from Chapter 22 to find the magnetic **field** that I_1 creates at the field point, and then apply the right-hand rule from Chapter 21 to find the magnetic **force** exerted on I_2.

- Apply the right-hand rule for magnetic **field** (Chapter 22) to I_1. Grab I_1 with your thumb out of the page ($\odot$), along I_1, and your fingers wrapped around I_1. Your fingers are headed upward ($\uparrow$) at the field point ($\bigstar$). The magnetic field ($\vec{\mathbf{B}}_1$) that I_1 makes at the field point ($\bigstar$) is upward ($\uparrow$).
- **Note**: In the diagram above on the right, we drew a circular magnetic field line created by I_1. The magnetic field ($\vec{\mathbf{B}}_1$) that I_1 makes at the field point ($\bigstar$) is tangent to that circular magnetic field line, as we learned in Chapter 22.
- Now apply the right-hand rule for magnetic force (Chapter 21) to I_2. Point your fingers into the page ($\otimes$), along I_2. At the same time, face your palm upward ($\uparrow$), along $\vec{\mathbf{B}}_1$. Your thumb points to the right ($\rightarrow$), along the magnetic force ($\vec{\mathbf{F}}_1$).
- The left current (I_1) pushes the right current (I_2) to the <u>**right**</u> ($\rightarrow$). These anti-parallel currents repel.

(I) What is the direction of the magnetic force that the left current (I_1) exerts on the right current (I_2) at point C in the diagram below?

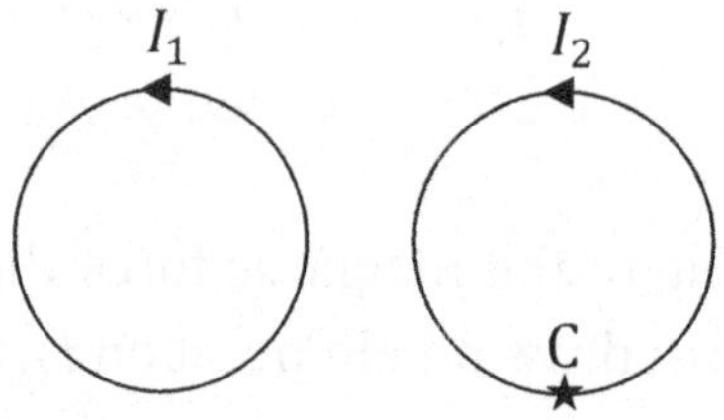

Solution. This question is thinking of the magnetic force that is exerted by I_1 and which is pushing or pulling I_2 at point C. Therefore, point C is the field point.

First apply the right-hand rule from Chapter 22 to find the magnetic **field** that I_1 creates at the field point, and then apply the right-hand rule from Chapter 21 to find the magnetic **force** exerted on I_2 at point C.

- Apply the right-hand rule for magnetic **field** (Chapter 22) to I_1. Grab I_1 at the **right** of the left loop, with your thumb upward (↑), since I_1 runs upward at the **right** of the left loop. (We're using the right side of the left loop since this point is nearest to the right loop, as it will have the dominant effect.) Your fingers are going into the page (⊗) at the field point (★). The magnetic field ($\vec{\mathbf{B}}_1$) that I_1 makes at the field point (★) is into the page (⊗).
- Now apply the right-hand rule for magnetic force (Chapter 21) to I_2 at point C. Point your fingers to the right (→), since I_2 runs to the right at point C. At the same time, face your palm into the page (⊗), along $\vec{\mathbf{B}}_1$. Your thumb points up (↑), along the magnetic force ($\vec{\mathbf{F}}_1$).
- The left current (I_1) pushes the right (I_2) **upward** (↑) at point C. More generally, the left loop pushes the right loop inward (meaning that at any point on the right loop, the magnetic force that I_1 exerts on I_2 is toward the center of the right loop).

24 MAGNETIC FORCE

Magnetic Force on a Moving Charge	Magnetic Force on a Current-Carrying Wire
$\vec{\mathbf{F}}_m = q\vec{\boldsymbol{v}} \times \vec{\mathbf{B}}$	$\vec{\mathbf{F}}_m = I\vec{\mathbf{L}} \times \vec{\mathbf{B}}$
Magnitude of the Magnetic Force on a Moving Charge	**Magnitude of the Magnetic Force on a Current-Carrying Wire**
$F_m = \lvert q \rvert vB \sin\theta$	$F_m = ILB \sin\theta$

Vector Product

$$\vec{\mathbf{A}} \times \vec{\mathbf{B}} = \begin{vmatrix} \hat{\mathbf{x}} & \hat{\mathbf{y}} & \hat{\mathbf{z}} \\ A_x & A_y & A_z \\ B_x & B_y & B_z \end{vmatrix} = \hat{\mathbf{x}} \begin{vmatrix} A_y & A_z \\ B_y & B_z \end{vmatrix} - \hat{\mathbf{y}} \begin{vmatrix} A_x & A_z \\ B_x & B_z \end{vmatrix} + \hat{\mathbf{z}} \begin{vmatrix} A_x & A_y \\ B_x & B_y \end{vmatrix}$$

$$\vec{\mathbf{A}} \times \vec{\mathbf{B}} = (A_y B_z - A_z B_y)\hat{\mathbf{x}} - (A_x B_z - A_z B_x)\hat{\mathbf{y}} + (A_x B_y - A_y B_x)\hat{\mathbf{z}}$$

$$\vec{\mathbf{A}} \times \vec{\mathbf{B}} = A_y B_z \hat{\mathbf{x}} - A_z B_y \hat{\mathbf{x}} + A_z B_x \hat{\mathbf{y}} - A_x B_z \hat{\mathbf{y}} + A_x B_y \hat{\mathbf{z}} - A_y B_x \hat{\mathbf{z}}$$

Uniform Circular Motion	Torque on a Current Loop
$\sum F_{in} = ma_c \quad , \quad a_c = \frac{v^2}{R}$	$\tau_{net} = IAB \sin\theta$

Source	Field	Force
mass (m)	gravitational field ($\vec{\mathbf{g}}$)	$\vec{\mathbf{F}}_g = m\vec{\mathbf{g}}$
charge (q)	electric field ($\vec{\mathbf{E}}$)	$\vec{\mathbf{F}}_e = q\vec{\mathbf{E}}$
current (I)	magnetic field ($\vec{\mathbf{B}}$)	$\vec{\mathbf{F}}_m = I\vec{\mathbf{L}} \times \vec{\mathbf{B}}$

Elementary Charge
$e = 1.60 \times 10^{-19}$ C

Symbol	Meaning
$\otimes$	into the page
$\odot$	out of the page
p	proton
n	neutron
e^-	electron

Symbol	Name	Units
$\vec{\mathbf{B}}$	magnetic field	T
$\vec{\mathbf{F}}_m$	magnetic force	N
q	electric charge	C
$\vec{v}$	velocity	m/s
v	speed	m/s
I	current	A
$\vec{\mathbf{L}}$	displacement vector along the current	m
L	length of the wire	m
θ	angle between $\vec{v}$ and $\vec{\mathbf{B}}$ or between I and $\vec{\mathbf{B}}$	° or rad
τ	torque	Nm
A	area of the loop	m^2
a_c	centripetal acceleration	m/s^2
R	radius	m

Example 84. There is a uniform magnetic field of 0.50 T directed along the positive y-axis.

(A) Determine the force exerted on a 4.0-A current in a 3.0-m long wire heading along the negative z-axis.

Solution. Make a list of the known quantities:

- The magnetic field is $\vec{\mathbf{B}} = 0.50\,\hat{\mathbf{y}}$ in vector form: It has a magnitude of $B = 0.50$ T and a direction along $\hat{\mathbf{y}}$.
- The current is $I = 4.0$ A.
- The displacement vector for the current is $\vec{\mathbf{L}} = -3\,\hat{\mathbf{z}}$ in vector form: It has a magnitude of $L = 3.0$ m (the length of the wire) and a direction along $-\hat{\mathbf{z}}$ (the direction of the current).
- The angle between $\vec{L}$ and $\vec{\mathbf{B}}$ is $\theta = 90°$ because the current and magnetic field are perpendicular: $\vec{L}$ is along $-\hat{\mathbf{z}}$, while $\vec{\mathbf{B}}$ is along $\hat{\mathbf{y}}$ (see the coordinate system below).

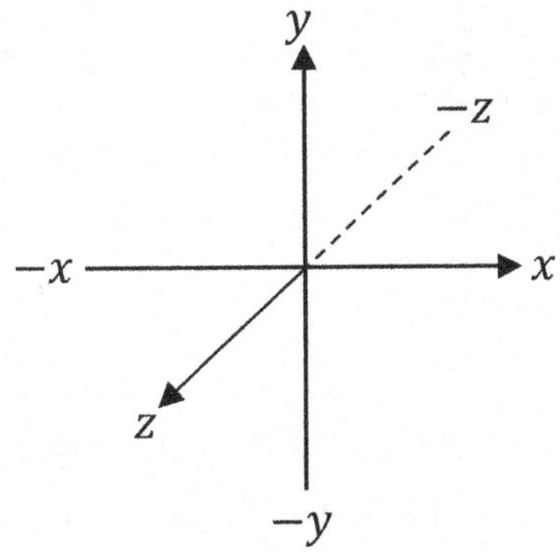

Use the appropriate trig equation to find the magnitude of the magnetic force.

$$F_m = ILB\sin\theta = (4)(3)(0.5)\sin 90° = (4)(3)(0.5)(1) = 6.0 \text{ N}$$

Recall from trig that $\sin 90° = 1$. The magnitude of the magnetic force is $F_m = 6.0$ N, but we still need to find the direction of the force. One way to find the direction of the magnetic force is to apply the right-hand rule for magnetic force (Chapter 21). Draw a right-handed, three-dimensional coordinate system with x, y, and z: We will use the one shown above.

- Point your fingers into the page (⊗), along the current (I), which is along $-\hat{\mathbf{z}}$.
- At the same time, face your palm upward (↑), along the magnetic field ($\vec{\mathbf{B}}$), which is along $\hat{\mathbf{y}}$.
- Your thumb points to the right (→), along the magnetic force ($\vec{\mathbf{F}}_m$), which is along the $+x$-axis (along $\hat{\mathbf{x}}$).

Alternatively, we could substitute $I = 4.0$ A, $\vec{\mathbf{L}} = -3\,\hat{\mathbf{z}}$, and $\vec{\mathbf{B}} = 0.50\,\hat{\mathbf{y}}$ into the vector product formula $\vec{\mathbf{F}}_m = I\vec{\mathbf{L}} \times \vec{\mathbf{B}}$ (similar to Example 85) to determine that $\vec{\mathbf{F}}_m = 6\,\hat{\mathbf{x}}$. Either way, the magnetic force has a magnitude of $F_m = 6.0$ N and a diretion along $\hat{\mathbf{x}}$.

(B) Determine the force exerted on a 200-μC charge moving 60 km/s at an angle of 30° below the $+x$-axis.

Solution. Make a list of the known quantities:

- The magnetic field is still $\vec{\mathbf{B}} = 0.50\,\hat{\mathbf{y}}$ in vector form: It has a magnitude of $B = 0.50$ T and a direction along $\hat{\mathbf{y}}$. This applies to parts (A), (B), and (C).
- The charge is $q = 200$ μC. Convert this to SI units: $q = 2.00 \times 10^{-4}$ C. Recall that the metric prefix μ stands for 10^{-6}.
- The speed is $v = 60$ km/s. Convert this to SI units: $v = 6.0 \times 10^4$ m/s. Recall that the metric prefix k stands for 1000.
- The angle between $\vec{v}$ and $\vec{\mathbf{B}}$ is $\theta = 120°$ because $\vec{v}$ is 30° below the $+x$-axis, while $\vec{\mathbf{B}}$ is 90° above x-axis, as shown in the diagram below: $30° + 90° = 120°$.

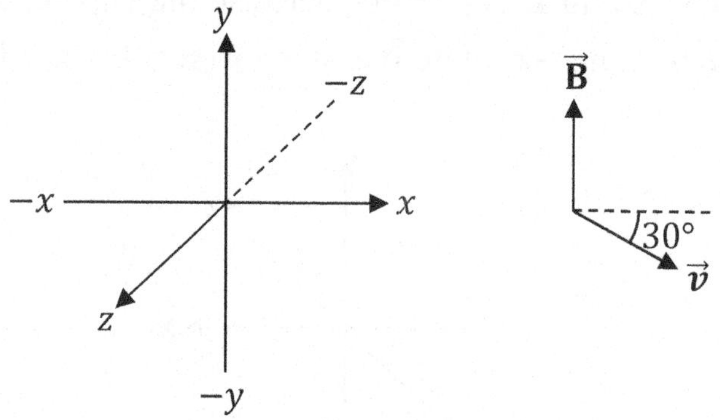

Use the appropriate trig equation to find the magnitude of the magnetic force.

$$F_m = |q|vB\sin\theta = (2 \times 10^{-4})(6 \times 10^4)(0.5)\sin 120°$$

$$= (2 \times 10^{-4})(6 \times 10^4)(0.5)\left(\frac{\sqrt{3}}{2}\right) = 3\sqrt{3}\text{ N}$$

Note that $10^{-4} \times 10^4 = 1$. The magnitude of the magnetic force is $F_m = 3\sqrt{3}$ N, but we still need to find the direction of the force. One way to find the direction of the magnetic force is to apply the right-hand rule for magnetic force (Chapter 21). Draw a right-handed, three-dimensional coordinate system with x, y, and z: We will use the one shown above.

- Point your fingers diagonally right and downward (↘), along the velocity ($\vec{v}$).
- At the same time, face your palm upward (↑), along the magnetic field ($\vec{\mathbf{B}}$), which is along $\hat{\mathbf{y}}$.
- Your thumb points out of the page (⊙), along the magnetic force ($\vec{\mathbf{F}}_m$), which is along the $+z$-axis (along $\hat{\mathbf{z}}$).

Alternatively, we could substitute $q = 2.00 \times 10^{-4}$ C, $\vec{v} = 30{,}000\sqrt{3}\,\hat{\mathbf{x}} - 30{,}000\,\hat{\mathbf{y}}$, and $\vec{\mathbf{B}} = 0.50\,\hat{\mathbf{y}}$ into the vector product formula $\vec{\mathbf{F}}_m = q\vec{v} \times \vec{\mathbf{B}}$ (similar to Example 85) to determine that $\vec{\mathbf{F}}_m = 3\sqrt{3}\,\hat{\mathbf{z}}$. Either way, the magnetic force has a magnitude of $F_m = 3\sqrt{3}$ N and a diretion along $\hat{\mathbf{z}}$. The equation $\vec{v} = 30{,}000\sqrt{3}\,\hat{\mathbf{x}} - 30{,}000\,\hat{\mathbf{y}}$ comes from $\vec{v} = v_x\,\hat{\mathbf{x}} - v_y\,\hat{\mathbf{y}}$, where $v_x = v\cos 330°$ and $v_y = v\sin 330°$. The angle 330° is equivalent to −30° (that is, 30° below the x-axis): Whether you go 330° counterclockwise from $+x$ or go 30° clockwise from $+x$, you arrive at the same place.

(C) In which direction(s) could a proton travel and experience zero magnetic force?

Solution. Apply the equation $F_m = |q|vB \sin\theta$. The angle θ represents the direction of the proton's velocity relative to the magnetic field. Which value(s) of θ make $\sin\theta = 0$? The answer is $\theta = 0°$ or $180°$, since $\sin 0° = 0$ and $\sin 180° = 0$. This means that the velocity ($\vec{v}$) must either be parallel or anti-parallel to the magnetic field ($\vec{\mathbf{B}}$). Since the magnetic field points along $+\hat{\mathbf{y}}$, the velocity must point along $+\hat{\mathbf{y}}$ or $-\hat{\mathbf{y}}$.

Example 85. A 200-μC charge has a velocity of $\vec{v} = 3\,\hat{\mathbf{x}} - 2\,\hat{\mathbf{y}} - \hat{\mathbf{z}}$ in a region where the magnetic field is $\vec{\mathbf{B}} = 5\,\hat{\mathbf{x}} + \hat{\mathbf{y}} - 4\,\hat{\mathbf{z}}$, where SI units have been suppressed. Find the magnetic force exerted on the charge.

Solution. Compare $\vec{v} = 3\,\hat{\mathbf{x}} - 2\,\hat{\mathbf{y}} - \hat{\mathbf{z}}$ to $\vec{v} = v_x\hat{\mathbf{x}} + v_y\hat{\mathbf{y}} + v_z\hat{\mathbf{z}}$ to see that $v_x = 3$, $v_y = -2$, $v_z = -1$. Similarly, compare $\vec{\mathbf{B}} = 5\,\hat{\mathbf{x}} + \hat{\mathbf{y}} - 4\,\hat{\mathbf{z}}$ to $\vec{\mathbf{B}} = B_x\hat{\mathbf{x}} + B_y\hat{\mathbf{y}} + B_z\hat{\mathbf{z}}$ to see that $B_x = 5$, $B_y = 1$, and $B_z = -4$. Convert the charge to SI units using $\mu = 10^{-6}$: $q = 2.00 \times 10^{-4}$ C. Recall that the metric prefix μ stands for 10^{-6}. Plug these values into the determinant form of the vector product:

$$\vec{\mathbf{F}}_m = q\vec{v} \times \vec{\mathbf{B}} = q\begin{vmatrix} \hat{\mathbf{x}} & \hat{\mathbf{y}} & \hat{\mathbf{z}} \\ v_x & v_y & v_z \\ B_x & B_y & B_z \end{vmatrix} = (2 \times 10^{-4})\begin{vmatrix} \hat{\mathbf{x}} & \hat{\mathbf{y}} & \hat{\mathbf{z}} \\ 3 & -2 & -1 \\ 5 & 1 & -4 \end{vmatrix}$$

It may be helpful to review the **vector product** from Chapter 19. Note that the charge ($q = 2.00 \times 10^{-4}$ C) is multiplying the determinant.

$$\vec{\mathbf{F}}_m = (2 \times 10^{-4})\,\hat{\mathbf{x}}\begin{vmatrix} -2 & -1 \\ 1 & -4 \end{vmatrix} - (2 \times 10^{-4})\,\hat{\mathbf{y}}\begin{vmatrix} 3 & -1 \\ 5 & -4 \end{vmatrix} + (2 \times 10^{-4})\,\hat{\mathbf{z}}\begin{vmatrix} 3 & -2 \\ 5 & 1 \end{vmatrix}$$

$$\vec{\mathbf{F}}_m = (2 \times 10^{-4})\,\hat{\mathbf{x}}[(-2)(-4) - (-1)(1)] - (2 \times 10^{-4})\,\hat{\mathbf{y}}[(3)(-4) - (-1)(5)]$$

$$+ (2 \times 10^{-4})\,\hat{\mathbf{z}}[(3)(1) - (-2)(5)]$$

$$\vec{\mathbf{F}}_m = (2 \times 10^{-4})\,\hat{\mathbf{x}}(8 + 1) - (2 \times 10^{-4})\,\hat{\mathbf{y}}(-12 + 5) + (2 \times 10^{-4})\,\hat{\mathbf{z}}(3 + 10)$$

$$\vec{\mathbf{F}}_m = (2 \times 10^{-4})\,\hat{\mathbf{x}}(9) - (2 \times 10^{-4})\,\hat{\mathbf{y}}(-7) + (2 \times 10^{-4})\,\hat{\mathbf{z}}(13)$$

$$\vec{\mathbf{F}}_m = 18 \times 10^{-4}\,\hat{\mathbf{x}} + 14 \times 10^{-4}\,\hat{\mathbf{y}} + 26 \times 10^{-4}\,\hat{\mathbf{z}}$$

$$\vec{\mathbf{F}}_m = 1.8 \times 10^{-3}\,\hat{\mathbf{x}} + 1.4 \times 10^{-3}\,\hat{\mathbf{y}} + 2.6 \times 10^{-3}\,\hat{\mathbf{z}}$$

Recall that two minus signs make a plus sign: For example, $-(2 \times 10^{-4})(-7) = +14 \times 10^{-4}$. Also note that $14 \times 10^{-4} = 1.4 \times 10^{-3}$. The magnetic force is $\vec{\mathbf{F}}_m = 1.8 \times 10^{-3}\,\hat{\mathbf{x}} + 1.4 \times 10^{-3}\,\hat{\mathbf{y}} + 2.6 \times 10^{-3}\,\hat{\mathbf{z}}$ in Newtons.

Example 86. As illustrated below, a 0.25-g object with a charge of −400 μC travels in a circle with a constant speed of 4000 m/s in an approximately zero-gravity region where there is a uniform magnetic field of 200,000 G perpendicular to the page.

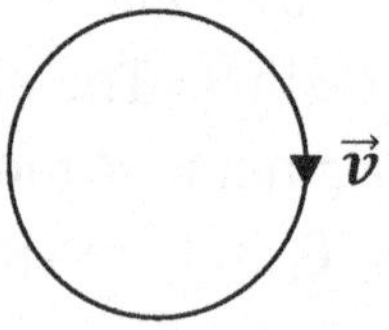

(A) What is the direction of the magnetic field?

Solution. The magnetic force ($\vec{\mathbf{F}}_m$) is a **centripetal** force: It pushes the charged particle towards the center of the circle. Invert the right-hand rule for magnetic force (Chapter 21) in order to find the magnetic field. We'll do this for the point when the charged particle is at the very rightmost position in the circle, when it is headed downward (this position corresponds to the arrow shown in the diagram above).

- Point your fingers down (↓), along the instantaneous velocity ($\vec{v}$), which is **tangent** to the circle.
- At the same time, point your **thumb** (it's **not** your palm in this example) to the left (←), along the magnetic force ($\vec{\mathbf{F}}_m$).
- Your **palm** faces out of the page (⊙). However, the electron has **negative** charge, so the answer is **backwards**: The magnetic field ($\vec{\mathbf{B}}$) is into the page (⊗). It may help to review parts C and D of Example 81 in Chapter 21.

(B) What is the radius of the circle?

Solution. Apply Newton's second law to the particle. Since the particle travels with uniform circular motion (meaning constant speed in a circle), the acceleration is **centripetal** (towards the center of the circle). Therefore, we sum the inward components of the forces and set that equal to mass times centripetal acceleration.

$$\sum F_{in} = ma_c$$

The magnetic force supplies the needed centripetal force. Apply the equation $F_m = |q|vB\sin\theta$. Since $\vec{\mathbf{F}}_m$ lies in the plane of the page (it points towards the center of the circle) and since we found in part (A) that $\vec{\mathbf{B}}$ points into the page (⊗), the angle is $\theta = 90°$.

$$|q|vB\sin 90° = ma_c$$

In Volume 1 of this series, we learned that $a_c = \frac{v^2}{R}$. Recall from trig that $\sin 90° = 1$.

$$|q|vB = m\frac{v^2}{R}$$

Divide both sides of the equation by the speed. Note that $\frac{v^2}{v} = v$.

$$|q|B = m\frac{v}{R}$$

Multiply both sides of the equation by R.

$$|q|BR = mv$$

Divide both sides of the equation by $|q|B$.

$$R = \frac{mv}{|q|B}$$

Convert the charge, mass, and magnetic field to SI units, using $\mu = 10^{-6}$, $1 \text{ g} = 10^{-3}$ kg, and $1 \text{ G} = 10^{-4}$ T (which is equivalent to $1 \text{ T} = 10^4$ G).

$$q = -400 \ \mu\text{C} = -4.00 \times 10^{-4} \text{ C}$$
$$m = 0.25 \text{ g} = 2.5 \times 10^{-4} \text{ kg}$$
$$B = 200{,}000 \text{ G} = 20 \text{ T}$$

Substitute these values, along with the speed ($v = 4000$ m/s), into the previous equation. Note the absolute values.

$$R = \frac{(2.5 \times 10^{-4})(4000)}{|-4 \times 10^{-4}|(20)} = \frac{(2.5 \times 10^{-4})(4000)}{(4 \times 10^{-4})(20)} = \frac{(2.5)(4000)}{(4)(20)} = 125 \text{ m}$$

Note that the 10^{-4}'s cancel out. The radius is $R = 125$ m. (That may seem like a huge circle, but charged particles can travel in even larger circles at a high-energy collider, such as the Large Hadron Collider.)

Example 87. A 30-A current runs through the rectangular loop of wire illustrated below. There is a uniform magnetic field of 8000 G directed downward. The width (which is horizontal) of the rectangle is 50 cm and the height (which is vertical) of the rectangle is 25 cm.

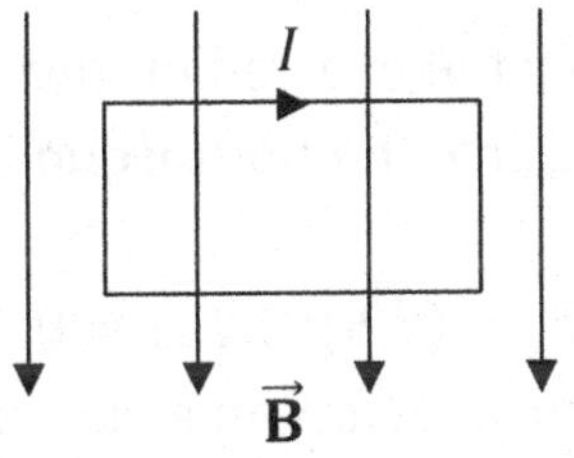

(A) Find the magnitude of the net force exerted on the loop.

Solution. We saw conceptual problems like this in Chapter 21. If you apply the right-hand rule for magnetic force to each side of the loop, you will see that the loop will rotate about the dashed horizontal axis shown on the following page.

- <u>Top side</u>: Point your fingers to the right (→) along the current (I) and your palm down (↓) along the magnetic field ($\vec{\mathbf{B}}$). The magnetic force ($\vec{\mathbf{F}}_{top}$) is into the page (⊗).
- <u>Right side</u>: The current (I) points down (↓) and the magnetic field ($\vec{\mathbf{B}}$) also points down (↓). Since I and $\vec{\mathbf{B}}$ are parallel, the magnetic force ($\vec{\mathbf{F}}_{right}$) is **<u>zero</u>**.
- <u>Bottom side</u>: Point your fingers to the left (←) along the current (I) and your palm down (↓) along the magnetic field ($\vec{\mathbf{B}}$). The magnetic force ($\vec{\mathbf{F}}_{bot}$) is out of the page (⊙).

- **Left side**: The current (I) points up ($\uparrow$) and the magnetic field ($\vec{\mathbf{B}}$) points down ($\downarrow$). Since I and $\vec{\mathbf{B}}$ are anti-parallel, the magnetic force ($\vec{\mathbf{F}}_{left}$) is **zero**.
- The top side of the loop is pushed into the page, while the bottom side of the loop is pulled out of the page. The loop rotates about the dashed axis shown below.

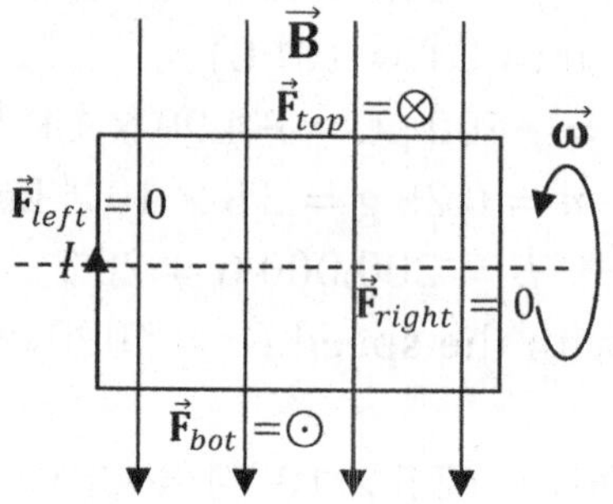

Calculate the magnitudes of the two nonzero forces, F_{top} and F_{bot}. Convert the magnetic field to SI units using $1\text{ G} = 10^{-4}\text{ T}$: $B = 8000\text{ G} = 0.80\text{ T}$. Note that $L = 50\text{ cm} = 0.50\text{ m}$ (the correct length is the "width," which is horizontal).

$$F_{top} = IL_{top}B\sin\theta = (30)(0.5)(0.8)\sin 90° = 12\text{ N}$$

$$F_{bot} = IL_{bot}B\sin\theta = (30)(0.5)(0.8)\sin 90° = 12\text{ N}$$

Since $\vec{\mathbf{F}}_{top}$ is into the page ($\otimes$) and $\vec{\mathbf{F}}_{bot}$ is out of the page ($\odot$), and since the two forces have equal magnitudes, they **cancel** out: $F_{net} = 0$. The net magnetic force exerted on the loop is **zero**.

(B) Find the magnitude of the net torque exerted on the loop.

Solution. Although the net **force** is zero, the net **torque** **isn't** zero. First find the area of the rectangular loop.

$$A = WH = (0.5)(0.25) = 0.125\text{ m}^2$$

Apply the equation for the net torque exerted on a current loop. Note that $\theta = 90°$ because the magnetic field is perpendicular to the axis of the loop, since axis of the loop is perpendicular to the plane of the loop. (Also note that the axis of the loop is **not** the same thing as the axis of rotation. In this example, the axis of the **loop** is out of the page, perpendicular to the plane of the loop, whereas the axis of **rotation** is horizontal, as indicated above.)

$$\tau_{net} = IAB\sin\theta = (30)(0.125)(0.8)\sin 90° = 3.0\text{ Nm}$$

The net torque is $\tau_{net} = 3.0$ Nm.

Another way to find the net torque is to apply the equation $\tau = rF\sin\theta$ to both F_{top} and F_{bot}, separately, using $r = \frac{H}{2}$ (see the dashed line in the figure above). Then add the two torques together: $\tau_{net} = \tau_1 + \tau_2$. You would get the same answer, $\tau_{net} = 3.0$ Nm.

25 MAGNETIC FIELD

Long Straight Current	Center of a Circular Loop
$B = \frac{\mu_0 I}{2\pi r_c}$	$B = \frac{\mu_0 I}{2a}$
Center of a Solenoid	**Force on a Straight Current**
$B = \frac{\mu_0 N I}{L} = \mu_0 n I$	$F_m = ILB \sin\theta$

Permeability of Free Space
$\mu_0 = 4\pi \times 10^{-7} \frac{\text{T} \cdot \text{m}}{\text{A}}$

Symbol	Name	Units
B	magnitude of the magnetic field	T
I	current	A
μ_0	permeability of free space	$\frac{\text{T}\cdot\text{m}}{\text{A}}$
r_c	distance from a long, straight wire	m
a	radius of a loop	m
N	number of loops (or turns)	unitless
n	number of turns per unit length	$\frac{1}{\text{m}}$
L	length of a wire or length of a solenoid	m
F_m	magnitude of the magnetic force	N
θ	angle between $\vec{v}$ and $\vec{\mathbf{B}}$ or between I and $\vec{\mathbf{B}}$	° or rad

Symbol	Meaning
⊗	into the page
⊙	out of the page

Example 88. Three currents are shown below. The currents run perpendicular to the page in the directions indicated. The triangle, which is **not** equilateral, has a base of 4.0 m and a height of 4.0 m. Find the magnitude and direction of the net magnetic field at the midpoint of the base.

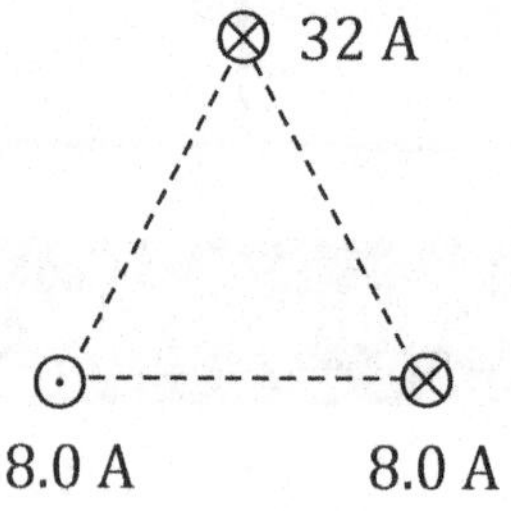

Solution. Draw the field point at the midpoint of the base, as shown below.

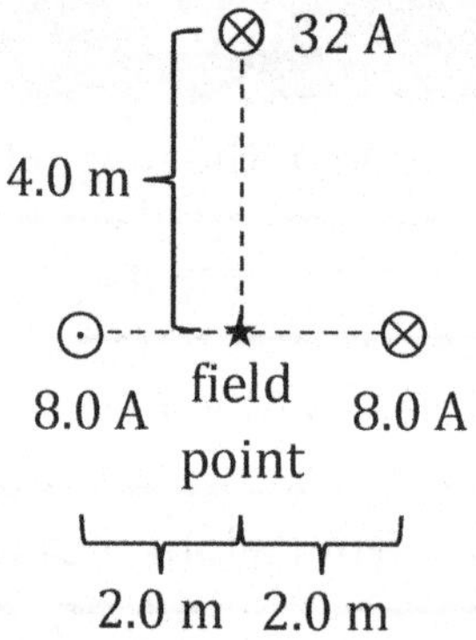

There are three separate magnetic fields to find: One magnetic field is created by each current at the field point. Note that $r_{left} = 2.0$ m, $r_{top} = 4.0$ m, and $r_{right} = 2.0$ m. (See the diagram above.) Recall that the permeability of free space is $\mu_0 = 4\pi \times 10^{-7} \frac{\text{T·m}}{\text{A}}$.

$$B_{left} = \frac{\mu_0 I_{left}}{2\pi r_{left}} = \frac{(4\pi \times 10^{-7})(8)}{2\pi(2)} = 8.0 \times 10^{-7}\text{ T}$$

$$B_{top} = \frac{\mu_0 I_{top}}{2\pi r_{top}} = \frac{(4\pi \times 10^{-7})(32)}{2\pi(4)} = 16.0 \times 10^{-7}\text{ T}$$

$$B_{right} = \frac{\mu_0 I_{right}}{2\pi r_{right}} = \frac{(4\pi \times 10^{-7})(8)}{2\pi(2)} = 8.0 \times 10^{-7}\text{ T}$$

Before we can determine how to combine these magnetic fields, we must apply the right-hand rule for magnetic field (Chapter 22) in order to determine the direction of each of these magnetic fields.

- When you grab I_{left} with your thumb out of the page (⊙), along I_{left}, and your fingers wrapped around I_{left}, your fingers are going up (↑) at the field point (★). The magnetic field ($\vec{\mathbf{B}}_{left}$) that I_{left} makes at the field point (★) is straight up (↑).
- When you grab I_{top} with your thumb into the page (⊗), along I_{top}, and your fingers wrapped around I_{top}, your fingers are going to the left (←) at the field point (★). The magnetic field ($\vec{\mathbf{B}}_{top}$) that I_{top} makes at the field point (★) is to the left (←).
- When you grab I_{right} with your thumb into the page (⊗), along I_{right}, and your fingers wrapped around I_{right}, your fingers are going up (↑) at the field point (★). The magnetic field ($\vec{\mathbf{B}}_{right}$) that I_{right} makes at the field point (★) is straight up (↑).

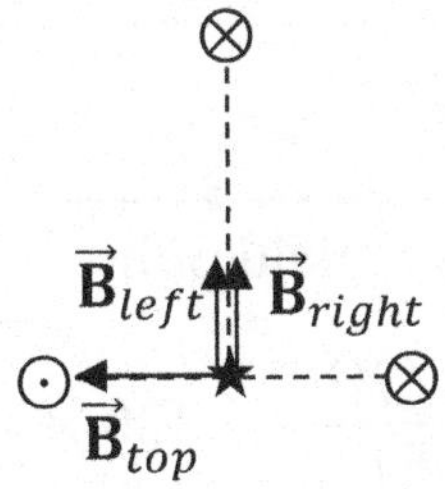

Since $\vec{\mathbf{B}}_{left}$ and $\vec{\mathbf{B}}_{right}$ both point straight up (↑), we add their magnitudes together:

$$B_y = B_{left} + B_{right} = 8.0 \times 10^{-7} + 8.0 \times 10^{-7} = 16.0 \times 10^{-7} \text{ T}$$

Since $\vec{\mathbf{B}}_{top}$ is perpendicular to $\vec{\mathbf{B}}_{left}$ and $\vec{\mathbf{B}}_{right}$ (since $\vec{\mathbf{B}}_{top}$ points to the left, while $\vec{\mathbf{B}}_{left}$ and $\vec{\mathbf{B}}_{right}$ both point up), we use the Pythagorean theorem, with B_y representing the combination of $\vec{\mathbf{B}}_{left}$ and $\vec{\mathbf{B}}_{right}$.

$$B_{net} = \sqrt{B_{top}^2 + B_y^2} = \sqrt{(16.0 \times 10^{-7})^2 + (16.0 \times 10^{-7})^2}$$

It's convenient to **factor** out the 16.0×10^{-7}.

$$B_{net} = \sqrt{(16.0 \times 10^{-7})^2(1^2 + 1^2)} = \sqrt{(16.0 \times 10^{-7})^2 2} = 16\sqrt{2} \times 10^{-7} \text{ T}$$

We applied the rules from algebra that $\sqrt{xy} = \sqrt{x}\sqrt{y}$ and $\sqrt{x^2} = x$. We can find the direction of the net magnetic field with an inverse tangent. We choose $+x$ to point right and $+y$ to point up, such that $B_x = -B_{top}$ (since $\vec{\mathbf{B}}_{top}$ points left).

$$\theta_B = \tan^{-1}\left(\frac{B_y}{B_x}\right) = \tan^{-1}\left(\frac{16.0 \times 10^{-7}}{-16.0 \times 10^{-7}}\right) = \tan^{-1}(-1) = 135°$$

The reference angle is 45° and the Quadrant II angle is $\theta_B = 180° - \theta_{ref} = 135°$. The magnitude of the net magnetic field at the field point is $B_{net} = 16\sqrt{2} \times 10^{-7}$ T and its direction is $\theta_B = 135°$. If you use a calculator, the magnitude of the net magnetic field at the field point works out to $B_{net} = 23 \times 10^{-7} \text{ T} = 2.3 \times 10^{-6}$ T.

Example 89. In the diagram below, the top wire carries a current of 8.0 A, the bottom wire carries a current of 5.0 A, each wire is 3.0 m long, and the distance between the wires is 0.050 m. What are the magnitude and direction of the magnetic force that the top current (I_1) exerts on the bottom current (I_2)?

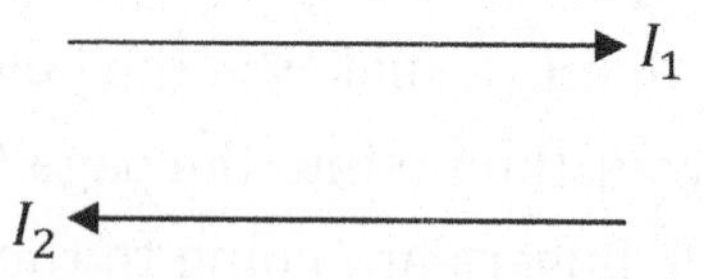

Solution. First imagine a field point (★) at the location of I_2 (since the force specified in the problem is exerted on I_2), and find the magnetic field at the field point (★) created by I_1. When we do this, we use $I_1 = 8.0$ A and $r_c = d = 0.050$ m (since the field point is 0.050 m from I_1).

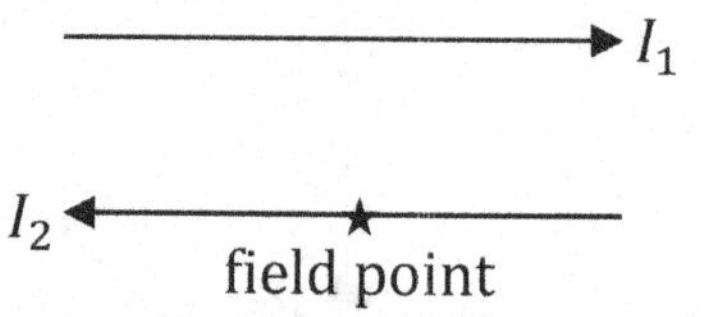

$$B_1 = \frac{\mu_0 I_1}{2\pi d} = \frac{(4\pi \times 10^{-7})(8)}{2\pi(0.05)} = 320 \times 10^{-7}\ \text{T} = 3.2 \times 10^{-5}\ \text{T}$$

Now we can find the force exerted on I_2. When we do this, we use $I_2 = 5.0$ A (since I_2 is experiencing the force specified in the problem) and $\theta = 90°$ (since I_2 is to the right and $\vec{\mathbf{B}}_1$ is into the page – as discussed below). What we're calling F_1 is the magnitude of the force that I_1 exerts on I_2.

$$F_1 = I_2 L_2 B_1 \sin\theta = (5)(3)(3.2 \times 10^{-5}) \sin 90° = 48 \times 10^{-5}\ \text{N} = 4.8 \times 10^{-4}\ \text{N}$$

Recall from trig that $\sin 90° = 1$. To find the direction of this force, apply the technique from Chapter 23. First apply the right-hand rule for magnetic **field** (Chapter 22) to I_1. Grab I_1 with your thumb along I_1 and your fingers wrapped around I_1. What are your fingers doing at the field point (★)? They are going into the page (⊗) at the field point (★). The magnetic field ($\vec{\mathbf{B}}_1$) that I_1 makes at the field point (★) is into the page (⊗).

Now apply the right-hand rule for magnetic **force** (Chapter 21) to I_2. Point your fingers to the left (←), along I_2. At the same time, face your palm into the page (⊗), along $\vec{\mathbf{B}}_1$. Your thumb points down (↓), along the magnetic force ($\vec{\mathbf{F}}_1$). The top current (I_1) pushes the bottom current (I_2) downward (↓).

The magnetic force that I_1 exerts on I_2 has a magnitude of $F_1 = 4.8 \times 10^{-4}$ N and a direction that is straight downward (↓). (You might recall from Chapter 23 that **anti-parallel** currents **repel** one another.) The magnitude of the magnetic force can also be expressed as $F_1 = 48 \times 10^{-5}$ N, $F_1 = 0.00048$ N, or $F_1 = 0.48$ mN.

Example 90. In the diagram below, the left wire carries a current of 3.0 A, the middle wire carries a current of 4.0 A, the right wire carries a current of 6.0 A, each wire is 5.0 m long, and the distance between neighboring wires is 0.25 m. What are the magnitude and direction of the net magnetic force exerted on the right current (I_3)?

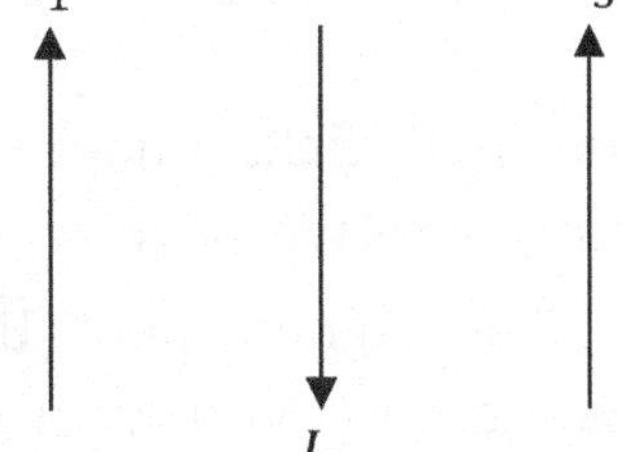

Solution. Here is how we will solve this example:

- We'll find the force that I_1 exerts on I_3 the way that we solved the previous example.
- We'll similarly find the force that I_2 exerts on I_3.
- Once we know the magnitudes and directions of both forces, we will know how to combine them.

First imagine a field point (★) at the location of I_3 (since the force specified in the problem is exerted on I_3), and find the magnetic fields at the field point (★) created by I_1 and I_2. When we do this, note that $d_1 = 0.25 + 0.25 = 0.50$ m and $d_2 = 0.25$ m (since these are the distances from I_1 to I_3 and from I_2 to I_3, respectively).

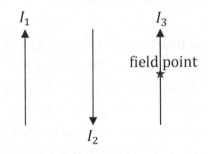

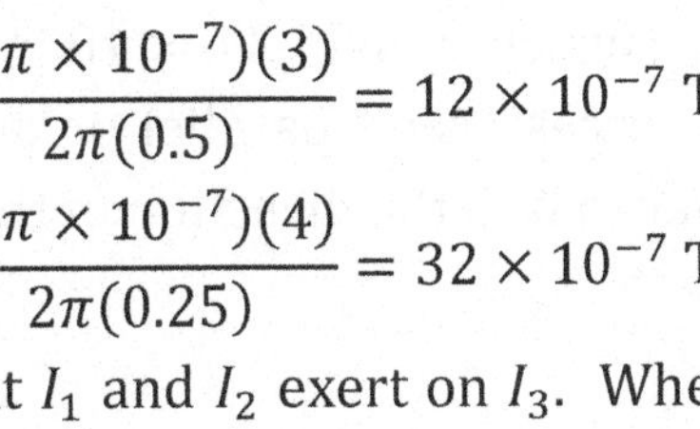

$$B_1 = \frac{\mu_0 I_1}{2\pi d_1} = \frac{(4\pi \times 10^{-7})(3)}{2\pi(0.5)} = 12 \times 10^{-7} \text{ T} = 1.2 \times 10^{-6} \text{ T}$$

$$B_2 = \frac{\mu_0 I_2}{2\pi d_2} = \frac{(4\pi \times 10^{-7})(4)}{2\pi(0.25)} = 32 \times 10^{-7} \text{ T} = 3.2 \times 10^{-6} \text{ T}$$

Now we can find the forces that I_1 and I_2 exert on I_3. When we do this, we use $I_3 = 6.0$ A (since I_3 is experiencing the force specified in the problem) and $\theta = 90°$ (since I_3 is up and since $\vec{\mathbf{B}}_1$ and $\vec{\mathbf{B}}_2$ are perpendicular to the page – as discussed on the following page). What we're calling F_1 is the magnitude of the force that I_1 exerts on I_3, and what we're calling F_2 is the magnitude of the force that I_2 exerts on I_3.

$$F_1 = I_3 L_3 B_1 \sin\theta = (6)(5)(1.2 \times 10^{-6}) \sin 90° = 36 \times 10^{-6} \text{ N} = 3.6 \times 10^{-5} \text{ N}$$
$$F_2 = I_3 L_3 B_2 \sin\theta = (6)(5)(3.2 \times 10^{-6}) \sin 90° = 96 \times 10^{-6} \text{ N} = 9.6 \times 10^{-5} \text{ N}$$

Recall from trig that $\sin 90° = 1$.

Before we can determine how to combine these magnetic forces, we must apply the technique from Chapter 23 in order to determine the direction of each of these forces.

First apply the right-hand rule for magnetic **field** (Chapter 22) to I_1 and I_2. When you grab I_1 with your thumb along I_1 and your fingers wrapped around I_1, your fingers are going into the page (⊗) at the field point (★). The magnetic field ($\vec{\mathbf{B}}_1$) that I_1 makes at the field point (★) is into the page (⊗). When you grab I_2 with your thumb along I_2 and your fingers wrapped around I_2, your fingers are coming out of the page (⊙) at the field point (★). The magnetic field ($\vec{\mathbf{B}}_2$) that I_2 makes at the field point (★) is out of the page (⊙).

Now apply the right-hand rule for magnetic **force** (Chapter 21) to I_3. Point your fingers up (↑), along I_3. At the same time, face your palm into the page (⊗), along $\vec{\mathbf{B}}_1$. Your thumb points to the left (←), along the magnetic force ($\vec{\mathbf{F}}_1$) that I_1 exerts on I_3. The current I_1 pulls I_3 to the left (←). Once again, point your fingers up (↑), along I_3. At the same time, face your palm out of the page (⊙), along $\vec{\mathbf{B}}_2$. Now your thumb points to the right (→), along the magnetic force ($\vec{\mathbf{F}}_2$) that I_2 exerts on I_3. The current I_2 pushes I_3 to the right (→). (You might recall from Chapter 23 that parallel currents, like I_1 and I_3, attract one another and **anti-parallel** currents, like I_2 and I_3, **repel** one another.)

Since $\vec{\mathbf{F}}_1$ and $\vec{\mathbf{F}}_2$ point in opposite directions, as $\vec{\mathbf{F}}_1$ points left (←) and $\vec{\mathbf{F}}_2$ points right (→), we subtract their magnitudes in order to find the magnitude of the net magnetic force. We use absolute values because the magnitude of the net magnetic force can't be negative.

$$F_{net} = |F_1 - F_2| = |3.6 \times 10^{-5} - 9.6 \times 10^{-5}| = |-6.0 \times 10^{-5}| = 6.0 \times 10^{-5} \text{ N}$$

The net magnetic force that I_1 and I_2 exert on I_3 has a magnitude of $F_{net} = 6.0 \times 10^{-5}$ N and a direction that is to the right (→). The reason that it's to the right is that F_2 (which equals 9.6×10^{-5} N) is greater than F_1 (which equals 3.6×10^{-5} N), and the dominant force $\vec{\mathbf{F}}_2$ points to the right. (If the two forces aren't parallel or anti-parallel, you would need to apply trig, as in Chapter 3, in order to find the direction of the net magnetic force.)

Example 91. The three currents below lie at the three corners of a square. The currents run perpendicular to the page in the directions indicated. The currents run through 3.0-m long wires, while the square has 25-cm long edges. Find the magnitude and direction of the net magnetic force exerted on the 2.0-A current.

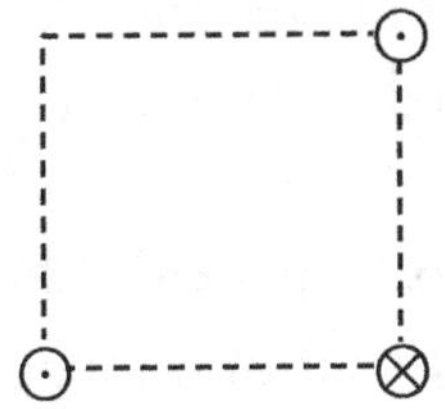

Solution. We will solve this example with the same strategy as the previous example:

- We'll find the force that left current exerts on the 2.0-A current.
- We'll similarly find the force that top current exerts on the 2.0-A current.
- Once we know the magnitudes and directions of both forces, we will know how to combine them.

First imagine a field point (★) at the location of the 2.0-A current (since the force specified in the problem is exerted on the 2.0-A current), and find the magnetic fields at the field point (★) created by each of the 4.0-A currents. When we do this, note that $d_{left} = d_{right} = 0.25$ m (the distance from each 4.0-A current to the 2.0-A current).

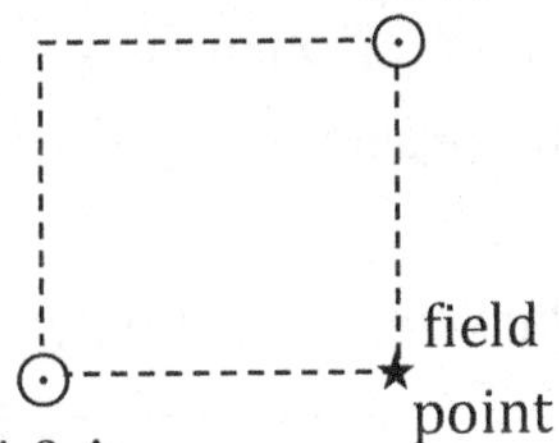

$$B_{left} = \frac{\mu_0 I_{left}}{2\pi d_{left}} = \frac{(4\pi \times 10^{-7})(4)}{2\pi(0.25)} = 32 \times 10^{-7}\text{ T} = 3.2 \times 10^{-6}\text{ T}$$

$$B_{top} = \frac{\mu_0 I_{top}}{2\pi d_{top}} = \frac{(4\pi \times 10^{-7})(4)}{2\pi(0.25)} = 32 \times 10^{-7}\text{ T} = 3.2 \times 10^{-6}\text{ T}$$

Now we can find the forces that the 4.0-A currents exert on the 2.0-A current. When we do this, we use $I_{br} = 2.0$ A (since the bottom right current, I_{br}, is experiencing the force specified in the problem) and $\theta = 90°$ (since I_{br} is into the page and since $\vec{\mathbf{B}}_{left}$ and $\vec{\mathbf{B}}_{top}$ lie in the plane of the rectangle – as discussed on the following page). What we're calling F_{left}

is the magnitude of the force that I_{left} exerts on I_{br}, and what we're calling F_{right} is the magnitude of the force that I_{right} exerts on I_{br}.

$$F_{left} = I_{br}L_{br}B_{left}\sin\theta = (2)(3)(3.2 \times 10^{-6})\sin 90° = 19.2 \times 10^{-6}\text{ N} = 1.92 \times 10^{-5}\text{ N}$$
$$F_{top} = I_{br}L_{br}B_{top}\sin\theta = (2)(3)(3.2 \times 10^{-6})\sin 90° = 19.2 \times 10^{-6}\text{ N} = 1.92 \times 10^{-5}\text{ N}$$

Before we can determine how to combine these magnetic forces, we must apply the technique from Chapter 23 in order to determine the direction of each of these forces.

First apply the right-hand rule for magnetic **field** (Chapter 22) to I_{left} and I_{top}. When you grab I_{left} with your thumb along I_{left} (out of the page) and your fingers wrapped around I_{left}, your fingers are going up (↑) at the field point (★). The magnetic field ($\vec{\mathbf{B}}_{left}$) that I_{left} makes at the field point (★) is up (↑). When you grab I_{top} with your thumb along I_{top} (out of the page) and your fingers wrapped around I_{top}, your fingers are going to the right (→) at the field point (★). The magnetic field ($\vec{\mathbf{B}}_{top}$) that I_{top} makes at the field point (★) is to the right (→).

Now apply the right-hand rule for magnetic **force** (Chapter 21) to I_{br}. (Remember: "br" stands for bottom right.) Point your fingers into the page (⊗), along I_{br}. At the same time, face your palm up (↑), along $\vec{\mathbf{B}}_{left}$. Your thumb points to the right (→), along the magnetic force ($\vec{\mathbf{F}}_{left}$) that I_{left} exerts on I_{br}. The current I_{left} pushes I_{br} to the right (→). Once again, point your fingers into the page (⊗), along I_{br}. At the same time, face your palm right (→), along $\vec{\mathbf{B}}_{top}$. Now your thumb points down (↓), along the magnetic force ($\vec{\mathbf{F}}_{top}$) that I_{top} exerts on I_{br}. The current I_{top} pushes I_{br} down (↓). (You might recall from Chapter 23 that **anti-parallel** currents **repel** one another.)

Since $\vec{\mathbf{F}}_{left}$ and $\vec{\mathbf{F}}_{top}$ are perpendicular, we apply the Pythagorean theorem.

$$F_{net} = \sqrt{F_{left}^2 + F_{top}^2} = \sqrt{(1.92 \times 10^{-5})^2 + (1.92 \times 10^{-5})^2}$$

It's convenient to factor out the 1.92×10^{-5}.

$$F_{net} = \sqrt{(1.92 \times 10^{-5})^2(1^2 + 1^2)} = \sqrt{(1.92 \times 10^{-5})^2(2)} = 1.92\sqrt{2} \times 10^{-5}\text{ N}$$

We applied the rules from algebra that $\sqrt{xy} = \sqrt{x}\sqrt{y}$ and $\sqrt{x^2} = x$. We can find the direction of the net magnetic force with an inverse tangent. We choose $+x$ to point right and $+y$ to point up, such that $F_y = -F_{top}$ (since $\vec{F}_{top}$ points down).

$$\theta_F = \tan^{-1}\left(\frac{F_y}{F_x}\right) = \tan^{-1}\left(\frac{-F_{top}}{F_{left}}\right) = \tan^{-1}\left(\frac{-1.92 \times 10^{-5}}{1.92 \times 10^{-5}}\right) = \tan^{-1}(-1) = 315°$$

The reference angle is 45° and the Quadrant IV angle is $\theta_F = 180° - \theta_{ref} = 315°$. (We know that the angle is in Quadrant IV because $F_y = -F_{top}$ points down while $F_x = F_{left}$

points right.) The magnitude of the net magnetic force exerted on the bottom right wire is $F_{net} = 192\sqrt{2} \times 10^{-7}$ N, which can also be expressed as $F_{net} = 1.92\sqrt{2} \times 10^{-5}$ N, and its direction is $\theta_F = 315°$. If you use a calculator, the magnitude of the net magnetic force works out to $F_{net} = 272 \times 10^{-7}$ N $= 2.7 \times 10^{-5}$ N.

Example 92. In the diagram below, the long straight wire is 5.0 m long and carries a current of 6.0 A, while the rectangular loop carries a current of 8.0 A. What are the magnitude and direction of the net magnetic force that the top current (I_1) exerts on the rectangular loop (I_2)?

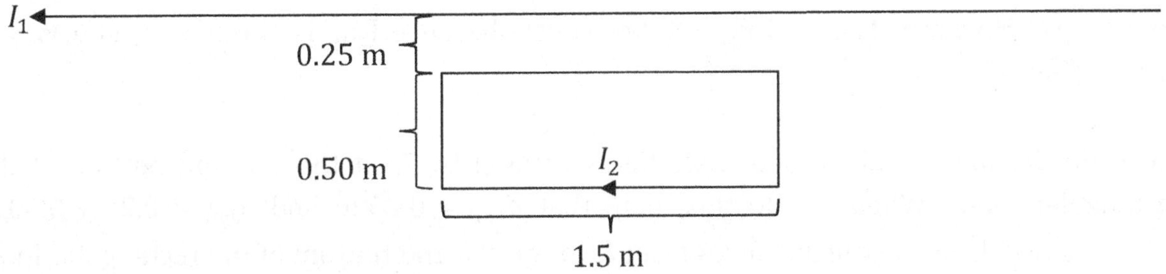

Solution. The first step is to determine the direction of the magnetic force that I_1 exerts on each side of the rectangular loop. Apply the right-hand rule for magnetic **field** (Chapter 22) to I_1. Grab I_1 with your thumb along I_1 (to the left) and your fingers wrapped around I_1. What are your fingers doing below I_1, where the rectangular loop is? They are coming out of the page (⊙) where the rectangular loop is. The magnetic field ($\vec{\mathbf{B}}_1$) that I_1 makes at the rectangular loop is out of the page (⊙).

Now apply the right-hand rule for magnetic **force** (Chapter 21) to each side of the loop.

- <u>Bottom side</u>: Point your fingers to the left (←) along the current (I_2) and your palm out of the page (⊙) along the magnetic field ($\vec{\mathbf{B}}_1$). The magnetic force ($\vec{\mathbf{F}}_{bot}$) is up (↑).
- <u>Left side</u>: Point your fingers up (↑) along the current (I_2) and your palm out of the page (⊙) along the magnetic field ($\vec{\mathbf{B}}_1$). The magnetic force ($\vec{\mathbf{F}}_{left}$) is right (→).
- <u>Top side</u>: Point your fingers to the right (→) along the current (I_2) and your palm out of the page (⊙) along the magnetic field ($\vec{\mathbf{B}}_1$). The magnetic force ($\vec{\mathbf{F}}_{top}$) is down (↓).
- <u>Right side</u>: Point your fingers down (↓) along the current (I_2) and your palm out of the page (⊙) along the magnetic field ($\vec{\mathbf{B}}_1$). The magnetic force ($\vec{\mathbf{F}}_{right}$) is left (←).

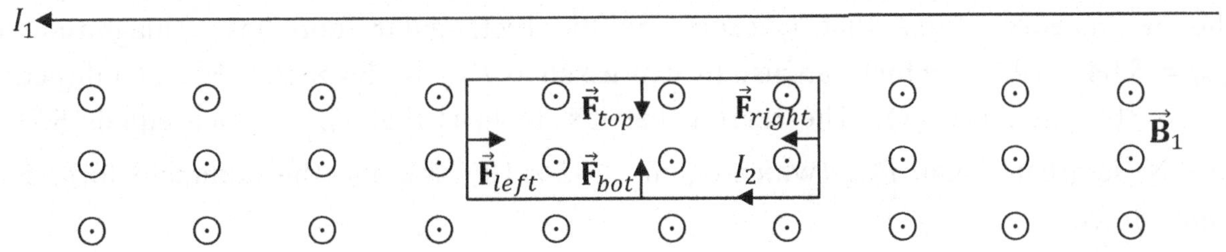

Study the diagram at the bottom of the previous page. You should see that $\vec{\mathbf{F}}_{left}$ and $\vec{\mathbf{F}}_{right}$ cancel out:

- $\vec{\mathbf{F}}_{left}$ and $\vec{\mathbf{F}}_{right}$ point in **opposite** directions: One points right, the other points left.
- $\vec{\mathbf{F}}_{left}$ and $\vec{\mathbf{F}}_{right}$ have **equal** magnitudes: They are the same distance from I_1.

We don't need to calculate $\vec{\mathbf{F}}_{left}$ and $\vec{\mathbf{F}}_{right}$ because they will cancel out later when we find the magnitude of the net force. (These would also be a challenge to calculate since they are perpendicular to I_1: That problem involves calculus.)

You might note that $\vec{\mathbf{F}}_{top}$ and $\vec{\mathbf{F}}_{bot}$ also have opposite directions (one points down, the other points up). However, $\vec{\mathbf{F}}_{top}$ and $\vec{\mathbf{F}}_{bot}$ do <u>**not**</u> cancel because $\vec{\mathbf{F}}_{top}$ is closer to I_1 and $\vec{\mathbf{F}}_{bot}$ is further from I_1.

To begin the math, find the magnetic fields created by I_1 at the top and bottom of the rectangular loop. When we do this, note that $d_{top} = 0.25$ m and $d_{bot} = 0.25 + 0.50 = 0.75$ m (since these are the distances from I_1 to the top and bottom of the rectangular loop, respectively). We use $I_1 = 6.0$ A in each case because I_1 is creating these two magnetic fields.

$$B_{top} = \frac{\mu_0 I_1}{2\pi d_{top}} = \frac{(4\pi \times 10^{-7})(6)}{2\pi(0.25)} = 48 \times 10^{-7} \text{ T} = 4.8 \times 10^{-6} \text{ T}$$

$$B_{bot} = \frac{\mu_0 I_1}{2\pi d_{bot}} = \frac{(4\pi \times 10^{-7})(6)}{2\pi(0.75)} = 16 \times 10^{-7} \text{ T} = 1.6 \times 10^{-6} \text{ T}$$

Now we can find the forces that I_1 exerts on the top and bottom sides of the rectangular loop. When we do this, we use $I_2 = 8.0$ A (since I_2 is experiencing the force specified in the problem) and $\theta = 90°$ (since we already determined that $\vec{\mathbf{B}}_1$, which points out of the page, is perpendicular to the loop). We also use the width of the rectangle, $L_2 = 1.5$ m, since that is the distance that I_2 travels in the top and bottom sides of the rectangular loop.

$$F_{top} = I_2 L_2 B_{top} \sin\theta = (8)(1.5)(4.8 \times 10^{-6}) \sin 90° = 57.6 \times 10^{-6} \text{ N} = 5.76 \times 10^{-5} \text{ N}$$

$$F_{bot} = I_2 L_2 B_{bot} \sin\theta = (8)(1.5)(1.6 \times 10^{-6}) \sin 90° = 19.2 \times 10^{-6} \text{ N} = 1.92 \times 10^{-5} \text{ N}$$

Since $\vec{\mathbf{F}}_{top}$ and $\vec{\mathbf{F}}_{bot}$ point in opposite directions, as $\vec{\mathbf{F}}_{top}$ points down (↓) and $\vec{\mathbf{F}}_{bot}$ points up (↑), we subtract their magnitudes in order to find the magnitude of the net magnetic force. We use absolute values because the magnitude of the net magnetic force can't be negative.

$$F_{net} = |F_{top} - F_{bot}| = |5.76 \times 10^{-5} - 1.92 \times 10^{-5}| = 3.84 \times 10^{-5} \text{ N}$$

The net magnetic force that I_1 exerts on the rectangular loop has a magnitude of $F_{net} = 3.84 \times 10^{-5}$ N, which can also be expressed as $F_{net} = 384 \times 10^{-7}$ N, and a direction that is straight down (↓). The reason that it's down is that F_{top} (which equals 5.76×10^{-5} N) is greater than F_{bot} (which equals 1.92×10^{-5} N), and the dominant force $\vec{\mathbf{F}}_{top}$ points down.

26 THE LAW OF BIOT-SAVART

Law of Biot-Savart for Current		
$\vec{\mathbf{B}} = \frac{\mu_0}{4\pi}\int \frac{I\, d\vec{\mathbf{s}} \times \hat{\mathbf{R}}}{R^2}$ (filamentary current)	$\vec{\mathbf{B}} = \frac{\mu_0}{4\pi}\int \frac{\vec{\mathbf{K}}\, dA \times \hat{\mathbf{R}}}{R^2}$ $I = \int \vec{\mathbf{K}} \cdot d\vec{\boldsymbol{\ell}}$ (surface current)	$\vec{\mathbf{B}} = \frac{\mu_0}{4\pi}\int \frac{\vec{\mathbf{J}}\, dV \times \hat{\mathbf{R}}}{R^2}$ $I = \int \vec{\mathbf{J}} \cdot d\vec{\mathbf{A}}$ (through a volume)
Law of Biot-Savart for Moving Charge		
$\vec{\mathbf{B}} = \frac{\mu_0}{4\pi}\int \frac{\vec{\boldsymbol{v}}\, dq \times \hat{\mathbf{R}}}{R^2}$	$v = r_{rot}\omega$ (if rotating)	$Q = \int dq$
$dq = \lambda ds$ (line or thin arc)	$dq = \sigma dA$ (surface area)	$dq = \rho dV$ (volume)
Relation Among Coordinate Systems and Unit Vectors		
$x = r\cos\theta$ $y = r\sin\theta$ $\hat{\mathbf{r}} = \hat{\mathbf{x}}\cos\theta + \hat{\mathbf{y}}\sin\theta$ $\hat{\boldsymbol{\theta}} = -\hat{\mathbf{x}}\sin\theta + \hat{\mathbf{y}}\cos\theta$ (2D polar)	$x = r_c\cos\theta$ $y = r_c\sin\theta$ $\hat{\mathbf{r}}_c = \hat{\mathbf{x}}\cos\theta + \hat{\mathbf{y}}\sin\theta$ $\hat{\boldsymbol{\theta}} = -\hat{\mathbf{x}}\sin\theta + \hat{\mathbf{y}}\cos\theta$ (cylindrical)	$x = r\sin\theta\cos\varphi$ $y = r\sin\theta\sin\varphi$ $z = r\cos\theta$ $\hat{\mathbf{r}} = \hat{\mathbf{x}}\cos\varphi\sin\theta$ $+\hat{\mathbf{y}}\sin\varphi\sin\theta + \hat{\mathbf{z}}\cos\theta$ (spherical)

Permeability of Free Space
$\mu_0 = 4\pi \times 10^{-7}\ \frac{\mathrm{T}\cdot\mathrm{m}}{\mathrm{A}}$

Symbol	Meaning
$\otimes$	into the page
$\odot$	out of the page

Differential Arc Length		
$d\vec{\mathbf{s}} = \hat{\mathbf{x}}\,dx$ $ds = dx$ (along x)	$d\vec{\mathbf{s}} = \hat{\mathbf{y}}\,dy$ or $\hat{\mathbf{z}}\,dz$ $ds = dy$ or dz (along y or z)	$d\vec{\mathbf{s}} = \pm\hat{\boldsymbol{\theta}}\,a d\theta$ $ds = a d\theta$ (circular arc of radius a)
Differential Area Element		
$dA = dxdy$ (polygon in xy plane)	$dA = rdrd\theta$ (pie slice, disc, thick ring)	$dA = a^2 \sin\theta\, d\theta d\varphi$ (sphere of radius a)
Differential Volume Element		
$dV = dxdydz$ (bounded by flat sides)	$dV = r_c dr_c d\theta dz$ (cylinder or cone)	$dV = r^2 \sin\theta\, drd\theta d\varphi$ (spherical)

Vector Product

$$\vec{\mathbf{A}} \times \vec{\mathbf{B}} = \begin{vmatrix} \hat{\mathbf{x}} & \hat{\mathbf{y}} & \hat{\mathbf{z}} \\ A_x & A_y & A_z \\ B_x & B_y & B_z \end{vmatrix} = \hat{\mathbf{x}} \begin{vmatrix} A_y & A_z \\ B_y & B_z \end{vmatrix} - \hat{\mathbf{y}} \begin{vmatrix} A_x & A_z \\ B_x & B_z \end{vmatrix} + \hat{\mathbf{z}} \begin{vmatrix} A_x & A_y \\ B_x & B_y \end{vmatrix}$$

$$\vec{\mathbf{A}} \times \vec{\mathbf{B}} = (A_y B_z - A_z B_y)\hat{\mathbf{x}} - (A_x B_z - A_z B_x)\hat{\mathbf{y}} + (A_x B_y - A_y B_x)\hat{\mathbf{z}}$$

$$\vec{\mathbf{A}} \times \vec{\mathbf{B}} = A_y B_z \hat{\mathbf{x}} - A_z B_y \hat{\mathbf{x}} + A_z B_x \hat{\mathbf{y}} - A_x B_z \hat{\mathbf{y}} + A_x B_y \hat{\mathbf{z}} - A_y B_x \hat{\mathbf{z}}$$

Magnitude of the Vector Product

$$\|\vec{\mathbf{C}}\| = \|\vec{\mathbf{A}} \times \vec{\mathbf{B}}\| = AB\sin\theta = \sqrt{C_x^2 + C_y^2 + C_z^2}$$

Vector Products among Cartesian Unit Vectors

$$\hat{\mathbf{x}} \times \hat{\mathbf{x}} = 0 \quad , \quad \hat{\mathbf{y}} \times \hat{\mathbf{y}} = 0 \quad , \quad \hat{\mathbf{z}} \times \hat{\mathbf{z}} = 0$$

$$\hat{\mathbf{x}} \times \hat{\mathbf{y}} = \hat{\mathbf{z}} \quad , \quad \hat{\mathbf{y}} \times \hat{\mathbf{z}} = \hat{\mathbf{x}} \quad , \quad \hat{\mathbf{z}} \times \hat{\mathbf{x}} = \hat{\mathbf{y}}$$

$$\hat{\mathbf{y}} \times \hat{\mathbf{x}} = -\hat{\mathbf{z}} \quad , \quad \hat{\mathbf{z}} \times \hat{\mathbf{y}} = -\hat{\mathbf{x}} \quad , \quad \hat{\mathbf{x}} \times \hat{\mathbf{z}} = -\hat{\mathbf{y}}$$

Symbol	Name	SI Units
I	current	A
$\vec{\mathbf{B}}$	magnetic field	T
μ_0	the permeability of free space	$\frac{\text{T}\cdot\text{m}}{\text{A}}$
$\vec{\mathbf{R}}$	a vector from each differential element to the field point	m
$\hat{\mathbf{R}}$	a unit vector along $\vec{\mathbf{R}}$	unitless
R	the distance corresponding to $\vec{\mathbf{R}}$	m
x, y, z	Cartesian coordinates	m, m, m
$\hat{\mathbf{x}}, \hat{\mathbf{y}}, \hat{\mathbf{z}}$	unit vectors along the $+x$-, $+y$-, $+z$-axes	unitless
r, θ	2D polar coordinates	m, rad
r_c, θ, z	cylindrical coordinates	m, rad, m
r, θ, φ	spherical coordinates	m, rad, rad
$\hat{\mathbf{r}}, \hat{\boldsymbol{\theta}}, \hat{\boldsymbol{\varphi}}$	unit vectors along spherical coordinate axes	unitless
$\hat{\mathbf{r}}_c$	a unit vector pointing away from the $+z$-axis	unitless
$\vec{\mathbf{K}}$	surface current density (distributed over a surface)	A/m
$\vec{\mathbf{J}}$	current density (distributed throughout a volume)	A/m^2
ds	differential arc length	m
dA	differential area element	m^2
dV	differential volume element	m^3
$\vec{\boldsymbol{v}}$	velocity	m/s
ω	angular speed	rad/s
dq	differential charge element	C
Q	total charge of the object	C

Current Loop Example: A Prelude to Example 93. A filamentary current travels along the circular wire illustrated below. The radius of the circle is denoted by the symbol a. Derive an equation for the magnetic field at the center of the circle in terms of μ_0, I, a, and appropriate unit vectors.

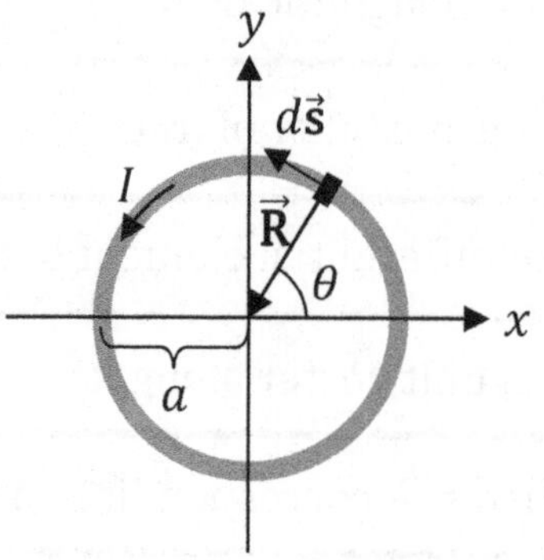

Solution. Begin with a labeled diagram. For a filamentary current, draw a representative $d\vec{s}$: Note that $d\vec{s}$ must lie on the circular wire and has a direction that is tangent to the current (I). Draw $\vec{R}$ from the source, $d\vec{s}$, to the field point (in this problem, it's at the origin). When we perform the integration, we effectively integrate over every $d\vec{s}$ that makes up the circle. Apply the law of Biot-Savart to the filamentary current.

$$\vec{B} = \frac{\mu_0}{4\pi}\int \frac{I\, d\vec{s} \times \hat{R}}{R^2}$$

Examine the picture above:

- The vector $\vec{R}$ has the same length for each dq that makes up the circle. Its magnitude, R, equals the radius of the circle: $R = a$.
- $\vec{R}$ points **inward**, toward the center of the circle. Since the unit vector $\hat{r}$ of 2D polar coordinates points **outward**, we can write $\hat{R} = -\hat{r}$.

For a circular filamentary current, write $d\vec{s} = \hat{\theta}\, a d\theta$ (it's positive because I is counter-clockwise in this example). See page 282. Substitute the expressions $R = a$, $\hat{R} = -\hat{r}$, and $d\vec{s} = \hat{\theta}\, a d\theta$ into the magnetic field integral. The current I and radius a are constants and may come out of the integral. Note that $\frac{a}{a^2} = \frac{1}{a}$. Also note that the $(-\hat{r})$ is actually part of the integration. The limits of integration are from $\theta = 0$ to $\theta = 2\pi$ **radians** for a full circle.

$$\vec{B} = \frac{\mu_0}{4\pi}\int \frac{I\, d\vec{s} \times \hat{R}}{R^2} = \frac{\mu_0}{4\pi}\int_{\theta=0}^{2\pi} \frac{I\left(\hat{\theta}\, a d\theta\right) \times (-\hat{r})}{a^2} = -\frac{\mu_0 I}{4\pi a}\int_{\theta=0}^{2\pi} \hat{\theta} \times \hat{r}\, d\theta$$

Now we perform the vector product $\hat{\theta} \times \hat{r}$. There are two ways to do this. One way is to insert the expressions $\hat{r} = \hat{x}\cos\theta + \hat{y}\sin\theta$ and $\hat{\theta} = -\hat{x}\sin\theta + \hat{y}\cos\theta$ (see page 281) into the determinant form of the vector product (Chapter 19):

$$\hat{\theta} \times \hat{r} = \begin{vmatrix} \hat{x} & \hat{y} & \hat{z} \\ -\sin\theta & \cos\theta & 0 \\ \cos\theta & \sin\theta & 0 \end{vmatrix} = \hat{z}(-\sin^2\theta - \cos^2\theta) = -\hat{z}(\sin^2\theta + \cos^2\theta) = -\hat{z}$$

We used the trigonometric identity $\sin^2\theta + \cos^2\theta = 1$.

There is an alternative method to determine that $\hat{\boldsymbol{\theta}} \times \hat{\mathbf{r}} = -\hat{\mathbf{z}}$: Apply the right-hand rule as follows.

Point your fingers along $\hat{\boldsymbol{\theta}}$ (a counterclockwise tangent along $d\vec{\mathbf{s}}$) and face your palm along $\hat{\mathbf{r}}$ (opposite to $\hat{\mathbf{R}}$, since $\hat{\mathbf{r}}$ points outward). Your thumb points into the page ($\otimes$), along $-\hat{\mathbf{z}}$ (since $+z$ comes out of the page). Since $\hat{\boldsymbol{\theta}}$ is perpendicular to $\hat{\mathbf{r}}$, and since these are unit vectors, the vector product $\hat{\boldsymbol{\theta}} \times \hat{\mathbf{r}}$ has a magnitude of $\|\hat{\boldsymbol{\theta}} \times \hat{\mathbf{r}}\| = (1)(1)\sin 90° = 1$ according to the formula $\|\vec{\mathbf{A}} \times \vec{\mathbf{B}}\| = AB\sin\theta$ from Chapter 19. Thus, $\hat{\boldsymbol{\theta}} \times \hat{\mathbf{r}} = (1)(-\hat{\mathbf{z}}) = -\hat{\mathbf{z}}$.

Substitute $\hat{\boldsymbol{\theta}} \times \hat{\mathbf{r}} = -\hat{\mathbf{z}}$ into our previous expression for the magnetic field integral. Note that the two minus signs make a plus sign.

$$\vec{\mathbf{B}} = -\frac{\mu_0 I}{4\pi a} \int_{\theta=0}^{2\pi} \hat{\boldsymbol{\theta}} \times \hat{\mathbf{r}}\, d\theta = \frac{\mu_0 I}{4\pi a} \int_{\theta=0}^{2\pi} \hat{\mathbf{z}}\, d\theta$$

The unit vector $\hat{\mathbf{z}}$ is a constant because it always points one unit along the $+z$-axis. (Note that the unit vectors $\hat{\boldsymbol{\theta}}$ and $\hat{\mathbf{r}}$ are **<u>not</u>** constants, since their directions are different for each point on the circle. It may help to study the diagram on the previous page and draw these unit vectors for a few different points on the circle.) Since $\hat{\mathbf{z}}$ is constant, we may pull it out of the integral. The remaining integral is trivial.

$$\vec{\mathbf{B}} = \frac{\mu_0 I}{4\pi a}\hat{\mathbf{z}} \int_{\theta=0}^{2\pi} d\theta = \frac{\mu_0 I}{4\pi a}\hat{\mathbf{z}}[\theta]_{\theta=0}^{2\pi} = \frac{\mu_0 I}{4\pi a}\hat{\mathbf{z}}(2\pi - 0) = \frac{\mu_0 I}{2a}\hat{\mathbf{z}}$$

The magnetic field at the origin is $\vec{\mathbf{B}} = \frac{\mu_0 I}{2a}\hat{\mathbf{z}}$. It has a magnitude of $B = \frac{\mu_0 I}{2a}$ and a direction of $\hat{\mathbf{B}} = \hat{\mathbf{z}}$. If you apply the right-hand rule for magnetic **field** (Chapter 22) to the diagram on the previous page, you should see that your fingers are coming out of the page ($\odot$) along $\hat{\mathbf{z}}$ at the origin, which agrees with the direction of our answer above.

Tips: It's a good habit to check that the direction of your answer from applying the law of Biot-Savart agrees with the right-hand rule for magnetic field. It's easy to make a mistake with the vector product or to make a sign mistake in the algebra, so it's good to have a quick way to help check your answer for consistency.

Another thing that you can check is the units. Your final expression for magnetic field should include the units of μ_0 times the unit of current divided by the unit of length. Our answer, $B = \frac{\mu_0 I}{2a}$, meets this criteria.

Circular Arc Example: Another Prelude to Example 93. A filamentary current travels along the path illustrated below, which includes a semicircular arc with radius a. Derive an equation for the magnetic field at the origin in terms of μ_0, I, a, and appropriate unit vectors.

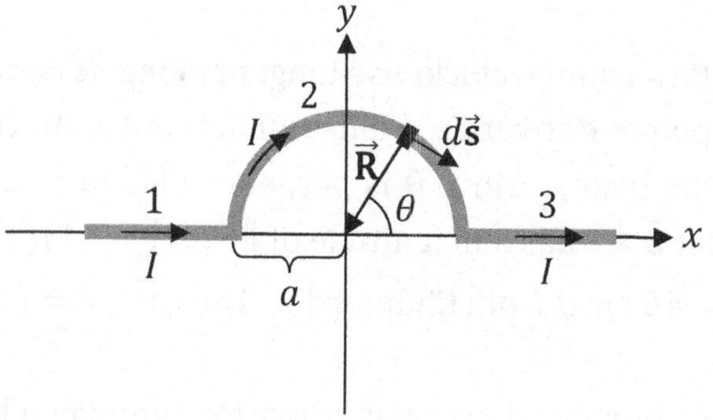

Solution. One "trick" to problems like this, which involve a circular arc and straight sections, is to divide the shape up into pieces:

- Piece 1 is the straight current on the left.
- Piece 2 is the semicircular current in the middle.
- Piece 3 is the straight current on the right.

A second "trick" is to realize that $\vec{\mathbf{B}}_1 = 0$ and $\vec{\mathbf{B}}_3 = 0$. For these two straight sections, $\hat{\mathbf{R}}$ points along $\hat{\mathbf{x}}$ or $-\hat{\mathbf{x}}$, while $d\vec{\mathbf{s}}$ points along $\hat{\mathbf{x}}$. For these two sections, $d\vec{\mathbf{s}} \times \hat{\mathbf{R}} = 0$ because $\hat{\mathbf{x}} \times \hat{\mathbf{x}} = 0$ (see Chapter 19). Whenever two vectors are parallel or anti-parallel, their vector product equals zero (since $\theta = 0°$ or $180°$, so that $\left\|\vec{\mathbf{A}} \times \vec{\mathbf{B}}\right\| = AB \sin\theta = AB \sin 0° = 0$).

Therefore, we only need to find the magnetic field ($\vec{\mathbf{B}}_2$) created by the semicircular section at the origin. The math for this is nearly identical to the previous example. The only differences between these two examples are:

- The limits of integration will now be from $\theta = 0$ to $\theta = \pi$ **radians** for the semicircle (instead of from $\theta = 0$ to $\theta = 2\pi$ for a full circle).
- The current in this example runs clockwise instead of counterclockwise, such that $(-\hat{\boldsymbol{\theta}}) \times \hat{\mathbf{r}} = \hat{\mathbf{z}}$ (instead of $\hat{\boldsymbol{\theta}} \times \hat{\mathbf{r}} = -\hat{\mathbf{z}}$). The different direction of the current in this example will simply change the overall sign of the answer.

The solution to this example will be the same as for the previous problem, except that the magnetic field integral will become:

$$\vec{\mathbf{B}}_2 = -\frac{\mu_0 I}{4\pi a}\hat{\mathbf{z}} \int_{\theta=0}^{\pi} d\theta = -\frac{\mu_0 I}{4a}\hat{\mathbf{z}}$$

The magnetic field at the origin is $\vec{\mathbf{B}} = \vec{\mathbf{B}}_2 = -\frac{\mu_0 I}{4a}\hat{\mathbf{z}}$. It has a magnitude of $B = \frac{\mu_0 I}{4a}$ and a direction of $\hat{\mathbf{B}} = -\hat{\mathbf{z}}$. If you apply the right-hand rule for magnetic **field** (Chapter 22) to the diagram above, you should see that your fingers are going into the page ($\otimes$) along $-\hat{\mathbf{z}}$, which agrees with the direction of our answer above.

Example 93. A filamentary current travels along the path illustrated below, which includes two circular arcs with radii of $2a$ and $3a$ (where the "complete" circle would be centered about the origin) and three straight line segments. The angles are in radians, measured counterclockwise from the $+x$-axis. Derive an equation for the magnetic field at the origin in terms of μ_0, I, a, and appropriate unit vectors.

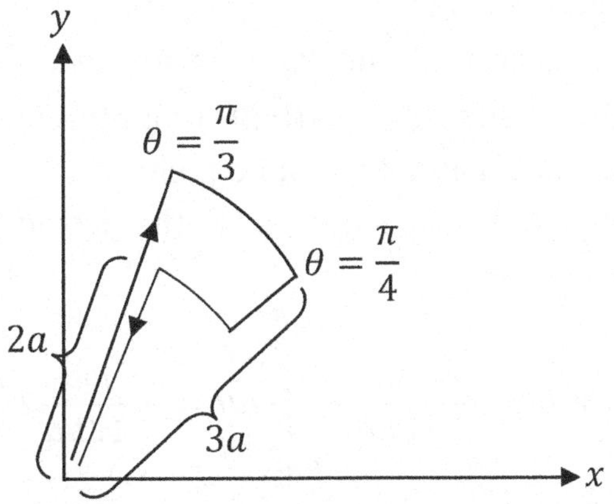

Solution. As we did in the previous example, we first divide the shape up into pieces: See the sections numbered below on the left. Draw $d\vec{\mathbf{s}}$ tangent to I and draw $\vec{\mathbf{R}}$ from the source, $d\vec{\mathbf{s}}$, to the field point (at the origin). See the diagram below on the right for section 2. From the diagram, you should see that $\hat{\mathbf{R}}$ points inward, along $-\hat{\mathbf{r}}$.

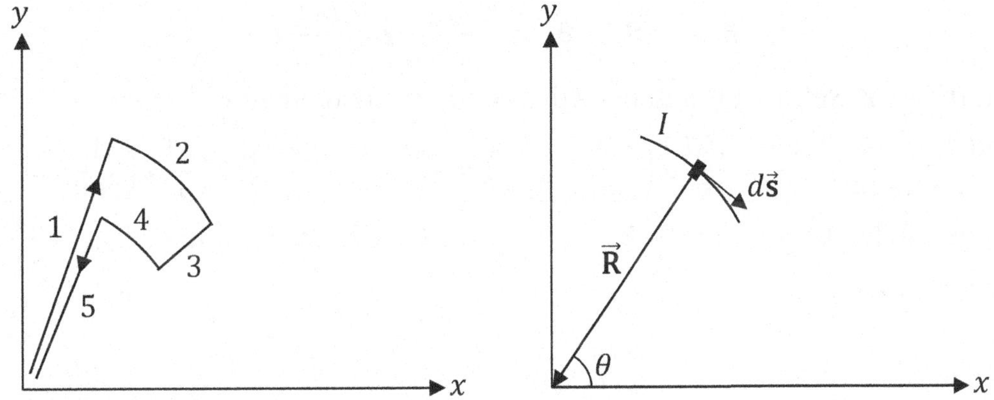

The magnetic field at the origin is zero for the straight sections (1, 3, and 5) because $d\vec{\mathbf{s}}$ is either parallel or anti-parallel to $\hat{\mathbf{R}}$ for these sections, such that $d\vec{\mathbf{s}} \times \hat{\mathbf{R}} = 0$ (because $\|\vec{\mathbf{A}} \times \vec{\mathbf{B}}\| = AB\sin\theta = AB\sin 0° = 0$ or $AB\sin 180° = 0$). Therefore, $\vec{\mathbf{B}}_1 = 0$, $\vec{\mathbf{B}}_3 = 0$, and $\vec{\mathbf{B}}_5 = 0$. We just need to apply the law of Biot-Savart to the curved sections (2 and 4).

Therefore, we only need to find the magnetic field ($\vec{\mathbf{B}}_2$) created by the curved sections at the origin. The math for this is nearly identical to the two previous examples. The only differences between this example and the previous examples are:

- The limits of integration will now be from $\theta = \frac{\pi}{4}$ to $\theta = \frac{\pi}{3}$ **radians** for the two circular arc lengths (sections 2 and 4).
- The current runs clockwise along section 2 and counterclockwise along section 4, such that $(-\hat{\boldsymbol{\theta}}) \times \hat{\mathbf{r}} = \hat{\mathbf{z}}$ for section 2 and $\hat{\boldsymbol{\theta}} \times \hat{\mathbf{r}} = -\hat{\mathbf{z}}$ for section 4. The different direction of the current along each circular arc will result in a different sign for the magnetic field created by each arc.
- Note that $R_2 = 3a$ for section 2 and $R_4 = 2a$ for section 4. Write $d\vec{\mathbf{s}}_2 = -3a\hat{\boldsymbol{\theta}}\,d\theta$ (clockwise) for section 2 and $d\vec{\mathbf{s}}_4 = 2a\hat{\boldsymbol{\theta}}\,d\theta$ (counterclockwise) for section 4, since the radii of these arcs are $3a$ and $2a$, respectively.

The solution to this example will be the same as for the previous example, except that the magnetic field integrals will become:

$$\vec{\mathbf{B}}_2 = \frac{\mu_0 I}{12\pi a} \int_{\theta=\pi/4}^{\pi/3} \hat{\boldsymbol{\theta}} \times \hat{\mathbf{r}}\,d\theta = -\frac{\mu_0 I}{12\pi a}\hat{\mathbf{z}} \int_{\theta=\frac{\pi}{4}}^{\frac{\pi}{3}} d\theta = -\frac{\mu_0 I}{12\pi a}\hat{\mathbf{z}}\left(\frac{\pi}{3} - \frac{\pi}{4}\right) = -\frac{\mu_0 I}{144a}\hat{\mathbf{z}}$$

$$\vec{\mathbf{B}}_4 = -\frac{\mu_0 I}{8\pi a} \int_{\theta=\pi/4}^{\pi/3} \hat{\boldsymbol{\theta}} \times \hat{\mathbf{r}}\,d\theta = \frac{\mu_0 I}{8\pi a}\hat{\mathbf{z}} \int_{\theta=\pi/4}^{\pi/3} d\theta = \frac{\mu_0 I}{8\pi a}\hat{\mathbf{z}}\left(\frac{\pi}{3} - \frac{\pi}{4}\right) = \frac{\mu_0 I}{96a}\hat{\mathbf{z}}$$

Note that $\frac{\pi}{3} - \frac{\pi}{4} = \frac{4\pi}{12} - \frac{3\pi}{12} = \frac{4\pi - 3\pi}{12} = \frac{\pi}{12}$ (subtract fractions with a **common denominator**).

Combine $\vec{\mathbf{B}}_2$ and $\vec{\mathbf{B}}_4$ together to form the net magnetic field, $\vec{\mathbf{B}}_{net}$.

$$\vec{\mathbf{B}}_{net} = \vec{\mathbf{B}}_2 + \vec{\mathbf{B}}_4 = -\frac{\mu_0 I}{144a}\hat{\mathbf{z}} + \frac{\mu_0 I}{96a}\hat{\mathbf{z}}$$

Factor out the $\frac{\mu_0 I}{a}\hat{\mathbf{z}}$. Subtract fractions with a **common denominator**.

$$\vec{\mathbf{B}}_{net} = \frac{\mu_0 I}{a}\hat{\mathbf{z}}\left(\frac{-1}{144} + \frac{1}{96}\right) = \frac{\mu_0 I}{a}\hat{\mathbf{z}}\left(\frac{-2}{288} + \frac{3}{288}\right) = \frac{\mu_0 I}{a}\hat{\mathbf{z}}\left(\frac{-2+3}{288}\right) = \frac{\mu_0 I}{a}\hat{\mathbf{z}}\left(\frac{1}{288}\right) = \frac{\mu_0 I}{288a}\hat{\mathbf{z}}$$

The magnetic field at the origin is $\vec{\mathbf{B}}_{net} = \frac{\mu_0 I}{288a}\hat{\mathbf{z}}$. It has a magnitude of $B_{net} = \frac{\mu_0 I}{288a}$ and a direction of $\hat{\mathbf{B}} = \hat{\mathbf{z}}$.

Example 94. A filamentary current travels along a finite length of wire on the z-axis with endpoints at $\left(0,0,-\frac{L}{2}\right)$ and $\left(0,0,\frac{L}{2}\right)$, as illustrated below. Derive an equation for the magnetic field at the point $(a,0,0)$, where a is a constant, in terms of μ_0, I, L, a, and appropriate unit vectors.

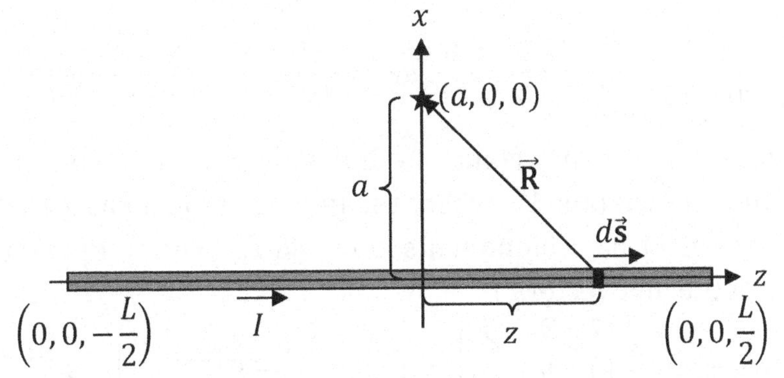

Solution. Begin with a labeled diagram. For a filamentary current, draw a representative $d\vec{\mathbf{s}}$ along the current (I). Draw $\vec{\mathbf{R}}$ from the source, $d\vec{\mathbf{s}}$, to the field point $(a,0,0)$. When we perform the integration, we effectively integrate over every $d\vec{\mathbf{s}}$ that makes up the current. Apply the law of Biot-Savart to the filamentary current.

$$\vec{\mathbf{B}} = \frac{\mu_0}{4\pi}\int \frac{I\, d\vec{\mathbf{s}} \times \hat{\mathbf{R}}}{R^2}$$

Examine the picture above:

- The vector $\vec{\mathbf{R}}$ extends a units up, along the x-axis (along $\hat{\mathbf{x}}$), and z units to the left, along the negative z-axis (along $-\hat{\mathbf{z}}$). Therefore, $\vec{\mathbf{R}} = a\hat{\mathbf{x}} - z\hat{\mathbf{z}}$.
- Apply the Pythagorean theorem to find the magnitude of $\vec{\mathbf{R}}$.

$$R = \sqrt{a^2 + z^2}$$

- Divide $\vec{\mathbf{R}}$ by R to find the direction of $\vec{\mathbf{R}}$.

$$\hat{\mathbf{R}} = \frac{\vec{\mathbf{R}}}{R} = \frac{a\hat{\mathbf{x}} - z\hat{\mathbf{z}}}{\sqrt{a^2 + z^2}}$$

For a straight filamentary current along the z-axis, write $d\vec{\mathbf{s}} = \hat{\mathbf{z}}\, dz$. See page 282. The limits of integration are from $z = -\frac{L}{2}$ to $z = \frac{L}{2}$ (the endpoints of the wire). Substitute the expressions for R, $\hat{\mathbf{R}}$, and $d\vec{\mathbf{s}}$ into the integral. Watch how we rewrite the magnetic field integral in order to manage the substitutions:

$$\vec{\mathbf{B}} = \frac{\mu_0}{4\pi}\int \frac{I\, d\vec{\mathbf{s}} \times \hat{\mathbf{R}}}{R^2} = \frac{\mu_0}{4\pi}\int I\, d\vec{\mathbf{s}}\left(\frac{1}{R^2}\right) \times \hat{\mathbf{R}}$$

Now substitute the expressions for R, $\hat{\mathbf{R}}$, and $d\vec{\mathbf{s}}$ into the integral. Compare the previous line to the following line.

$$\vec{\mathbf{B}} = \frac{\mu_0}{4\pi}\int_{z=-L/2}^{L/2} I(\hat{\mathbf{z}}\,dz)\frac{1}{a^2+z^2}\times\frac{a\hat{\mathbf{x}}-z\hat{\mathbf{z}}}{\sqrt{a^2+z^2}}$$

Note that $R^2 = \left(\sqrt{a^2+z^2}\right)^2 = a^2+z^2$. In the next step, we will apply the rule from algebra that $(a^2+z^2)\sqrt{a^2+z^2} = (a^2+z^2)^1(a^2+z^2)^{1/2} = (a^2+z^2)^{3/2}$.

$$\vec{\mathbf{B}} = \frac{\mu_0}{4\pi}\int_{z=-L/2}^{L/2}\frac{I\,\hat{\mathbf{z}}\times(a\hat{\mathbf{x}}-z\hat{\mathbf{z}})}{(a^2+z^2)^{3/2}}dz = \frac{\mu_0 I}{4\pi}\int_{z=-L/2}^{L/2}\frac{\hat{\mathbf{z}}\times(a\hat{\mathbf{x}}-z\hat{\mathbf{z}})}{(a^2+z^2)^{3/2}}dz$$

Note that I is a constant, which may come out of the integral. Work out the vector product $\hat{\mathbf{z}}\times(a\hat{\mathbf{x}}-z\hat{\mathbf{z}})$ according to Chapter 19, where the first vector just has a z-component (equal to 1) and the second vector has components a, 0, and z. (The components of the vectors are the coefficients of the unit vectors.)

$$\hat{\mathbf{z}}\times(a\hat{\mathbf{x}}-z\hat{\mathbf{z}}) = \begin{vmatrix}\hat{\mathbf{x}} & \hat{\mathbf{y}} & \hat{\mathbf{z}}\\ 0 & 0 & 1\\ a & 0 & -z\end{vmatrix} = \hat{\mathbf{x}}\begin{vmatrix}0 & 1\\ 0 & -z\end{vmatrix} - \hat{\mathbf{y}}\begin{vmatrix}0 & 1\\ a & -z\end{vmatrix} + \hat{\mathbf{z}}\begin{vmatrix}0 & 0\\ a & 0\end{vmatrix}$$

$$\hat{\mathbf{z}}\times(a\hat{\mathbf{x}}-z\hat{\mathbf{z}}) = \hat{\mathbf{x}}(0-0)-\hat{\mathbf{y}}(0-a)+\hat{\mathbf{z}}(0-0) = a\,\hat{\mathbf{y}}$$

Substitute this result into the magnetic field integral.

$$\vec{\mathbf{B}} = \frac{\mu_0 I}{4\pi}\int_{z=-L/2}^{L/2}\frac{a\,\hat{\mathbf{y}}}{(a^2+z^2)^{3/2}}dz$$

The constants a and $\hat{\mathbf{y}}$ may come out of the integral.

$$\vec{\mathbf{B}} = \frac{\mu_0 I a}{4\pi}\hat{\mathbf{y}}\int_{z=-L/2}^{L/2}\frac{dz}{(a^2+z^2)^{3/2}}$$

This integral can be performed via the following trigonometric substitution:

$$z = a\tan\theta$$

$$dz = a\sec^2\theta\,d\theta$$

Solving for θ, we get $\theta = \tan^{-1}\left(\frac{z}{a}\right)$. It's easier to ignore the new limits for now and deal with them later. Note that the denominator of the integral simplifies as follows, using the trig identity $1+\tan^2\theta = \sec^2\theta$:

$$(a^2+z^2)^{3/2} = [a^2+(a\tan\theta)^2]^{3/2} = [a^2(1+\tan^2\theta)]^{3/2} = (a^2\sec^2\theta)^{3/2} = a^3\sec^3\theta$$

In the last step, we applied the rule from algebra that $(a^2x^2)^{3/2} = (a^2)^{3/2}(x^2)^{3/2} = a^3x^3$. Substitute the above expressions for dz and $(a^2+z^2)^{3/2}$ into the previous integral.

$$\vec{\mathbf{B}} = \frac{\mu_0 I a}{4\pi}\hat{\mathbf{y}}\int\frac{a\sec^2\theta\,d\theta}{a^3\sec^3\theta} = \frac{\mu_0 I a}{4\pi}\hat{\mathbf{y}}\int\frac{d\theta}{a^2\sec\theta}$$

Recall that $\sec\theta = \frac{1}{\cos\theta}$. Pull the constant $\frac{1}{a^2}$ out of the integral. Note that $a\left(\frac{1}{a^2}\right) = \frac{1}{a}$.

$$\vec{\mathbf{B}} = \frac{\mu_0 I}{4\pi a}\hat{\mathbf{y}}\int\cos\theta\,d\theta = \frac{\mu_0 I}{4\pi a}\hat{\mathbf{y}}[\sin\theta]$$

Now we must evaluate $\sin\theta$ over the limits of integration. Recall that $\theta = \tan^{-1}\left(\frac{z}{a}\right)$. If we plug this into $\sin\theta$, we get the complicated looking expression $\sin\left[\tan^{-1}\left(\frac{z}{a}\right)\right]$. Whenever you find yourself taking the sine of an inverse tangent, it's simpler to draw a right triangle. Since $z = a\tan\theta$, which means that $\tan\theta = \frac{z}{a}$, we draw a right triangle with z opposite to θ and with a adjacent to θ.

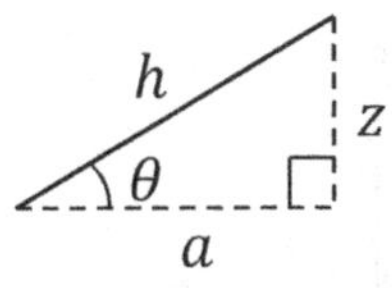

Apply the Pythagorean theorem to find the hypotenuse.

$$h = \sqrt{z^2 + a^2}$$

From the right triangle, you should see that $\sin\theta = \frac{z}{\sqrt{z^2+a^2}}$. Now we may evaluate $\sin\theta$ over the original limits from $z = -\frac{L}{2}$ to $z = \frac{L}{2}$.

$$\vec{\mathbf{B}} = \frac{\mu_0 I}{4\pi a}\hat{\mathbf{y}}[\sin\theta] = \frac{\mu_0 I}{4\pi a}\hat{\mathbf{y}}\left[\frac{z}{\sqrt{z^2+a^2}}\right]_{z=-L/2}^{L/2} = \frac{\mu_0 I}{4\pi a}\hat{\mathbf{y}}\left[\frac{\frac{L}{2}}{\sqrt{\frac{L^2}{4}+a^2}} - \left(\frac{-\frac{L}{2}}{\sqrt{\frac{L^2}{4}+a^2}}\right)\right]$$

Two minus signs make a plus sign.

$$\vec{\mathbf{B}} = \frac{\mu_0 I}{4\pi a}\hat{\mathbf{y}}\left(\frac{L/2}{\sqrt{L^2/4+a^2}} + \frac{L/2}{\sqrt{L^2/4+a^2}}\right)$$

Note that $\frac{L}{2} + \frac{L}{2} = L$.

$$\vec{\mathbf{B}} = \frac{\mu_0 I}{4\pi a}\hat{\mathbf{y}}\left(\frac{L}{\sqrt{L^2/4+a^2}}\right) = \frac{\mu_0 I L}{4\pi a\sqrt{\frac{L^2}{4}+a^2}}\hat{\mathbf{y}}$$

The magnetic field at the point $(a, 0, 0)$ is $\vec{\mathbf{B}} = \frac{\mu_0 I L}{4\pi a\sqrt{\frac{L^2}{4}+a^2}}\hat{\mathbf{y}}$. It has a magnitude of $B = \frac{\mu_0 I L}{4\pi a\sqrt{\frac{L^2}{4}+a^2}}$ and a direction of $\hat{\mathbf{B}} = \hat{\mathbf{y}}$. If you apply the right-hand rule for magnetic **field** (Chapter 22) to the original diagram, you should see that your fingers are coming out of the page ($\odot$) along $\hat{\mathbf{y}}$, at the point $(a, 0, 0)$, which agrees with the direction of our answer above.

Three-dimensional Current Loop Example: A Prelude to Example 95. A filamentary current travels along the circular wire illustrated below. The circular loop lies in the xy plane, centered about the origin. The radius of the circle is denoted by the symbol a. Derive an equation for the magnetic field at the point $(0, 0, p)$ in terms of μ_0, I, a, p, and appropriate unit vectors.

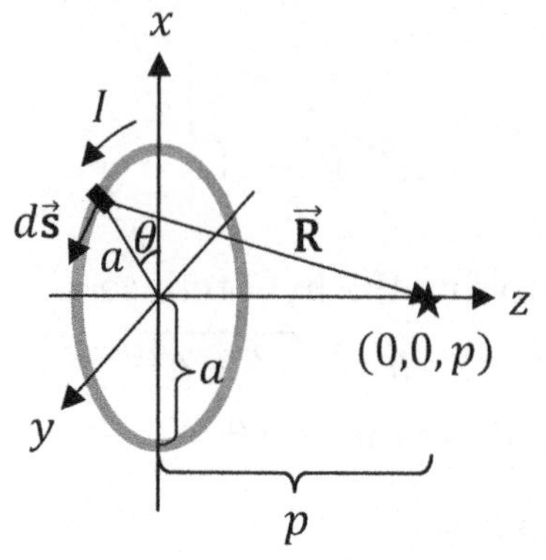

Solution. Begin with a labeled diagram. For a filamentary current, draw a representative $d\vec{s}$ that is tangent to the current (I). Draw $\vec{R}$ from the source, $d\vec{s}$, to the field point $(0, 0, p)$. When we perform the integration, we effectively integrate over every $d\vec{s}$ that makes up the current loop. Apply the law of Biot-Savart to the filamentary current.

$$\vec{B} = \frac{\mu_0}{4\pi}\int \frac{I\, d\vec{s} \times \hat{R}}{R^2}$$

Examine the picture above:

- The vector $\vec{R}$ extends a units inward, towards the z-axis (along $-\hat{r}_c$), and p units along the z-axis (along $\hat{z}$). Therefore, $\vec{R} = -a\hat{r}_c + p\hat{z}$. Note that this is the $\hat{r}_c$ of **cylindrical** coordinates.
- Apply the Pythagorean theorem to find the magnitude of $\vec{R}$.

$$R = \sqrt{a^2 + p^2}$$

- Divide $\vec{R}$ by R to find the direction of $\vec{R}$.

$$\hat{R} = \frac{\vec{R}}{R} = \frac{-a\hat{r}_c + p\hat{z}}{\sqrt{a^2 + p^2}}$$

For a circular filamentary current, write $d\vec{s} = \hat{\theta}\, a d\theta$ (it's positive because I is counter-clockwise, looking from the $+z$-axis, in this example). See page 282. The limits of integration are from $\theta = 0$ to $\theta = 2\pi$ <u>**radians**</u> for a full circle. Watch how we rewrite the magnetic field integral in order to manage the substitutions:

$$\vec{B} = \frac{\mu_0}{4\pi}\int \frac{I\, d\vec{s} \times \hat{R}}{R^2} = \frac{\mu_0}{4\pi}\int I\, d\vec{s}\left(\frac{1}{R^2}\right) \times \hat{R}$$

Now substitute the expressions for R, $\hat{R}$, and $d\vec{s}$ into the integral. Compare the previous line to the following line.

$$\vec{\mathbf{B}} = \frac{\mu_0}{4\pi}\int_{\theta=0}^{2\pi} I(\hat{\boldsymbol{\theta}}\, a d\theta)\frac{1}{a^2+p^2} \times \frac{-a\hat{\mathbf{r}}_c + p\hat{\mathbf{z}}}{\sqrt{a^2+p^2}}$$

Note that $R^2 = \left(\sqrt{a^2+p^2}\right)^2 = a^2 + p^2$. In the next step, we will apply the rule from algebra that $(a^2+p^2)\sqrt{a^2+p^2} = (a^2+p^2)^1(a^2+p^2)^{1/2} = (a^2+p^2)^{3/2}$. We will also pull the constants I and a out of the integral.

$$\vec{\mathbf{B}} = \frac{\mu_0 I a}{4\pi}\int_{\theta=0}^{2\pi} \frac{\hat{\boldsymbol{\theta}} \times (-a\hat{\mathbf{r}}_c + p\hat{\mathbf{z}})}{(a^2+p^2)^{3/2}} d\theta$$

(This power of 3/2 is very common in electric and magnetic field integrals. If you get a different power, you should check your work very carefully for a possible mistake.)

Work out the vector product $\hat{\boldsymbol{\theta}} \times (-a\hat{\mathbf{r}}_c + p\hat{\mathbf{z}})$ according to Chapter 19. It's convenient to work with cylindrical coordinates for the determinant. It works the same way as in Cartesian coordinates: The first row has the unit vectors and the subsequent rows have the components of the two vectors corresponding to those unit vectors. The cylindrical components correspond to the unit vectors $\hat{\mathbf{r}}_c$, $\hat{\boldsymbol{\theta}}$, and $\hat{\mathbf{z}}$. For the first vector ($\hat{\boldsymbol{\theta}}$), the cylindrical components are simply 0, 1, and 0 (since $\hat{\boldsymbol{\theta}}$ is one unit long), while for the second vector $(-a\hat{\mathbf{r}}_c + p\hat{\mathbf{z}})$ they are $-a$, 0, and p. (In either case, the components are the coefficients of the unit vectors.)

$$\hat{\boldsymbol{\theta}} \times (-a\hat{\mathbf{r}}_c + p\hat{\mathbf{z}}) = \begin{vmatrix} \hat{\mathbf{r}}_c & \hat{\boldsymbol{\theta}} & \hat{\mathbf{z}} \\ 0 & 1 & 0 \\ -a & 0 & p \end{vmatrix} = \hat{\mathbf{r}}_c \begin{vmatrix} 1 & 0 \\ 0 & p \end{vmatrix} - \hat{\boldsymbol{\theta}} \begin{vmatrix} 0 & 0 \\ -a & p \end{vmatrix} + \hat{\mathbf{z}} \begin{vmatrix} 0 & 1 \\ -a & 0 \end{vmatrix}$$

$$\hat{\boldsymbol{\theta}} \times (-a\hat{\mathbf{r}}_c + p\hat{\mathbf{z}}) = \hat{\mathbf{r}}_c(p-0) - \hat{\boldsymbol{\theta}}(0-0) + \hat{\mathbf{z}}(0+a) = p\,\hat{\mathbf{r}}_c + a\,\hat{\mathbf{z}}$$

Substitute this result into the magnetic field integral.

$$\vec{\mathbf{B}} = \frac{\mu_0 I a}{4\pi}\int_{\theta=0}^{2\pi} \frac{p\,\hat{\mathbf{r}}_c + a\,\hat{\mathbf{z}}}{(a^2+p^2)^{3/2}} d\theta = \frac{\mu_0 I a}{4\pi(a^2+p^2)^{3/2}}\int_{\theta=0}^{2\pi} (p\,\hat{\mathbf{r}}_c + a\,\hat{\mathbf{z}})\, d\theta$$

Note that $(a^2+p^2)^{3/2}$ is a constant and may come out of the integral. Separate the integral into two terms.

$$\vec{\mathbf{B}} = \frac{\mu_0 I a}{4\pi(a^2+p^2)^{3/2}}\int_{\theta=0}^{2\pi} p\,\hat{\mathbf{r}}_c\, d\theta + \frac{\mu_0 I a}{4\pi(a^2+p^2)^{3/2}}\int_{\theta=0}^{2\pi} a\,\hat{\mathbf{z}}\, d\theta$$

Note that p, a, and $\hat{\mathbf{z}}$ are constants, and thus may come out of their integrals, but $\hat{\mathbf{r}}_c$ is <u>**not**</u> constant: Since $\hat{\mathbf{r}}_c$ points away from the z-axis, its direction is different for each $d\vec{\mathbf{s}}$ that makes up the current loop. We may <u>**not**</u> pull $\hat{\mathbf{r}}_c$ out of the integral. Instead, we use the handy equation from page 281.

$$\hat{\mathbf{r}}_c = \hat{\mathbf{x}}\cos\theta + \hat{\mathbf{y}}\sin\theta$$

Substitute this equation into the previous expression for magnetic field.

$$\vec{\mathbf{B}} = \frac{\mu_0 I a p}{4\pi(a^2+p^2)^{3/2}} \int_{\theta=0}^{2\pi} (\hat{\mathbf{x}}\cos\theta + \hat{\mathbf{y}}\sin\theta)\, d\theta + \frac{\mu_0 I a^2}{4\pi(a^2+p^2)^{3/2}} \hat{\mathbf{z}} \int_{\theta=0}^{2\pi} d\theta$$

The first integral we again separate into two terms. The last integral is trivial.

$$\vec{\mathbf{B}} = \frac{\mu_0 I a p}{4\pi(a^2+p^2)^{3/2}} \int_{\theta=0}^{2\pi} \hat{\mathbf{x}}\cos\theta\, d\theta + \frac{\mu_0 I a p}{4\pi(a^2+p^2)^{3/2}} \int_{\theta=0}^{2\pi} \hat{\mathbf{y}}\sin\theta\, d\theta + \frac{\mu_0 I a^2}{4\pi(a^2+p^2)^{3/2}} \hat{\mathbf{z}}(2\pi)$$

The unit vectors $\hat{\mathbf{x}}$ and $\hat{\mathbf{y}}$ are constants, so they may come out of their integrals.

$$\vec{\mathbf{B}} = \frac{\mu_0 I a p}{4\pi(a^2+p^2)^{3/2}} \hat{\mathbf{x}} \int_{\theta=0}^{2\pi} \cos\theta\, d\theta + \frac{\mu_0 I a p}{4\pi(a^2+p^2)^{3/2}} \hat{\mathbf{y}} \int_{\theta=0}^{2\pi} \sin\theta\, d\theta + \frac{\mu_0 I a^2}{2(a^2+p^2)^{3/2}} \hat{\mathbf{z}}$$

$$\vec{\mathbf{B}} = \frac{\mu_0 I a p}{4\pi(a^2+p^2)^{3/2}} \hat{\mathbf{x}}\, [\sin\theta]_{\theta=0}^{2\pi} + \frac{\mu_0 I a p}{4\pi(a^2+p^2)^{3/2}} \hat{\mathbf{y}}\, [-\cos\theta]_{\theta=0}^{2\pi} + \frac{\mu_0 I a^2}{2(a^2+p^2)^{3/2}} \hat{\mathbf{z}}$$

$$\vec{\mathbf{B}} = \frac{\mu_0 I a p}{4\pi(a^2+p^2)^{3/2}} \hat{\mathbf{x}}\, (\sin 2\pi - \sin 0) + \frac{\mu_0 I a p}{4\pi(a^2+p^2)^{3/2}} \hat{\mathbf{y}}\, (-\cos 2\pi + \cos 0)$$

$$+ \frac{\mu_0 I a^2}{2(a^2+p^2)^{3/2}} \hat{\mathbf{z}}$$

$$\vec{\mathbf{B}} = \frac{\mu_0 I a p}{4\pi(a^2+p^2)^{3/2}} \hat{\mathbf{x}}\, (0-0) + \frac{\mu_0 I a p}{4\pi(a^2+p^2)^{3/2}} \hat{\mathbf{y}}\, (-1+1) + \frac{\mu_0 I a^2}{2(a^2+p^2)^{3/2}} \hat{\mathbf{z}}$$

$$\vec{\mathbf{B}} = \frac{\mu_0 I a p}{4\pi(a^2+p^2)^{3/2}} \hat{\mathbf{x}}\, (0) + \frac{\mu_0 I a p}{4\pi(a^2+p^2)^{3/2}} \hat{\mathbf{y}}\, (0) + \frac{\mu_0 I a^2}{2(a^2+p^2)^{3/2}} \hat{\mathbf{z}}$$

$$\vec{\mathbf{B}} = \frac{\mu_0 I a^2}{2(a^2+p^2)^{3/2}} \hat{\mathbf{z}}$$

Look at what happened: The integral $\int_{\theta=0}^{2\pi} p\, \hat{\mathbf{r}}_c\, d\theta$ turned out to be exactly zero. If you really understand the magnetic field lines from Chapter 22, you could have deduced this earlier and saved the trouble of doing that extra math. The magnetic field at the point $(0,0,p)$ will point along $\hat{\mathbf{z}}$, so only the integral $\int_{\theta=0}^{2\pi} a\, \hat{\mathbf{z}}\, d\theta$ mattered.

The magnetic field at the point $(0,0,p)$ is $\vec{\mathbf{B}} = \frac{\mu_0 I a^2}{2(a^2+p^2)^{3/2}} \hat{\mathbf{z}}$. It has a magnitude of $B = \frac{\mu_0 I a^2}{2(a^2+p^2)^{3/2}}$ and a direction of $\hat{\mathbf{B}} = \hat{\mathbf{z}}$.

Example 95. In the Helmholtz coils illustrated below, equal filamentary currents travel in the same direction along two parallel loops of wire. The circular loops are parallel to the xy plane, with one at $z = -\frac{a}{2}$ and the other at $z = \frac{a}{2}$ (such that the loops are separated by a distance a, which also equals the radius of each loop). Derive an equation for the magnetic field at the origin in terms of μ_0, I, a, and appropriate unit vectors.

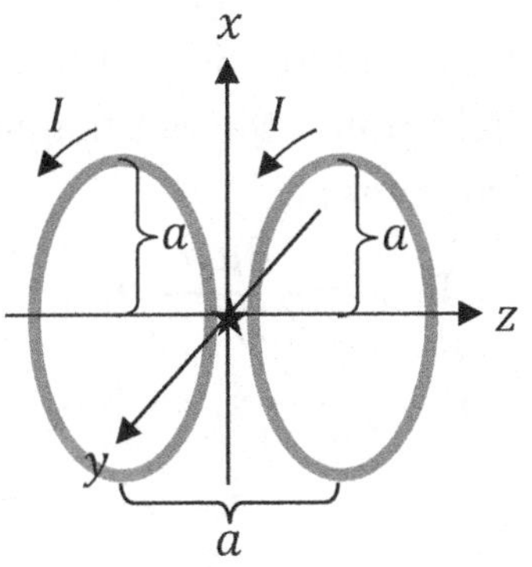

Solution. We don't need to perform a new integral for this example. We just need to apply the result from the previous example, which involves a circular current loop and a field point at $(0, 0, p)$. Recall from the previous example that the magnetic field a distance p along the axis of a circular loop of radius a is:

$$\vec{\mathbf{B}}_{one\ loop} = \frac{\mu_0 I a^2}{2(a^2 + p^2)^{3/2}} \hat{\mathbf{z}}$$

In our current example, note that the field point lies midway between the two loops, such that $p = \frac{a}{2}$ (since the two loops are a distance a apart in this example). Set $p = \frac{a}{2}$ in the previous expression for magnetic field. Note that $\left(\frac{a}{2}\right)^2 = \frac{a^2}{4}$.

$$\vec{\mathbf{B}}_{one\ loop} = \frac{\mu_0 I a^2}{2\left(a^2 + \frac{a^2}{4}\right)^{3/2}} \hat{\mathbf{z}}$$

Add fractions with a **common denominator**: $a^2 + \frac{a^2}{4} = \frac{4a^2}{4} + \frac{a^2}{4} = \frac{5a^2}{4}$.

$$\vec{\mathbf{B}}_{one\ loop} = \frac{\mu_0 I a^2}{2\left(\frac{5a^2}{4}\right)^{3/2}} \hat{\mathbf{z}}$$

Apply the rule from algebra that $\left(\frac{5a^2}{4}\right)^{3/2} = \frac{(5)^{3/2}(a^2)^{3/2}}{(4)^{3/2}}$. Note that $(a^2)^{3/2} = a^3$ according to the rule $(x^m)^n = x^{mn}$. Note that $(4)^{3/2} = \left(\sqrt{4}\right)^3 = 2^3 = 8$ (or enter 4^1.5 on your calculator).

$$\vec{\mathbf{B}}_{one\ loop} = \frac{\mu_0 I a^2}{2\frac{(5)^{3/2}a^3}{8}} \hat{\mathbf{z}}$$

To divide by a fraction, multiply by its **reciprocal**. Note that $\frac{a^2}{a^3} = \frac{1}{a}$.

$$\vec{\mathbf{B}}_{one\ loop} = \frac{\mu_0 I a^2}{2}\hat{\mathbf{z}}\frac{8}{(5)^{3/2}a^3} = \frac{4\mu_0 I}{(5)^{3/2}a}\hat{\mathbf{z}}$$

Both loops are equidistant from the field point, such that the magnitude of the magnetic field at the field point is the same for each loop. The direction of the magnetic field is also the same at the field point for both loops, which you can see by applying the right-hand rule for magnetic field (Chapter 22) to each loop. Since the magnitude and direction of both magnetic fields is the same at the field point, the net magnetic field is double the magnetic field created by one loop. The final answer is:

$$\vec{\mathbf{B}}_{net} = \frac{8\mu_0 I}{(5)^{3/2}a}\hat{\mathbf{z}}$$

The magnetic field at the origin is $\vec{\mathbf{B}} = \frac{8\mu_0 I}{(5)^{3/2}a}\hat{\mathbf{z}}$. It has a magnitude of $B = \frac{8\mu_0 I}{(5)^{3/2}a}$ and a direction of $\hat{\mathbf{B}} = \hat{\mathbf{z}}$.

Example 96. A solid disc lying in the xy plane and centered about the origin has positive charge Q, radius a, and uniform charge density σ. The solid disc rotates about the z-axis counterclockwise with constant angular speed ω. Derive an equation for the magnetic field at the origin in terms of μ_0, Q, a, ω, and appropriate unit vectors.

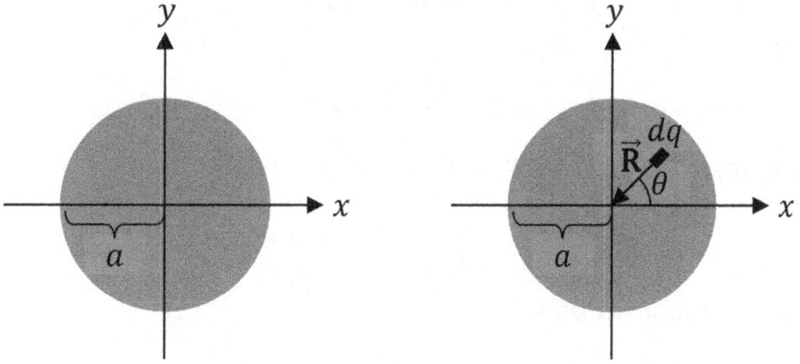

Solution. Begin with a labeled diagram. For a rotating charged disc, draw a representative dq within the area of the disc. Draw $\vec{\mathbf{R}}$ from the source, dq, to the field point (at the origin). When we perform the integration, we effectively integrate over every dq that makes up the solid disc. Apply the law of Biot-Savart to the rotating charged disc (see page 281).

$$\vec{\mathbf{B}} = \frac{\mu_0}{4\pi}\int \frac{\vec{v}\, dq \times \hat{\mathbf{R}}}{R^2}$$

Examine the diagram above on the right: The vector $\vec{\mathbf{R}}$ extends r units inward, towards the origin (along $-\hat{\mathbf{r}}$). Therefore, $R = r$ and $\hat{\mathbf{R}} = -\hat{\mathbf{r}}$ (using 2D polar coordinates). For a solid disc, $dq = \sigma dA$ and $dA = rdrd\theta$ (see pages 281-282), such that $dq = \sigma rdrd\theta$. The limits of integration are from $r = 0$ to $r = a$ and $\theta = 0$ to $\theta = 2\pi$ **radians**. Now substitute the expressions for R, $\hat{\mathbf{R}}$, and dq into the integral.

$$\vec{\mathbf{B}} = \frac{\mu_0}{4\pi}\int \frac{\vec{v}\, dq \times \hat{\mathbf{R}}}{R^2} = -\frac{\mu_0}{4\pi}\int_{r=0}^{a}\int_{\theta=0}^{2\pi} \frac{\vec{v} \times \hat{\mathbf{r}}}{r^2}\sigma rdrd\theta$$

Note that $\frac{r}{r^2} = \frac{1}{r}$.The charge density σ is constant since the disc is **uniform**, so it may come out of the integral.

$$\vec{\mathbf{B}} = -\frac{\mu_0\sigma}{4\pi}\int_{r=0}^{a}\int_{\theta=0}^{2\pi} \frac{\vec{v} \times \hat{\mathbf{r}}}{r}drd\theta$$

Since the velocity ($\vec{v}$) of each dq is tangential to the circle of rotation, and since $\hat{\boldsymbol{\theta}}$ is tangential to a circle lying in the xy plane, we write $\vec{v} = v\,\hat{\boldsymbol{\theta}}$. For a rotating charged object, we write $v = r_{rot}\omega$, where r_{rot} is the radius of rotation of each dq. Based on how this solid disc is rotating, $r_{rot} = r$. Substitute $\vec{v} = v\,\hat{\boldsymbol{\theta}} = r\omega\hat{\boldsymbol{\theta}}$ into the magnetic field integral. Note that the r cancels.

$$\vec{\mathbf{B}} = -\frac{\mu_0 \sigma \omega}{4\pi} \int_{r=0}^{a} \int_{\theta=0}^{2\pi} \hat{\boldsymbol{\theta}} \times \hat{\mathbf{r}} \, dr d\theta$$

Recall from page 284 that the vector product $\hat{\boldsymbol{\theta}} \times \hat{\mathbf{r}}$ works out to $\hat{\boldsymbol{\theta}} \times \hat{\mathbf{r}} = -\hat{\mathbf{z}}$.

$$\vec{\mathbf{B}} = \frac{\mu_0 \sigma \omega}{4\pi} \hat{\mathbf{z}} \int_{r=0}^{a} \int_{\theta=0}^{2\pi} dr \, d\theta = \frac{\mu_0 \sigma \omega}{4\pi} \hat{\mathbf{z}} [r]_{r=0}^{a} [\theta]_{\theta=0}^{2\pi}$$

$$\vec{\mathbf{B}} = \frac{\mu_0 \sigma a \omega}{2} \hat{\mathbf{z}}$$

Integrate $Q = \int dq$ using the same substitutions as before.

$$Q = \sigma \int_{r=0}^{a} \int_{\theta=0}^{2\pi} r \, dr d\theta = \sigma \pi a^2$$

Divide both sides of the equation by πa^2.

$$\sigma = \frac{Q}{\pi a^2}$$

Plug this expression for σ into the previous equation for magnetic field.

$$\vec{\mathbf{B}} = \frac{\mu_0 \sigma a \omega}{2} \hat{\mathbf{z}} = \frac{\mu_0 a \omega}{2} \hat{\mathbf{z}} \frac{Q}{\pi a^2} = \frac{\mu_0 Q \omega}{2\pi a} \hat{\mathbf{z}}$$

The magnetic field at the origin is $\vec{\mathbf{B}} = \frac{\mu_0 Q \omega}{2\pi a} \hat{\mathbf{z}}$. It has a magnitude of $B = \frac{\mu_0 Q \omega}{2\pi a}$ and a direction of $\hat{\mathbf{B}} = \hat{\mathbf{z}}$.

27 AMPÈRE'S LAW

Ampère's Law

$$\oint_C \vec{\mathbf{B}} \cdot d\vec{\mathbf{s}} = \mu_0 I_{enc}$$

Current Enclosed

$I = \int \vec{\mathbf{K}} \cdot d\vec{\boldsymbol{\ell}}$ (surface current)	$I = \int \vec{\mathbf{J}} \cdot d\vec{\mathbf{A}}$ (through a volume)

Differential Arc Length

$d\vec{\mathbf{s}} = \hat{\mathbf{x}}\, dx$ $ds = dx$ (along x)	$d\vec{\mathbf{s}} = \hat{\mathbf{y}}\, dy$ or $\hat{\mathbf{z}}\, dz$ $ds = dy$ or dz (along y or z)	$d\vec{\mathbf{s}} = \pm\hat{\boldsymbol{\theta}}\, r_c d\theta$ $ds = a d\theta$ (circular arc of radius a)

Differential Area Element

$dA = dxdy$ (polygon in xy plane)	$dA = rdrd\theta$ (pie slice, disc, thick ring)	$dA = a^2 \sin\theta\, d\theta d\varphi$ (sphere of radius a)

Differential Volume Element

$dV = dxdydz$ (bounded by flat sides)	$dV = r_c dr_c d\theta dz$ (cylinder or cone)	$dV = r^2 \sin\theta\, drd\theta d\varphi$ (spherical)

Permeability of Free Space

$$\mu_0 = 4\pi \times 10^{-7} \ \frac{\text{T} \cdot \text{m}}{\text{A}}$$

Symbol	Name	SI Units
I_{enc}	the current enclosed by the Ampèrian loop	A
I	the total current	A
$\vec{\mathbf{B}}$	magnetic field	T
μ_0	the permeability of free space	$\frac{\text{T}\cdot\text{m}}{\text{A}}$
x, y, z	Cartesian coordinates	m, m, m
$\hat{\mathbf{x}}, \hat{\mathbf{y}}, \hat{\mathbf{z}}$	unit vectors along the $+x$-, $+y$-, $+z$-axes	unitless
r, θ	2D polar coordinates	m, rad
r_c, θ, z	cylindrical coordinates	m, rad, m
r, θ, φ	spherical coordinates	m, rad, rad
$\hat{\mathbf{r}}, \hat{\boldsymbol{\theta}}, \hat{\boldsymbol{\varphi}}$	unit vectors along spherical coordinate axes	unitless
$\hat{\mathbf{r}}_c$	a unit vector pointing away from the $+z$-axis	unitless
$\vec{\mathbf{K}}$	surface current density (distributed over a surface)	A/m
$\vec{\mathbf{J}}$	current density (distributed throughout a volume)	A/m^2
$d\vec{\mathbf{s}}$	differential displacement along a filamentary current	m
dA	differential area element	m^2
dV	differential volume element	m^3
N	number of loops (or turns)	unitless

Symbol	Meaning
$\otimes$	into the page
$\odot$	out of the page

Infinite Wire Example: A Prelude to Example 97. An infinite solid cylindrical conductor coaxial with the z-axis has radius a, uniform current density $\vec{\mathbf{J}}$ along $\hat{\mathbf{z}}$, and carries total current I. Derive an expression for the magnetic field both inside and outside of the cylinder.

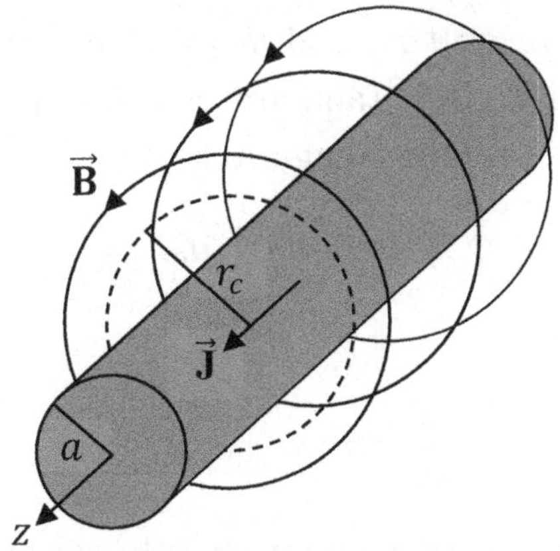

Solution. First sketch the magnetic field lines for the conducting cylinder. It's hard to draw, but the cylinder is perpendicular to the page with the current (I) coming out of the page ($\odot$). Apply the right-hand rule for magnetic field (Chapter 22): Grab the current with your thumb pointing out of the page ($\odot$), along the current (I). Your fingers make counter-clockwise (use the right-hand rule to see this) circles around the wire (toward your fingertips), as shown in the diagram above.

We choose our Ampèrian loop to be a circle (dashed line above) coaxial with the conducting cylinder such that $\vec{\mathbf{B}}$ and $d\vec{\mathbf{s}}$ (which is tangent to the Ampèrian loop) will be parallel and the magnitude of $\vec{\mathbf{B}}$ will be constant over the Ampèrian loop (since every point on the Ampèrian loop is equidistant from the axis of the conducting cylinder).

Note: In Chapter 26, when we applied the law of Biot-Savart, $d\vec{\mathbf{s}}$ was tangent to the current. However, in Chapter 27, when we apply Ampère's law, $d\vec{\mathbf{s}}$ is instead tangent to the Ampèrian loop, **<u>not</u>** tangent to the current.

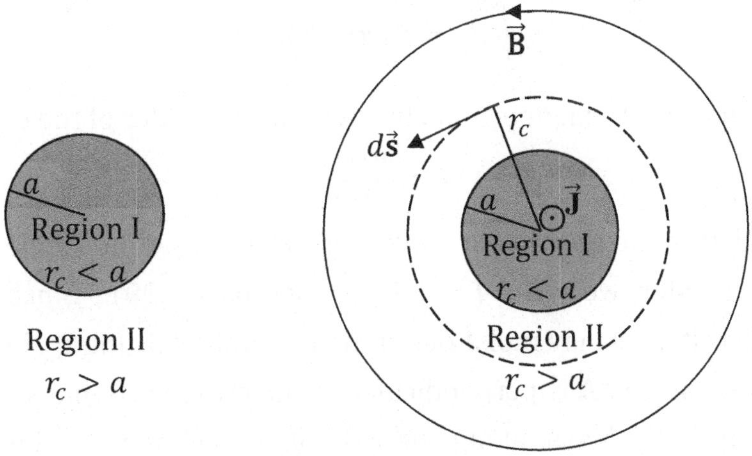

Write the formula for Ampère's law.

$$\oint_C \vec{\mathbf{B}} \cdot d\vec{\mathbf{s}} = \mu_0 I_{enc}$$

The scalar product is $\vec{\mathbf{B}} \cdot d\vec{\mathbf{s}} = B\cos\theta\, ds$, and $\theta = 0°$ since the magnetic field lines make circles coaxial with our Ampèrian circle (and are therefore parallel to $d\vec{\mathbf{s}}$, which is tangent to the Ampèrian circle). See the previous diagram.

$$\oint_C B\cos 0°\, ds = \mu_0 I_{enc}$$

Recall from trig that $\cos 0° = 1$.

$$\oint_C B\, ds = \mu_0 I_{enc}$$

The magnitude of the magnetic field is constant over the Ampèrian circle, since every point on the circle is equidistant from the axis of the conducting cylinder. Therefore, we may pull B out of the integral.

$$B\oint_C ds = \mu_0 I_{enc}$$

This integral is over the path of the Ampèrian circle of radius r_c, where r_c is a variable because the magnetic field depends on how close or far the field point is away from the conducting cylinder. We work with cylindrical coordinates (Chapter 6) and write $ds = r_c d\theta$ (see page 299). This is a purely angular integration (over the angle θ), and we treat the independent variable r_c as if it were a constant for the purpose of the integration because every point on the path of the Ampèrian circle has the same value of r_c.

$$\oint_C ds = \int_{\theta=0}^{2\pi} r_c\, d\theta = r_c \int_{\theta=0}^{2\pi} d\theta = 2\pi r_c$$

This is the circumference of the Ampèrian circle. Substitute this expression for circumference into the previous equation for magnetic field.

$$B\oint_C ds = B 2\pi r_c = \mu_0 I_{enc}$$

Isolate the magnitude of the magnetic field by dividing both sides of the equation by $2\pi r_c$.

$$B = \frac{\mu_0 I_{enc}}{2\pi r_c}$$

Now we need to determine how much current is enclosed by the Ampèrian circle. For a solid conducting cylinder, we write $I = \int \vec{\mathbf{J}} \cdot d\vec{\mathbf{A}}$ (see page 299). Since the cylinder has uniform current density, we may pull $\vec{\mathbf{J}}$ out of the integral. This integral is over the area of the Ampèrian circle. Since $d\vec{\mathbf{A}}$ is perpendicular to the cross-sectional area, $\vec{\mathbf{J}}$ is parallel to $d\vec{\mathbf{A}}$ (both come out of the page, along the axis of the cylinder). Therefore, the scalar product is $\vec{\mathbf{J}} \cdot d\vec{\mathbf{A}} = J \cos 0° \, dA = JdA$.

$$I_{enc} = \int \vec{\mathbf{J}} \cdot d\vec{\mathbf{A}} = \int J \cos 0° \, dA = \int J \, dA = J \int dA$$

We work with cylindrical coordinates and write the differential area element as $dA = r_c dr_c d\theta$ (see page 299 or Chapter 6). Since we are now integrating over area (not arc length like before), r_c is a variable and we will have a double integral.

We must consider two different regions:

- The Ampèrian circle could be smaller than the conducting cylinder. This will help us find the magnetic field in region I.
- The Ampèrian circle could be larger than the conducting cylinder. This will help us find the magnetic field in region II.

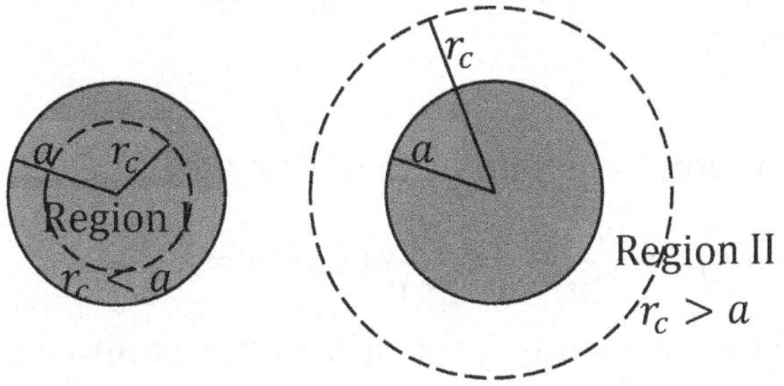

Region I: $r_c < a$.

Inside of the conducting cylinder, only a fraction of the cylinder's current is enclosed by the Ampèrian circle. In this region, the upper limit of the r_c-integration is the variable r_c: The larger the Ampèrian circle, the more current it encloses, up to a maximum radius of a (the radius of the conducting cylinder).

$$I_{enc} = J \int dA = J \int_{r_c=0}^{r_c} \int_{\theta=0}^{2\pi} r_c \, dr_c \, d\theta$$

This integral is separable:

$$I_{enc} = J \int_{r_c=0}^{r_c} r_c \, dr_c \int_{\theta=0}^{2\pi} d\theta = J \left[\frac{r_c^2}{2}\right]_{r_c=0}^{r_c} [\theta]_{\theta=0}^{2\pi} = J\left(\frac{r_c^2}{2}\right)(2\pi) = \pi r_c^2 J$$

You should recognize that πr_c^2 is the area of a circle. For a conducting cylinder with uniform current density, current equals J times area. (For a non-uniform current density, you would need to integrate: Then you couldn't just multiply J by area. Although we could have skipped the calculus in this example, we performed the integral so that it would be easier for you to adapt the solution to a non-uniform current density.) Substitute this expression for the current enclosed into the previous equation for magnetic field.

$$B_I = \frac{\mu_0 I_{enc}}{2\pi r_c} = \frac{\mu_0}{2\pi r_c}(\pi r_c^2 J) = \frac{\mu_0 J r_c}{2}$$

Since the magnetic field lines circulate around the conducting cylinder, we can include a direction with the magnetic field by adding on the cylindrical unit vector $\hat{\boldsymbol{\theta}}$.

$$\vec{\mathbf{B}}_I = \frac{\mu_0 J r_c}{2} \hat{\boldsymbol{\theta}}$$

The answer is different outside of the conducting cylinder. We will explore that next.

<u>Region II</u>: $r_c > a$.

Outside of the conducting cylinder, 100% of the current is enclosed by the Ampèrian circle. This changes the upper limit of the r_c-integration to a (since all of the current lies inside a cylinder of radius a).

$$I_{enc} = I = J \int dA = J \int_{r_c=0}^{r_c} \int_{\theta=0}^{2\pi} r_c \, dr_c \, d\theta$$

We don't need to work out the entire double integral again. We'll get the same expression as before, but with a in place of r_c.

$$I_{enc} = I = \pi a^2 J$$

Substitute this into the equation for magnetic field that we obtained from Ampère's law.

$$B_{II} = \frac{\mu_0 I_{enc}}{2\pi r_c} = \frac{\mu_0}{2\pi r_c}(\pi a^2 J) = \frac{\mu_0 J a^2}{2 r_c}$$

We can turn this into a vector by including the appropriate unit vector.

$$\vec{\mathbf{B}}_{II} = \frac{\mu_0 J a^2}{2 r_c} \hat{\boldsymbol{\theta}}$$

<u>Alternate forms of the answers in regions I and II</u>.

Since the total current is $I = \pi a^2 J$ (we found this equation for region II above), we can alternatively express the magnetic field in terms of the total current (I) of the conducting cylinder instead of the current density (J).

<u>Region I</u>: $r_c < a$.

$$\vec{\mathbf{B}}_I = \frac{\mu_0 J r_c}{2} \hat{\boldsymbol{\theta}} = \frac{\mu_0 I r_c}{2\pi a^2} \hat{\boldsymbol{\theta}}$$

<u>Region II</u>: $r_c > a$.

$$\vec{\mathbf{B}}_{II} = \frac{\mu_0 J a^2}{2 r_c} \hat{\boldsymbol{\theta}} = \frac{\mu_0 I}{2\pi r_c} \hat{\boldsymbol{\theta}}$$

Note that the magnetic field in region II is identical to the magnetic field created by an infinite filamentary current (see Chapter 26, but note that a has a different meaning there; you can also find this formula in Chapter 25 for a long straight wire). Note also that the expressions for the magnetic field in the two different regions both agree at the boundary: That is, in the limit that r_c approaches a, both expressions approach $\frac{\mu_0 I}{2\pi a} \hat{\boldsymbol{\theta}}$.

Example 97. An infinite solid cylindrical conductor coaxial with the z-axis has radius a and carries total current I. The current density is non-uniform: $\vec{\mathbf{J}} = \beta r_c \hat{\mathbf{z}}$, where β is a positive constant, Derive an expression for the magnetic field both inside and outside of the cylinder.

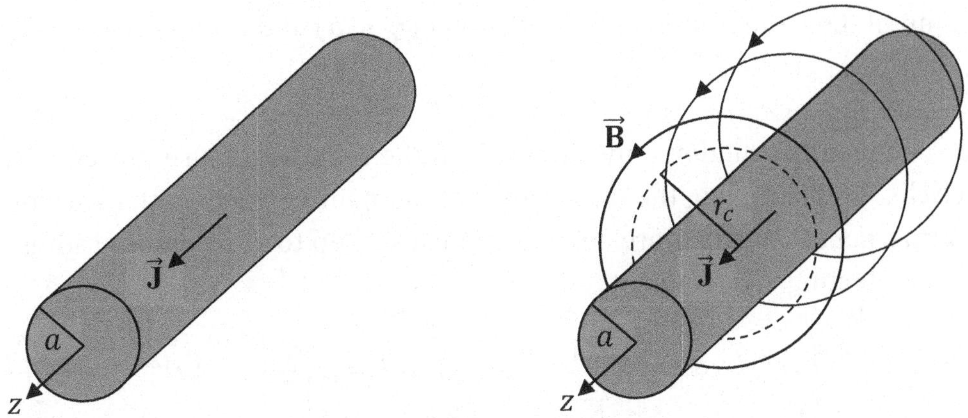

Solution. This problem is just like the previous example, except that the current density is non-uniform. The magnetic field lines and Ampèrian loop are the same as before. To save time, we'll simply repeat the steps that are identical to the previous example, and pick up from where this solution deviates from the earlier one. It would be a good exercise to see if you can understand each step (if not, review the previous example for the explanation).

$$\oint_C \vec{\mathbf{B}} \cdot d\vec{\mathbf{s}} = \mu_0 I_{enc}$$

$$\oint_C B \cos 0° \, ds = \mu_0 I_{enc}$$

$$\oint_C B \, ds = \mu_0 I_{enc}$$

$$B \oint_C ds = \mu_0 I_{enc}$$

$$B \int_{\theta=0}^{2\pi} r_c \, d\theta = B r_c \int_{\theta=0}^{2\pi} d\theta = B 2\pi r_c = \mu_0 I_{enc}$$

$$B = \frac{\mu_0 I_{enc}}{2\pi r_c}$$

Now we have reached the point where it will matter that the current density is non-uniform. This time, we can't pull the charge density ($\vec{\mathbf{J}}$) out of the integral. Instead, we must apply the equation $\vec{\mathbf{J}} = \beta r_c \hat{\mathbf{z}}$ that was given in the problem. The symbol β, however, is constant and may come out of the integral.

$$I_{enc} = \int \vec{\mathbf{J}} \cdot d\vec{\mathbf{A}} = \int J \cos 0° \, dA = \int J \, dA = \int \beta r_c \, dA = \beta \int r_c \, dA$$

$$I_{enc} = \beta \int_{r_c=0}^{r_c \text{ or } a} \int_{\theta=0}^{2\pi} r_c \, (r_c dr_c \, d\theta) = \beta \int_{r_c=0}^{r_c \text{ or } a} \int_{\theta=0}^{2\pi} r_c^2 \, dr_c \, d\theta$$

The upper limit of the r_c-integration is r_c in region I ($r_c < a$) and is a in region II ($r_c > a$).

Region I: $r_c < a$.

Inside of the conducting cylinder, only a fraction of the cylinder's current is enclosed by the Ampèrian circle. In this region, the upper limit of the r_c-integration is the variable r_c: The larger the Ampèrian circle, the more current it encloses, up to a maximum radius of a (the radius of the conducting cylinder).

$$I_{enc} = \beta \int_{r_c=0}^{r_c} \int_{\theta=0}^{2\pi} r_c^2 \, dr_c \, d\theta = \beta \int_{r_c=0}^{r_c} r_c^2 \, dr_c \int_{\theta=0}^{2\pi} d\theta = \beta \left[\frac{r_c^3}{3}\right]_{r_c=0}^{r_c} [\theta]_{\theta=0}^{2\pi} = \frac{2\pi\beta r_c^3}{3}$$

We must eliminate the constant β from our answer. To do this, perform the integral for the current enclosed again, but this time integrate up to $r_c = a$ and call this the total current, I.

$$I = \beta \int_{r_c=0}^{a} \int_{\theta=0}^{2\pi} r_c^2 \, dr_c \, d\theta = \beta \int_{r_c=0}^{a} r_c^2 \, dr_c \int_{\theta=0}^{2\pi} d\theta = \beta \left[\frac{r_c^3}{3}\right]_{r_c=0}^{a} [\theta]_{\theta=0}^{2\pi} = \frac{2\pi\beta a^3}{3}$$

Divide the previous two equations.

$$\frac{I_{enc}}{I} = \frac{r_c^3}{a^3}$$

(Everything else cancels out.) Multiply both sides of the equation by I.

$$I_{enc} = I \frac{r_c^3}{a^3}$$

Substitute this expression for the current enclosed into the previous equation for magnetic field.

$$B_I = \frac{\mu_0 I_{enc}}{2\pi r_c} = \frac{\mu_0}{2\pi r_c}\left(I \frac{r_c^3}{a^3}\right) = \frac{\mu_0 I r_c^2}{2a^3}$$

Since the magnetic field lines circulate around the conducting cylinder, we can include a direction with the magnetic field by adding on the cylindrical unit vector $\hat{\boldsymbol{\theta}}$.

$$\vec{\mathbf{B}}_I = \frac{\mu_0 I r_c^2}{2a^3} \hat{\boldsymbol{\theta}}$$

The answer is different outside of the conducting cylinder. We will explore that next.

Region II: $r_c > a$.

Outside of the conducting cylinder, 100% of the current is enclosed by the Ampèrian circle.

$$I_{enc} = I$$

Substitute this into the equation for magnetic field that we obtained from Ampère's law.

$$B_I = \frac{\mu_0 I_{enc}}{2\pi r_c} = \frac{\mu_0 I}{2\pi r_c}$$

We can turn this into a vector by including the appropriate unit vector.

$$\vec{\mathbf{B}}_{II} = \frac{\mu_0 I}{2 r_c}\hat{\boldsymbol{\theta}}$$

Example 98. An infinite solid cylindrical conductor coaxial with the z-axis has radius a and uniform current density $\vec{\mathbf{J}}$. Coaxial with the solid cylindrical conductor is a thick infinite cylindrical conducting shell of inner radius b, outer radius c, and uniform current density $\vec{\mathbf{J}}$. The conducting cylinder carries total current I coming out of the page, while the conducting cylindrical shell carries the same total current I, except that its current is going into the page. Derive an expression for the magnetic field in each region. (This is a **coaxial cable**.)

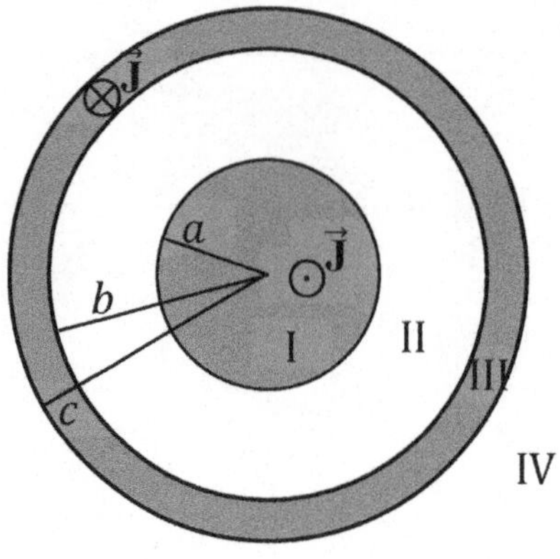

Solution. This problem is similar to the previous example except that there are four regions (as indicated in the diagram above).

Note: If you recall how we treated the conducting shell in the context of Gauss's law (Chapter 8), you'll want to note that the solution to a conducting shell problem in the context of Ampère's law is significantly different because current is **not** an electrostatic situation (since current involves a flow of charge). Most notably, the magnetic field is **not** zero in region III, like the electric field would be in an electrostatic Gauss's law problem. (As we will see, in this particular problem the magnetic field is instead zero in region IV.)

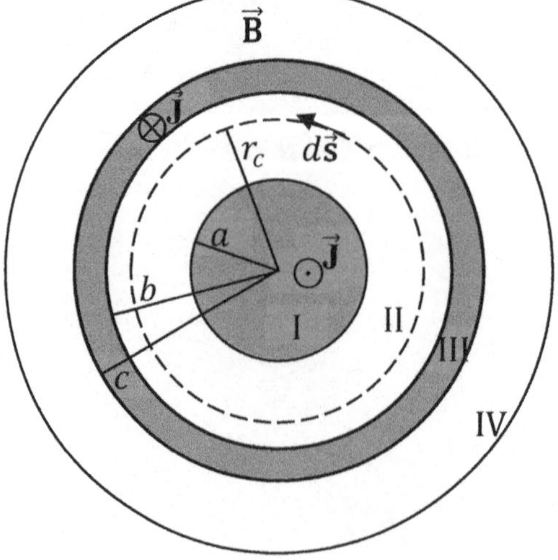

Region I: $r_c < a$.

The conducting shell does **not** matter for region I, since none of its current will reside in an Ampèrian circle with $r_c < a$. The answer is exactly the same as the example with the infinite solid cylinder on pages 301-304. We choose to use one of the alternate forms of the equation on page 304.

$$\vec{\mathbf{B}}_I = \frac{\mu_0 I r_c}{2\pi a^2}\hat{\boldsymbol{\theta}}$$

Region II: $a < r_c < b$.

The conducting shell also does **not** matter for region II, since again none of its current will reside in an Ampèrian circle with $r_c < b$. (Recall that Ampère's law involves the current enclosed, I_{enc}, by the Ampèrian loop.) We obtain the same result as for region II of the previous examples (see, for example, the bottom of page 304).

$$\vec{\mathbf{B}}_{II} = \frac{\mu_0 I}{2\pi r_c}\hat{\boldsymbol{\theta}}$$

Region III: $b < r_c < c$.

Now the conducting shell matters. There is a "trick" to finding the current enclosed in region III: We must consider both conductors for this region, plus the fact that the two currents run in opposite directions. The current enclosed in region III includes 100% of the solid cylinder's current, $+I$, which comes out of the page (⨀) **plus** a fraction of the cylindrical shell's current, $-I$, which goes into the page (⨂):

$$I_{enc} = I - I_{shell}$$

Note that I_{shell} refers to the current enclosed within the shell in region III from b to r_c, which will be a fraction of the cylindrical shell's total current (I). The following integral gives the current enclosed within the shell in region III (see page 299).

$$I_{shell} = \int \vec{\mathbf{J}} \cdot d\vec{\mathbf{A}}$$

Since the cylinder has uniform current density, we may pull $\vec{\mathbf{J}}$ out of the integral. This integral is over the area of the Ampèrian circle. Since $d\vec{\mathbf{A}}$ is perpendicular to the cross-sectional area, $\vec{\mathbf{J}}$ is parallel to $d\vec{\mathbf{A}}$ (both go into the page, along the axis of the cylindrical shell). Therefore, the scalar product is $\vec{\mathbf{J}} \cdot d\vec{\mathbf{A}} = J\cos 0°\, dA = JdA$.

$$I_{shell} = \int \vec{\mathbf{J}} \cdot d\vec{\mathbf{A}} = \int J\cos 0°\, dA = \int J\, dA = J\int dA$$

We work with cylindrical coordinates and write the differential area element as $dA = r_c dr_c d\theta$ (see page 299 or Chapter 6). The limits of the r_c-integration are from $r_c = b$ to $r_c = r_c$ in region III ($b < r_c < c$).

$$I_{shell} = J\int_{r_c=b}^{r_c}\int_{\theta=0}^{2\pi} r_c\, dr_c\, d\theta = J\left[\frac{r_c^2}{2}\right]_{r_c=b}^{r_c}[\theta]_{\theta=0}^{2\pi} = \pi J(r_c^2 - b^2)$$

Now we need an expression for the total current (I). If we do the same integral with the limits of the r_c-integration from $r_c = b$ to $r_c = c$, we will get the total current:

$$I = J \int_{r_c=b}^{c} \int_{\theta=0}^{2\pi} r_c \, dr_c \, d\theta = J \left[\frac{r_c^2}{2}\right]_{r_c=b}^{c} [\theta]_{\theta=0}^{2\pi} = \pi J(c^2 - b^2)$$

Divide the two previous equations. Note that the πJ cancels out.

$$\frac{I_{shell}}{I} = \frac{r_c^2 - b^2}{c^2 - b^2}$$

Multiply both sides of the equation by I.

$$I_{shell} = I\left(\frac{r_c^2 - b^2}{c^2 - b^2}\right)$$

Plug this expression for I_{shell} into the previous equation $I_{enc} = I - I_{shell}$. The minus **sign** represents that the shell's current runs **opposite** to the solid cylinder's current.

$$I_{enc} = I - I\left(\frac{r_c^2 - b^2}{c^2 - b^2}\right) = I\left(1 - \frac{r_c^2 - b^2}{c^2 - b^2}\right)$$

We **factored** out the I. Subtract the fractions using a **common denominator**.

$$I_{enc} = I\left(\frac{c^2 - b^2}{c^2 - b^2} - \frac{r_c^2 - b^2}{c^2 - b^2}\right) = I\left(\frac{c^2 - b^2 - r_c^2 + b^2}{c^2 - b^2}\right) = I\left(\frac{c^2 - r_c^2}{c^2 - b^2}\right)$$

Note that when you **distribute** the minus sign ($-$) to the second term, the two minus signs combine to make $+b^2$. That is, $-(r_c^2 - b^2) = -r_c^2 + b^2$. The b^2's cancel out. Plug this expression for I_{enc} into the equation for magnetic field from Ampère's law.

$$B_{III} = \frac{\mu_0 I_{enc}}{2\pi r_c} = \frac{\mu_0 I(c^2 - r_c^2)}{2\pi r_c(c^2 - b^2)}$$

<u>**Check**</u>: We can verify that our solution matches our answer for regions II and IV at the boundaries. As r_c approaches b, the $c^2 - b^2$ will cancel out and the magnetic field approaches $\frac{\mu_0 I}{2\pi b}$, which matches region II at the boundary. As r_c approaches c, the magnetic field approaches 0, which matches region IV at the boundary. Our solution checks out.

Note: It's possible to get a seemingly much different answer that turns out to be exactly the same (even if it doesn't seem like it's the same) if you approach the enclosed current a different way. (Of course, if your answer is different, it could also just be <u>**wrong**</u>. There are some common mistakes that students tend to make when attempting to solve this problem. We will mention two of these later.)

A common way that students arrive at a different (yet equivalent) answer to this problem is to obtain the total current for the solid conductor instead of the conducting shell. We can find the total current for the solid cylinder by performing the following integral:

$$I = J \int_{r_c=0}^{a} \int_{\theta=0}^{2\pi} r_c \, dr_c \, d\theta = J\pi a^2$$

In this case, find the shell current by dividing $I_{shell} = \pi J(r_c^2 - b^2)$ by $I = J\pi a^2$.

$$I_{shell} = I\left(\frac{r_c^2 - b^2}{a^2}\right)$$

Recall that the current enclosed equals $I_{enc} = I - I_{shell}$.

$$I_{enc} = I - I\left(\frac{r_c^2 - b^2}{a^2}\right) = I\left(1 - \frac{r_c^2 - b^2}{a^2}\right) = I\left(\frac{a^2}{a^2} - \frac{r_c^2 - b^2}{a^2}\right) = I\left(\frac{a^2 - r_c^2 + b^2}{a^2}\right)$$

In this case, the magnetic field in region III is:

$$B_{III} = \frac{\mu_0 I_{enc}}{2\pi r_c} = \frac{\mu_0 I(a^2 + b^2 - r_c^2)}{2\pi r_c a^2}$$

You can verify that this expression matches region II as r_c approaches b, but it's much more difficult to see that it matches region IV as r_c approaches c. With this in mind, it's arguably better to go with our previous answer, which is the **same** as the answer shown above:

$$B_{III} = \frac{\mu_0 I(c^2 - r_c^2)}{2\pi r_c(c^2 - b^2)}$$

As usual, we can turn this into a vector by adding the appropriate unit vector.

$$\vec{\mathbf{B}}_{III} = \frac{\mu_0 I(c^2 - r_c^2)}{2\pi r_c(c^2 - b^2)}\hat{\boldsymbol{\theta}}$$

Another way to have a correct, but seemingly different, answer to this problem is to express your answer in terms of the current density (J) instead of the total current (I). For example, $B_{III} = \frac{\mu_0 J(c^2 - r_c^2)}{2r_c}$ or $B_{III} = \frac{\mu_0 J(a^2 - r_c^2 + b^2)}{2r_c}$. (There are yet other ways to get a correct, yet different, answer: You could mix and match our substitutions for I_{shell} and I, for example.)

Two common ways for students to arrive at an **incorrect** answer to this problem are to integrate from $r_c = 0$ to $r_c = r_c$ for I_{shell} (when the lower limit should be $r_c = b$) or to forget to include the solid cylinder's current in $I_{enc} = I - I_{shell}$.

Region IV: $r_c < c$.

Outside both of the conducting cylinders, the net current enclosed is zero: $I_{enc} = I - I = 0$. That's because one current equal to $+I$ comes out of page for the solid cylinder, while another current equal to $-I$ goes into page for the cylindrical shell. Therefore, the net magnetic field in region IV is zero:

$$\vec{\mathbf{B}}_{IV} = 0$$

Infinite Current Sheet Example: A Prelude to Example 99. The infinite current sheet illustrated below is a very thin infinite conducting plane with uniform current density* $\vec{\mathbf{K}}$ coming out of the page. The infinite current sheet lies in the xy plane at $z = 0$. Derive an expression for the magnetic field on either side of the infinite current sheet.

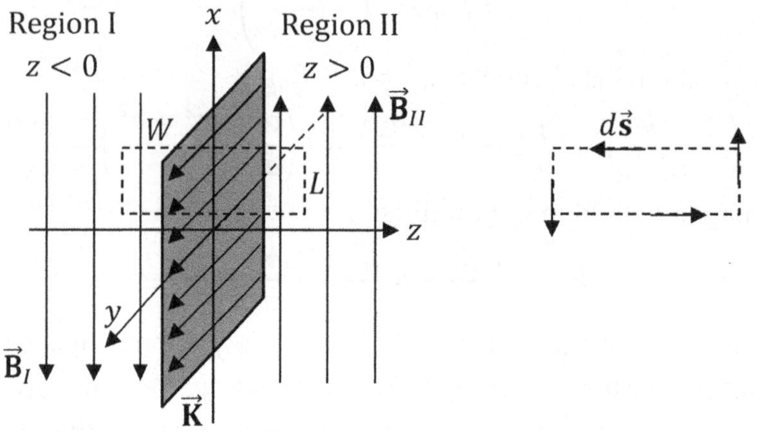

Solution. First sketch the magnetic field lines for the current sheet. Note that the current (I) is coming out of the page ($\odot$) along $\vec{\mathbf{K}}$. Apply the right-hand rule for magnetic field (Chapter 22): Grab the current with your thumb pointing out of the page ($\odot$), along the current (I). As shown above, the magnetic field ($\vec{\mathbf{B}}$) lines are straight up ($\uparrow$) to the right of the sheet and straight down ($\downarrow$) to the left of the sheet.

We choose our Ampèrian loop to be a rectangle lying in the zx plane (the same as the plane of this page) such that $\vec{\mathbf{B}}$ and $d\vec{\mathbf{s}}$ (which is tangent to the Ampèrian loop) will be parallel or perpendicular at each side of the rectangle. The Ampèrian rectangle is centered about the current sheet, as shown above on the left (we redrew the Ampèrian rectangle again on the right in order to make it easier to visualize $d\vec{\mathbf{s}}$).

- Along the top and bottom sides, $\vec{\mathbf{B}}$ is vertical and $d\vec{\mathbf{s}}$ is horizontal, such that $\vec{\mathbf{B}}$ and $d\vec{\mathbf{s}}$ are perpendicular.
- Along the right and left sides, $\vec{\mathbf{B}}$ and $d\vec{\mathbf{s}}$ are both parallel (they either both point up or both point down).

Write the formula for Ampère's law.

$$\oint_C \vec{\mathbf{B}} \cdot d\vec{\mathbf{s}} = \mu_0 I_{enc}$$

The closed integral on the left-hand side of the equation involves integrating over the complete path of the Ampèrian rectangle. The rectangle includes four sides: the right, top, left, and bottom edges.

* Note that a few textbooks may use different symbols for the current densities, $\vec{\mathbf{K}}$ and $\vec{\mathbf{J}}$.

$$\int_{right} \vec{\mathbf{B}} \cdot d\vec{\mathbf{s}} + \int_{top} \vec{\mathbf{B}} \cdot d\vec{\mathbf{s}} + \int_{left} \vec{\mathbf{B}} \cdot d\vec{\mathbf{s}} + \int_{bot} \vec{\mathbf{B}} \cdot d\vec{\mathbf{s}} = \mu_0 I_{enc}$$

The scalar product is $\vec{\mathbf{B}} \cdot d\vec{\mathbf{s}} = B \cos\theta \, ds$, where θ is the angle between $\vec{\mathbf{B}}$ and $d\vec{\mathbf{s}}$. Also recall that the direction of $d\vec{\mathbf{s}}$ is tangential to the Ampèrian loop. Study the direction of $\vec{\mathbf{B}}$ and $d\vec{\mathbf{s}}$ at each side of the rectangle in the previous diagram.

- For the right and left sides, $\theta = 0°$ because $\vec{\mathbf{B}}$ and $d\vec{\mathbf{s}}$ either both point up or both point down.
- For the top and bottom sides, $\theta = 90°$ because $\vec{\mathbf{B}}$ and $d\vec{\mathbf{s}}$ are perpendicular.

$$\int_{right} B \cos 0° \, ds + \int_{top} B \cos 90° \, ds + \int_{left} B \cos 0° \, ds + \int_{bot} B \cos 90° \, ds = \mu_0 I_{enc}$$

Recall from trig that $\cos 0° = 1$ and $\cos 90° = 0$.

$$\int_{right} Bds + 0 + \int_{left} Bds + 0 = \mu_0 I_{enc}$$

Over the right or left side of the Ampèrian rectangle, the magnitude of the magnetic field is constant, since every point on either side of the rectangle is equidistant from the infinite sheet. Therefore, we may pull B out of the integrals. (We choose our Ampèrian rectangle to be centered about the infinite sheet such that the value of B is the same† at both ends.)

$$B \int_{right} ds + B \int_{left} ds = \mu_0 I_{enc}$$

The remaining integrals are trivial: $\int ds = L$ for the left and right sides of the rectangle. Note that L is the height of the Ampèrian rectangle, <u>**not**</u> the length of the current sheet (which is infinite).

$$BL + BL = 2BL = \mu_0 I_{enc}$$

Isolate the magnitude of the magnetic field by dividing both sides of the equation by $2L$.

$$B = \frac{\mu_0 I_{enc}}{2L}$$

Now we need to determine how much current is enclosed by the Ampèrian rectangle. For a current sheet, we write $I = \int \vec{\mathbf{K}} \cdot d\vec{\boldsymbol{\ell}}$ (see page 299). Since the current sheet has uniform current density, we may pull $\vec{\mathbf{K}}$ out of the integral. This integral is over the length L of the Ampèrian rectangle (since that's the direction that encompasses some current). Note that the direction of $d\vec{\boldsymbol{\ell}}$ is perpendicular to the length L (consistent with how $d\vec{\mathbf{A}}$ is perpendicular to the cross-sectional area when working with $\vec{\mathbf{J}} \cdot d\vec{\mathbf{A}}$), which is parallel to $\vec{\mathbf{K}}$. Therefore, the scalar product is $\vec{\mathbf{K}} \cdot d\vec{\boldsymbol{\ell}} = K \cos 0° \, d\ell = K d\ell$.

† Once we reach our final answer, we will see that this doesn't matter: It turns out that the magnetic field is independent of the distance from the infinite current sheet.

$$I_{enc} = \int \vec{\mathbf{K}} \cdot d\vec{\boldsymbol{\ell}} = \int K \cos 0^\circ \, d\ell = \int K \, d\ell = K \int d\ell = KL$$

Substitute this expression for the current enclosed into the previous equation for magnetic field.

$$B = \frac{\mu_0 I_{enc}}{2L} = \frac{\mu_0}{2L}(KL) = \frac{\mu_0 K}{2}$$

The magnitude of the magnetic field is $B = \frac{\mu_0 K}{2}$, which is a constant. Thus, the magnetic field created by an infinite current sheet is uniform. We can use a unit vector to include the direction of the magnetic field with our answer: $\vec{\mathbf{B}} = \frac{\mu_0 K}{2}\hat{\mathbf{x}}$ for $z > 0$ (to the right of the sheet) and $\vec{\mathbf{B}} = -\frac{\mu_0 K}{2}\hat{\mathbf{x}}$ for $z < 0$ (to the left of the sheet), since $\hat{\mathbf{x}}$ points one unit along the $+x$-axis (which is upward in our original diagram). There is a clever way to combine both results into a single equation: We can simply write $\vec{\mathbf{B}} = \frac{\mu_0 K}{2}\frac{|z|}{z}\hat{\mathbf{x}}$, since $\frac{|z|}{z} = +1$ if $z > 0$ and $\frac{|z|}{z} = -1$ if $z < 0$. (Note carefully the combination of z and $\hat{\mathbf{x}}$: The z-coordinate tells us whether we're on the right or left side of the sheet, while the unit vector $\hat{\mathbf{x}}$ points upward.)

Example 99. An infinite conducting slab with thickness T is parallel to the xy plane, centered about $z = 0$, and has uniform current density $\vec{\mathbf{J}}$ coming out of the page. Derive an expression for the magnetic field in each region.

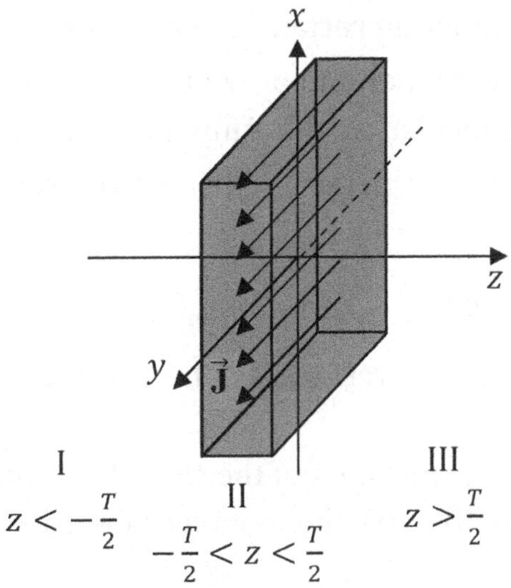

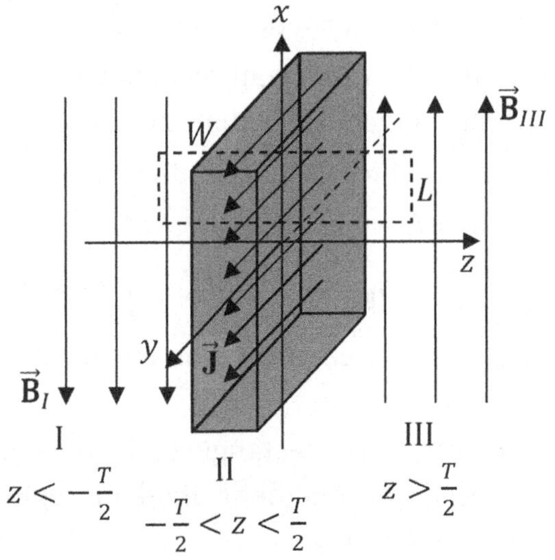

Solution. This problem is similar to the previous example. The difference is that the infinite slab has thickness, unlike the infinite plane. The magnetic field lines and Ampèrian loop are the same as before. To save time, we'll simply repeat the steps that are identical to the previous example, and pick up from where this solution deviates from the earlier one. It would be a good exercise to see if you can understand each step (if not, review the previous example for the explanation).

$$\oint_C \vec{\mathbf{B}} \cdot d\vec{\mathbf{s}} = \mu_0 I_{enc}$$

$$\int_{right} \vec{\mathbf{B}} \cdot d\vec{\mathbf{s}} + \int_{top} \vec{\mathbf{B}} \cdot d\vec{\mathbf{s}} + \int_{left} \vec{\mathbf{B}} \cdot d\vec{\mathbf{s}} + \int_{bot} \vec{\mathbf{B}} \cdot d\vec{\mathbf{s}} = \mu_0 I_{enc}$$

$$\int_{right} B \cos 0° \, ds + \int_{top} B \cos 90° \, ds + \int_{left} B \cos 0° \, ds + \int_{bot} B \cos 90° \, ds = \mu_0 I_{enc}$$

$$\int_{right} Bds + 0 + \int_{left} Bds + 0 = \mu_0 I_{enc}$$

$$B \int_{right} ds + B \int_{left} ds = \mu_0 I_{enc}$$

$$BL + BL = 2BL = \mu_0 I_{enc}$$

$$B = \frac{\mu_0 I_{enc}}{2L}$$

Now we have reached the point where it will matter that the slab has thickness.

We need to determine how much current is enclosed by the Ampèrian rectangle. For a solid thick slab, we write $I_{enc} = \int \vec{\mathbf{J}} \cdot d\vec{\mathbf{A}}$ (see page 299). Since the slab has uniform current density, we may pull $\vec{\mathbf{J}}$ out of the integral. This integral is over the length L of the Ampèrian rectangle (since that's the direction that encompasses some current). Note that the direction of $d\vec{\mathbf{A}}$ is perpendicular to the area of the Ampèrian rectangle, which is parallel to $\vec{\mathbf{J}}$. (Note that the Ampèrian rectangle in the right diagram on the previous page lies in the plane of the paper, such that the direction perpendicular to the Ampèrian rectangle is parallel to $\vec{\mathbf{J}}$, which is perpendicular to the plane of the paper.) The scalar product is therefore $\vec{\mathbf{J}} \cdot d\vec{\mathbf{A}} = J \cos 0° \, dA = JdA$.

$$I_{enc} = \int \vec{\mathbf{J}} \cdot d\vec{\mathbf{A}} = \int J \cos 0° \, dA = \int J \, dA = J \int dA = JA_{enc}$$

Here, A_{enc} is the area of the slab enclosed by the Ampèrian rectangle. There are two cases to consider:

- The Ampèrian rectangle could be wider than the thickness of the slab. This will help us find the magnetic field in regions I and III (see the regions labeled on the previous page).
- The Ampèrian rectangle could be narrower than the thickness of the slab. This will help us find the magnetic field in region II.

<u>Regions I and III</u>: $z < -\frac{T}{2}$ and $z > \frac{T}{2}$.

When the Ampèrian rectangle is wider than the thickness of the slab, the area of current enclosed equals the intersection of the rectangle and the slab: It is a rectangle with a length equal to the thickness of the slab. The area of this rectangle equals the thickness of the slab times the length of the rectangle: $A_{enc} = TL$.

$$I_{enc} = JA_{enc} = JTL$$

Substitute this expression into the previous equation for magnetic field.

$$B_I = \frac{\mu_0 I_{enc}}{2L} = \frac{\mu_0 JT}{2}$$

<u>Region II</u>: $-\frac{T}{2} < z < \frac{T}{2}$.

When the Ampèrian rectangle is narrower than the thickness of the slab, the area of current enclosed equals the area of the Ampèrian rectangle. The width of the Ampèrian rectangle is $2|z|$ (since the Ampèrian rectangle extends from $-z$ to $+z$), where $-\frac{T}{2} < z < \frac{T}{2}$. The narrower the Ampèrian rectangle, the less current it encloses. The area of the Ampèrian rectangle equals $2|z|$ times the length of the rectangle: $A_{enc} = 2|z|L$.

$$I_{enc} = JA_{enc} = J2|z|L$$

Substitute this expression into the previous equation for magnetic field.

$$B_{II} = \frac{\mu_0 I_{enc}}{2L} = \frac{\mu_0 J2|z|}{2L} = \frac{\mu_0 J|z|}{L}$$

Infinite Solenoid Example. The infinite tightly wound solenoid illustrated below is coaxial with the z-axis and carries an insulated filamentary current I. Derive an expression for the magnetic field inside of the solenoid.

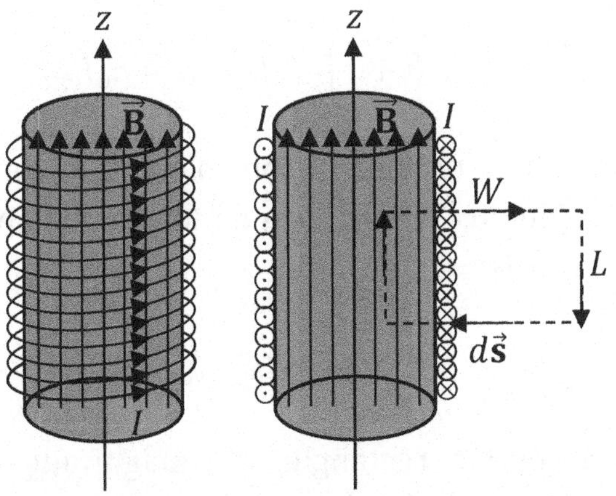

Solution. First sketch the magnetic field lines for the solenoid. Recall that we have already done this in Chapter 22: As shown above, the magnetic field ($\vec{\mathbf{B}}$) lines are straight up (↑) inside of the solenoid. Review Example 82, part (H) in Chapter 22 (but it's backwards here).

We choose our Ampèrian loop to be a rectangle with one edge perpendicular to the axis of the solenoid such that $\vec{\mathbf{B}}$ and $d\vec{\mathbf{s}}$ (which is tangent to the Ampèrian loop) will be parallel or perpendicular at each side of the rectangle. See the right diagram above.

- Along the top and bottom sides, $\vec{\mathbf{B}}$ is vertical and $d\vec{\mathbf{s}}$ is horizontal, such that $\vec{\mathbf{B}}$ and $d\vec{\mathbf{s}}$ are perpendicular.
- Along the right and left sides, $\vec{\mathbf{B}}$ and $d\vec{\mathbf{s}}$ are both parallel (they either both point up or both point down).

Write the formula for Ampère's law.

$$\oint_C \vec{\mathbf{B}} \cdot d\vec{\mathbf{s}} = \mu_0 I_{enc}$$

The closed integral on the left-hand side of the equation involves integrating over the complete path of the Ampèrian rectangle. The rectangle includes four sides.

$$\int_{right} \vec{\mathbf{B}} \cdot d\vec{\mathbf{s}} + \int_{top} \vec{\mathbf{B}} \cdot d\vec{\mathbf{s}} + \int_{left} \vec{\mathbf{B}} \cdot d\vec{\mathbf{s}} + \int_{bot} \vec{\mathbf{B}} \cdot d\vec{\mathbf{s}} = \mu_0 I_{enc}$$

The scalar product is $\vec{\mathbf{B}} \cdot d\vec{\mathbf{s}} = B \cos\theta \, ds$, where θ is the angle between $\vec{\mathbf{B}}$ and $d\vec{\mathbf{s}}$. Also recall that the direction of $d\vec{\mathbf{s}}$ is tangential to the Ampèrian loop. Study the direction of $\vec{\mathbf{B}}$ and $d\vec{\mathbf{s}}$ at each side of the rectangle in the previous diagram.

- For the right and left sides, $\theta = 0°$ because $\vec{\mathbf{B}}$ and $d\vec{\mathbf{s}}$ either both point up or both point down.
- For the top and bottom sides, $\theta = 90°$ because $\vec{\mathbf{B}}$ and $d\vec{\mathbf{s}}$ are perpendicular.

$$\int_{right} B\cos 0°\, ds + \int_{top} B\cos 90°\, ds + \int_{left} B\cos 0°\, ds + \int_{bot} B\cos 90°\, ds = \mu_0 I_{enc}$$

Recall from trig that $\cos 0° = 1$ and $\cos 90° = 0$.

$$\int_{right} Bds + 0 + \int_{left} Bds + 0 = \mu_0 I_{enc}$$

It turns out that the magnetic field outside of the solenoid is very weak compared to the magnetic field inside of the solenoid. With this in mind, we will approximate $B \approx 0$ outside of the solenoid, such that $\int_{right} Bds \approx 0$.

$$\int_{left} Bds = \mu_0 I_{enc}$$

Over the left side of the Ampèrian rectangle, the magnitude of the magnetic field is constant, since every point on the left side of the rectangle is equidistant from the axis of the infinite solenoid. Therefore, we may pull B out of the integral.

$$B \int_{left} ds = \mu_0 I_{enc}$$

The remaining integral is trivial: $\int ds = L$ for the left side of the rectangle.

$$BL = \mu_0 I_{enc}$$

Isolate the magnitude of the magnetic field by dividing both sides of the equation by L.

$$B = \frac{\mu_0 I_{enc}}{L}$$

Now we need to determine how much current is enclosed by the Ampèrian rectangle. The answer is simple: The current passes through the Ampèrian rectangle N times. Therefore, the current enclosed by the Ampèrian rectangle is $I_{enc} = NI$. Substitute this expression for the current enclosed into the previous equation for magnetic field.

$$B = \frac{\mu_0 I_{enc}}{L} = \frac{\mu_0 NI}{L} = \mu_0 nI$$

Note that lowercase n is the number of turns (or loops) per unit length: $n = \frac{N}{L}$. For a truly infinite solenoid, N and L would each be infinite, yet n is finite. The magnitude of the magnetic field inside of the solenoid is $B = \mu_0 nI$, which is a constant. Thus, the magnetic field inside of an infinitely long, tightly wound solenoid is uniform. We can use a unit vector to include the direction of the magnetic field with our answer: $\vec{\mathbf{B}} = \mu_0 nI\hat{\mathbf{z}}$.

Toroidal Coil Example. The tightly wound toroidal coil illustrated below carries an insulated filamentary current I. Derive an expression for the magnetic field in the region that is shaded gray in the diagram below.

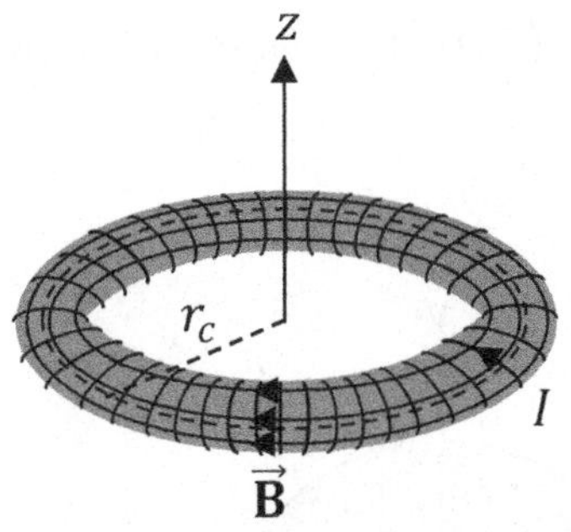

Solution. Although some students struggle to visualize or draw the toroidal coil, you shouldn't be afraid of the math: As we will see, the mathematics involved in this application of Ampère's law is very simple.

First sketch the magnetic field lines for the toroidal coil. Apply the right-hand rule for magnetic field (Chapter 22): As shown above, the magnetic field ($\vec{\mathbf{B}}$) lines are circles running along the (circular) axis of the toroidal coil. We choose our Ampèrian loop to be a circle inside of the toroid (dashed circle above) such that $\vec{\mathbf{B}}$ and $d\vec{\mathbf{s}}$ (which is tangent to the Ampèrian loop) will be parallel. Write the formula for Ampère's law.

$$\oint_C \vec{\mathbf{B}} \cdot d\vec{\mathbf{s}} = \mu_0 I_{enc}$$

The scalar product is $\vec{\mathbf{B}} \cdot d\vec{\mathbf{s}} = B \cos\theta \, ds$, where $\theta = 0°$, and the magnitude of the magnetic field is constant over the Ampèrian circle. Ampère's law thus reduces to:

$$B \oint_C ds = \mu_0 I_{enc}$$

This integral is over the path of the Ampèrian circle. Note that r_c is constant for a circle.

$$B \oint_C ds = B \oint_{\theta=0}^{2\pi} r_c \, d\theta = B r_c \oint_{\theta=0}^{2\pi} d\theta = B 2\pi r_c = \mu_0 I_{enc}$$

Isolate the magnitude of the magnetic field by dividing both sides of the equation by $2\pi r_c$.

$$B = \frac{\mu_0 I_{enc}}{2\pi r_c}$$

Similar to the previous example, the current enclosed by the Ampèrian loop is $I_{enc} = NI$.

$$B = \frac{\mu_0 N I}{2\pi r_c}$$

The magnitude of the magnetic field inside of the toroidal coil is $B = \frac{\mu_0 N I}{2\pi r_c}$. We can use a unit vector to include the direction of the magnetic field with our answer: $\vec{\mathbf{B}} = \frac{\mu_0 N I}{2\pi r_c} \hat{\boldsymbol{\theta}}$.

Example 100. An infinite solid cylindrical conductor coaxial with the z-axis has radius a, uniform current density $\vec{\mathbf{J}}$ along $\hat{\mathbf{z}}$,, and carries total current I. As illustrated below, the solid cylinder has a cylindrical cavity with radius $\frac{a}{2}$ centered about a line parallel to the z-axis and passing through the point $\left(\frac{a}{2}, 0, 0\right)$. Determine the magnitude and direction of the magnetic field at the point $\left(\frac{3a}{2}, 0, 0\right)$.

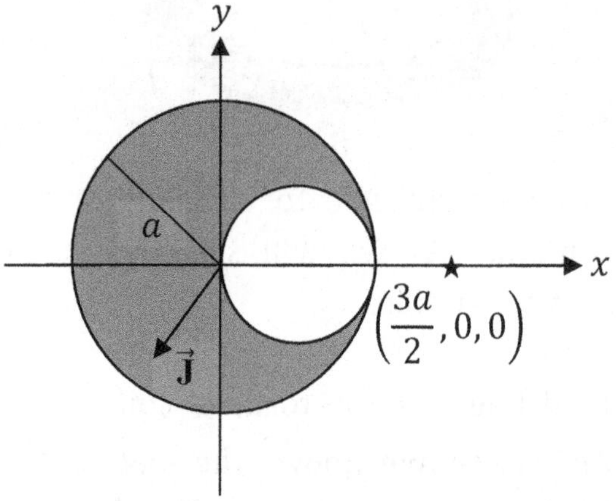

Solution. There are a couple of "tricks" involved in this solution. The first "trick" is to apply the principle of superposition (vector addition), which we learned in Chapter 3. Another "trick" is to express the current correctly for each object (we'll clarify this "trick" when we reach that stage of the solution).

Geometrically, we could make a complete solid cylinder (which we will call the "**big**" cylinder) by adding the given shape (which we will call the "**given**" shape, and which includes the cavity) to a solid cylinder that is the same size as the cavity (which we will call the "**small**" cylinder). See the diagram below.

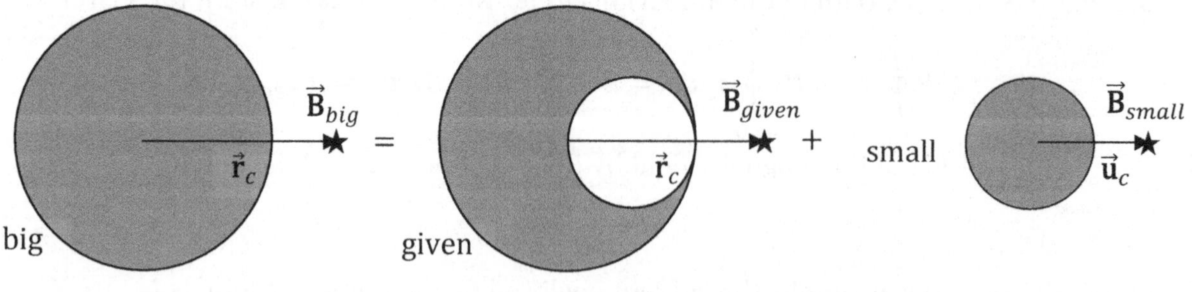

First find the magnetic field due to the big cylinder (ignoring the cavity) at a distance $r_c = \frac{3a}{2}$ away from the axis of the cylinder. Since the field point $\left(\frac{3a}{2}, 0, 0\right)$ lies outside of the cylinder, use the equation for the magnetic field in region II from the example with the infinite cylinder. We will use one of the alternate forms from the bottom of page 304. Note that this corresponds to region II with $r_c = \frac{3a}{2}$. Apply the right-hand rule for magnetic field (Chapter 22) to see that the magnetic field points up (↑) along $\hat{\mathbf{y}}$ at the field point $\left(\frac{3a}{2}, 0, 0\right)$.

$$\vec{\mathbf{B}}_{big} = \frac{\mu_0 I_{big}}{2\pi r_c}\hat{\mathbf{y}} = \frac{\mu_0 I_{big}}{3\pi a}\hat{\mathbf{y}}$$

Next find the magnetic field due to the small cylinder (the same size as the cavity) at a distance $u_c = a$ away from the axis of this small cylinder. Note that the field point is closer to the axis of the small cylinder (which "fits" into the cavity) than it is to the axis of the big cylinder: That's why $u_c = a$ for the small cylinder, whereas $r_c = \frac{3a}{2}$ for the big cylinder. The equation is the same, except for the distance being smaller.

$$\vec{\mathbf{B}}_{small} = \frac{\mu_0 I_{small}}{2\pi u_c}\hat{\mathbf{y}} = \frac{\mu_0 I_{small}}{2\pi a}\hat{\mathbf{y}}$$

Note that the two currents are different. More current would pass through the big cylinder than would proportionately pass through the small cylinder: $I_{big} > I_{small}$.

Find the magnetic field of the given shape (the cylinder with the cavity) through the principle of **superposition**. We subtract, as shown geometrically on the previous page.

$$\vec{\mathbf{B}}_{given} = \vec{\mathbf{B}}_{big} - \vec{\mathbf{B}}_{small}$$

$$\vec{\mathbf{B}}_{given} = \frac{\mu_0 I_{big}}{3\pi a}\hat{\mathbf{y}} - \frac{\mu_0 I_{small}}{2\pi a}\hat{\mathbf{y}}$$

Factor out the $\frac{\mu_0}{\pi a}\hat{\mathbf{y}}$.

$$\vec{\mathbf{B}}_{given} = \frac{\mu_0}{\pi a}\hat{\mathbf{y}}\left(\frac{I_{big}}{3} - \frac{I_{small}}{2}\right)$$

Use the equation $I = \int \vec{\mathbf{J}} \cdot d\vec{\mathbf{A}}$ (see page 299) to find the current corresponding to each shape. Since the cylinder has uniform current density, we may pull $\vec{\mathbf{J}}$ out of the integral. This integral is over the area of the Ampèrian circle. Since $d\vec{\mathbf{A}}$ is perpendicular to the cross-sectional area, $\vec{\mathbf{J}}$ is parallel to $d\vec{\mathbf{A}}$ (both come out of the page, along the axis of the cylinder). Therefore, the scalar product is $\vec{\mathbf{J}} \cdot d\vec{\mathbf{A}} = J \cos 0° \, dA = J dA$.

$$I_{enc} = \int \vec{\mathbf{J}} \cdot d\vec{\mathbf{A}} = \int J \cos 0° \, dA = \int J \, dA = J \int dA = JA_{enc}$$

In region II (outside of the cylinder), the area enclosed corresponds to the cross-sectional area of the cylinder (which is a circle): $A_{enc} = \pi(\text{radius})^2$.

- $I_{big} = J\pi a^2$. (The big cylinder has a radius equal to a.)
- $I_{small} = \frac{J\pi a^2}{4}$. (The small cylinder has a radius equal to $\frac{a}{2}$. When you square $\frac{a}{2}$ in the formula πr^2 for the area of a circle, you get $\frac{a^2}{4}$).
- $I_{given} = \frac{3J\pi a^2}{4}$. (Get this by subtracting: $I_{big} - I_{small}$. Note that $1 - \frac{1}{4} = \frac{3}{4}$).

Divide the equation for I_{big} by the equation for I_{given}.

$$\frac{I_{big}}{I_{given}} = \frac{J\pi a^2}{\frac{3J\pi a^2}{4}} = J\pi a^2 \div \frac{3J\pi a^2}{4} = J\pi a^2 \times \frac{4}{3J\pi a^2} = \frac{4}{3}$$

To divide by a fraction, multiply by its **reciprocal**. Multiply both sides by I_{given}.

$$I_{big} = \frac{4}{3} I_{given}$$

Now divide the equation for I_{small} by the equation for I_{given}.

$$\frac{I_{small}}{I_{given}} = \frac{\frac{J\pi a^2}{4}}{\frac{3J\pi a^2}{4}} = \frac{J\pi a^2}{4} \div \frac{3J\pi a^2}{4} = \frac{J\pi a^2}{4} \times \frac{4}{3J\pi a^2} = \frac{1}{3}$$

To divide by a fraction, multiply by its **reciprocal**. Multiply both sides by I_{given}.

$$I_{small} = \frac{1}{3} I_{given}$$

The second "trick" to this solution is to realize that what the problem is calling the total current (I) refers to the current through the "**given**" shape (the cylinder with the cavity in it). It's a common conceptual mistake for students to want to associate the total current (I) with the big cylinder. However, since the actual shape in this problem is the given shape with the cavity in it, the total current (I) is the actual current in the given shape, which is less than the propoportional amount of current which would run through the big cylinder.

Set $I_{given} = I$ in the above equations.

$$I_{big} = \frac{4}{3} I$$

$$I_{small} = \frac{1}{3} I$$

Substitute these equations into the previous equation for $\vec{\mathbf{B}}_{given}$.

$$\vec{\mathbf{B}}_{given} = \frac{\mu_0}{\pi a}\hat{\mathbf{y}}\left(\frac{I_{big}}{3} - \frac{I_{small}}{2}\right) = \frac{\mu_0}{\pi a}\hat{\mathbf{y}}\left[\frac{1}{3}\left(\frac{4}{3}I\right) - \frac{1}{2}\left(\frac{1}{3}I\right)\right] = \frac{\mu_0}{\pi a}\hat{\mathbf{y}}\left(\frac{4I}{9} - \frac{I}{6}\right)$$

Factor out the current (I). Subtract fractions with a **common denominator**.

$$\vec{\mathbf{B}}_{given} = \frac{\mu_0 I}{\pi a}\hat{\mathbf{y}}\left(\frac{4}{9} - \frac{1}{6}\right) = \frac{\mu_0 I}{\pi a}\hat{\mathbf{y}}\left(\frac{8}{18} - \frac{3}{18}\right) = \frac{\mu_0 I}{\pi a}\hat{\mathbf{y}}\left(\frac{8-3}{18}\right) = \frac{5\mu_0 I}{18\pi a}\hat{\mathbf{y}}$$

The magnetic field at the point $\left(\frac{a}{2}, 0, 0\right)$ is $\vec{\mathbf{B}}_{given} = \frac{5\mu_0 I}{18\pi a}\hat{\mathbf{y}}$. It has a magnitude of $B_{given} = \frac{5\mu_0 I}{18\pi a}$ and a direction of $\hat{\mathbf{B}}_{given} = \hat{\mathbf{y}}$.

28 LENZ'S LAW

Lenz's Law

According to **Faraday's law**, a **current** (I_{ind}) is induced in a loop of wire when there is a **changing magnetic flux** (Φ_m) through the loop. Lenz's law provides the direction of the induced current, while Faraday's law provides the **emf** (ε_{ind}) induced in the loop (from which the induced current can be determined). We'll explore Faraday's law in Chapter 29.

According to **Lenz's law**, the induced current runs through the loop in a direction such that the induced magnetic field created by the induced current opposes the change in the magnetic flux. That's the "fancy" way of explaining Lenz's law. Following is what the "fancy" definition really means:

- If the magnetic flux (Φ_m) through the area of the loop is **increasing**, the induced magnetic field ($\vec{\mathbf{B}}_{ind}$) created by the induced current (I_{ind}) will be **opposite** to the external magnetic field ($\vec{\mathbf{B}}_{ext}$).
- If the magnetic flux (Φ_m) through the area of the loop is **decreasing**, the induced magnetic field ($\vec{\mathbf{B}}_{ind}$) created by the induced current (I_{ind}) will be **parallel** to the external magnetic field ($\vec{\mathbf{B}}_{ext}$).

We will break Lenz's law down into four steps as follows.

1. What is the direction of the **external magnetic field** ($\vec{\mathbf{B}}_{ext}$) in the area of the loop? Draw an arrow to show the direction of $\vec{\mathbf{B}}_{ext}$.
   - What is creating a magnetic field in the area of the loop to begin with?
   - If it is a bar magnet, you'll need to know what the magnetic field lines of a bar magnet look like (see Chapter 20).
   - You want to know the direction of $\vec{\mathbf{B}}_{ext}$ specifically in the area of the loop.
   - If $\vec{\mathbf{B}}_{ext}$ points in multiple directions within the area of the loop, ask yourself which way $\vec{\mathbf{B}}_{ext}$ points on average.
2. Is the **magnetic flux** (Φ_m) through the loop increasing or decreasing? Write one of the following words: increasing, decreasing, or constant.
   - Is the relative number of magnetic field lines passing through the loop increasing or decreasing?
   - Is a magnet or external current getting closer or further from the loop?
   - Is something rotating? If so, ask yourself if more or fewer magnetic field lines will pass through the loop while it rotates.
3. What is the direction of the **induced magnetic field** ($\vec{\mathbf{B}}_{ind}$) created by the induced current? Determine this from your answer for Φ_m in Step 2 and your answer for $\vec{\mathbf{B}}_{ext}$ in Step 1 as follows. Draw an arrow to show the direction of $\vec{\mathbf{B}}_{ind}$.

- If Φ_m is **increasing**, draw $\vec{\mathbf{B}}_{ind}$ **opposite** to your answer for Step 1.
- If Φ_m is **decreasing**, draw $\vec{\mathbf{B}}_{ind}$ **parallel** to your answer for Step 1.
- If Φ_m is **constant**, then $\vec{\mathbf{B}}_{ind} = 0$ and there won't be any induced current.

4. Apply the **right-hand rule for magnetic field** (Chapter 22) to determine the direction of the **induced current** (I_{ind}) from your answer to Step 3 for the induced magnetic field ($\vec{\mathbf{B}}_{ind}$). Draw and label your answer for I_{ind} on the diagram.
   - Be sure to use your answer to Step 3 and **not** your answer to Step 1.
   - You're actually inverting the right-hand rule. In Chapter 22, we knew the direction of the current and applied the right-hand rule to determine the direction of the magnetic field. With Lenz's law, we know the direction of the magnetic field (from Step 3), and we're applying the right-hand rule to determine the direction of the induced current.
   - The right-hand rule for magnetic field still works the same way. The difference is that now you'll need to grab the loop both ways (try it both with your thumb clockwise and with your thumb counterclockwise to see which way points your fingers correctly inside of the loop).
   - What matters is which way your fingers are pointing inside of the loop (they must match your answer to Step 3) and which way your thumb points (this is your answer for the direction of the induced current).

Right-hand Rule for Magnetic Field

A long straight wire creates magnetic field lines that circulate around the current, as shown below.

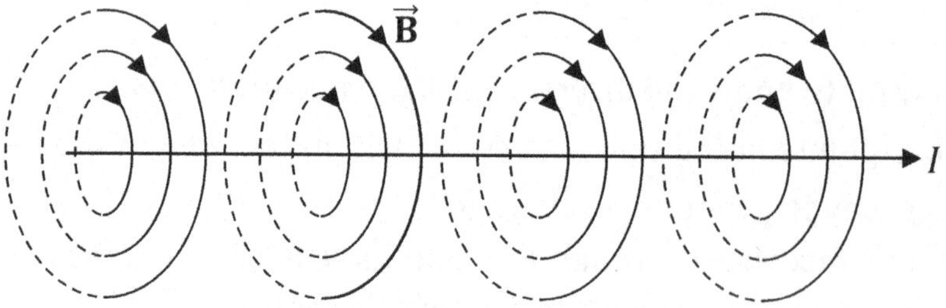

Recall the right-hand rule for magnetic **field** from Chapter 22: This right-hand rule gives you the direction of the magnetic field ($\vec{\mathbf{B}}$) created by a current (I) or moving charge:

- Imagine grabbing the wire with your right hand. **Tip**: You can use a pencil to represent the wire and actually grab the pencil.
- Grab the wire with your thumb pointing along the current (I).
- Your fingers represent **circular** magnetic field lines traveling around the wire **toward your fingertips**. At a given point, the direction of the magnetic field is **tangent** to these circles (your fingers). See the diagram on the following page.

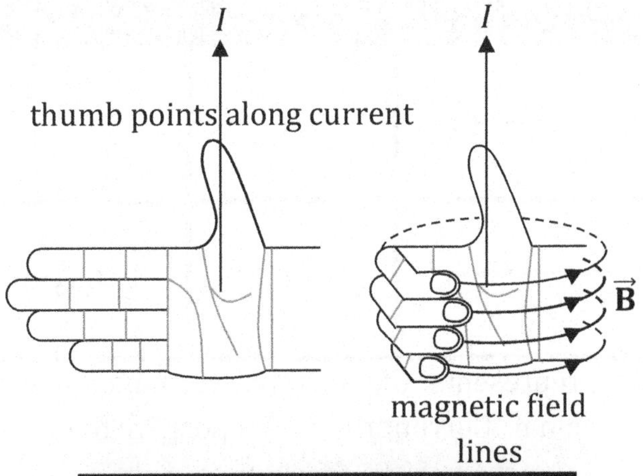

Symbol	Meaning
$\otimes$	into the page
$\odot$	out of the page
N	north pole
S	south pole
p	proton
n	neutron
e^-	electron

Symbol	Name	SI Units
$\vec{\mathbf{B}}_{ext}$	external magnetic field (that is, external to the loop)	T
$\vec{\mathbf{B}}_{ind}$	induced magnetic field (created by the induced current)	T
Φ_m	magnetic flux through the area of the loop	T·m^2 or Wb
I_{ind}	induced current (that is, induced in the loop)	A
$\vec{v}$	velocity	m/s

Schematic Representation	Symbol	Name
	R	resistor
	ΔV	battery or DC power supply

Recall that the long line represents the positive terminal, while the small rectangle represents the negative terminal. Current runs from positive to negative.

Example 101. If the magnetic field is increasing in the diagram below, what is the direction of the current induced in the loop?

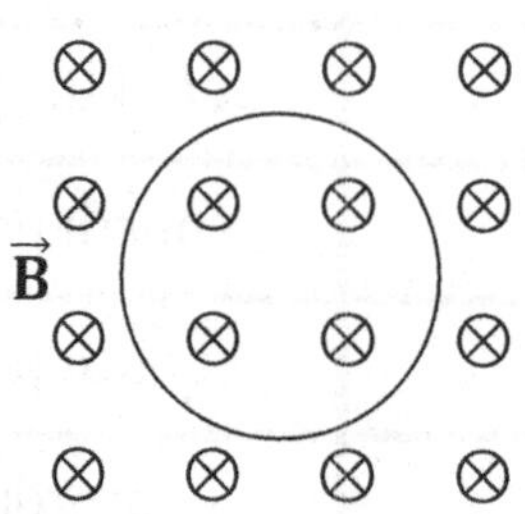

Solution. Apply the four steps of Lenz's law.

1. The **external** magnetic field ($\vec{\mathbf{B}}_{ext}$) is into the page ($\otimes$). It happens to already be drawn in the problem.
2. The magnetic flux (Φ_m) is **increasing** because the problem states that the external magnetic field is increasing.
3. The **induced** magnetic field ($\vec{\mathbf{B}}_{ind}$) is out of the page ($\odot$). Since Φ_m is increasing, the direction of $\vec{\mathbf{B}}_{ind}$ is **opposite** to the direction of $\vec{\mathbf{B}}_{ext}$ from Step 1.
4. The **induced current** (I_{ind}) is counterclockwise, as drawn below. If you grab the loop with your thumb pointing counterclockwise and your fingers wrapped around the wire, inside the loop your fingers will come out of the page ($\odot$). Remember, you want your fingers to match Step 3 inside the loop: You **<u>don't</u>** want your fingers to match Step 1 or the magnetic field lines originally drawn in the problem.

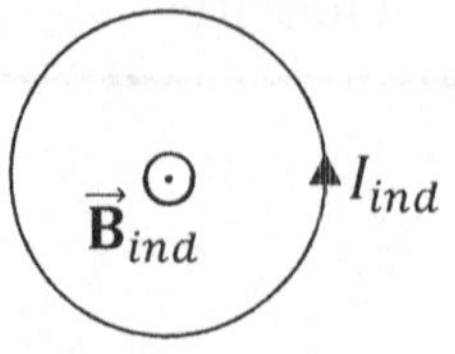

Example 102. If the magnetic field is decreasing in the diagram below, what is the direction of the current induced in the loop?

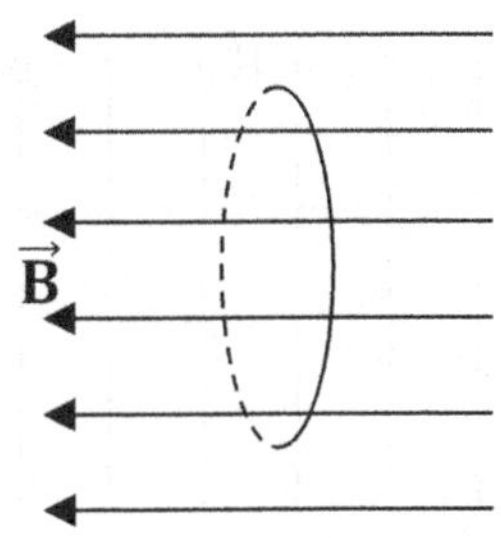

Solution. Apply the four steps of Lenz's law.

1. The **external** magnetic field ($\vec{\mathbf{B}}_{ext}$) is to the left (←). It happens to already be drawn in the problem.
2. The magnetic flux (Φ_m) is **decreasing** because the problem states that the external magnetic field is decreasing.
3. The **induced** magnetic field ($\vec{\mathbf{B}}_{ind}$) is to the left (←). Since Φ_m is decreasing, the direction of $\vec{\mathbf{B}}_{ind}$ is the **same** as the direction of $\vec{\mathbf{B}}_{ext}$ from Step 1.
4. The **induced current** (I_{ind}) runs up the front of the loop (and therefore runs down the back of the loop), as drawn below. Note that this loop is vertical. If you grab the front of the loop with your thumb pointing up and your fingers wrapped around the wire, inside the loop your fingers will go to the left (←).

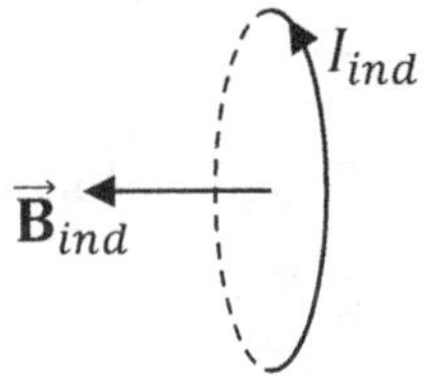

Example 103. If the magnetic field is increasing in the diagram below, what is the direction of the current induced in the loop?

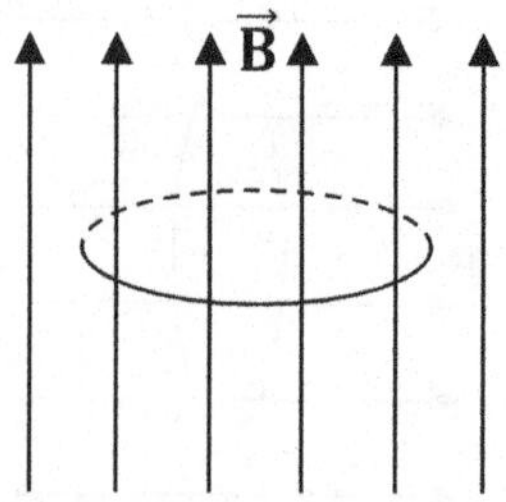

Solution. Apply the four steps of Lenz's law.

1. The **external** magnetic field ($\vec{\mathbf{B}}_{ext}$) is up (↑). It happens to already be drawn in the problem.
2. The magnetic flux (Φ_m) is **increasing** because the problem states that the external magnetic field is increasing.
3. The **induced** magnetic field ($\vec{\mathbf{B}}_{ind}$) is down (↓). Since Φ_m is increasing, the direction of $\vec{\mathbf{B}}_{ind}$ is **opposite** to the direction of $\vec{\mathbf{B}}_{ext}$ from Step 1.
4. The **induced current** (I_{ind}) runs to the left in the front of the loop (and therefore runs to the right in the back of the loop), as drawn below. Note that this loop is horizontal. If you grab the front of the loop with your thumb pointing to the left and your fingers wrapped around the wire, inside the loop your fingers will go down (↓). Remember, you want your fingers to match Step 3 inside the loop (**<u>not</u>** Step 1).

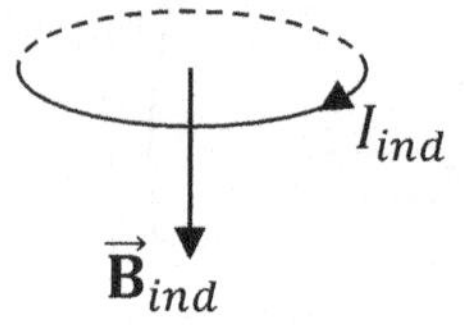

Example 104. As the magnet is moving towards the loop in the diagram below, what is the direction of the current induced in the loop?

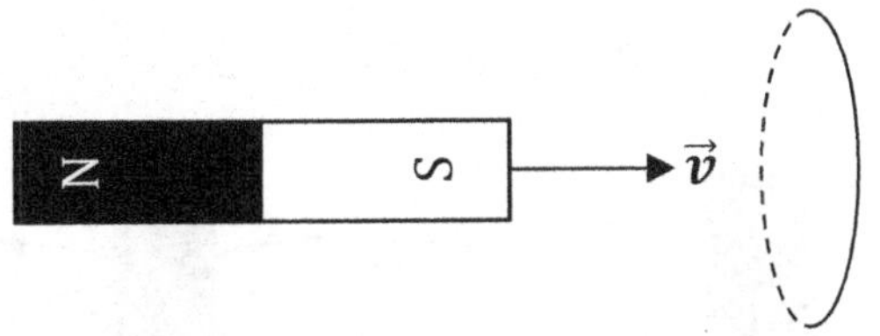

Solution. Apply the four steps of Lenz's law.

1. The **external** magnetic field ($\vec{\mathbf{B}}_{ext}$) is to the left (←). This is because the magnetic field lines of the magnet are going to the left, towards the south (S) pole, in the area of the loop as illustrated below. **Tip**: When you view the diagram below, ask yourself which way, on average, the magnetic field lines would be headed if you extend the diagram to where the loop is. (**<u>Don't</u>** use the velocity in Step 1.)

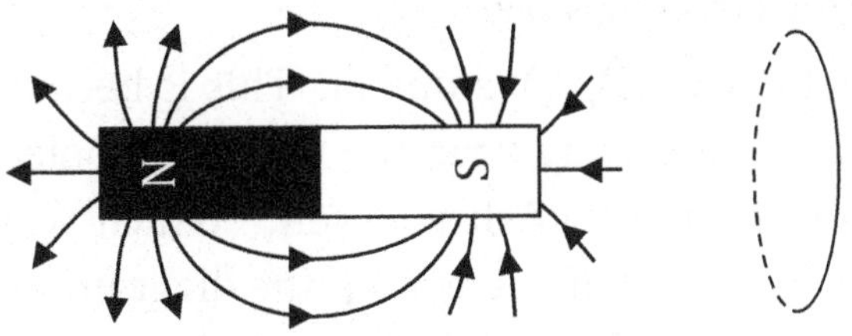

2. The magnetic flux (Φ_m) is **increasing** because the magnet is getting closer to the loop. This is the step where the direction of the velocity ($\vec{v}$) matters.
3. The **induced** magnetic field ($\vec{\mathbf{B}}_{ind}$) is to the right (→). Since Φ_m is increasing, the direction of $\vec{\mathbf{B}}_{ind}$ is **opposite** to the direction of $\vec{\mathbf{B}}_{ext}$ from Step 1.
4. The **induced current** (I_{ind}) runs down the front of the loop (and therefore runs up in the back of the loop), as drawn below. Note that this loop is vertical. If you grab the front of the loop with your thumb pointing down and your fingers wrapped around the wire, inside the loop your fingers will go to the right (→). Remember, you want your fingers to match Step 3 inside the loop (<u>**not**</u> Step 1).

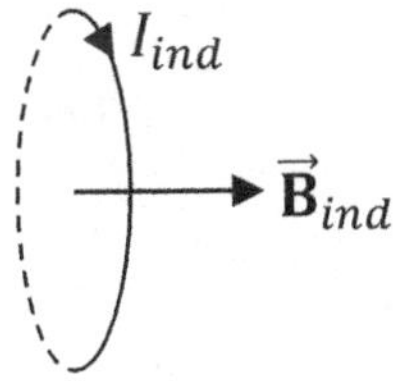

Example 105. As the magnet is moving away from the loop in the diagram below, what is the direction of the current induced in the loop?

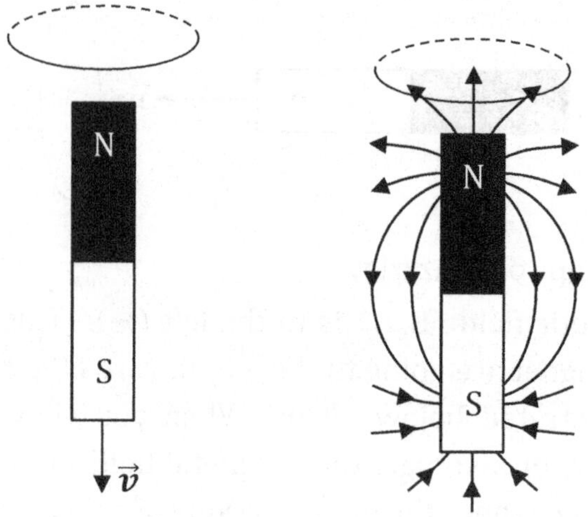

Solution. Apply the four steps of Lenz's law.

1. The **external** magnetic field ($\vec{\mathbf{B}}_{ext}$) is up (↑). This is because the magnetic field lines of the magnet are going up, away from the north (N) pole, in the area of the loop as illustrated above on the right. Ask yourself: Which way, on average, would the magnetic field lines be headed if you extend the diagram to where the loop is?
2. The magnetic flux (Φ_m) is **decreasing** because the magnet is getting further away from the loop. This is the step where the direction of the velocity ($\vec{v}$) matters.
3. The **induced** magnetic field ($\vec{\mathbf{B}}_{ind}$) is up (↑). Since Φ_m is decreasing, the direction of $\vec{\mathbf{B}}_{ind}$ is the **same** as the direction of $\vec{\mathbf{B}}_{ext}$ from Step 1.
4. The **induced current** (I_{ind}) runs to the right in the front of the loop (and therefore runs to the left in the back of the loop), as drawn below. Note that this loop is horizontal. If you grab the front of the loop with your thumb pointing right and your fingers wrapped around the wire, inside the loop your fingers will go up (↑).

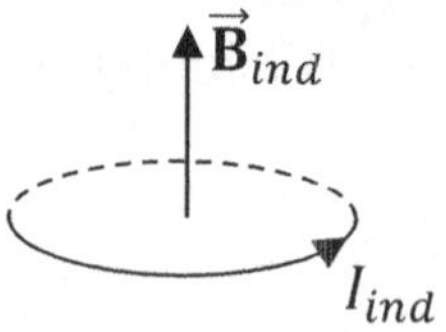

Example 106. As the magnet (which is perpendicular to the page) is moving into the page and towards the loop in the diagram below, what is the direction of the current induced in the loop? (Unlike the magnet, the loop lies in the plane of the page.)

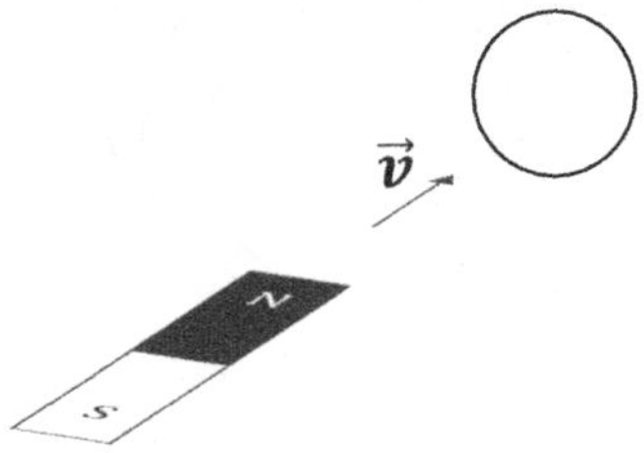

Solution. Apply the four steps of Lenz's law.

1. The **external** magnetic field ($\vec{\mathbf{B}}_{ext}$) is into the page ($\otimes$). This is because the magnetic field lines of the magnet are going into the page (towards the loop), away from the north (N) pole, in the area of the loop as illustrated below. Ask yourself: Which way, on average, would the magnetic field lines be headed if you extend the diagram to where the loop is?

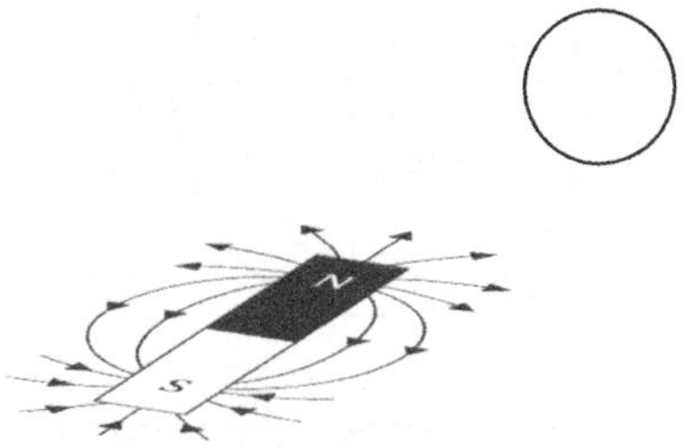

2. The magnetic flux (Φ_m) is **increasing** because the magnet is getting closer to the loop. This is the step where the direction of the velocity ($\vec{v}$) matters.
3. The **induced** magnetic field ($\vec{\mathbf{B}}_{ind}$) is out of the page ($\odot$). Since Φ_m is increasing, the direction of $\vec{\mathbf{B}}_{ind}$ is **opposite** to the direction of $\vec{\mathbf{B}}_{ext}$ from Step 1.
4. The **induced current** (I_{ind}) is counterclockwise, as drawn below. If you grab the loop with your thumb pointing counterclockwise and your fingers wrapped around the wire, inside the loop your fingers will come out of the page ($\odot$). Remember, you want your fingers to match Step 3 inside the loop (**<u>not</u>** Step 1).

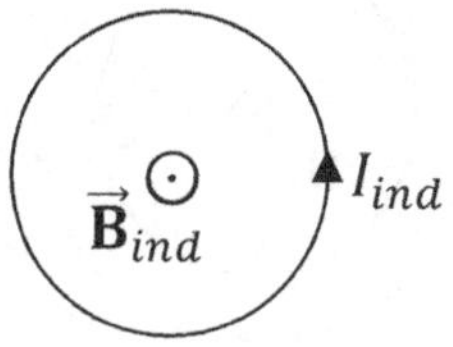

Example 107. As the current increases in the outer loop in the diagram below, what is the direction of the current induced in the inner loop?

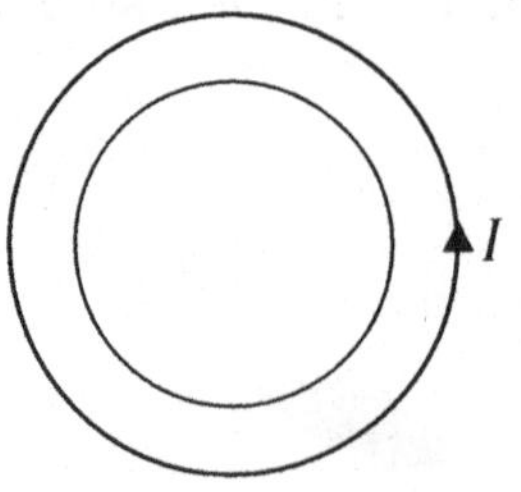

Solution. Apply the four steps of Lenz's law.

1. The **external** magnetic field ($\vec{\mathbf{B}}_{ext}$) is out of the page ($\odot$). This is because the outer loop creates a magnetic field that is out of the page in the region where the inner loop is. Get this from the right-hand rule for magnetic field (Chapter 22). When you grab the outer loop with your thumb pointed counterclockwise (along the given current) and your fingers wrapped around the wire, inside the outer loop (because the inner loop is inside of the outer loop) your fingers point out of the page.

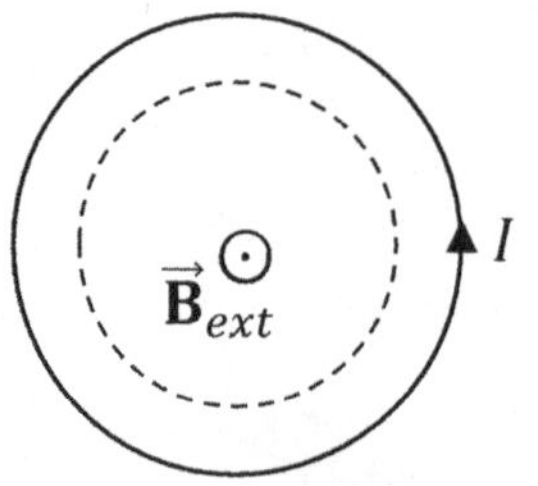

2. The magnetic flux (Φ_m) is **increasing** because the problem states that the given current is increasing.
3. The **induced** magnetic field ($\vec{\mathbf{B}}_{ind}$) is into the page ($\otimes$). Since Φ_m is increasing, the direction of $\vec{\mathbf{B}}_{ind}$ is **opposite** to the direction of $\vec{\mathbf{B}}_{ext}$ from Step 1.
4. The **induced current** (I_{ind}) is clockwise, as drawn below. If you grab the loop with your thumb pointing clockwise and your fingers wrapped around the wire, inside the loop your fingers will go into the page ($\otimes$). Remember, you want your fingers to match Step 3 inside the loop (**<u>not</u>** Step 1).

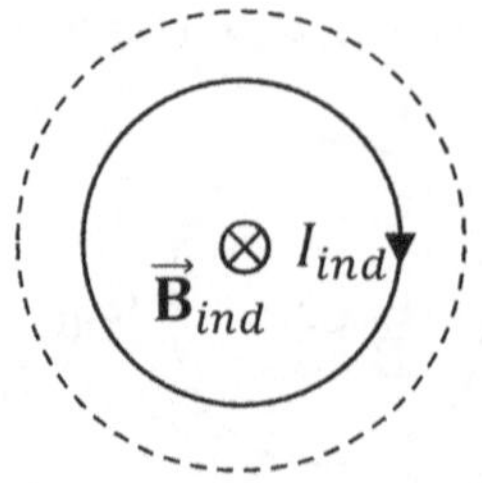

Example 108. As the current in the straight conductor decreases in the diagram below, what is the direction of the current induced in the rectangular loop?

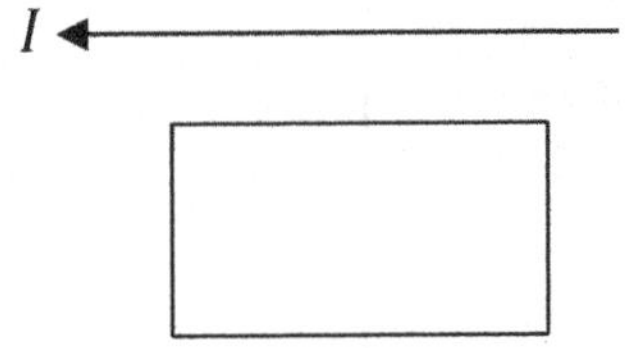

Solution. Apply the four steps of Lenz's law.

1. The **external** magnetic field ($\vec{\mathbf{B}}_{ext}$) is out of the page ($\odot$). This is because the given current creates a magnetic field that is out of the page in the region where the loop is. Get this from the right-hand rule for magnetic field (Chapter 22). When you grab the given current with your thumb pointed to the left and your fingers wrapped around the wire, below the given current (because the loop is below the straight wire) your fingers point out of the page.

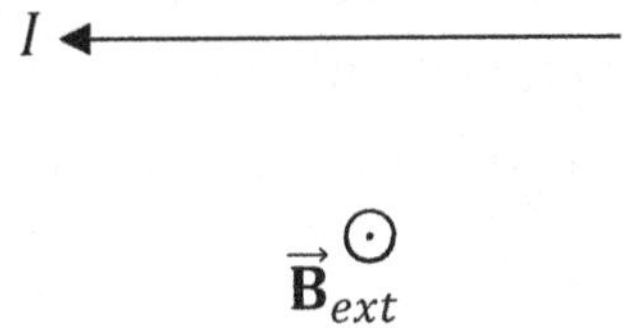

2. The magnetic flux (Φ_m) is **decreasing** because the problem states that the given current is decreasing.
3. The **induced** magnetic field ($\vec{\mathbf{B}}_{ind}$) is out of the page ($\odot$). Since Φ_m is decreasing, the direction of $\vec{\mathbf{B}}_{ind}$ is the **same** as the direction of $\vec{\mathbf{B}}_{ext}$ from Step 1.
4. The **induced current** (I_{ind}) is counterclockwise, as drawn below. If you grab the loop with your thumb pointing counterclockwise and your fingers wrapped around the wire, inside the loop your fingers will come out of the page ($\odot$).

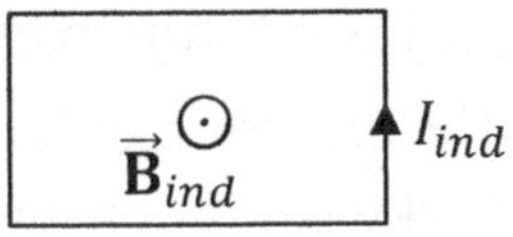

Example 109. As the potential difference of the DC power supply increases in the diagram below, what is the direction of the current induced in the right loop?

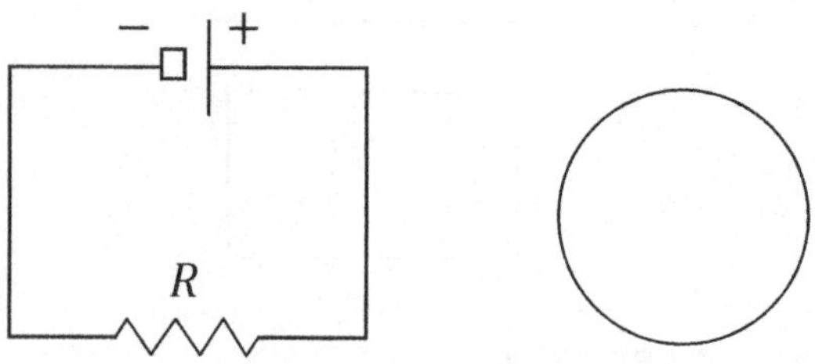

Solution. Apply the four steps of Lenz's law.

1. The **external** magnetic field ($\vec{\mathbf{B}}_{ext}$) is out of the page ($\odot$). First of all, an external current runs clockwise through the left loop, from the positive (+) terminal of the battery to the negative (−) terminal, as shown below. Secondly, the left loop creates a magnetic field that is out of the page in the region where the right loop is. Get this from the right-hand rule for magnetic field (Chapter 22). When you grab the **right side** of the **left loop** (because that side is nearest to the right loop) with your thumb pointed down (since the given current runs down at the right side of the left loop), outside of the left loop (because the right loop is outside of the left loop) your fingers point out of the page.

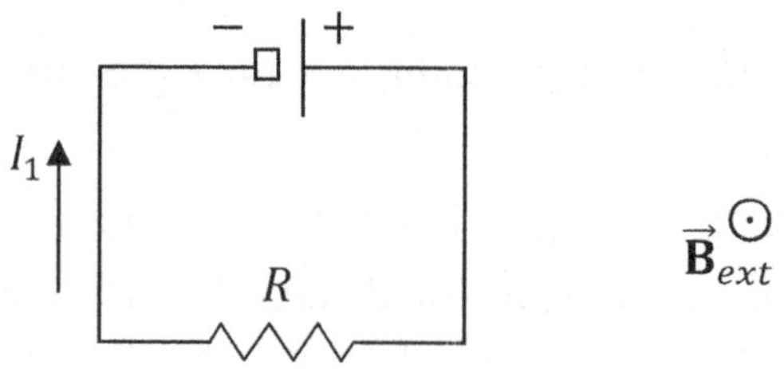

2. The magnetic flux (Φ_m) is **increasing** because the problem states that the potential difference in the battery is increasing, which increases the external current.
3. The **induced** magnetic field ($\vec{\mathbf{B}}_{ind}$) is into the page ($\otimes$). Since Φ_m is increasing, the direction of $\vec{\mathbf{B}}_{ind}$ is **opposite** to the direction of $\vec{\mathbf{B}}_{ext}$ from Step 1.
4. The **induced current** (I_{ind}) is clockwise, as drawn below. If you grab the loop with your thumb pointing clockwise and your fingers wrapped around the wire, inside the loop your fingers will go into the page ($\otimes$). Remember, you want your fingers to match Step 3 inside the loop (**not** Step 1).

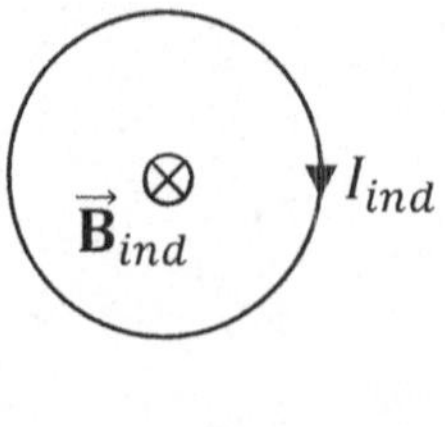

Example 110. As the conducting bar illustrated below slides to the right along the rails of a bare U-channel conductor, what is the direction of the current induced in the conducting bar?

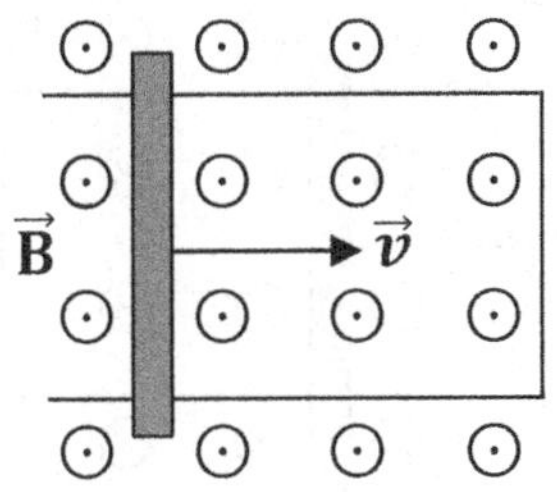

Solution. Apply the four steps of Lenz's law.

1. The **external** magnetic field ($\vec{\mathbf{B}}_{ext}$) is out of the page (⊙). It happens to already be drawn in the problem.
2. The magnetic flux (Φ_m) is **decreasing** because the area of the loop is getting smaller as the conducting bar travels to the right. Note that the conducting bar makes electrical contract where it touches the bare U-channel conductor. The dashed (---) line below illustrates how the area of the loop is getting smaller as the conducting bar travels to the right.

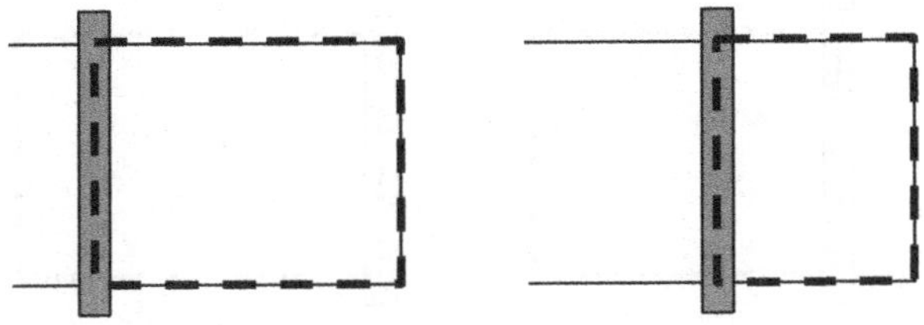

3. The **induced** magnetic field ($\vec{\mathbf{B}}_{ind}$) is out of the page (⊙). Since Φ_m is decreasing, the direction of $\vec{\mathbf{B}}_{ind}$ is the **same** as the direction of $\vec{\mathbf{B}}_{ext}$ from Step 1.
4. The **induced current** (I_{ind}) is counterclockwise, as drawn below. If you grab the loop with your thumb pointing counterclockwise and your fingers wrapped around the wire, inside the loop your fingers will come out of the page (⊙). Since the induced current is counterclockwise in the loop, the induced current runs down (↓) the conducting bar.

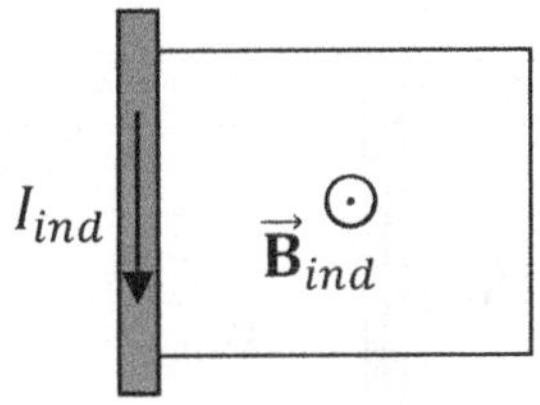

Example 111. As the conducting bar illustrated below slides to the left along the rails of a bare U-channel conductor, what is the direction of the current induced in the conducting bar?

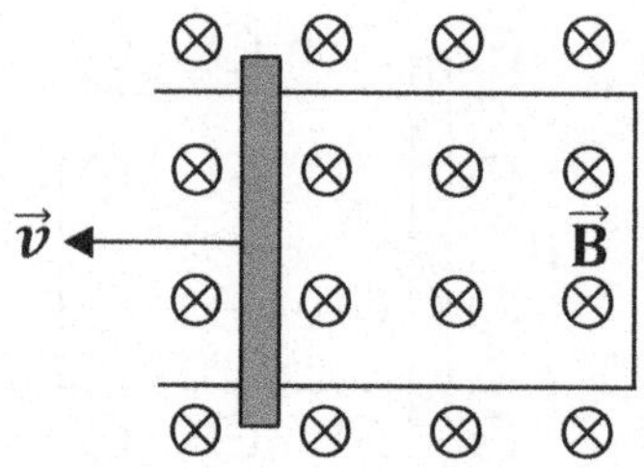

Solution. Apply the four steps of Lenz's law.

1. The **external** magnetic field ($\vec{\mathbf{B}}_{ext}$) is into the page (⊗). It happens to already be drawn in the problem.
2. The magnetic flux (Φ_m) is **increasing** because the area of the loop is getting larger as the conducting bar travels to the left. Note that the conducting bar makes electrical contract where it touches the bare U-channel conductor. The dashed (---) line below illustrates how the area of the loop is getting larger as the conducting bar travels to the left.

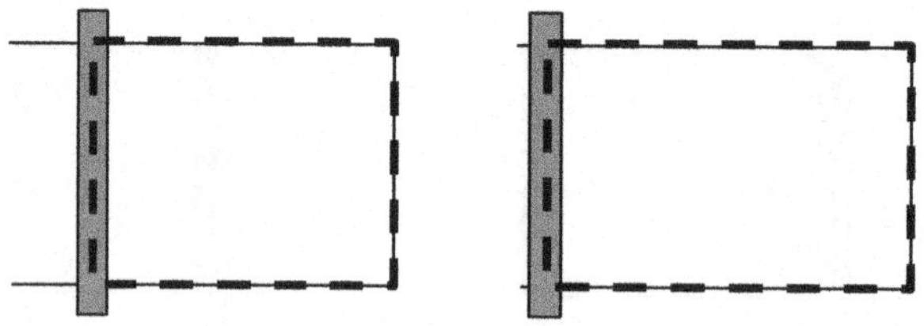

3. The **induced** magnetic field ($\vec{\mathbf{B}}_{ind}$) is out of the page (⊙). Since Φ_m is increasing, the direction of $\vec{\mathbf{B}}_{ind}$ is **opposite** to the direction of $\vec{\mathbf{B}}_{ext}$ from Step 1.
4. The **induced current** (I_{ind}) is counterclockwise, as drawn below. If you grab the loop with your thumb pointing counterclockwise and your fingers wrapped around the wire, inside the loop your fingers will come out of the page (⊙). Since the induced current is counterclockwise in the loop, the induced current runs down (↓) the conducting bar. Remember, you want your fingers to match Step 3 inside the loop (<u>**not**</u> Step 1).

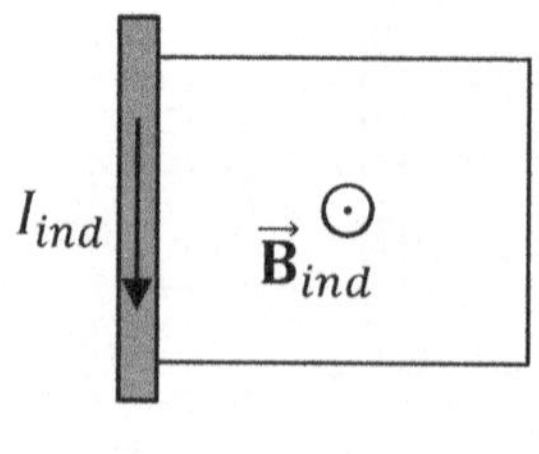

Example 112. As the vertex of the triangle illustrated below is pushed downward from point A to point C, what is the direction of the current induced in the triangular loop?

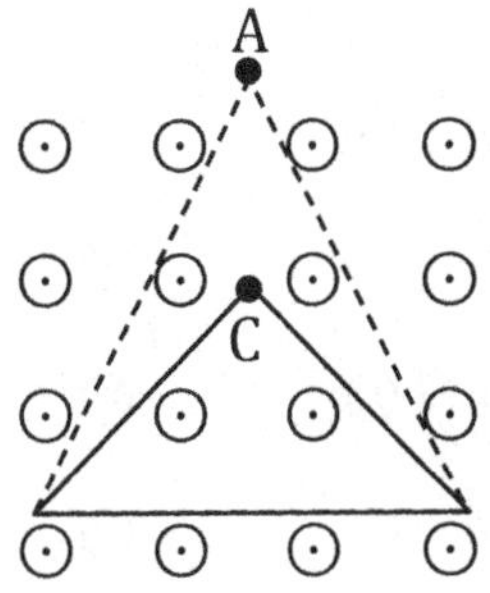

Solution. Apply the four steps of Lenz's law.

1. The **external** magnetic field ($\vec{\mathbf{B}}_{ext}$) is out of the page ($\odot$). It happens to already be drawn in the problem.
2. The magnetic flux (Φ_m) is **decreasing** because the area of the loop is getting smaller as the vertex of the triangle is pushed down from point A to point C. The formula for the area of a triangle is $A = \frac{1}{2}bh$, and the height (h) is getting shorter.
3. The **induced** magnetic field ($\vec{\mathbf{B}}_{ind}$) is out of the page ($\odot$). Since Φ_m is decreasing, the direction of $\vec{\mathbf{B}}_{ind}$ is the **same** as the direction of $\vec{\mathbf{B}}_{ext}$ from Step 1.
4. The **induced current** (I_{ind}) is counterclockwise, as drawn below. If you grab the loop with your thumb pointing counterclockwise and your fingers wrapped around the wire, inside the loop your fingers will come out of the page ($\odot$).

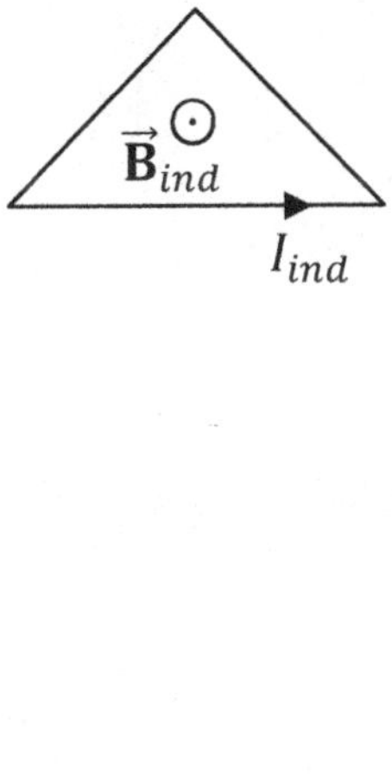

Example 113. The loop in the diagram below rotates with the right side of the loop coming out of the page. What is the direction of the current induced in the loop during this 90° rotation?

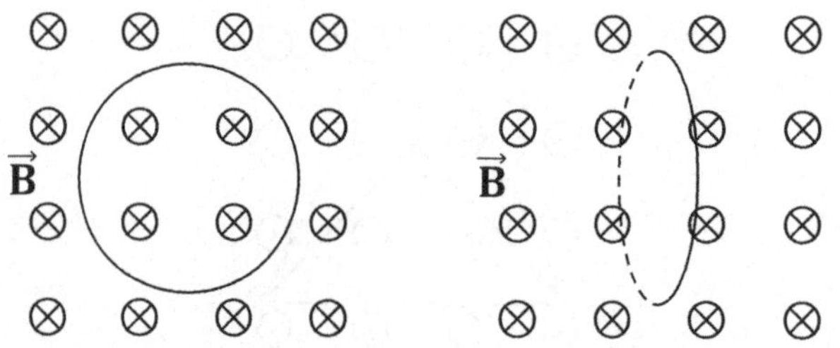

Solution. Apply the four steps of Lenz's law.

1. The **external** magnetic field ($\vec{\mathbf{B}}_{ext}$) is into the page ($\otimes$). It happens to already be drawn in the problem.
2. The magnetic flux (Φ_m) is **decreasing** because fewer magnetic field lines pass through the loop as it rotates. At the end of the described 90° rotation, the loop is vertical and the final magnetic flux is zero.
3. The **induced** magnetic field ($\vec{\mathbf{B}}_{ind}$) is into the page ($\otimes$). Since Φ_m is decreasing, the direction of $\vec{\mathbf{B}}_{ind}$ is the **same** as the direction of $\vec{\mathbf{B}}_{ext}$ from Step 1.
4. The **induced current** (I_{ind}) is clockwise, as drawn below. If you grab the loop with your thumb pointing clockwise and your fingers wrapped around the wire, inside the loop your fingers will go into the page ($\otimes$).

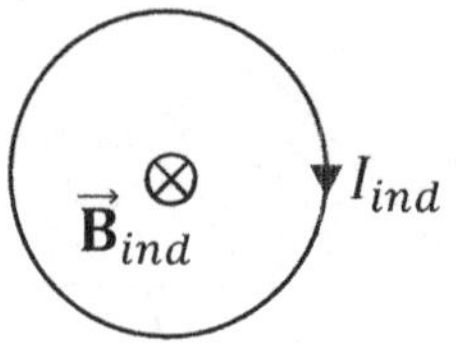

Example 114. The magnetic field in the diagram below rotates 90° clockwise. What is the direction of the current induced in the loop during this 90° rotation?

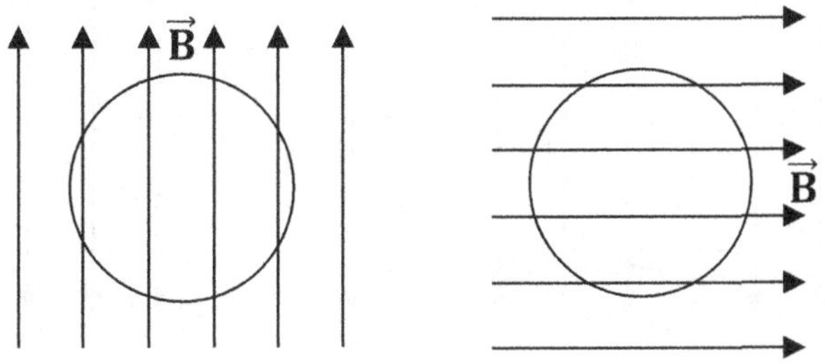

Solution. Apply the four steps of Lenz's law.

1. The **external** magnetic field ($\vec{\mathbf{B}}_{ext}$) is initially upward (↑). It happens to already be drawn in the problem.
2. The magnetic flux (Φ_m) is **constant** because the number of magnetic field lines passing through the loop doesn't change. The magnetic flux through the loop is zero at all times. This will be easier to see when you study Faraday's law in Chapter 29: The angle between the axis of the loop (which is perpendicular to the page) and the magnetic field (which remains in the plane of the page throughout the rotation) is 90° such that $\Phi_m = BA\cos\theta = BA\cos 90° = 0$ (since $\cos 90° = 0$).
3. The **induced** magnetic field ($\vec{\mathbf{B}}_{ind}$) is **zero** because the magnetic flux (Φ_m) through the loop is **constant**.
4. The **induced current** (I_{ind}) is also **zero** because the magnetic flux (Φ_m) through the loop is **constant**.

Example 115. The loop in the diagram below rotates with the top of the loop coming out of the page. What is the direction of the current induced in the loop during this 90° rotation?

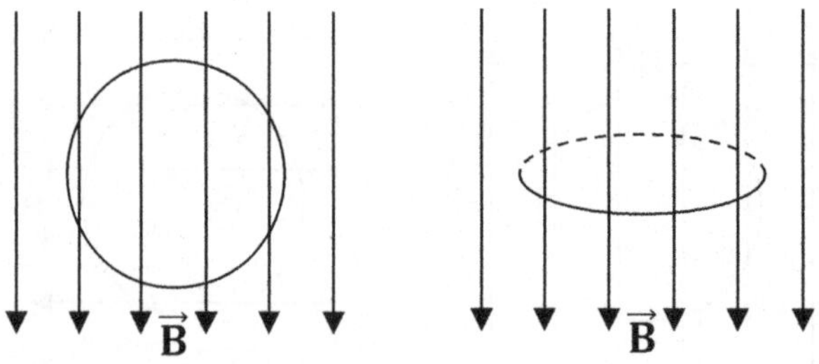

Solution. Apply the four steps of Lenz's law.

1. The **external** magnetic field ($\vec{\mathbf{B}}_{ext}$) is downward (↓). It happens to already be drawn in the problem.
2. The magnetic flux (Φ_m) is **increasing** because more magnetic field lines pass through the loop as it rotates. At the beginning of the described 90° rotation, the initial magnetic flux is zero. This will be easier to see when you study Faraday's law in Chapter 29: The angle between the axis of the loop and the magnetic field is initially 90° such that $\Phi_m = BA\cos\theta = BA\cos 90° = 0$ (since $\cos 90° = 0$). As the loop rotates, θ decreases from 90° to 0° and $\cos\theta$ increases from 0 to 1. Therefore, the magnetic flux increases during the described 90° rotation.
3. The **induced** magnetic field ($\vec{\mathbf{B}}_{ind}$) is upward (↑). Since Φ_m is increasing, the direction of $\vec{\mathbf{B}}_{ind}$ is **opposite** to the direction of $\vec{\mathbf{B}}_{ext}$ from Step 1.
4. The **induced current** (I_{ind}) runs to the right in the front of the loop (and therefore runs to the left in the back of the loop), as drawn below. Note that this loop is horizontal at the end of the described 90° rotation. If you grab the front of the loop with your thumb pointing right and your fingers wrapped around the wire, inside the loop your fingers will go up (↑).

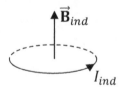

29 FARADAY'S LAW

Magnetic Flux	Change in Magnetic Flux
$\Phi_m = \int_S \vec{\mathbf{B}} \cdot d\vec{\mathbf{A}}$	$\Delta\Phi_m = \Phi_m - \Phi_{m0}$
Faraday's Law	**Average Emf Induced**
$\varepsilon_{ind} = -N\frac{d\Phi_m}{dt}$	$\varepsilon_{ave} = -N\frac{\Delta\Phi_m}{\Delta t}$
Motional Emf	**Relation to Induced Current**
$\varepsilon_{ind} = -B\ell v$	$\varepsilon_{ind} = I_{ind}R_{loops}$
Resistance	**Uniform Magnetic Field**
$R_{loops} = \frac{\rho\ell}{A_{wire}}$	$\Phi_m = BA\cos\theta$
Magnetic Field of a Long Straight Wire	**Magnetic Field Near the Center of a Solenoid**
$B = \frac{\mu_0 I}{2\pi r_c}$	$B = \frac{\mu_0 NI}{L} = \mu_0 nI$
Area of a Circle	**Area of a Triangle**
$A = \pi a^2$	$A = \frac{1}{2}bh$
Area of a Square	**Area of a Rectangle**
$A = L^2$	$A = LW$

Symbol	Name	Units
Φ_m	magnetic flux	$T{\cdot}m^2$ or Wb
B	magnitude of the external magnetic field	T
A	area of the loop	m^2
A_{wire}	area of the wire	m^2
θ	angle between $\vec{\mathbf{B}}$ and the axis of the loop, or the angle between $\vec{\mathbf{B}}$ and $d\vec{\mathbf{A}}$	° or rad
ε_{ind}	induced emf	V
I_{ind}	induced current	A
R_{loops}	resistance of the loops	Ω
ρ	resistivity	Ω·m
N	number of loops (or turns)	unitless
n	number of turns per unit length	$\frac{1}{m}$
t	time	s
μ_0	permeability of free space	$\frac{T{\cdot}m}{A}$
r_c	distance from a long, straight wire	m
a	radius of a loop	m
a_{wire}	radius of a wire	m
ℓ	length of the wire or length of conducting bar	m
L	length of a solenoid	m
v	speed	m/s
ω	angular speed	rad/s

Note: The symbol Φ is the uppercase Greek letter phi.

Permeability of Free Space
$\mu_0 = 4\pi \times 10^{-7} \dfrac{\text{T} \cdot \text{m}}{\text{A}}$

Converting Gauss to Tesla
$1 \text{ G} = 10^{-4} \text{ T}$

Symbol	Meaning
$\otimes$	into the page
$\odot$	out of the page

Example 116. A circular loop of wire with a diameter of 8.0 cm lies in the zx plane, centered about the origin. A uniform magnetic field of 2,500 G is oriented along the $+y$ direction. What is the magnetic flux through the area of the loop?

Solution. First convert the magnetic field from Gauss (G) to Tesla (T). The conversion factor is $1 \text{ G} = 10^{-4}$ T.

$$B = 2500 \text{ G} = 0.25 \text{ T}$$

We need to find the area of the loop before we can find the magnetic flux. Use the formula for the area of a circle. The radius is one-half the diameter: $a = \frac{D}{2} = \frac{8}{2} = 4.0 \text{ cm} = 0.040 \text{ m}$. (We're using a for radius to avoid possible confusion with resistance.)

$$A = \pi a^2 = \pi(0.04)^2 = 0.0016\pi \text{ m}^2$$

If you use a calculator, the area comes out to $A = 0.0050 \text{ m}^2$. Use the equation for the magnetic flux for a **uniform** magnetic field through a planar surface (since the circular loop lies in a plane - specifically, the zx plane). Note that the y-axis is the axis of the loop because the axis of a loop is defined as the line that is perpendicular to the loop and passing through the center of the loop. The magnetic field also lies on the y-axis. Since θ is the angle between the magnetic field and the axis of the loop, in this example $\theta = 0°$.

$$\Phi_m = BA\cos\theta = (0.25)(0.0016\pi)\cos 0° = (0.25)(0.0016\pi)(1) = 0.0004\pi \text{ T·m}^2$$

Recall from trig that $\cos 0° = 1$. The magnetic flux through the loop is $\Phi_m = 0.0004\pi \text{ T·m}^2$, which can also be expressed as $\Phi_m = 4\pi \times 10^{-4} \text{ T·m}^2$. If you use a calculator, it is $\Phi_m = 0.0013 \text{ T·m}^2 = 1.3 \times 10^{-3} \text{ T·m}^2$.

Example 117. The vertical loop illustrated below has the shape of a square with 2.0-m long edges. The uniform magnetic field shown makes a 30° angle with the plane of the square loop and has a magnitude of 7.0 T. What is the magnetic flux through the area of the loop?

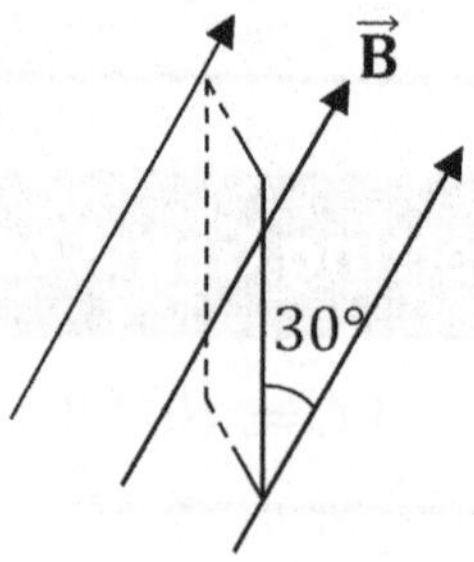

Solution. We need to find the area of the loop before we can find the magnetic flux. Use the formula for the area of a square.

$$A = L^2 = 2^2 = 4.0 \text{ m}^2$$

Use the equation for the magnetic flux for a **uniform** magnetic field through a planar surface (since the square loop lies in a plane). Recall that the axis of a loop is defined as the line that is perpendicular to the loop and passing through the center of the loop. Since the square loop is **vertical**, its axis is **horizontal** (see the dashed line below). Also, since the diagram above labels 30° as the angle between the magnetic field and the **vertical**, the angle we need in the magnetic flux equation is the complement to this: We need the angle between the magnetic field and the **horizontal**, which is $\theta = 60°$ (**not** 30°) as shown below.

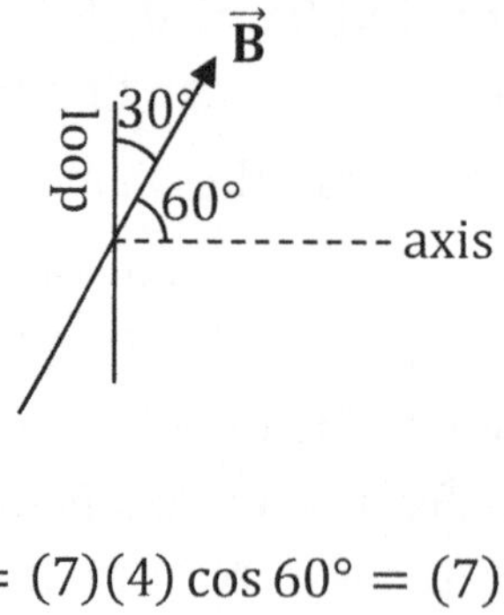

$$\Phi_m = BA\cos\theta = (7)(4)\cos 60° = (7)(4)\left(\frac{1}{2}\right) = 14 \text{ T}\cdot\text{m}^2$$

The magnetic flux through the loop is $\Phi_m = 14 \text{ T}\cdot\text{m}^2$.

Example 118. A rectangular loop of wire with a length of 50 cm, a width of 30 cm, and a resistance of 4.0 Ω lies in the xy plane. A magnetic field is oriented along the $+z$ direction and is uniform throughout the loop at any given moment. The magnetic field increases from 6,000 G to 8,000 G in 500 ms.

(A) Find the initial magnetic flux through the loop.

Solution. First convert the given quantities to SI units. Recall that $1 \text{ G} = 10^{-4}$ T.

$$L = 50 \text{ cm} = 0.50 \text{ m}$$
$$W = 30 \text{ cm} = 0.30 \text{ m}$$
$$B_0 = 6000 \text{ G} = 0.60 \text{ T}$$
$$B = 8000 \text{ G} = 0.80 \text{ T}$$
$$t = 500 \text{ ms} = 0.500 \text{ s}$$

We need to find the area of the loop before we can find the magnetic flux. Use the formula for the area of a rectangle.

$$A = LW = (0.5)(0.3) = 0.15 \text{ m}^2$$

Use the equation for the magnetic flux through a loop in a uniform magnetic field. Note that $\theta = \theta_0 = 0°$ because the z-axis is the axis of the loop (because the axis of a loop is defined as the line that is perpendicular to the loop and passing through the center of the loop) and the magnetic field lines run parallel to the z-axis. Recall from trig that $\cos 0° = 1$. Since the area of this loop isn't changing, both A_0 and A equal 0.15 m^2.

$$\Phi_{m0} = B_0 A_0 \cos\theta_0 = (0.6)(0.15)\cos 0° = 0.090 \text{ T·m}^2$$

The initial magnetic flux through the loop is $\Phi_{m0} = 0.090 \text{ T·m}^2$.

(B) Find the final magnetic flux through the loop.

Solution. Use the same equation that we used to find the initial magnetic flux. The only difference is that the final magnetic field is 0.80 T instead of 0.60 T.

$$\Phi_m = BA\cos\theta = (0.8)(0.15)\cos 0° = 0.120 \text{ T·m}^2$$

The final magnetic flux through the loop is $\Phi_m = 0.120 \text{ T·m}^2$.

(C) What is the average emf induced in the loop during this time?

Solution. Note that this problem asked for the **average** emf induced (as opposed to the instantaneous value of the emf). The distinction is that the formula for the average emf involves the change in magnetic flux over the elapsed time, whereas the formula for the instantaneous emf involves a derivative. Since the formula for average emf involves the **change** in the magnetic flux, we must first subtract the initial magnetic flux from the final magnetic flux.

$$\Delta\Phi_m = \Phi_m - \Phi_m = 0.120 - 0.090 = 0.030 \text{ T·m}^2$$

Use the equation for the average emf induced. There is just one loop in this example such that $N = 1$. Recall that $t = 0.500$ s.

$$\varepsilon_{ave} = -N\frac{\Delta\Phi_m}{\Delta t} = -(1)\frac{0.03}{0.5} = -0.060 \text{ V}$$

The average emf induced in the loop is $\varepsilon_{ave} = -0.060$ V, which equates to $\varepsilon_{ave} = -60$ mV.

(D) What is the average current induced in the loop during this time?

Solution. Apply Ohm's law.

$$\varepsilon_{ave} = I_{ave}R_{loop}$$

Divide both sides of the equation by the resistance of the loop.

$$I_{ave} = \frac{\varepsilon_{ave}}{R_{loop}} = -\frac{0.06}{4} = -0.015 \text{ A}$$

The average current induced in the loop is $I_{ave} = -0.015$ A, which can also be expressed as $I_{ave} = -15$ mA, where the metric prefix milli (m) stands for 10^{-3}.

Example 119. The top vertex of the triangular loops of wire illustrated below is pushed up in 250 ms. The base of the triangle is 50 cm, the initial height is 25 cm, and the final height is 50 cm. The magnetic field has a magnitude of 4000 G. There are 200 loops.

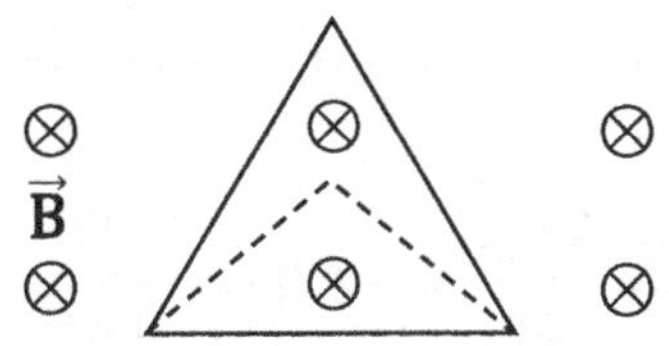

(A) Find the initial magnetic flux through each loop.

Solution. First convert the given quantities to SI units. Recall that $1 \text{ G} = 10^{-4}$ T.

$$b = 50 \text{ cm} = 0.50 \text{ m}$$
$$h_0 = 25 \text{ cm} = 0.25 \text{ m}$$
$$h = 50 \text{ cm} = 0.50 \text{ m}$$
$$B = 4000 \text{ G} = 0.40 \text{ T}$$
$$t = 250 \text{ ms} = 0.250 \text{ s}$$

We need to find the area of the loop before we can find the magnetic flux. Use the formula for the area of a triangle: one-half of the base (b) times the height (h).

$$A_0 = \frac{1}{2}bh_0 = \frac{1}{2}(0.5)(0.25) = 0.0625 \text{ m}^2$$

Use the equation for the magnetic flux through a loop in a uniform magnetic field. Note that $\theta = \theta_0 = 0°$ because the axis of a loop is defined as the line that is perpendicular to the loop and passing through the center of the loop, which is parallel to the magnetic field lines. Recall from trig that $\cos 0° = 1$.

$$\Phi_{m0} = B_0 A_0 \cos\theta_0 = (0.4)(0.0625)\cos 0° = (0.4)(0.0625)(1) = 0.025 \text{ T·m}^2$$

The initial magnetic flux through each loop is $\Phi_{m0} = 0.025$ T·m^2.

(B) Find the final magnetic flux through each loop.

Solution. The initial and final areas are different since the height of the triangle changes.

$$A = \frac{1}{2}bh = \frac{1}{2}(0.5)(0.5) = 0.125 \text{ m}^2$$

Use the same equation that we used to find the initial magnetic flux. The only difference is that the area is 0.125 m^2 instead of 0.0625 m^2. In this problem, the magnetic field remains constant (but the magnetic **flux** changes).

$$\Phi_m = BA\cos\theta = (0.4)(0.125)\cos 0° = (0.4)(0.125)(1) = 0.050 \text{ T·m}^2$$

The final magnetic flux through each loop is $\Phi_m = 0.050$ T·m^2.

(C) What is the average emf induced in the loop during this time?

Solution. Since the formula for average emf involves the **change** in the magnetic flux, we must first subtract the initial magnetic flux from the final magnetic flux.

$$\Delta\Phi_m = \Phi_m - \Phi_m = 0.050 - 0.025 = 0.025 \text{ T·m}^2$$

Use the equation for the average emf induced. There are $N = 200$ loops in this example. Recall that $t = 0.250$ s.

$$\varepsilon_{ave} = -N\frac{\Delta\Phi_m}{\Delta t} = -(200)\frac{0.025}{0.25} = -20 \text{ V}$$

The average emf induced in the loops is $\varepsilon_{ave} = -20$ V.

(D) Find the direction of the induced current.

Solution. Apply the four steps of Lenz's law (Chapter 28).

1. The **external** magnetic field ($\vec{\mathbf{B}}_{ext}$) is into the page (⊗). It happens to already be drawn in the problem.
2. The magnetic flux (Φ_m) is **increasing** because the area of the loop is getting larger as the vertex of the triangle is pushed up. The formula for the area of a triangle is $A = \frac{1}{2}bh$, and the height (h) is getting taller.
3. The **induced** magnetic field ($\vec{\mathbf{B}}_{ind}$) is out of the page (⊙). Since Φ_m is increasing, the direction of $\vec{\mathbf{B}}_{ind}$ is **opposite** to the direction of $\vec{\mathbf{B}}_{ext}$ from Step 1.
4. The **induced current** (I_{ind}) is counterclockwise, as drawn below. If you grab the loop with your thumb pointing counterclockwise and your fingers wrapped around the wire, inside the loop your fingers will come out of the page (⊙). Remember, you want your fingers to match Step 3 inside the loop (<u>**not**</u> Step 1).

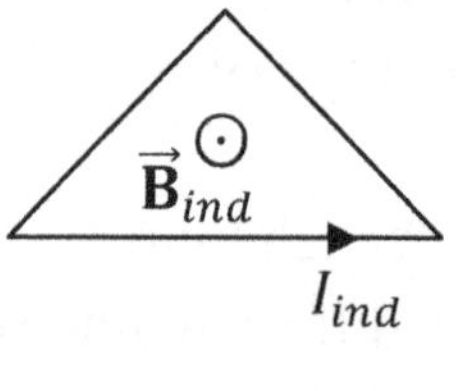

The **induced current** (I_{ind}) runs **counterclockwise** through the loops, as shown above.

Example 120. Consider the loop in the figure below, which initially has the shape of a square. Each side is 2.0 m long and has a resistance of 5.0 Ω. There is a uniform magnetic field of $5000\sqrt{3}$ G directed into the page. The loop is hinged at each vertex. Monkeys pull corners K and M apart until corners L and N are 2.0 m apart, changing the shape to that of a rhombus. The duration of this process is $(2 - \sqrt{3})$ s.

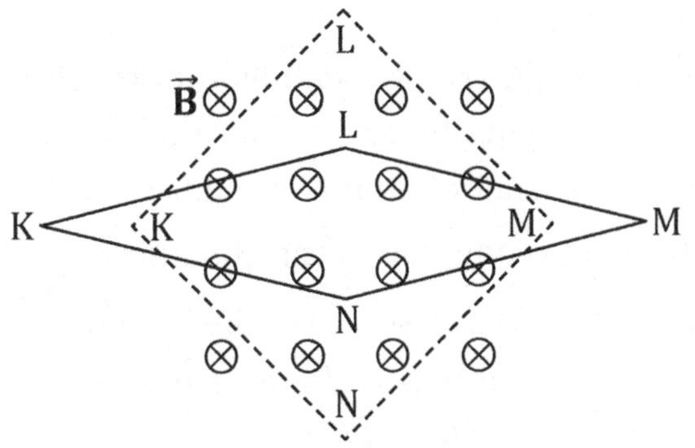

(A) Find the initial magnetic flux through the loop.

Solution. First convert the magnetic field to Tesla using $1 \text{ G} = 10^{-4}$ T.

$$B = 5000\sqrt{3} \text{ G} = 0.5\sqrt{3} \text{ T} = \frac{\sqrt{3}}{2} \text{ T}$$

Note that $5000 \times 10^{-4} = 0.5 = \frac{1}{2}$. If you use a calculator, the magnetic field works out to $B = 0.87$ T. We need to find the area of the loop before we can find the magnetic flux. Use the formula for the area of a square, since the loop initially has the shape of a square with 2.0-m long edges.

$$A_0 = L^2 = 2^2 = 4.0 \text{ m}^2$$

Use the equation for the magnetic flux through a loop in a uniform magnetic field. Note that $\theta = \theta_0 = 0°$ because the axis of a loop is defined as the line that is perpendicular to the loop and passing through the center of the loop, which is parallel to the magnetic field lines. Recall from trig that $\cos 0° = 1$.

$$\Phi_{m0} = B_0 A_0 \cos\theta_0 = \left(\frac{\sqrt{3}}{2}\right)(4)\cos 0° = \left(\frac{\sqrt{3}}{2}\right)(4)(1) = 2\sqrt{3} \text{ T·m}^2$$

The initial magnetic flux through the loop is $\Phi_{m0} = 2\sqrt{3}$ T·m². If you use a calculator, this works out to $\Phi_{m0} = 3.5$ T·m².

(B) Find the final magnetic flux through the loop.

Solution. The initial and final areas are different since the shape of the loop changes from a square to a rhombus. (You don't need to guess about this: We will do the math and calculate the area of the rhombus, and then you will see that it is different from the area of the square.) It isn't necessary to look up the formula for the area of a rhombus: We can simply divide the rhombus up into four identical triangles, as shown on the next page.

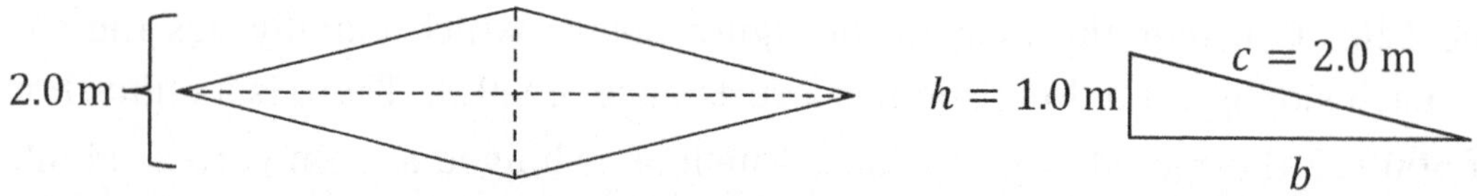

We will determine the area of the triangle (A_{tri}), and then multiply this times 4 in order to obtain the area of the rhombus, since the rhombus is made up of 4 identical triangles. Let's find the base and height of the right triangle:

- The hypotenuse of each right triangle equals the length of each side of the square: $c = 2.0$ m. However, the hypotenuse is neither the base nor the height.
- The problem states that corners L and N are 2.0 m apart, and as shown above, the height of the right triangle is one-half of this distance: $h = \frac{2}{2} = 1.0$ m.
- Apply the Pythagorean theorem to solve for the base of each right triangle. The base of each right triangle comes out to $b = \sqrt{3}$ m, as worked out below.

$$b^2 + h^2 = c^2$$
$$b^2 = c^2 - h^2$$
$$b = \sqrt{c^2 - h^2} = \sqrt{2^2 - 1^2} = \sqrt{4 - 1} = \sqrt{3}\text{ m}$$

Plug the base ($b = \sqrt{3}$ m) and height ($h = 1.0$ m) into the formula for the area of a triangle.

$$A_{tri} = \frac{1}{2}bh = \frac{1}{2}(\sqrt{3})(1) = \frac{\sqrt{3}}{2}\text{ m}^2$$

The area of the rhombus is 4 times the area of one of the right triangles.

$$A = 4A_{tri} = 4\left(\frac{\sqrt{3}}{2}\right) = 2\sqrt{3}\text{ m}^2$$

If you use a calculator, the area of the rhombus works out to $A = 3.5\text{ m}^2$. Use the same equation that we used to find the initial magnetic flux. The only difference is that the area is $2\sqrt{3}\text{ m}^2$ instead of 4.0 m^2. In this problem, the magnetic field remains constant (but the magnetic **flux** changes). Recall that the magnetic field is $B = \frac{\sqrt{3}}{2}$ T.

$$\Phi_m = BA\cos\theta = \left(\frac{\sqrt{3}}{2}\right)(2\sqrt{3})\cos 0° = \left(\frac{\sqrt{3}}{2}\right)(2\sqrt{3})(1) = 3.0\text{ T·m}^2$$

Note that $\left(\frac{1}{2}\right)(2) = 1$ and that $\sqrt{3}\sqrt{3} = 3$. The final magnetic flux through the loop is $\Phi_m = 3.0\text{ T·m}^2$.

(C) What is the average current induced in the loop during this time?

Solution. Before we can find the induced current, we must apply Faraday's law in order to determine the induced emf. Since the formula for average emf involves the **change** in the magnetic flux, we must first subtract the initial magnetic flux from the final magnetic flux.

$$\Delta\Phi_m = \Phi_m - \Phi_m = (3 - 2\sqrt{3})\text{ T·m}^2$$

Use the equation for the average emf induced. There is just one loop in this example such that $N = 1$. Recall that time is given in the problem as $t = \left(2 - \sqrt{3}\right)$ s.

$$\varepsilon_{ave} = -N\frac{\Delta\Phi_m}{\Delta t} = -(1)\frac{3 - 2\sqrt{3}}{2 - \sqrt{3}} = -\frac{3 - 2\sqrt{3}}{2 - \sqrt{3}}$$

If you're not using a calculator, there is a "trick" to the math. First we will **distribute** the minus sign: Note that $-\left(3 - 2\sqrt{3}\right) = -3 + 2\sqrt{3}$, which is the same thing as $2\sqrt{3} - 3$.

$$\varepsilon_{ave} = -\frac{3 - 2\sqrt{3}}{2 - \sqrt{3}} = \frac{2\sqrt{3} - 3}{2 - \sqrt{3}}$$

Next we will multiply the numerator and denominator both by $\sqrt{3}$.

$$\varepsilon_{ave} = \frac{2\sqrt{3} - 3}{2 - \sqrt{3}}\left(\frac{\sqrt{3}}{\sqrt{3}}\right) = \frac{\left(2\sqrt{3} - 3\right)\left(\sqrt{3}\right)}{\left(2 - \sqrt{3}\right)\left(\sqrt{3}\right)}$$

Now we will **distribute** the $\sqrt{3}$ in the denominator only. Recall from algebra that $\sqrt{3}\sqrt{3} = 3$.

$$\varepsilon_{ave} = \frac{\left(2\sqrt{3} - 3\right)\left(\sqrt{3}\right)}{\left(2 - \sqrt{3}\right)\left(\sqrt{3}\right)} = \frac{\left(2\sqrt{3} - 3\right)\left(\sqrt{3}\right)}{2\sqrt{3} - \sqrt{3}\sqrt{3}} = \frac{\left(2\sqrt{3} - 3\right)\left(\sqrt{3}\right)}{2\sqrt{3} - 3}$$

It's like "magic": The $2\sqrt{3} - 3$ in the numerator cancels the $2\sqrt{3} - 3$ in the denominator, so all that's left is $\sqrt{3}$.

$$\varepsilon_{ave} = \frac{\left(2\sqrt{3} - 3\right)\left(\sqrt{3}\right)}{2\sqrt{3} - 3} = \sqrt{3}\text{ V}$$

The average emf induced in the loops is $\varepsilon_{ave} = \sqrt{3}$ V. If you use a calculator, $\varepsilon_{ave} = 1.7$ V.

Apply Ohm's law in order to determine the induced current.

$$\varepsilon_{ave} = I_{ave}R_{loop}$$

Divide both sides of the equation by the resistance of the loop. Note that $R_{loop} = 5 + 5 + 5 + 5 = 20\ \Omega$ (since the loop has 4 sides and each side has a resistance of 5.0 Ω).

$$I_{ave} = \frac{\varepsilon_{ave}}{R_{loop}} = \frac{\sqrt{3}}{20}\text{ A}$$

The average current induced in the loop is $I_{ave} = \frac{\sqrt{3}}{20}$ A. If you use a calculator, it comes out to $I_{ave} = 0.087$ A, which can also be expressed as $I_{ave} = 87$ mA since the metric prefix milli (m) stands for $\text{m} = 10^{-3}$.

(D) Find the direction of the induced current.

Solution. Apply the four steps of Lenz's law (Chapter 28).

1. The **external** magnetic field ($\vec{\mathbf{B}}_{ext}$) is into the page (⊗). It happens to already be drawn in the problem.
2. The magnetic flux (Φ_m) is **decreasing** because the area of the loop is getting smaller. One way to see this is to compare $A = 2\sqrt{3}\text{ m}^2 = 3.5\text{ m}^2$ to $A_0 = 4.0\text{ m}^2$. Another way is to imagine the extreme case where points L and N finally touch, for which the

area will equal zero. Either way, the area of the loop is getting smaller.

3. The **induced** magnetic field ($\vec{\mathbf{B}}_{ind}$) is into the page ($\otimes$). Since Φ_m is decreasing, the direction of $\vec{\mathbf{B}}_{ind}$ is the **same** as the direction of $\vec{\mathbf{B}}_{ext}$ from Step 1.
4. The **induced current** (I_{ind}) is clockwise, as drawn below. If you grab the loop with your thumb pointing clockwise and your fingers wrapped around the wire, inside the loop your fingers will go into the page ($\otimes$).

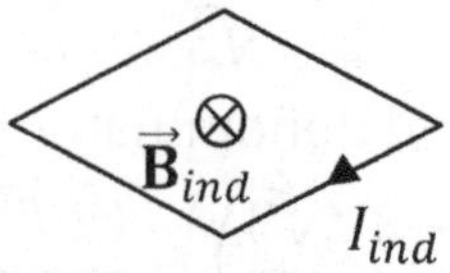

The **induced current** (I_{ind}) runs **clockwise** through the loop, as shown above.

Example 121. A solenoid that is not connected to any power supply has 300 turns, a length of 18 cm, and a radius of 4.0 cm. A uniform magnetic field with a magnitude of 500,000 G is initially oriented along the axis of the solenoid. The axis of the solenoid rotates through an angle of 30° relative to the magnetic field with a constant angular speed of 20 rad/s.

(A) What instantaneous emf is induced across the solenoid as the angle reaches 30°?
Solution. First convert the radius and magnetic field to SI units. Recall that $1 \text{ G} = 10^{-4}$ T. Also convert $\Delta\theta$ from 30° to radians using $180° = \pi$ rad in order to be consistent with the units of the angular speed (which is given as $\omega = 20$ rad/s).

$$a = 4.0 \text{ cm} = 0.040 \text{ m}$$

$$B = 500{,}000 \text{ G} = 50 \text{ T}$$

$$\Delta\theta = 30° = 30° \times \frac{\pi \text{ rad}}{180°} = \frac{\pi}{6} \text{ rad}$$

(We're using a for radius in order to avoid possible confusion with resistance.) Note that the length of the solenoid (18 cm) is **irrelevant** to the solution. Just ignore it. (In the lab, you can measure almost anything you want, such as the temperature, so it's a valuable skill to be able to tell which quantities you do or don't need.)

We need to find the area of each loop before we can find the magnetic flux. Use the formula for the area of a circle, since a solenoid has circular loops.

$$A = \pi a^2 = \pi(0.04)^2 = 0.0016\pi \text{ m}^2$$

If you use a calculator, the area works out to $A = 0.0050 \text{ m}^2$.

Apply Faraday's law. For **instantaneous** emf, this involves a **derivative**.

$$\varepsilon_{ind} = -N\frac{d\Phi_m}{dt}$$

Plug in the expression $\Phi_m = BA\cos\theta$ for a **uniform** magnetic field.

$$\varepsilon_{ind} = -N\frac{d\Phi_m}{dt} = -N\frac{d(BA\cos\theta)}{dt}$$

Both A and B are constants in this example: Neither the area nor the magnetic field is changing. (However, the magnetic **flux** is changing.) Therefore, we may factor BA out of the derivative.

$$\varepsilon_{ind} = -NBA\frac{d(\cos\theta)}{dt}$$

Apply the **chain rule,** $\frac{df}{dx} = \frac{df}{du}\frac{du}{dx}$, identifying $f = \cos\theta, u = \theta$, and $x = t$.

$$\varepsilon_{ind} = -NBA\frac{df}{du}\frac{du}{dx} = -NBA\frac{d(\cos\theta)}{d\theta}\frac{d\theta}{dt}$$

Note that a derivative of angle with respect to time is angular velocity: $\omega = \frac{d\theta}{dt}$. (This is analogous to how a derivative of position with respect to time equals velocity.)

$$\varepsilon_{ind} = -NBA\omega\frac{d(\cos\theta)}{d\theta}$$

Recall from calculus that the derivative of cosine with respect to theta is $\frac{d\cos\theta}{d\theta} = -\sin\theta$.

$$\varepsilon_{ind} = -NBA\omega(-\sin\theta)$$

The two minus signs make a plus sign.

$$\varepsilon_{ind} = NBA\omega\sin\theta$$

Plug in the known values. Note that $N = 300$, $B = 50$ T, $\omega = 20$ rad/s, and we previously found $A = 0.0016\pi$ m^2 $= 0.0050$ m^2 and $\Delta\theta = \frac{\pi}{6}$ rad. If you're using a calculator, be sure that it is set in radians mode rather than degrees mode. (You could simply use $\Delta\theta = 30°$ instead of $\Delta\theta = \frac{\pi}{6}$ rad and then leave your calculator in degrees mode, but in part (B) we'll need to use radians instead of degrees.) Recall from trig that $\sin\left(\frac{\pi}{6}\right) = \frac{1}{2}$.

$$\varepsilon_{ind} = (300)(50)(0.0016\pi)(20)\sin\left(\frac{\pi}{6}\right) = (300)(50)(0.0016\pi)(20)\left(\frac{1}{2}\right) = 240\pi \text{ V}$$

The **instantaneous** emf induced when θ reaches 30° is $\varepsilon_{ind} = 240\pi$ V. Using a calculator, this comes out to $\varepsilon_{ind} = 754$ V.

(B) What average emf is induced during the 30° rotation?

Solution. The formula for **average** emf involves a **ratio**, whereas the formula for **instantaneous** emf involves a **derivative.** Unlike part (A), for **average** emf, we will need to determine both the initial and final magnetic flux. Note that $\theta_0 = 0°$, whereas $\Delta\theta = \frac{\pi}{6}$ rad. Only the angle changes: Both B and A are constant in this example.

$$\Phi_{m0} = BA\cos\theta_0 = (50)(0.0016\pi)\cos(0) = (50)(0.0016\pi)(1) = 0.08\pi \text{ T·m}^2$$

Recall from trig that $\cos(0) = 1$. The initial magnetic flux is $\Phi_{m0} = 0.08\pi$ T·m^2. Using a calculator, $\Phi_{m0} = 0.251$ T·m^2.

$$\Phi_m = BA\cos\theta = (50)(0.0016\pi)\cos\left(\frac{\pi}{6}\right) = (50)(0.0016\pi)\left(\frac{\sqrt{3}}{2}\right) = 0.04\pi\sqrt{3}\text{ T}\cdot\text{m}^2$$

Recall from trig that $\cos\left(\frac{\pi}{6}\right) = \frac{\sqrt{3}}{2}$. The final magnetic flux is $\Phi_m = 0.04\pi\sqrt{3}\text{ T}\cdot\text{m}^2$. Using a calculator, $\Phi_m = 0.218\text{ T}\cdot\text{m}^2$.

Subtract the initial magnetic flux from the final magnetic flux in order to determine the change in the magnetic flux during the 30° rotation.

$$\Delta\Phi_m = \Phi_m - \Phi_{m0} = 0.04\pi\sqrt{3} - 0.08\pi$$

If you're not using a calculator, it's convenient to **factor** out the 0.04π.

$$\Delta\Phi_m = 0.04\pi\left(\sqrt{3} - 2\right)\text{ T}\cdot\text{m}^2$$

The change in the magnetic flux is $\Delta\Phi_m = 0.04\pi\left(\sqrt{3} - 2\right)\text{ T}\cdot\text{m}^2$. If you use a calculator, this works out to $\Delta\Phi_m = -0.034\text{ T}\cdot\text{m}^2$. It's negative because $\sqrt{3} \approx 1.73$ is less than 2.

Before we can use the formula for average emf, we also need to determine the time interval (Δt). Since the angular speed is constant, we can set the angular speed equal to the change in the angle divided by the time interval. (This is analogous to the equation for constant speed, where speed equals distance over time.)

$$\omega = \frac{\Delta\theta}{\Delta t}$$

Multiply both sides of the equation by Δt.

$$\omega\Delta t = \Delta\theta$$

Divide both sides of the equation by ω in order to solve for the time interval. The following equation is where it's important to use radians (**<u>not</u>** degrees). Recall that the angular speed is given in the problem: $\omega = 20$ rad/s. Also recall that we found $\Delta\theta = \frac{\pi}{6}$ rad in part (A).

$$\Delta t = \frac{\Delta\theta}{\omega} = \frac{\frac{\pi}{6}}{20}$$

To divide a fraction by 20, multiply by the **reciprocal** of 20, which is $\frac{1}{20}$.

$$\Delta t = \frac{\pi}{6} \div 20 = \frac{\pi}{6} \times \frac{1}{20} = \frac{\pi}{120}\text{ s}$$

The time interval is $\Delta t = \frac{\pi}{120}$ s. Using a calculator, this comes out to $\Delta t = 0.026$ s.

Plug $\Delta\Phi_m = 0.04\pi\left(\sqrt{3} - 2\right)\text{ T}\cdot\text{m}^2$ and $\Delta t = \frac{\pi}{120}$ s into the equation for average emf. There are $N = 300$ loops (called turns) in this example.

$$\varepsilon_{ave} = -N\frac{\Delta\Phi_m}{\Delta t} = -(300)\frac{0.04\pi\left(\sqrt{3} - 2\right)}{\pi/120}$$

To divide by a fraction, multiply by its **reciprocal**.

$$\varepsilon_{ave} = -(300)(0.04\pi)(\sqrt{3}-2) \div \frac{\pi}{120} = -(300)(0.04\pi)(\sqrt{3}-2) \times \frac{120}{\pi}$$

Note that the π's cancel.

$$\varepsilon_{ave} = -(300)(0.04\pi)(\sqrt{3}-2)(120) = -1440(\sqrt{3}-2)$$

If you're not using a calculator, let's **distribute** the minus sign: $-(\sqrt{3}-2) = 2-\sqrt{3}$.

$$\varepsilon_{ave} = 1440(2-\sqrt{3})\text{ V}$$

The **average** emf induced from 0° to 30° is $\varepsilon_{ave} = 1440(2-\sqrt{3})$ V. If you use a calculator, this comes out to $\varepsilon_{ave} = 386$ V.

(C) Why should our answer for the "average" emf that we found in part (B) be about one-half of the "instantaneous" emf that we found in part (A)?

Solution. Consider the instantaneous emf again, which we found in part (A) using a derivative: $\varepsilon_{ind} = NBA\omega \sin\theta$. We found that $\varepsilon_{ind} = 754$ V when θ reaches 30°. According to the equation $\varepsilon_{ind} = NBA\omega \sin\theta$, the instantaneous emf is initially **zero**: When θ is 0°, we get $\varepsilon_{ind} = 0$ (since $\sin 0° = 0$).

Think about this: The **instantaneous** emf grows from 0 to 754 as θ varies from 0° to 30°. We should expect the **average** to be about $\frac{0+754}{2} = \frac{754}{2} = 357$ V. Indeed, the average emf turned out to be $\varepsilon_{ave} = 386$ V in part (B). (The reason it isn't "exactly" 357 V is because the emf is a **non-linear** function of θ.)

Example 122. In the diagram below, a conducting bar with a length of 12 cm slides to the right with a constant speed of 3.0 m/s along the rails of a bare U-channel conductor in the presence of a uniform magnetic field of 25 T. The conducting bar has a resistance of 3.0 Ω and the U-channel conductor has negligible resistance.

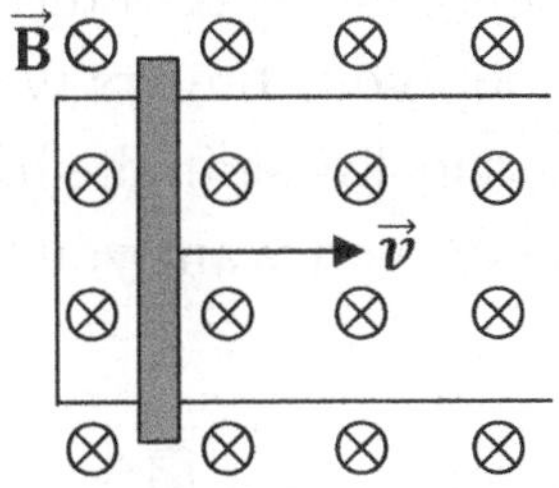

(A) What emf is induced across the ends of the conducting bar?

Solution. First convert the length to meters.

$$\ell = 12\text{ cm} = 0.12\text{ m}$$

Use the equation for motional emf.

$$\varepsilon_{ind} = -B\ell v = -(25)(0.12)(3) = -9.0\text{ V}$$

The emf induced across the conducting bar is $\varepsilon_{ind} = -9.0$ V.

(B) How much current is induced in the loop?

Solution. Apply Ohm's law.

$$\varepsilon_{ave} = I_{ave} R_{loop}$$

Divide both sides of the equation by the resistance of the loop.

$$I_{ave} = \frac{\varepsilon_{ave}}{R_{loop}} = -\frac{9}{3} = -3.0 \text{ A}$$

The average current induced in the loop is $I_{ave} = -3.0$ A.

(C) Find the direction of the induced current.

Solution. Apply the four steps of Lenz's law (Chapter 28).

1. The **external** magnetic field ($\vec{\mathbf{B}}_{ext}$) is into the page (⊗). It happens to already be drawn in the problem.
2. The magnetic flux (Φ_m) is **increasing** because the area of the loop is getting larger as the conducting bar travels to the right. Note that the conducting bar makes electrical contract where it touches the bare U-channel conductor. The dashed (---) line below illustrates how the area of the loop is getting larger as the conducting bar travels to the right.

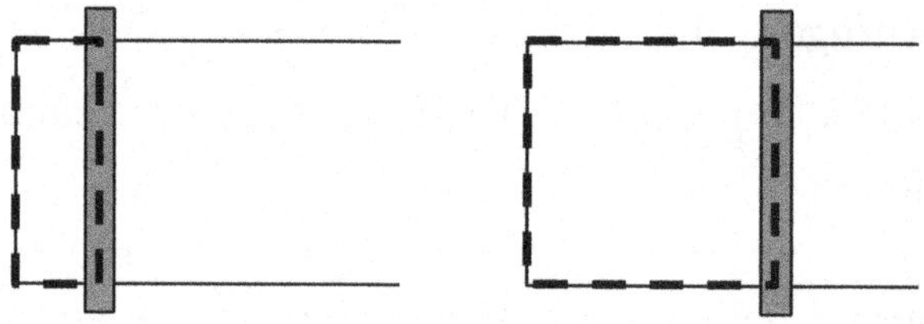

3. The **induced** magnetic field ($\vec{\mathbf{B}}_{ind}$) is out of the page (⊙). Since Φ_m is increasing, the direction of $\vec{\mathbf{B}}_{ind}$ is **opposite** to the direction of $\vec{\mathbf{B}}_{ext}$ from Step 1.
4. The **induced current** (I_{ind}) is counterclockwise, as drawn on the next page. If you grab the loop with your thumb pointing counterclockwise and your fingers wrapped around the wire, inside the loop your fingers will come out of the page (⊙). Since the induced current is counterclockwise in the loop, the induced current runs up (↑) the conducting bar. Remember, you want your fingers to match Step 3 inside the loop (<u>**not**</u> Step 1).

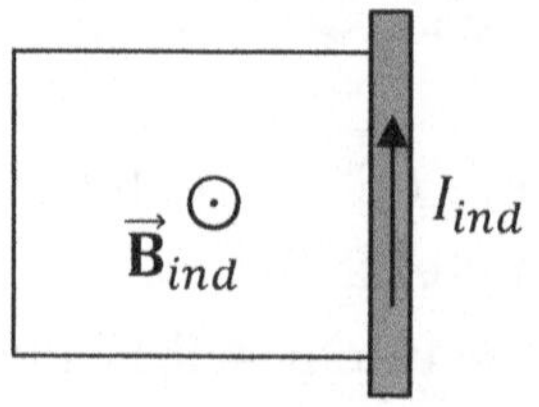

The **induced current** (I_{ind}) runs **counterclockwise** through the loop, as shown above.

Example 123. The conducting bar illustrated below has a length of 8.0 cm and travels with a constant speed of 25 m/s through a uniform magnetic field of 40,000 G. What emf is induced across the ends of the conducting bar?

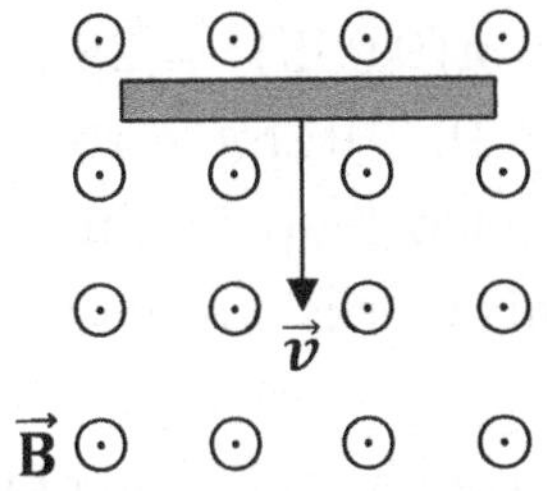

Solution. First convert the length and magnetic field to SI units. Recall that 1 G = 10^{-4} T.

$$\ell = 8.0 \text{ cm} = 0.080 \text{ m}$$
$$B = 40{,}000 \text{ G} = 4.0 \text{ T}$$

Use the equation for motional emf.

$$\varepsilon_{ind} = -B\ell v = -(4)(0.08)(25) = -8.0 \text{ V}$$

The emf induced across the conducting bar is $\varepsilon_{ind} = -8.0$ V.

Example 124. A solenoid has 3,000 turns, a length of 50 cm, and a radius of 5.0 mm. The current running through the solenoid is increased from 3.0 A to 7.0 A in 16 ms. A second solenoid coaxial with the first solenoid has 2,000 turns, a length of 20 cm, a radius of 5.0 mm, and a resistance of 5.0 Ω. The second solenoid is not connected to the first solenoid or any power supply.

(A) What initial magnetic field does the first solenoid create inside of its coils?

Solution. Use subscripts 1 and 2 to distinguish between the two different solenoids.

- $N_1 = 3000$, $L_1 = 0.50$ m, and $a_1 = 0.0050$ m for the first solenoid.
- $N_2 = 2000$, $L_2 = 0.20$ m, and $a_2 = 0.0050$ m for the second solenoid.
- $I_{10} = 3.0$ A and $I_1 = 7.0$ A for the initial and final current in the first solenoid.
- $R_2 = 5.0\ \Omega$ for the resistance of the second solenoid.
- $\Delta t = 0.016$ s since the metric prefix milli (m) stands for 10^{-3}.

Note that this question specifies magnetic **field** (B). We're <u>**not**</u> finding magnetic **flux** yet. Use the equation for the magnetic field created near the center of the first solenoid to find the initial magnetic field created by the first solenoid. (You can find this equation on page 341, or in Chapter 25.) Recall that the permeability of free space is $\mu_0 = 4\pi \times 10^{-7} \frac{\text{T·m}}{\text{A}}$.

$$B_{10} = \frac{\mu_0 N_1 I_{10}}{L_1} = \frac{(4\pi \times 10^{-7})(3000)(3)}{0.5} = 72\pi \times 10^{-4} \text{ T}$$

The initial magnetic field created by the first solenoid is $B_{10} = 72\pi \times 10^{-4}$ T. If you use a calculator, this works out to $B_{10} = 0.023$ T.

(B) What is the initial magnetic flux through each loop of the second solenoid?
Solution. Use the equation for magnetic flux, where $\theta = 0°$ and A_2 is the area of the loops for the **second** solenoid. Use the formula for the area of a circle since a solenoid has circular loops. Recall from trig that $\cos 0° = 1$.

$$A_2 = \pi a_2^2 = \pi(0.005)^2 = 25\pi \times 10^{-6} \text{ m}^2$$

$$\Phi_{2m0} = B_{10} A_2 \cos\theta = (72\pi \times 10^{-4})(25\pi \times 10^{-6}) \cos 0° = 18\pi^2 \times 10^{-8} \text{ T·m}^2$$

The initial magnetic flux through each loop of the second solenoid is $\Phi_{2m0} = 18\pi^2 \times 10^{-8}$ T·m^2. If you use a calculator, this works out to $\Phi_{2m0} = 1.8 \times 10^{-6}$ T·m^2.

(C) What is the final magnetic flux through each loop of the second solenoid?
Solution. First find the final magnetic field created by the first solenoid. This is the same as part (A), except for using the final current ($I_1 = 7.0$ A) instead of the initial current ($I_{10} = 3.0$ A).

$$B_1 = \frac{\mu_0 N_1 I_1}{L_1} = \frac{(4\pi \times 10^{-7})(3000)(7)}{0.5} = 168\pi \times 10^{-4} \text{ T}$$

If you use a calculator, the final magnetic field comes out to $B_1 = 0.053$ T. Plug the final magnetic field into the equation for magnetic flux.

$$\Phi_{2m} = B_1 A_2 \cos\theta = (168\pi \times 10^{-4})(25\pi \times 10^{-6}) \cos 0° = 42\pi^2 \times 10^{-8} \text{ T·m}^2$$

The final magnetic flux through each loop of the second solenoid is $\Phi_{2m} = 42\pi^2 \times 10^{-8}$ T·m^2. If you use a calculator, this works out to $\Phi_{2m} = 4.1 \times 10^{-6}$ T·m^2.

(D) What is the average current induced in the second solenoid during this time?
Solution. First subtract the initial magnetic flux from the final magnetic flux in order to determine the change in the magnetic flux.

$$\Delta\Phi_{2m} = \Phi_{2m} - \Phi_{2m0} = 42\pi^2 \times 10^{-8} - 18\pi^2 \times 10^{-8} = 24\pi^2 \times 10^{-8} \text{ T·m}^2$$

If you use a calculator,* this works out to $\Delta\Phi_{2m} = 2.4 \times 10^{-6}$ T·m^2. Plug the change in the magnetic flux into the equation for average emf. Be sure to use $N_2 = 2000$ (**not** $N_1 = 3000$) because we want the emf induced in the **second** solenoid.

$$\varepsilon_{ave} = -N_2 \frac{\Delta\Phi_{2m}}{\Delta t} = -(2000)\frac{24\pi^2 \times 10^{-8}}{0.016} = -0.03\pi^2 \text{ V}$$

The average emf induced in the loop is $\varepsilon_{ave} = -0.03\pi^2$ V. Using a calculator, $\varepsilon_{ave} = -0.296$ V. (Since $\pi^2 \approx 10$, that's essentially how the decimal point moved.) Apply Ohm's law to find the induced current.

$$I_{ave} = \frac{\varepsilon_{ave}}{R_{solenoid2}} = -\frac{0.03\pi^2}{5} = -6\pi^2 \times 10^{-3} \text{ A}$$

The average current induced in the loop is $I_{ave} = -6\pi^2 \times 10^{-3}$ A $= -6\pi^2$ mA. If you use a calculator, this works out to $I_{ave} = -0.059$ A $= -59$ mA.

* If you subtract the rounded decimal values, 4.1×10^{-6} T·m^2 $- 1.8 \times 10^{-6}$ T·m^2, you instead get 2.3×10^{-6} T·m^2. However, you shouldn't round to two decimal places before the final value is obtained: You should keep extra digits on your calculator throughout your calculation, and only round your final answers.

Example 125. A rectangular loop of wire with a length of 150 cm and a width of 80 cm lies in the xy plane. A magnetic field is oriented along the $+z$ direction and its magnitude is given by the following equation, where SI units have been suppressed.

$$B = 15e^{-t/3}$$

(A) What is the instantaneous emf induced in the loop at $t = 3.0$ s?

Solution. First convert the distances to meters.

$$L = 150 \text{ cm} = 1.50 \text{ m} \quad , \quad W = 80 \text{ cm} = 0.80 \text{ m}$$

Find the area of the rectangular loop.

$$A = LW = (1.5)(0.8) = 1.2 \text{ m}^2$$

Apply Faraday's law. For **instantaneous** emf, this involves a **derivative**.

$$\varepsilon_{ind} = -N\frac{d\Phi_m}{dt}$$

Plug in the expression for magnetic flux: Note that we can use the equation $\Phi_m = BA\cos\theta$ instead of $\Phi_m = \int_S \vec{\mathbf{B}} \cdot d\vec{\mathbf{A}}$ since B only depends on time (not x or y).

$$\varepsilon_{ind} = -N\frac{d\Phi_m}{dt} = -N\frac{d(BA\cos\theta)}{dt} = -NA\cos\theta\frac{dB}{dt}$$

Note that A and θ are constants. The derivative is $\frac{dB}{dt} = \frac{d}{dt}(15e^{-t/3}) = -5e^{-t/3}$ (see Chapter 17). Plug $\frac{dB}{dt} = -5e^{-t/3}$ into the equation from Faraday's law.

$$\varepsilon_{ind} = -NA\cos\theta\frac{dB}{dt} = -NA\cos\theta\left(-5e^{-t/3}\right) = 5NA\cos\theta\, e^{-t/3}$$

The two minus signs make a plus sign. Plug numbers into this equation. Note that $N = 1$, $t = 3.0$ s, $\theta = 0°$, and we previously found $A = 1.2 \text{ m}^2$.

$$\varepsilon_{ind} = 5(1)(1.2)\cos 0°\, e^{-3/3} = 6e^{-1} = \frac{6}{e} \text{ V}$$

Note that Note that $(5)(1.2) = 6$ and $e^{-1} = \frac{1}{e}$. The **instantaneous** emf induced at $t = 3.0$ s is $\varepsilon_{ind} = \frac{6}{e}$ V. Using a calculator, this comes out to $\varepsilon_{ind} = 2.2$ V.

(B) What is the average emf induced in the loop from $t = 0$ to $t = 3.0$ s?

Solution. Use the equation for average emf. For **average** emf, this involves a **ratio.**

$$\varepsilon_{ave} = -N\frac{\Delta\Phi_m}{\Delta t}$$

First plug $t = 0$ into $B = 15e^{-t/3}$ to find the initial magnetic field (B_0).

$$B_0 = 15e^{-0/3} = 15e^0 = 15 \text{ T}$$

Note that $e^0 = 1$ (see Chapter 17). Now find the initial magnetic flux. Note that $\theta = 0°$.

$$\Phi_{m0} = B_0 A\cos\theta = (15)(1.2)\cos 0° = 18 \text{ T·m}^2$$

Now plug $t = 3.0$ s into $B = 15e^{-t/3}$ to find the final magnetic field (B).

$$B = 15e^{-\frac{3}{3}} = 15e^{-1} = \frac{15}{e} \text{ T}$$

The final magnetic field is $B = \frac{15}{e}$ T. Using a calculator with $e \approx 2.718$, we get $B = 5.52$ T.

Now find the final magnetic flux.

$$\Phi_m = BA\cos\theta = \left(\frac{15}{e}\right)(1.2)\cos 0° = \frac{18}{e}\ \text{T·m}^2$$

The final magnetic flux is $\Phi_m = \frac{18}{e}$ T·m^2. Using a calculator, $\Phi_m = 6.6$ T·m^2.

Subtract the initial magnetic flux from the final magnetic flux in order to find the change in the magnetic flux.

$$\Delta\Phi_m = \Phi_m - \Phi_{m0} = \frac{18}{e} - 18 = \left(\frac{18}{e} - 18\right)\ \text{T·m}^2$$

The change in magnetic flux is $\Delta\Phi_m = \left(\frac{18}{e} - 18\right)$ T·m^2. If you use a calculator, this works out to $\Delta\Phi_m = 11.4$ T·m^2. (You could be off by a little round-off error.)

Plug numbers into the equation for average emf. Recall that $N = 1$ and $\Delta t = 3.0$ s.

$$\varepsilon_{ave} = -N\frac{\Delta\Phi_m}{\Delta t} = -(1)\frac{\left(\frac{18}{e} - 18\right)}{3} = -\left(\frac{6}{e} - 6\right) = \left(6 - \frac{6}{e}\right)\ \text{V}$$

Note that we **distributed** the minus sign: $-\left(\frac{6}{e} - 6\right) = -\frac{6}{e} + 6 = 6 - \frac{6}{e}$. The **average** emf induced from $t = 0$ to $t = 3.0$ s is $\varepsilon_{ave} = \left(6 - \frac{6}{e}\right)$ V. Using a calculator, this comes out to $\varepsilon_{ave} = 3.8$ V.

(C) Compare the values from (A) and (B) and comment on the difference between them.

Solution. Consider the instantaneous emf, which we found in part (A) to be $\varepsilon_{ind} = 2.2$ V when $t = 3.0$ s. Plug $t = 0$ into $\varepsilon_{ind} = 5NA\cos\theta\, e^{-t/3}$ in order to find the initial emf: $\varepsilon_0 = 6.0$ V.

Think about this: The **instantaneous** emf drops from 6.0 V to 2.2 V as t increases from 0 to 3.0 s. We should expect the **average** to be about $\frac{6+2.2}{2} = \frac{8.2}{2} = 4.1$ V.

The average emf turned out to be $\varepsilon_{ave} = 3.8$ V in part (B), which is just 7% off. (The reason it isn't "exactly" 4.1 V because B is a **non-linear** function of t.)

30 INDUCTANCE

Self-Inductance	Self-induced Emf
$L = \frac{N\Phi_m}{I}$	$\varepsilon_L = -L\frac{dI}{dt}$
Magnetic Flux	**Magnetic Flux (Uniform $\vec{\mathbf{B}}$)**
$\Phi_m = \int_S \vec{\mathbf{B}} \cdot d\vec{\mathbf{A}}$	$\Phi_m = BA\cos\theta$

Mutual Inductance
$\varepsilon_1 = -M_{12}\frac{dI_2}{dt}$, $\varepsilon_2 = -M_{21}\frac{dI_1}{dt}$, $M_{12} = \frac{N_1\Phi_{12}}{I_2}$, $M_{21} = \frac{N_2\Phi_{21}}{I_1}$

Magnetic Energy	Magnetic Field in a Solenoid
$U_L = \frac{1}{2}LI^2$	$B = \frac{\mu_0 NI}{L} = \mu_0 nI$

Schematic Representation	Symbol	Name
	R	resistor
	C	capacitor
	L	inductor
	ΔV	battery or DC power supply

RL Circuit (No Battery)

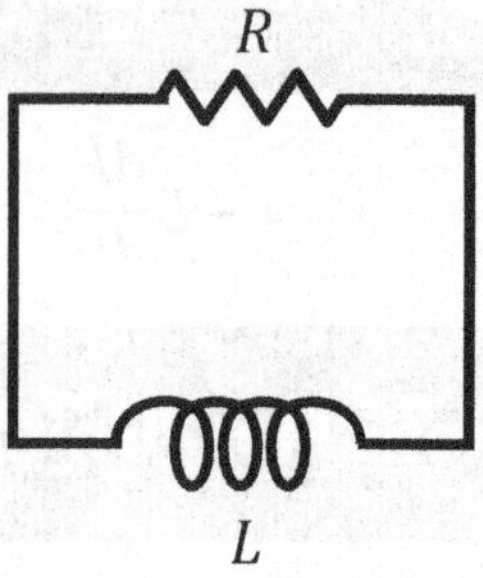

$$I = I_m e^{-t/\tau}$$

$$\tau = \frac{L}{R} \quad , \quad t_{½} = \tau \ln(2)$$

$$\Delta V_R = IR \quad , \quad \varepsilon_L = -L\frac{dI}{dt}$$

RL Circuit (with DC Power Supply)

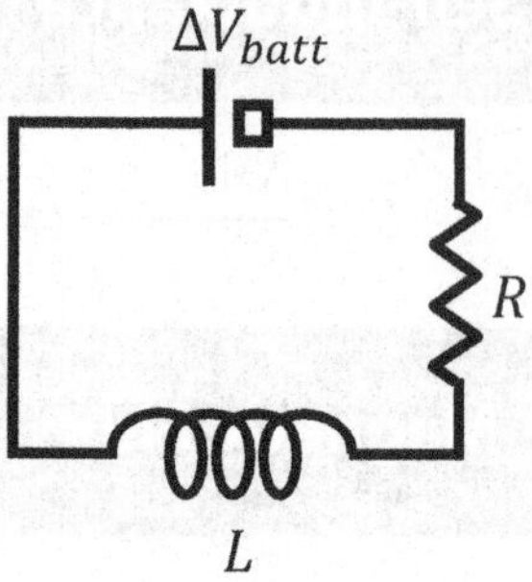

$$I = I_m(1 - e^{-t/\tau})$$

$$\tau = \frac{L}{R} \quad , \quad t_{½} = \tau \ln(2)$$

$$\Delta V_R = IR \quad , \quad \varepsilon_L = -L\frac{dI}{dt}$$

LC Circuit (No Battery)

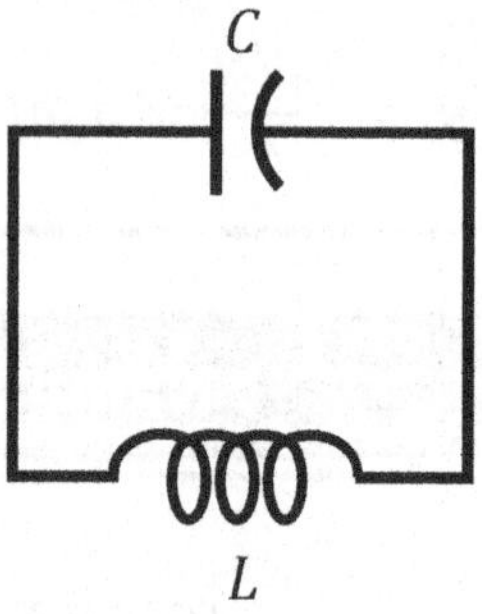

$$I = -I_m \sin(\omega t + \varphi)$$

$$Q = Q_m \cos(\omega t + \varphi)$$

$$\omega = \frac{1}{\sqrt{LC}} = 2\pi f$$

$$I_m = \omega Q_m$$

RLC Circuit (No Battery)

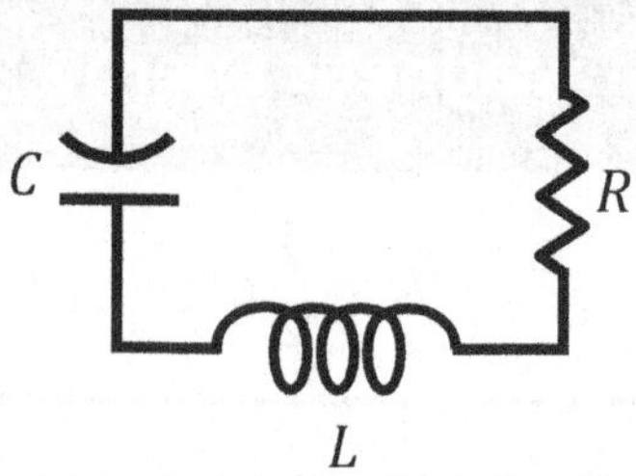

$$Q = Q_m e^{-Rt/2L} \cos(\omega_d t + \varphi)$$

$$I = \frac{dQ}{dt}$$

$$\omega_d = \sqrt{\frac{1}{LC} - \left(\frac{R}{2L}\right)^2}$$

$$R_c = \sqrt{\frac{4L}{C}}$$

$$Q = C\Delta V_C \quad , \quad \Delta V_R = IR \quad , \quad \varepsilon_L = -L\frac{dI}{dt}$$

Deriving an Equation for Inductance		
$L = \frac{N\Phi_m}{I}$	$\Phi_m = \int_S \vec{\mathbf{B}} \cdot d\vec{\mathbf{A}}$	$I = \int \vec{\mathbf{K}} \cdot d\vec{\boldsymbol{\ell}}$ (surface) $I = \int \vec{\mathbf{J}} \cdot d\vec{\mathbf{A}}$ (volume)

Ampère's Law
$\oint_C \vec{\mathbf{B}} \cdot d\vec{\mathbf{s}} = \mu_0 I_{enc}$

Relation Among Coordinate Systems and Unit Vectors		
$x = r\cos\theta$ $y = r\sin\theta$ $\hat{\mathbf{r}} = \hat{\mathbf{x}}\cos\theta + \hat{\mathbf{y}}\sin\theta$ $\hat{\boldsymbol{\theta}} = -\hat{\mathbf{x}}\sin\theta + \hat{\mathbf{y}}\cos\theta$ (2D polar)	$x = r_c\cos\theta$ $y = r_c\sin\theta$ $\hat{\mathbf{r}}_c = \hat{\mathbf{x}}\cos\theta + \hat{\mathbf{y}}\sin\theta$ $\hat{\boldsymbol{\theta}} = -\hat{\mathbf{x}}\sin\theta + \hat{\mathbf{y}}\cos\theta$ (cylindrical)	$x = r\sin\theta\cos\varphi$ $y = r\sin\theta\sin\varphi$ $z = r\cos\theta$ $\hat{\mathbf{r}} = \hat{\mathbf{x}}\cos\varphi\sin\theta$ $+\hat{\mathbf{y}}\sin\varphi\sin\theta + \hat{\mathbf{z}}\cos\theta$ (spherical)

Differential Arc Length		
$d\vec{\mathbf{s}} = \hat{\mathbf{x}}\,dx$ $ds = dx$ (along x)	$d\vec{\mathbf{s}} = \hat{\mathbf{y}}\,dy$ or $\hat{\mathbf{z}}\,dz$ $ds = dy$ or dz (along y or z)	$d\vec{\mathbf{s}} = \pm\hat{\boldsymbol{\theta}}\,r_c d\theta$ $ds = a d\theta$ (circular arc of radius a)

Differential Area Element		
$dA = dxdy$ (polygon in xy plane)	$dA = rdrd\theta$ (pie slice, disc, thick ring)	$dA = a^2\sin\theta\, d\theta d\varphi$ (sphere of radius a)

Differential Volume Element		
$dV = dxdydz$ (bounded by flat sides)	$dV = r_c dr_c d\theta dz$ (cylinder or cone)	$dV = r^2\sin\theta\, dr d\theta d\varphi$ (spherical)

Symbol	Name	Units
L	inductance (or self-inductance)	H
M	mutual inductance	H
ε_L	self-induced emf	V
ε_{ind}	induced emf	V
I	current	A
t	time	s
Φ_m	magnetic flux	$\text{T}\cdot\text{m}^2$ or Wb
N	number of loops (or turns)	unitless
n	number of turns per unit length	$\frac{1}{\text{m}}$
B	magnitude of the external magnetic field	T
A	area of the loop	m^2
θ	angle between $\vec{\mathbf{B}}$ and the axis of the loop, or the angle between $\vec{\mathbf{B}}$ and $d\vec{\mathbf{A}}$	° or rad
R	resistance	Ω
C	capacitance	F
U_L	magnetic energy stored in an inductor	J
Q	the charge stored on the positive plate of a capacitor	C
ΔV	the potential difference between two points in a circuit	V
$t_{1/2}$	half-life	s
τ	time constant	s
ω	angular frequency	rad/s
φ	phase angle	rad

Note: The symbols Φ and φ are the uppercase and lowercase forms of the Greek letter phi.

Permeability of Free Space
$\mu_0 = 4\pi \times 10^{-7} \frac{\text{T}\cdot\text{m}}{\text{A}}$

Prefix	Name	Power of 10
k	kilo	10^3
m	milli	10^{-3}
μ	micro	10^{-6}
n	nano	10^{-9}
p	pico	10^{-12}

Example 126. The emf induced in an 80-mH inductor is -0.72 V when the current through the inductor increases at a constant rate. What is the rate at which the current increases?

Solution. First convert the inductance from milliHenry (mH) to Henry (H) using $\text{m} = 10^{-3}$.

$$L = 80 \text{ mH} = 0.080 \text{ H}$$

Note that $\frac{dI}{dt}$ is the rate at which the current increases, and that's what we're solving for: Simply solve for $\frac{dI}{dt}$ in the equation that relates the self-induced emf to the inductance. (There is no need to apply calculus: We're not taking a derivative, since we aren't given current as a function of time. Rather, we're just solving for what the derivative equals.*)

$$\varepsilon_L = -L\frac{dI}{dt}$$

$$-0.72 = -(0.08)\frac{dI}{dt}$$

Divide both sides of the equation by -0.08. The minus signs cancel.

$$\frac{dI}{dt} = \frac{0.72}{0.08} = 9.0 \text{ A/s}$$

The current increases at a rate of $\frac{dI}{dt} = 9.0$ A/s.

* You've actually done this many times in physics without even realizing it! For example, consider the one-dimensional uniform acceleration equation, $v_x = v_{x0} + a_x t$. Acceleration is a derivative of velocity with respect to time: $a_x = \frac{dv_x}{dt}$. Therefore, we could rewrite the previous one-dimensional uniform acceleration equation as $v_x = v_{x0} + \frac{dv_x}{dt}t$. It's the exact same equation. If you want to know the "rate at which velocity increases" (which means the same thing as "acceleration") in a uniform acceleration problem, you could solve for $\frac{dv_x}{dt}$ in this equation the same way that you would ordinarily solve for a_x. Whether we call it $\frac{dv_x}{dt}$ or call it a_x makes no difference to how we would perform the algebra to solve for it.

Example 127. A 50.0-Ω resistor is connected in series with a 20-mH inductor. What is the time constant?

Solution. First convert the inductance from milliHenry (mH) to Henry (H) using m $= 10^{-3}$.

$$L = 20 \text{ mH} = 0.020 \text{ H}$$

Use the equation for the time constant of an RL circuit.

$$\tau = \frac{L}{R} = \frac{0.020}{50} = 0.00040 \text{ s} = 0.40 \text{ ms}$$

The time constant is $\tau = 0.40$ ms. (That's in milliseconds.)

Example 128. A 40.0-μF capacitor that is initially charged is connected in series with a 16.0-mH inductor. What is the angular frequency of oscillations in the current?

Solution. First convert the given values to SI units using m $= 10^{-3}$ and μ $= 10^{-6}$.

$$L = 16 \text{ mH} = 0.016 \text{ H}$$

$$C = 40.0 \text{ μF} = 4.00 \times 10^{-5} \text{ F}$$

Use the equation for the angular frequency in an LC circuit.

$$\omega = \frac{1}{\sqrt{LC}} = \frac{1}{\sqrt{(0.016)(4.0 \times 10^{-5})}} = \frac{1}{\sqrt{0.064 \times 10^{-5}}} = \frac{1}{\sqrt{64 \times 10^{-8}}}$$

Note that $6.4 \times 10^{-7} = 64 \times 10^{-8}$. It's simpler to take a square root of an even power of 10, which makes it preferable to work with 10^{-8} instead of 10^{-7}. Apply the rule from algebra that $\sqrt{ax} = \sqrt{a}\sqrt{x}$.

$$\omega = \frac{1}{\sqrt{64}} \frac{1}{\sqrt{10^{-8}}} = \frac{1}{8} \frac{1}{10^{-4}} = \frac{1}{8} \times 10^4 = 1250 \text{ rad/s} = 1.25 \times 10^3 \text{ rad/s}$$

The angular frequency is $\omega = 1250$ rad/s $= 1.25 \times 10^3$ rad/s.

Solenoid Inductance Example: A Prelude to Example 129. The long, tightly wound solenoid illustrated below (both diagrams show the same solenoid) is coaxial with the z-axis and carries an insulated filamentary current I. Derive an expression for the self-inductance of the solenoid.

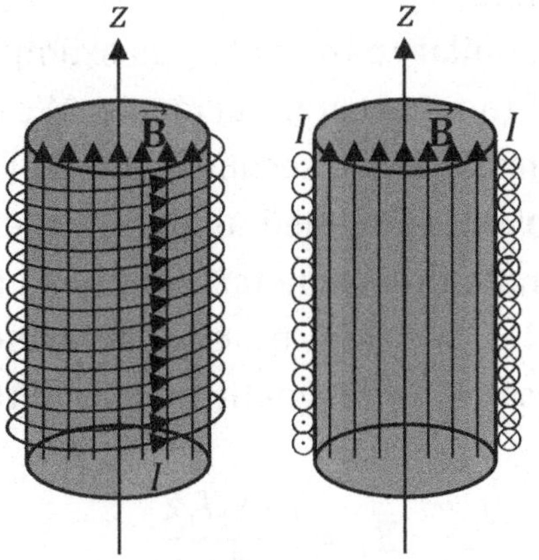

Solution. We applied Ampère's law to derive an equation for the magnetic field inside of a similar solenoid in Chapter 27. The result was:

$$\vec{\mathbf{B}} = \mu_0 n I \hat{\mathbf{z}} = \frac{\mu_0 N I \hat{\mathbf{z}}}{\ell}$$

Here, n is the number of turns per unit length, N is the number of turns, and ℓ is the length of the solenoid. Plug this equation for magnetic field into the magnetic flux integral.

$$\Phi_m = \int_S \vec{\mathbf{B}} \cdot d\vec{\mathbf{A}} = \int_S \frac{\mu_0 N I \hat{\mathbf{z}}}{\ell} \cdot d\vec{\mathbf{A}} = \frac{\mu_0 N I}{\ell} \int_S \hat{\mathbf{z}} \cdot d\vec{\mathbf{A}}$$

The surface is the area of one loop. The corresponding differential area element has a magnitude of $dA = r_c dr_c d\theta$ (see page 363). The direction of $d\vec{\mathbf{A}}$ is along the z-axis, since $d\vec{\mathbf{A}}$ is in general perpendicular to the surface and the z-axis is perpendicular to each loop. Combining this together, we get $d\vec{\mathbf{A}} = \hat{\mathbf{z}}\, dA = \hat{\mathbf{z}}\, r_c dr_c d\theta$. The limits are from $r_c = 0$ to $r_c = a$ (the radius of the solenoid) and from $\theta = 0$ to $\theta = 2\pi$.

$$\Phi_m = \frac{\mu_0 N I}{\ell} \int_{r_c=0}^{a} \int_{\theta=0}^{2\pi} \hat{\mathbf{z}} \cdot \hat{\mathbf{z}}\, r_c\, dr_c\, d\theta$$

The scalar product is $\hat{\mathbf{z}} \cdot \hat{\mathbf{z}} = 1$ (see Chapter 19).

$$\Phi_m = \frac{\mu_0 N I}{\ell} \int_{r_c=0}^{a} \int_{\theta=0}^{2\pi} r_c dr_c d\theta = \frac{\mu_0 N I}{\ell} \int_{r_c=0}^{a} r_c\, dr_c \int_{\theta=0}^{2\pi} d\theta$$

$$\Phi_m = \frac{\mu_0 N I}{\ell} \left[\frac{r_c^2}{2}\right]_{r_c=0}^{a} [\theta]_{\theta=0}^{2\pi} = \frac{\pi \mu_0 N I a^2}{\ell}$$

Plug this expression into the equation for self-inductance.

$$L = \frac{N\Phi_m}{I} = \frac{N}{I}\left(\frac{\pi \mu_0 N I a^2}{\ell}\right) = \frac{\pi \mu_0 N^2 a^2}{\ell}$$

Note that it's the same as $L = \frac{\mu_0 N^2 A}{\ell}$, where $A = \pi a^2$ is the area of each loop.

Mutual Inductance Example: Another Prelude to Example 129. Two long, tightly wound solenoids are coaxial with the z-axis, have the same radius (a), and carry insulated filamentary currents I_1 and I_2. Derive an expression for the mutual inductance (M_{21}) of solenoid 2 with respect to solenoid 1.

Solution. This example is very similar to the previous example, except that we will need to use subscripts to keep track of the mutual inductance. We begin with the magnetic field that the first solenoid creates inside of the second solenoid (since Φ_{21}, which corresponds to M_{21} as specified in the problem, is defined as the magnetic flux created by solenoid 1 through solenoid 2 – although note that some textbooks call this the magnetic flux through solenoid 2 created by solenoid 1, which is exactly the **same thing** worded in a different order). We use the same expression for magnetic field as in the previous example, but with subscripts.

$$\vec{\mathbf{B}}_1 = \frac{\mu_0 N_1 I_1 \hat{\mathbf{z}}}{\ell_1}$$

Next, we find the magnetic flux (Φ_{21}) through solenoid 2 created by solenoid 1.

$$\Phi_{21} = \int_{S_2} \vec{\mathbf{B}}_1 \cdot d\vec{\mathbf{A}}$$

We use the magnetic field ($\vec{\mathbf{B}}_1$) created by solenoid 1, but the integral is over the surface (S_2) of solenoid 2 (which is the area of one of solenoid 2's loops) – although in this problem it doesn't matter because the two solenoids happen to have the same radius (a). Just as in the previous example, $d\vec{\mathbf{A}} = \hat{\mathbf{z}}\, dA = \hat{\mathbf{z}}\, r_c dr_c d\theta$ and $\hat{\mathbf{z}} \cdot \hat{\mathbf{z}} = 1$. The limits are from $r_c = 0$ to $r_c = a$ (technically, the radius of solenoid 2, since it's the flux through solenoid 2, though in this problem the two solenoids have the same radius) and from $\theta = 0$ to $\theta = 2\pi$.

$$\Phi_{21} = \int_{S_2} \vec{\mathbf{B}}_1 \cdot d\vec{\mathbf{A}} = \int_{S_2} \frac{\mu_0 N_1 I_1 \hat{\mathbf{z}}}{\ell_1} \cdot d\vec{\mathbf{A}} = \frac{\mu_0 N_1 I_1}{\ell_1} \int_{r_c=0}^{a} \int_{\theta=0}^{2\pi} \hat{\mathbf{z}} \cdot \hat{\mathbf{z}}\, r_c\, dr_c\, d\theta$$

$$\Phi_{21} = \frac{\mu_0 N_1 I_1}{\ell_1} \int_{r_c=0}^{a} r_c\, dr_c \int_{\theta=0}^{2\pi} d\theta = \frac{\mu_0 N_1 I_1}{\ell_1} \left[\frac{r_c^2}{2}\right]_{r_c=0}^{a} [\theta]_{\theta=0}^{2\pi} = \frac{\pi \mu_0 N_1 I_1 a^2}{\ell_1}$$

Plug this expression into the equation for mutual inductance.

$$M = \frac{N_2 \Phi_{21}}{I_1} = \frac{N_2}{I_1}\left(\frac{\pi \mu_0 N_1 I_1 a^2}{\ell_1}\right) = \frac{\pi \mu_0 N_1 N_2 a^2}{\ell_1}$$

Note that it's the same as $M = \frac{\mu_0 N_1 N_2 A_2}{\ell_1}$, where $A_2 = \pi a_2^2 = \pi a^2$ is the area of each loop in solenoid 2.

Example 129. The tightly wound (even though the diagram is not drawn to look that way) toroidal coil illustrated below has square cross section and carries an insulated filamentary current I. Derive an expression for the self-inductance of the toroidal coil.

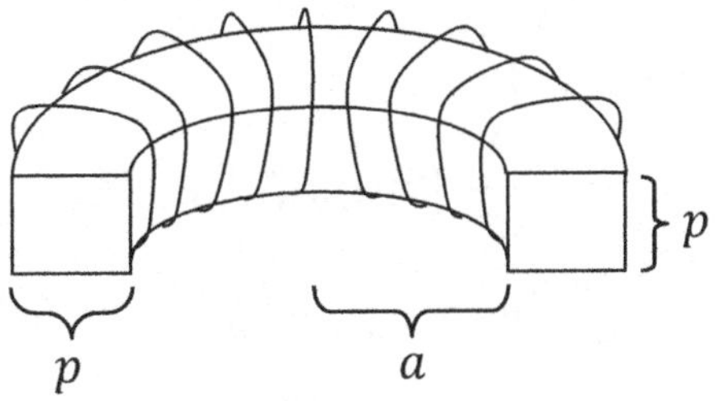

Solution. Note that the square cross section (compared to a toroid with a circular cross section) actually makes the integral for magnetic flux simpler.

First find the magnetic field inside of the toroidal coil. The answer for this step is the same as for a toroidal coil with circular cross section. We found this using Ampère's law in Chapter 27. See page 319.

$$\vec{\mathbf{B}} = \frac{\mu_0 NI}{2\pi r_c}\hat{\boldsymbol{\theta}}$$

Substitute the above expression for magnetic field into the integral for magnetic flux.

$$\Phi_m = \int_S \vec{\mathbf{B}} \cdot d\vec{\mathbf{A}}$$

$$\Phi_m = \int_S \frac{\mu_0 NI}{2\pi r_c}\hat{\boldsymbol{\theta}} \cdot d\vec{\mathbf{A}} = \frac{\mu_0 NI}{2\pi}\int_S \frac{1}{r_c}\hat{\boldsymbol{\theta}} \cdot d\vec{\mathbf{A}}$$

The differential area element is along $\hat{\boldsymbol{\theta}}$ (which is perpendicular to each square loop), such that $d\vec{\mathbf{A}} = \hat{\boldsymbol{\theta}}\, dA$. The integral is over the area of one of the square loops. For a square lying in the $r_c z$ plane, we write $dA = dr_c dz$ and $d\vec{\mathbf{A}} = \hat{\boldsymbol{\theta}}\, dr_c dz$. Note that r_c varies outward from $r_c = a$ to $r_c = a + p$, while z varies upward from $z = 0$ to $z = p$ (if you put the bottom of the toroidal coil in the xy plane). This is the r_c of cylindrical coordinates (see Chapter 6). Note that $\hat{\boldsymbol{\theta}} \cdot \hat{\boldsymbol{\theta}} = 1$ (this is true of any unit vector dotted into itself.)

$$\Phi_m = \int_S \frac{\mu_0 NI}{2\pi r_c}\hat{\boldsymbol{\theta}} \cdot d\vec{\mathbf{A}} = \frac{\mu_0 NI}{2\pi}\int_{r_c=a}^{a+p}\int_{z=0}^{p} \frac{\hat{\boldsymbol{\theta}}}{r_c} \cdot \hat{\boldsymbol{\theta}}\, dr_c\, dz = \frac{\mu_0 NI}{2\pi}\int_{r_c=a}^{a+p}\int_{z=0}^{p} \frac{dr_c}{r_c} dz$$

(Note that it's <u>**incorrect**</u> to write $r_c dr_c d\theta$ for this problem, even though we ordinarily write dA this way for a cylindrically shaped object. That's because we're integrating over a square loop that extends outward along r_c and upward along z, whereas when we write $r_c dr_c d\theta$, we would be integrating over the area of a circle.)

When integrating over z, we treat the independent variable r_c as if it were a constant. This means that we may pull r_c out of the z integral.

$$\Phi_m = \frac{\mu_0 NI}{2\pi} \int_{r_c=a}^{a+p} \frac{dr_c}{r_c} \int_{z=0}^{p} dz$$

Recall from Chapter 17 that $\int \frac{dx}{x} = \ln(x)$.

$$\Phi_m = \frac{\mu_0 NI}{2\pi} [\ln(r_c)]_{r_c=a}^{a+p} [z]_{z=0}^{p}$$

$$\Phi_m = \frac{\mu_0 NI}{2\pi} [\ln(a+p) - \ln(a)](p-0)$$

Apply the logarithm identity $\ln(x) - \ln(y) = \ln\left(\frac{x}{y}\right)$ to write $\ln(a+p) - \ln(a) = \ln\left(\frac{a+p}{a}\right)$.

$$\Phi_m = \frac{\mu_0 NI}{2\pi} \ln\left(\frac{a+p}{a}\right) p = \frac{\mu_0 NIp}{2\pi} \ln\left(\frac{a+p}{a}\right)$$

Substitute the expression for magnetic flux into the equation for inductance.

$$L = \frac{N\Phi_m}{I} = \frac{\mu_0 N^2 p}{2\pi} \ln\left(\frac{a+p}{a}\right)$$

31 AC CIRCUITS

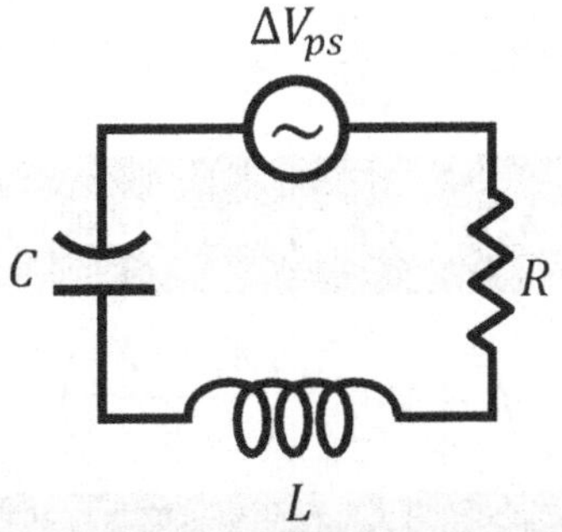

Instantaneous Potential Difference in an AC Circuit

$$\Delta V_{ps} = \Delta V_m \sin(\omega t) \quad , \quad \Delta V_R = \Delta V_{Rm} \sin(\omega t)$$
$$\Delta V_L = \Delta V_{Lm} \cos(\omega t) \quad , \quad \Delta V_C = -\Delta V_{Cm} \cos(\omega t)$$

Amplitudes of Potential Difference in an AC Circuit

$$\Delta V_m = I_m Z \quad , \quad \Delta V_{Rm} = I_m R \quad , \quad \Delta V_{Lm} = I_m X_L \quad , \quad \Delta V_{Cm} = I_m X_C$$

Rms Values of Potential Difference in an AC Circuit

$$\Delta V_{rms} = I_{rms} Z \quad , \quad \Delta V_{Rrms} = I_{rms} R \quad , \quad \Delta V_{Lrms} = I_{rms} X_L \quad , \quad \Delta V_{Crms} = I_{rms} X_C$$

Current in an AC Circuit

$$I = I_m \sin(\omega t - \varphi)$$

Inductive Reactance in an AC Circuit

$$X_L = \omega L \quad , \quad \Delta V_{Lm} = I_m X_L$$

Capacitive Reactance in an AC Circuit

$$X_C = \frac{1}{\omega C} \quad , \quad \Delta V_{Cm} = I_m X_C$$

Impedance for an RLC Circuit

$$\Delta V_m = I_m Z \quad , \quad Z = \sqrt{R^2 + (X_L - X_C)^2}$$

Root-mean-square (rms) Values in an AC Circuit

$$\Delta V_{rms} = \frac{\Delta V_m}{\sqrt{2}} \quad , \quad I_{rms} = \frac{I_m}{\sqrt{2}}$$

Phase Angle for an RLC Circuit

$$\varphi = \tan^{-1}\left(\frac{X_L - X_C}{R}\right)$$

Angular Frequency, Frequency, and Period

$$\omega = 2\pi f = \frac{2\pi}{T} \quad , \quad f = \frac{1}{T}$$

Resonance Angular Frequency and Resonance Frequency

$$\omega_0 = \frac{1}{\sqrt{LC}} = 2\pi f_0$$

Instantaneous Power in an RLC Circuit

$$P = I\Delta V_{ps} = I_m \Delta V_m \sin(\omega t)\sin(\omega t - \varphi)$$

Average Power in an RLC Circuit

$$P_{av} = I_{rms}\Delta V_{rms}\cos\varphi = I_{rms}^2 R$$

Quality Factor for an RLC Circuit

$$Q_0 = \frac{\omega_0}{\Delta\omega} = \frac{\omega_0 L}{R}$$

Instantaneous Current and Charge

$$I = \frac{dQ}{dt}$$

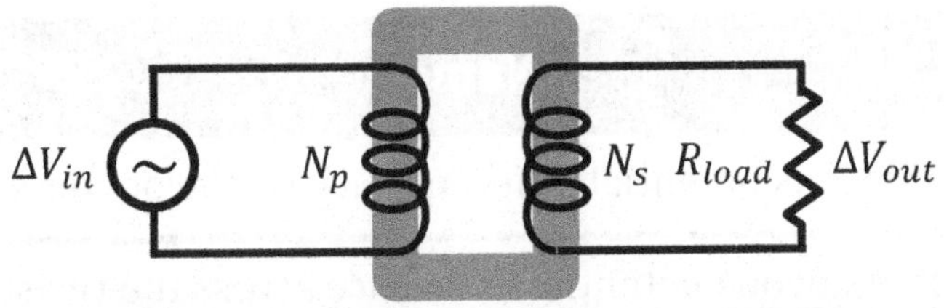

Transformers

$$\frac{\Delta V_{out}}{\Delta V_{in}} = \frac{N_s}{N_p} \quad , \quad R_{eq} = R_{load}\left(\frac{N_p}{N_s}\right)^2$$

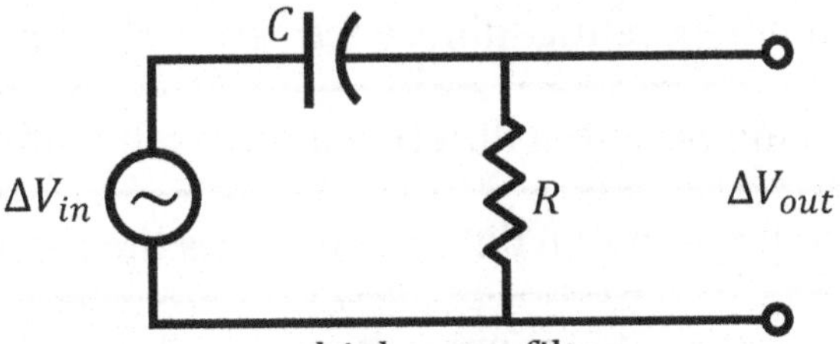

high-pass filter

High-Pass Filters

$$\frac{\Delta V_{out}}{\Delta V_{in}} = \frac{I_m R}{I_m\sqrt{R^2 + \left(\frac{1}{\omega C}\right)^2}} = \frac{R}{\sqrt{R^2 + \left(\frac{1}{\omega C}\right)^2}}$$

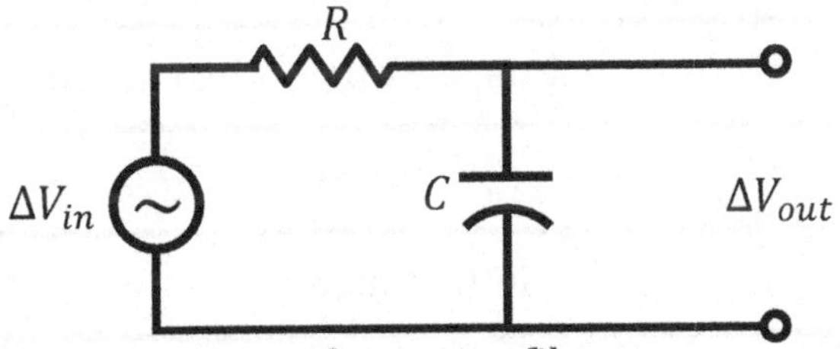

low-pass filter

Low-Pass Filters

$$\frac{\Delta V_{out}}{\Delta V_{in}} = \frac{I_m X_C}{I_m\sqrt{R^2 + \left(\frac{1}{\omega C}\right)^2}} = \frac{\frac{1}{\omega C}}{\sqrt{R^2 + \left(\frac{1}{\omega C}\right)^2}}$$

Symbol	Name	SI Units
ΔV_{ps}	instantaneous potential difference across the power supply	V
ΔV_R	instantaneous potential difference across the resistor	V
ΔV_L	instantaneous potential difference across the inductor	V
ΔV_C	instantaneous potential difference across the capacitor	V
ΔV_m	maximum potential difference across the power supply	V
ΔV_{rms}	root-mean-square value of the potential difference	V
ΔV_{Rm}	maximum potential difference across the resistor	V
ΔV_{Lm}	maximum potential difference across the inductor	V
ΔV_{Cm}	maximum potential difference across the capacitor	V
ΔV_{in}	input voltage	V
ΔV_{out}	output voltage	V
N_p	number of loops in the primary inductor	unitless
N_s	number of loops in the secondary inductor	unitless
I	instantaneous current	A
I_m	maximum value (amplitude) of the current	A
I_{rms}	root-mean-square value of the current	A
R	resistance	Ω
L	inductance	H
C	capacitance	F
Z	impedance	Ω
X_L	inductive reactance	Ω
X_C	capacitive reactance	Ω
P	instantaneous power	W

P_{av}	average power	W
t	time	s
ω	angular frequency	rad/s
ω_0	resonance (angular) frequency	rad/s
$\Delta\omega$	full-width at half maximum (FWHM)	rad/s
f	frequency	Hz
f_0	resonance frequency	Hz
T	period	s
φ	phase angle	rad
Q_0	quality factor (**not** charge)	unitless
Q	instantaneous charge stored on the capacitor	C

Schematic Representation	**Symbol**	**Name**
	R	resistor
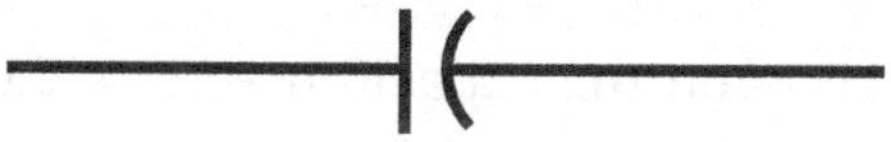	C	capacitor
	L	inductor
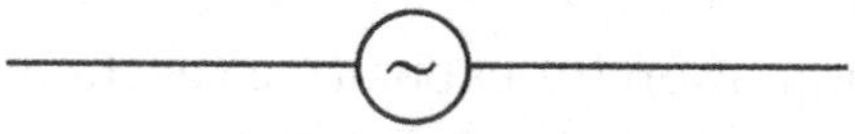	ΔV	AC power supply
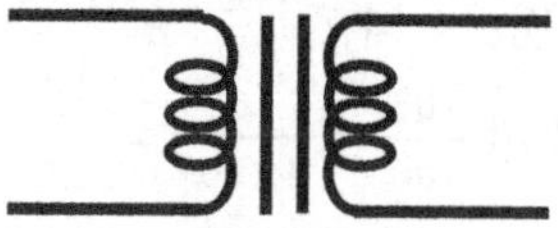		transformer

Example 130. The current varies as a sine wave in an AC circuit with an amplitude of 2.0 A. What is the rms current?

Solution. We are given the amplitude of the current: $I_m = 2.0$ A. We are looking for the root-mean-square (rms) value of the current (I_{rms}). Divide by the squareroot of 2.

$$I_{rms} = \frac{I_m}{\sqrt{2}} = \frac{2}{\sqrt{2}} = \frac{2}{\sqrt{2}}\frac{\sqrt{2}}{\sqrt{2}} = \frac{2\sqrt{2}}{2} = \sqrt{2}\text{ A}$$

We multiplied the numerator and denominator both by $\sqrt{2}$ in order to **rationalize** the denominator. The rms current is $I_{rms} = \sqrt{2}$ A, which is the same as $I_{rms} = \frac{2}{\sqrt{2}}$ A. If you use a calculator, this works out to $I_{rms} = 1.4$ A.

Example 131. An AC circuit where the power supply's frequency is set to 50 Hz includes a 30-mH inductor. What is the inductive reactance?

Solution. We are given the frequency ($f = 50$ Hz) and the inductance ($L = 30$ mH). We are looking for the inductive reactance (X_L). First convert the inductance to Henry.

$$L = 30\text{ mH} = 0.030\text{ H}$$

The equation for inductive reactance involves angular frequency (ω). Multiply the frequency by 2π to find the angular frequency.

$$\omega = 2\pi f = 2\pi(50) = 100\pi\text{ rad/s}$$

Plug angular frequency and inductance into the equation for inductive reactance.

$$X_L = \omega L = (100\pi)(0.03) = 3\pi\ \Omega$$

The inductive reactance is $X_L = 3\pi\ \Omega$. If you use a calculator, this comes out to $X_L = 9.4\ \Omega$.

Example 132. An AC circuit where the power supply's frequency is set to 100 Hz includes a 2.0-μF capacitor. What is the capacitive reactance?

Solution. We are given the frequency ($f = 100$ Hz) and the capacitance ($C = 2.0$ μF). We are looking for the capacitive reactance (X_C). First convert the capacitance to Farads.

$$C = 2.0\ \mu\text{F} = 2.0 \times 10^{-6}\text{ F}$$

The equation for capacitive reactance involves angular frequency (ω). Multiply the frequency by 2π to find the angular frequency.

$$\omega = 2\pi f = 2\pi(100) = 200\pi\text{ rad/s}$$

Plug angular frequency and capacitance into the equation for capacitive reactance.

$$X_C = \frac{1}{\omega C} = \frac{1}{(200\pi)(2.0 \times 10^{-6})} = \frac{10^6}{400\pi} = \frac{10^4}{4\pi} = \frac{2500}{\pi}\ \Omega$$

Note that $10^{-6} = \frac{1}{10^6}$ such that $\frac{1}{10^{-6}} = 10^6$. Also note that $\frac{10^6}{400} = \frac{10^6}{4\times 10^2} = \frac{10^4}{4} = \frac{10{,}000}{4} = 2500$. The capacitive reactance is $X_C = \frac{2500}{\pi}\ \Omega$. If you use a calculator, this works out to $X_C = 796\ \Omega$, which is $800\ \Omega$ or 0.80 kΩ to two significant figures.

Example 133. An AC power supply that operates at an angular frequency of 25 rad/s is connected in series with a $100\sqrt{3}$-Ω resistor, a 6.0-H inductor, and an 800-μF capacitor.

(A) What is the impedance for this RLC circuit?

Solution. Begin by making a list of the given quantities in SI units:

- The angular frequency is $\omega = 25$ rad/s. **Note**: It's **not** ω_0 (that is, it isn't resonance).
- The resistance is $R = 100\sqrt{3}\ \Omega$.
- The inductance is $L = 6.0$ H.
- The capacitance is $C = 8.00 \times 10^{-4}$ F. The metric prefix micro (μ) stands for 10^{-6}.

First use the equations for inductive and capacitive reactance.

$$X_L = \omega L = (25)(6) = 150\ \Omega$$

$$X_C = \frac{1}{\omega C} = \frac{1}{(25)(8 \times 10^{-4})} = \frac{1}{200 \times 10^{-4}} = \frac{1}{2 \times 10^{-2}} = \frac{10^2}{2} = \frac{100}{2} = 50\ \Omega$$

The inductive reactance is $X_L = 150\ \Omega$ and the capacitive reactance is $X_C = 50\ \Omega$. Plug the resistance, inductive reactance, and capacitive reactance into the equation for impedance.

$$Z = \sqrt{R^2 + (X_L - X_C)^2} = \sqrt{\left(100\sqrt{3}\right)^2 + (150 - 50)^2} = \sqrt{(100)^2\left(\sqrt{3}\right)^2 + (100)^2}$$

Factor out the $(100)^2$.

$$Z = \sqrt{(100)^2\left[\left(\sqrt{3}\right)^2 + 1\right]} = \sqrt{(100)^2(3 + 1)} = \sqrt{(100)^2(4)} = 100(2) = 200\ \Omega$$

We applied the rules that $\sqrt{xy} = \sqrt{x}\sqrt{y}$ and $\sqrt{3}\sqrt{3} = 3$. The impedance is $Z = 200\ \Omega$.

(B) What is the phase angle for the current with respect to the power supply voltage?

Solution. Use the equation that relates the phase angle to resistance and reactance.

$$\varphi = \tan^{-1}\left(\frac{X_L - X_C}{R}\right) = \tan^{-1}\left(\frac{150 - 50}{100\sqrt{3}}\right) = \tan^{-1}\left(\frac{100}{100\sqrt{3}}\right) = \tan^{-1}\left(\frac{1}{\sqrt{3}}\right)$$

Multiply the numerator and denominator by $\sqrt{3}$ in order to **rationalize** the denominator.

$$\varphi = \tan^{-1}\left(\frac{1}{\sqrt{3}}\frac{\sqrt{3}}{\sqrt{3}}\right) = \tan^{-1}\left(\frac{\sqrt{3}}{3}\right) = 30°$$

The phase angle is $\varphi = 30°$. Since the phase angle is **positive**, the current **lags** behind the power supply's potential difference by 30° (one-twelfth of a cycle, since 360° is full-cycle).

Example 134. An AC power supply that provides an rms voltage of 200 V and operates at an angular frequency of 50 rad/s is connected in series with a $50\sqrt{3}$-Ω resistor, a 3.0-H inductor, and a 100-μF capacitor.

(A) What would an AC ammeter measure for this RLC circuit?

Solution. Begin by making a list of the given quantities in SI units:

- The rms voltage of the power supply is $\Delta V_{rms} = 200$ V.
- The angular frequency is $\omega = 50$ rad/s. **Note**: It's <u>**not**</u> ω_0 (that is, it isn't resonance).
- The resistance is $R = 50\sqrt{3}\ \Omega$.
- The inductance is $L = 3.0$ H.
- The capacitance is $C = 1.00 \times 10^{-4}$ F. The metric prefix micro (μ) stands for 10^{-6}.

First use the equations for inductive and capacitive reactance.

$$X_L = \omega L = (50)(3) = 150\ \Omega$$

$$X_C = \frac{1}{\omega C} = \frac{1}{(50)(1 \times 10^{-4})} = \frac{1}{50 \times 10^{-4}} = \frac{1}{5 \times 10^{-3}} = \frac{10^3}{5} = \frac{1000}{5} = 200\ \Omega$$

The inductive reactance is $X_L = 150\ \Omega$ and the capacitive reactance is $X_C = 200\ \Omega$. Plug the resistance, inductive reactance, and capacitive reactance into the equation for impedance.

$$Z = \sqrt{R^2 + (X_L - X_C)^2} = \sqrt{\left(50\sqrt{3}\right)^2 + (150 - 200)^2} = \sqrt{(50)^2\left(\sqrt{3}\right)^2 + (-50)^2}$$

Factor out the $(50)^2$.

$$Z = \sqrt{(50)^2\left[\left(\sqrt{3}\right)^2 + (-1)^2\right]} = \sqrt{(50)^2(3+1)} = \sqrt{(50)^2(4)} = 50(2) = 100\ \Omega$$

We applied the rules that $\sqrt{xy} = \sqrt{x}\sqrt{y}$ and $\sqrt{3}\sqrt{3} = 3$. The impedance is $Z = 100\ \Omega$.

AC meters measure rms values. The rms voltage and rms current are related to one another by the impedance.

$$\Delta V_{rms} = I_{rms} Z$$

Divide both sides of the equation by the impedance.

$$I_{rms} = \frac{\Delta V_{rms}}{Z} = \frac{200}{100} = 2.0 \text{ A}$$

An AC ammeter would measure $I_{rms} = 2.0$ A.

(B) What would an AC voltmeter measure across the resistor?

Solution. Multiply the rms current by the resistance.

$$\Delta V_{Rrms} = I_{rms} R = (2)\left(50\sqrt{3}\right) = 100\sqrt{3}\text{ V}$$

An AC voltmeter would measure $\Delta V_{Rrms} = 100\sqrt{3}$ V across the resistor. If you use a calculator, this works out to $\Delta V_{Rrms} = 173 \text{ V} = 0.17 \text{ kV}$.

(C) What would an AC voltmeter measure across the inductor?

Solution. Multiply the rms current by the inductive reactance.

$$\Delta V_{Lrms} = I_{rms}X_L = (2)(150) = 300 \text{ V}$$

An AC voltmeter would measure $\Delta V_{Lrms} = 300$ V across the inductor.

(D) What would an AC voltmeter measure across the capacitor?

Solution. Multiply the rms current by the capacitive reactance.

$$\Delta V_{Crms} = I_{rms}X_C = (2)(200) = 400 \text{ V}$$

An AC voltmeter would measure $\Delta V_{Crms} = 400$ V across the capacitor.

(E) What would an AC voltmeter measure if the two probes were connected across the inductor-capacitor combination?

Solution. The phasors for the inductor and capacitor point in **opposite** directions: The phase angle for the inductor is 90°, while the phase angle for the capacitor is −90°. Subtract the previous two answers.

$$\Delta V_{LCrms} = |\Delta V_{Lrms} - \Delta V_{Crms}| = |300 - 400| = |-100| = 100 \text{ V}$$

An AC voltmeter would measure $\Delta V_{LCrms} = 100$ V across the inductor-capacitor combination.

(F) Show that the answers to parts (B) through (D) are consistent with the 200-V rms voltage supplied by the AC power supply.

Solution. Add the voltage phasors like you would add vectors. Subtract the voltages across the inductor and capacitor since the voltage across the inductor and voltage across the capacitor are 180° out of phase with one another (one leads the current by 90°, while the other lags the current by 90°). Combine this with the voltage across the resistor using the Pythagorean theorem, since the voltage across the resistor is 90° out of phase with the voltages across the inductor and capacitor (since the voltage across the resistor is in phase with the current).

$$\Delta V_{rms} = \sqrt{\Delta V_{Rrms}^2 + (\Delta V_{Lrms} - \Delta V_{Crms})^2} = \sqrt{\left(100\sqrt{3}\right)^2 + (300 - 400)^2}$$

$$\Delta V_{rms} = \sqrt{(100)^2\left(\sqrt{3}\right)^2 + (-100)^2}$$

Factor out the $(100)^2$.

$$\Delta V_{rms} = \sqrt{(100)^2\left[\left(\sqrt{3}\right)^2 + (-1)^2\right]} = \sqrt{(100)^2(3+1)} = \sqrt{(100)^2(4)} = 100(2) = 200 \text{ V}$$

The voltage across the AC power supply is $\Delta V_{rms} = 200$ V, which agrees with the value given in the problem. It's instructive to note that it would be **incorrect** to add the answers for ΔV_{Rrms}, ΔV_{Lrms}, and ΔV_{Crms} together. The correct way to combine these values is through phasor addition, which accounts for the direction (or phase angle) of each phasor, which is reflected in the formula $\Delta V_{rms} = \sqrt{\Delta V_{Rrms}^2 + (\Delta V_{Lrms} - \Delta V_{Crms})^2}$.

Example 135. An AC power supply is connected in series with a $100\sqrt{3}$-Ω resistor, an 80-mH inductor, and a 50-μF capacitor.

(A) What angular frequency would produce resonance for this RLC circuit?
Solution. Begin by making a list of the given quantities in SI units:

- The resistance is $R = 100\sqrt{3}\ \Omega$.
- The inductance is $L = 8.0 \times 10^{-2}$ H. The metric prefix milli (m) stands for 10^{-3}.
- The capacitance is $C = 5.0 \times 10^{-5}$ F. The metric prefix micro (μ) stands for 10^{-6}.

Use the equation for resonance angular frequency.

$$\omega_0 = \frac{1}{\sqrt{LC}} = \frac{1}{\sqrt{(8 \times 10^{-2})(5 \times 10^{-5})}} = \frac{1}{\sqrt{40 \times 10^{-7}}}$$

Note that $40 \times 10^{-7} = 4 \times 10^{-6}$. It's simpler to take a square root of an even power of 10, which makes it preferable to work with 10^{-6} instead of 10^{-7}. Apply the rule from algebra that $\sqrt{ax} = \sqrt{a}\sqrt{x}$.

$$\omega_0 = \frac{1}{\sqrt{4 \times 10^{-6}}} = \frac{1}{\sqrt{4}}\frac{1}{\sqrt{10^{-6}}} = \frac{1}{2}\frac{1}{10^{-3}} = \frac{10^3}{2} = \frac{1000}{2} = 500 \text{ rad/s}$$

Note that $\sqrt{10^{-6}} = 10^{-3}$ and that $\frac{1}{10^{-3}} = 10^3$. The resonance angular frequency for this RLC circuit is $\omega_0 = 500$ rad/s.

(B) What is the resonance frequency in Hertz?
Solution. Divide the resonance angular frequency by 2π.

$$f_0 = \frac{\omega_0}{2\pi} = \frac{500}{2\pi} = \frac{250}{\pi} \text{ Hz}$$

The resonance frequency in Hertz is $f_0 = \frac{250}{\pi}$ Hz. If you use a calculator, it comes out to $f_0 = 80$ Hz.

Example 136. An AC power supply that provides an rms voltage of 120 V is connected in series with a 30-Ω resistor, a 60-H inductor, and a 90-μF capacitor.

(A) What is the maximum possible rms current that the AC power supply could provide to this RLC circuit?

Solution. Begin by making a list of the given quantities in SI units:

- The rms voltage of the power supply is $\Delta V_{rms} = 120$ V.
- The resistance is $R = 30\ \Omega$.
- The inductance is $L = 60$ H.
- The capacitance is $C = 9.0 \times 10^{-5}$ F. The metric prefix micro (μ) stands for 10^{-6}.

The rms voltage and rms current are related by the impedance.

$$\Delta V_{rms} = I_{rms} Z$$

$$I_{rms} = \frac{\Delta V_{rms}}{Z}$$

Since the impedance is in the denominator of the above equation, the rms current is maximum when the impedance (Z) is minimum. Let's look at the equation for impedance.

$$Z = \sqrt{R^2 + (X_L - X_C)^2}$$

The inductive reactance ($X_L = \omega L$) and capacitive reactance ($X_C = \frac{1}{\omega C}$) depend on the angular frequency (ω). By varying the frequency of the AC power supply, the impedance will be minimum (and the rms current will therefore be a maximum) when $X_L - X_C = 0$. (This occurs at the resonance frequency, but as we will see, we don't actually need to find the resonance frequency as a number in order to answer the question.) Set $X_L - X_C = 0$ in the equation for impedance in order to find that the minimum possible impedance equals the resistance.

$$Z_{min} = R = 30\ \Omega$$

Plug the minimum impedance into the equation for the rms current in order to find the maximum possible rms current.

$$I_{rms,max} = \frac{\Delta V_{rms}}{R} = \frac{120}{30} = 4.0\text{ A}$$

The maximum possible rms current that the AC power supply could provide to this RLC circuit is $I_{rms,max} = 4.0$ A.

(B) What is the minimum possible impedance for this RLC circuit?

Solution. The maximum current and minimum impedance both occur at **resonance**. Less **impedance** results in more **current** (for a fixed rms power supply voltage). As we discussed in part (A), at resonance, the impedance equals the **resistance**. There is no math to do. The minimum possible impedance that could be obtained by adjusting the frequency is $Z_{min} = R = 30\ \Omega$.

Example 137. An AC power supply that provides an rms voltage of 3.0 kV and operates at an angular frequency of 25 rad/s is connected in series with a $500\sqrt{3}$-Ω resistor, an 80-H inductor, and an 80-μF capacitor.

(A) What is the rms current for this RLC circuit?

Solution. Begin by making a list of the given quantities in SI units:

- The rms voltage of the power supply is $\Delta V_{rms} = 3000$ V. The metric prefix kilo (k) stands for $10^3 = 1000$.
- The angular frequency is $\omega = 25$ rad/s. **Note**: It's **not** ω_0 (that is, it isn't resonance).
- The resistance is $R = 500\sqrt{3}\ \Omega$.
- The inductance is $L = 80$ H.
- The capacitance is $C = 8.0 \times 10^{-5}$ F. The metric prefix micro (μ) stands for 10^{-6}.

First use the equations for inductive and capacitive reactance.

$$X_L = \omega L = (25)(80) = 2000\ \Omega$$

$$X_C = \frac{1}{\omega C} = \frac{1}{(25)(8 \times 10^{-5})} = \frac{1}{200 \times 10^{-5}} = \frac{1}{2 \times 10^{-3}} = \frac{10^3}{2} = \frac{1000}{2} = 500\ \Omega$$

The inductive reactance is $X_L = 2000\ \Omega$ and the capacitive reactance is $X_C = 500\ \Omega$. Plug the resistance, inductive reactance, and capacitive reactance into the equation for impedance.

$$Z = \sqrt{R^2 + (X_L - X_C)^2} = \sqrt{\left(500\sqrt{3}\right)^2 + (2000 - 500)^2} = \sqrt{(500)^2\left(\sqrt{3}\right)^2 + (1500)^2}$$

Factor out $(500)^2$. Note that $(1500)^2 = (500)^2(3)^2$.

$$Z = \sqrt{(500)^2\left[\left(\sqrt{3}\right)^2 + (3)^2\right]} = \sqrt{(500)^2(3+9)} = \sqrt{(500)^2(12)} = 500\sqrt{12}$$

$$Z = 500\sqrt{(4)(3)} = 500\sqrt{4}\sqrt{3} = 500(2)\sqrt{3} = 1000\sqrt{3}\ \Omega$$

We applied the rules that $\sqrt{xy} = \sqrt{x}\sqrt{y}$ and $\sqrt{3}\sqrt{3} = 3$. The impedance is $Z = 1000\sqrt{3}\ \Omega$.

The rms voltage and rms current are related to one another by the impedance.

$$\Delta V_{rms} = I_{rms} Z$$

Divide both sides of the equation by the impedance.

$$I_{rms} = \frac{\Delta V_{rms}}{Z} = \frac{3000}{1000\sqrt{3}} = \frac{3}{\sqrt{3}} = \frac{3}{\sqrt{3}}\frac{\sqrt{3}}{\sqrt{3}} = \frac{3\sqrt{3}}{3} = \sqrt{3}\text{ A}$$

We multiplied by $\frac{\sqrt{3}}{\sqrt{3}}$ in order to **rationalize** the denominator. Note that $\sqrt{3}\sqrt{3} = 3$. The rms current for this RLC circuit is $I_{rms} = \sqrt{3}$ A. If you use a calculator, this comes out to $I_{rms} = 1.7$ A.

(B) What is the average power delivered by the AC power supply?

Solution. Use the appropriate equation for average power. Recall that we found $I_{rms} = \sqrt{3}$ A in part (A), while $R = 500\sqrt{3}\ \Omega$ was given in the problem.

$$P_{av} = I_{rms}^2 R = \left(\sqrt{3}\right)^2 \left(500\sqrt{3}\right) = (3)\left(500\sqrt{3}\right) = 1500\sqrt{3}\text{ W}$$

Note that $\sqrt{3}\sqrt{3} = 3$. The average power that the AC power supply delivers to the circuit is $P_{av} = 1500\sqrt{3}$ W. If you use a calculator, this comes out to $P_{av} = 2.6$ kW, where the metric prefix kilo (k) stands for k = 1000.

(C) What angular frequency would produce resonance for this RLC circuit?

Solution. Use the equation for resonance angular frequency.

$$\omega_0 = \frac{1}{\sqrt{LC}} = \frac{1}{\sqrt{(80)(8 \times 10^{-5})}} = \frac{1}{\sqrt{640 \times 10^{-5}}} = \frac{1}{\sqrt{64 \times 10^{-4}}} = \frac{1}{\sqrt{64}}\frac{1}{\sqrt{10^{-4}}}$$

$$\omega_0 = \frac{1}{8}\frac{1}{10^{-2}} = \frac{10^2}{8} = \frac{100}{8} = \frac{25}{2} = 12.5\text{ rad/s}$$

The resonance angular frequency for this RLC circuit is $\omega_0 = \frac{25}{2}$ rad/s = 12.5 rad/s (or 13 rad/s if you round to two significant figures, as you should in this problem).

(D) What is the maximum possible rms current that could be obtained by adjusting the frequency of the power supply?

Solution. Use the same equation and reasoning that we applied in Example 136, part (A).

$$I_{rms,max} = \frac{\Delta V_{rms}}{R} = \frac{3000}{500\sqrt{3}} = \frac{6}{\sqrt{3}} = \frac{6}{\sqrt{3}}\frac{\sqrt{3}}{\sqrt{3}} = \frac{6\sqrt{3}}{3} = 2\sqrt{3}\text{ A}$$

We multiplied by $\frac{\sqrt{3}}{\sqrt{3}}$ in order to **rationalize** the denominator. Note that $\sqrt{3}\sqrt{3} = 3$. The maximum possible rms current that the AC power supply could provide to this RLC circuit by adjusting the frequency is $I_{rms,max} = 2\sqrt{3}$ A.

Example 138. An AC power supply is connected in series with a 2.0-Ω resistor, a 30-mH inductor, and a 12-μF capacitor. What is the quality factor for this RLC circuit?

Solution. Begin by making a list of the given quantities in SI units:

- The resistance is $R = 2.0\ \Omega$.
- The inductance is $L = 3.0 \times 10^{-2}$ H. The metric prefix milli (m) stands for 10^{-3}.
- The capacitance is $C = 1.2 \times 10^{-5}$ F. The metric prefix micro (μ) stands for 10^{-6}.

First find the resonance angular frequency.

$$\omega_0 = \frac{1}{\sqrt{LC}} = \frac{1}{\sqrt{(3 \times 10^{-2})(1.2 \times 10^{-5})}} = \frac{1}{\sqrt{3.6 \times 10^{-7}}} = \frac{1}{\sqrt{36 \times 10^{-8}}} = \frac{1}{\sqrt{36}}\frac{1}{\sqrt{10^{-8}}}$$

$$\omega_0 = \frac{1}{6}\frac{1}{10^{-4}} = \frac{10^4}{6} = \frac{10{,}000}{6} = \frac{5000}{3} \text{ rad/s}$$

The resonance angular frequency for this RLC circuit is $\omega_0 = \frac{5000}{3}$ rad/s. Now use the equation for quality factor (Q_0) that involves the resonance angular frequency (ω_0), resistance (R), and inductance (L).

$$Q_0 = \omega_0 \frac{L}{R} = \left(\frac{5000}{3}\right)\frac{(3 \times 10^{-2})}{2} = 2500 \times 10^{-2} = 25$$

The quality factor for this RLC circuit is $Q_0 = 25$.

Example 139. What is the quality factor for the graph of average power shown below?

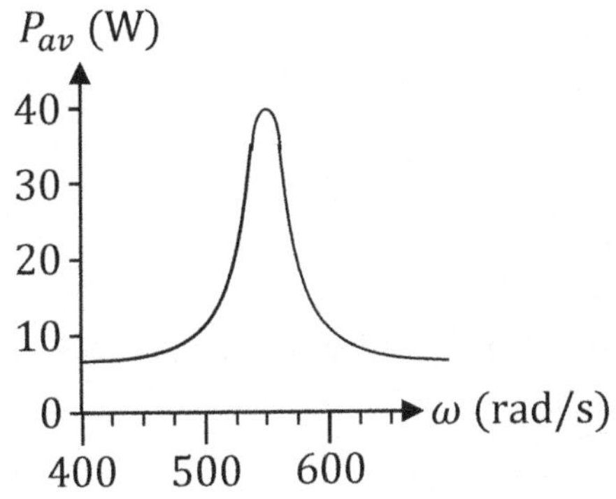

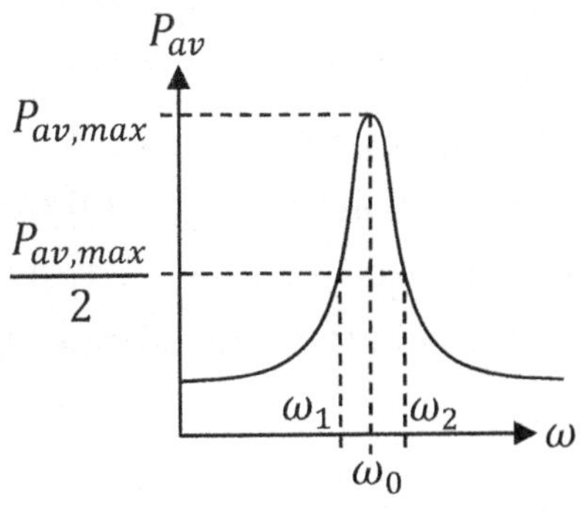

Solution. Read off the needed values from the graph:

- The resonance angular frequency is $\omega_0 = 550$ rad/s. This is the angular frequency for which the curve reaches its peak.
- The maximum average power is $P_{av,max} = 40$ W. This is the maximum vertical value of the curve.
- One-half of the maximum average power is $\frac{P_{av,max}}{2} = \frac{40}{2} = 20$ W.
- Draw a horizontal line on the graph where the vertical value of the curve is equal to 20 W, which corresponds to $\frac{P_{av,max}}{2}$.
- Draw two vertical lines on the graph where the curve intersects the horizontal line that you drew in the previous step. See the right figure above. These two vertical lines correspond to ω_1 and ω_2.
- Read off ω_1 and ω_2 from the graph: $\omega_1 = 525$ rad/s and $\omega_2 = 575$ rad/s.
- Subtract these values: $\Delta\omega = \omega_2 - \omega_1 = 575 - 525 = 50$ rad/s. This is the **full-width at half max**. Recall that the resonance frequency is $\omega_0 = 550$ rad/s.

Use the equation for quality factor that involves $\Delta\omega$ and ω_0.

$$Q_0 = \frac{\omega_0}{\Delta\omega} = \frac{550}{50} = 11$$

The quality factor for this graph is $Q_0 = 11$.

Example 140. An AC power supply with an rms voltage of 240 V is connected across the primary coil of a transformer. The transformer has 400 turns in the primary coil and 100 turns in the secondary coil. What is the rms output voltage?

Solution. Identify the given quantities.

- The rms input voltage is $\Delta V_{in} = 240$ V.
- The primary coil has $N_p = 400$ turns.
- The secondary coil has $N_s = 100$ turns.

Use the ratio equation for a **transformer** to solve for the output voltage (ΔV_{out}).

$$\frac{\Delta V_{out}}{\Delta V_{in}} = \frac{N_s}{N_p}$$

Cross-multiply.

$$\Delta V_{out} N_p = \Delta V_{in} N_s$$

Divide both sides of the equation by N_p.

$$\Delta V_{out} = \frac{\Delta V_{in} N_s}{N_p} = \frac{(240)(100)}{400} = 60 \text{ V}$$

The rms output voltage is $\Delta V_{out} = 60$ V.

Example 141. An AC power supply with an rms voltage of 40 V is connected across the primary coil of a transformer. The transformer has 200 turns in the primary coil. How many turns are in the secondary coil if the rms output voltage is 120 V?

Solution. Identify the given quantities.

- The rms input voltage is $\Delta V_{in} = 40$ V.
- The primary coil has $N_p = 200$ turns.
- The rms output voltage is $\Delta V_{out} = 120$ V.

Use the ratio equation for a **transformer** to solve for the output voltage (ΔV_{out}).

$$\frac{\Delta V_{out}}{\Delta V_{in}} = \frac{N_s}{N_p}$$

Cross-multiply.

$$\Delta V_{out} N_p = \Delta V_{in} N_s$$

Divide both sides of the equation by ΔV_{in}.

$$N_s = \frac{\Delta V_{out} N_p}{\Delta V_{in}} = \frac{(120)(200)}{40} = 600$$

The secondary coil has $N_s = 600$ turns (or loops).

32 MAXWELL'S EQUATIONS

Maxwell's Equations (Integral Form)

$$\oint_S \vec{\mathbf{E}} \cdot d\vec{\mathbf{A}} = \frac{q_{enc}}{\epsilon_0} \qquad \oint_S \vec{\mathbf{B}} \cdot d\vec{\mathbf{A}} = 0$$

$$\oint_C \vec{\mathbf{E}} \cdot d\vec{\mathbf{s}} = -\frac{d\Phi_m}{dt} \qquad \oint_C \vec{\mathbf{B}} \cdot d\vec{\mathbf{s}} = \mu_0 I_{enc} + \epsilon_0\mu_0 \frac{d\Phi_e}{dt}$$

Maxwell's Equations (Differential Form)

$$\vec{\nabla} \cdot \vec{\mathbf{E}} = \frac{\rho}{\epsilon_0} \quad , \quad \vec{\nabla} \times \vec{\mathbf{E}} = -\frac{\partial \vec{\mathbf{B}}}{\partial t} \qquad \vec{\nabla} \cdot \vec{\mathbf{B}} = 0 \quad , \quad \vec{\nabla} \times \vec{\mathbf{B}} = \mu_0 \vec{\mathbf{J}} + \epsilon_0\mu_0 \frac{\partial \vec{\mathbf{E}}}{\partial t}$$

Electric Flux	Magnetic Flux
$\Phi_e = \int_S \vec{\mathbf{E}} \cdot d\vec{\mathbf{A}}$	$\Phi_m = \int_S \vec{\mathbf{B}} \cdot d\vec{\mathbf{A}}$

Lorentz Force	Displacement Current
$\vec{\mathbf{F}}_{net} = q\vec{\mathbf{E}} + q\vec{v} \times \vec{\mathbf{B}}$	$I_d = \epsilon_0 \frac{d\Phi_e}{dt}$

Poynting Vector	Speed of Light in Vacuum
$\vec{\mathbf{S}} = \frac{1}{\mu_0}\vec{\mathbf{E}} \times \vec{\mathbf{B}} \quad , \quad I = S_{av} = \frac{E_m B_m}{2\mu_0}$	$c = \frac{1}{\sqrt{\epsilon_0\mu_0}}$

Gradient, Divergence, and Curl

$$\vec{\nabla} f = \hat{\mathbf{x}}\frac{\partial f}{\partial x} + \hat{\mathbf{y}}\frac{\partial f}{\partial y} + \hat{\mathbf{z}}\frac{\partial f}{\partial z} \quad , \quad \vec{\nabla} \cdot \vec{\mathbf{F}} = \frac{\partial F_x}{\partial x} + \frac{\partial F_y}{\partial y} + \frac{\partial F_z}{\partial z} \quad , \quad \vec{\nabla} \times \vec{\mathbf{F}} = \begin{vmatrix} \hat{\mathbf{x}} & \hat{\mathbf{y}} & \hat{\mathbf{z}} \\ \frac{\partial}{\partial x} & \frac{\partial}{\partial y} & \frac{\partial}{\partial z} \\ F_x & F_y & F_z \end{vmatrix}$$

Symbol	Name	SI Units
q	charge	C
I	current	A
$\vec{\mathbf{E}}$	electric field	N/C or V/m
$\vec{\mathbf{B}}$	magnetic field	T
$\vec{\mathbf{S}}$	the Poynting vector	W/m^2
ϵ_0	permittivity of free space	$\frac{C^2}{N \cdot m^2}$ or $\frac{C^2 \cdot s^2}{kg \cdot m^3}$
μ_0	permeability of free space	$\frac{T \cdot m}{A}$
$d\vec{\mathbf{s}}$	differential displacement (vector)	m
$d\vec{\mathbf{A}}$	differential area element (vector)	m^2
t	time	s
$\vec{\mathbf{F}}_e$	electric force	N
$\vec{\mathbf{F}}_m$	magnetic force	N
Φ_e	electric flux	$\frac{N \cdot m^2}{C}$ or $\frac{kg \cdot m^3}{C \cdot s^2}$
Φ_m	magnetic flux	$T \cdot m^2$ or Wb
ε_{ind}	induced emf	V
ρ	volume charge density	C/m^3
$\vec{\mathbf{J}}$	current density (distributed throughout a volume)	A/m^2
$\vec{v}$	velocity	m/s
c	speed of light in vacuum	m/s
U	energy	J
u	energy density	J/m^3

Conceptual Example. Briefly describe the significance of each of Maxwell's equations in integral form (in vacuum).

Solution. Gauss's law in electricity (Chapter 8) states that the **net electric flux** ($\Phi_{e,net} = \oint_S \vec{\mathbf{E}} \cdot d\vec{\mathbf{A}}$) is proportional to the **charge enclosed** by the closed Gaussian surface.

$$\oint_S \vec{\mathbf{E}} \cdot d\vec{\mathbf{A}} = \frac{q_{enc}}{\epsilon_0} \quad \text{(Gauss's law in electricity)}$$

Gauss's law in magnetism states that the **net magnetic flux** ($\Phi_{m,net} = \oint_S \vec{\mathbf{B}} \cdot d\vec{\mathbf{A}}$) always equals <u>**zero**</u>. That's because only moving charges create magnetic fields. Nobody has ever discovered a magnetic **monopole** – the hypothetical magnetic equivalent of electric charge.

$$\oint_S \vec{\mathbf{B}} \cdot d\vec{\mathbf{A}} = 0 \quad \text{(Gauss's law in magnetism)}$$

Faraday's law (Chapter 29) states that a **changing magnetic flux** $\left(\frac{d\Phi_m}{dt}\right)$ through a loop of wire induces an emf ($\varepsilon_{ind} = \oint_C \vec{\mathbf{E}} \cdot d\vec{\mathbf{s}}$) in the loop of wire.

$$\oint_C \vec{\mathbf{E}} \cdot d\vec{\mathbf{s}} = -\frac{d\Phi_m}{dt} \quad \text{(Faraday's law)}$$

Ampère's law (Chapter 27) states that the **line integral of magnetic field** ($\oint_C \vec{\mathbf{B}} \cdot d\vec{\mathbf{s}}$) is proportional to the **current enclosed** (I_{enc}) by the Ampèrian loop. Maxwell generalized Ampère's law to account for a **displacement current** (I_d), which is associated with a **changing electric flux** $\left(\frac{d\Phi_e}{dt}\right)$: $I_d = \epsilon_0 \frac{d\Phi_e}{dt}$.

$$\oint_C \vec{\mathbf{B}} \cdot d\vec{\mathbf{s}} = \mu_0 I_{enc} + \epsilon_0 \mu_0 \frac{d\Phi_e}{dt} \quad \text{(generalized Ampère's law)}$$

Maxwell's four equations, which we have written in integral form, describe electromagnetism. With Faraday's law, we see that a changing magnetic flux induces an **electric field** (which causes an emf and current to be induced in a loop of wire). Similarly, the generalized form of Ampère's law shows that a changing electric flux induces a **magnetic field**. Maxwell's equations can be applied to describe **electromagnetic waves** (or **light**).

Conceptual Example. Briefly describe the force(s) that a moving charge experiences in an electromagnetic field.

Solution. Recall from Chapter 2 that a charged particle in the presence of an electric field ($\vec{\mathbf{E}}$) experiences an **electric** force $\vec{\mathbf{F}}_e = q\vec{\mathbf{E}}$, and recall from Chapter 24 that a moving charge in the presence of a magnetic field ($\vec{\mathbf{B}}$) experiences a **magnetic** force $\vec{\mathbf{F}}_m = q\vec{v} \times \vec{\mathbf{B}}$. When a moving charge is in the presence of both electric and magnetic fields, it experiences a Lorentz force, which combines these two equations together.

$$\vec{\mathbf{F}}_{net} = q\vec{\mathbf{E}} + q\vec{v} \times \vec{\mathbf{B}}$$

Conceptual Example. Briefly describe the significance of Gauss's law in magnetism.
Solution. Gauss's law applies to both electricity and magnetism, except that the right-hand side is zero in the case of magnetism.

$$\oint_S \vec{\mathbf{E}} \cdot d\vec{\mathbf{A}} = \frac{q_{enc}}{\epsilon_0}$$

$$\oint_S \vec{\mathbf{B}} \cdot d\vec{\mathbf{A}} = 0$$

The reason for this is that there evidently is no such thing as a magnetic **monopole** – a hypothetical particle that would serve as sort of a "magnetic charge," analogous to electric charge. When you study magnets macroscopically, at first they appear to consist of two separate poles (north and south), but it turns out that such north and south poles always come in pairs. If you cut a magnet in half, you **don't** get two pieces that each have only one pole: Instead, you get two new smaller magnets, where each new magnet has both a north and south pole. (If you keep cutting the magnet in half forever, eventually you will have to split a single atom.) As explained in Chapter 20, **all magnetic fields**, including those created by magnets, **are produced by moving charges**. A magnet doesn't really have a north and south pole inside of it: Rather, each atom in the magnet produces its own tiny magnetic field, and these atomic magnetic fields create the net magnetic field of the magnet.

It is instructive to compare the magnetic field lines for a bar magnet (Chapter 20) to the electric field lines of an electric dipole (Chapter 4). Outside of a bar magnet, the magnetic field lines resemble the electric field lines of an electric dipole. However, inside of the magnet, the magnetic field lines look much different. According to Gauss's law in magnetism, **the net magnetic flux through any closed surface is always zero**: The same number of **magnetic** field lines always enter and exit any closed surface.

Compare the electric field map for the electric dipole shown below on the left with the magnetic field map for a bar magnet shown below on the right.

- In the electric field map, if a closed surface surrounds the negative charge (as in surface X drawn below), there is a net electric flux through the surface, but if a closed surface surrounds zero charge (or a net charge of zero), the net electric flux through the surface is zero (as in surface W drawn below).
- In the magnetic field map, the net magnetic flux through any closed surface is zero (see surfaces Y and Z below).
- Pay close attention to the distinction between surfaces X and Z in the two different diagrams: The net electric flux through surface X is negative, while the net magnetic flux through surface Z is zero. (The net flux through surfaces W and Y is also zero.)

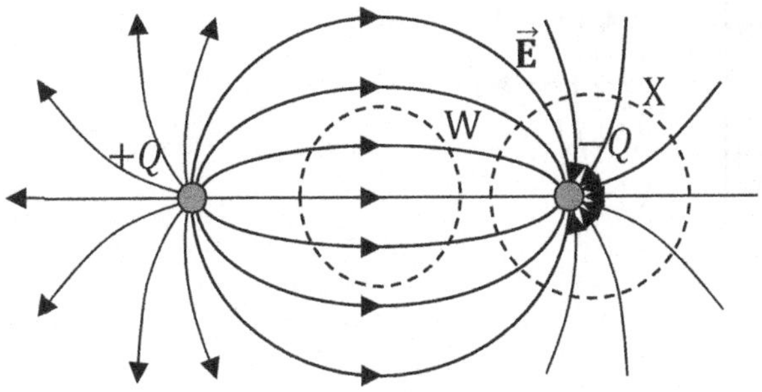

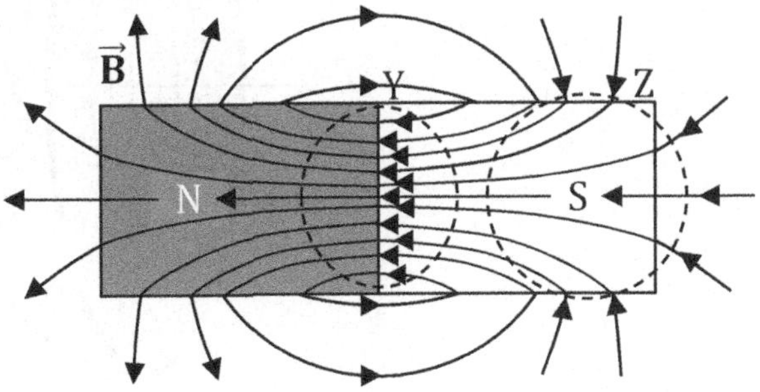

Because the right-hand side of Gauss's law is zero for a magnetic field, Gauss's law isn't as practical for deriving equations for magnetic field as it is for deriving equations for electric field (like we did in Chapter 8). However, if you want to derive an equation for magnetic field, we have Ampère's law (Chapter 27) for that, or the law of Biot-Savart (Chapter 26).

Conceptual Example. What would Maxwell's equations look like if there were a hypothetical magnetic **monopole** with "magnetic charge" q_m (to be distinguished from ordinary electric charge q_e) with SI units of Ampères times meters (Am)?

Solution. First replace I_{enc} with $\frac{dq_e}{dt}$ in Maxwell's equations, and then write the electric and magnetic flux in terms of the corresponding fields: $\Phi_e = \int_S \vec{\mathbf{E}} \cdot d\vec{\mathbf{A}}$, and $\Phi_m = \int_S \vec{\mathbf{B}} \cdot d\vec{\mathbf{A}}$. Now complete the near symmetry by adding $\mu_0 q_m$ to the right-hand side of Gauss's law in magnetism (corresponding to $\frac{q_{enc}}{\epsilon_0}$ in electricity), and by adding $-\mu_0 \frac{dq_m}{dt}$ to the right-hand side of the integral form of Faraday's law (corresponding to $\mu_0 \frac{dq_e}{dt}$ on the right-hand side of Ampère's law).

$$\oint_S \vec{\mathbf{E}} \cdot d\vec{\mathbf{A}} = \frac{q_{enc}}{\epsilon_0} \quad , \quad \oint_S \vec{\mathbf{B}} \cdot d\vec{\mathbf{A}} = \mu_0 q_m$$

$$\oint_C \vec{\mathbf{E}} \cdot d\vec{\mathbf{s}} = -\mu_0 \frac{dq_m}{dt} - \frac{d}{dt} \int_S \vec{\mathbf{B}} \cdot d\vec{\mathbf{A}} \quad , \quad \oint_C \vec{\mathbf{B}} \cdot d\vec{\mathbf{s}} = \mu_0 \frac{dq_e}{dt} + \epsilon_0 \mu_0 \frac{d}{dt} \int_S \vec{\mathbf{E}} \cdot d\vec{\mathbf{A}}$$

Conceptual Example. Briefly describe the significance of the generalized form of Ampère's law.
Solution. Maxwell generalized Ampère's law to account for a gap in a circuit, such as the space between the plates of a capacitor. For example, consider the very long straight (horizontal) wire illustrated below, carrying a steady current (I), which features a parallel-plate capacitor with two large, closely spaced plates of uniform charge density. We applied Ampère's law ($\oint_C \vec{\mathbf{B}} \cdot d\vec{\mathbf{s}} = \mu_0 I_{enc}$) to a very long straight wire in Chapter 27, where we found that the left-hand side was $2\pi r_c B$ for an Ampèrian circle coaxial with the current.

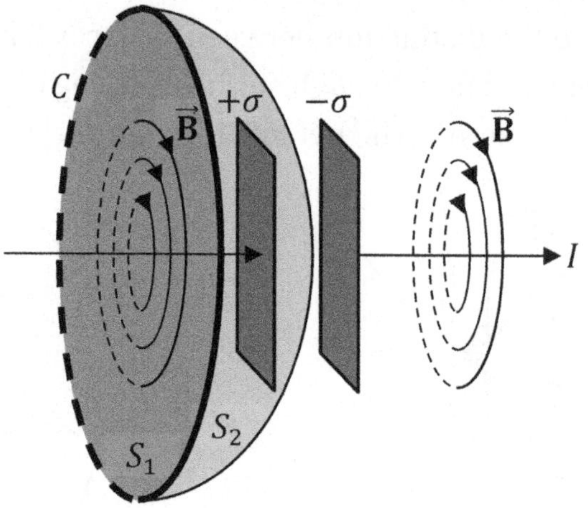

The right-hand side of Ampère's law includes $\mu_0 I_{enc}$. The current enclosed (I_{enc}) by the Ampèrian loop equals the current that passes through a surface that is bounded by the path C in the integral $\oint_C \vec{\mathbf{B}} \cdot d\vec{\mathbf{s}}$: The path C is the Ampèrian circle shown above. The need for Maxwell's displacement current (I_d) can be understood by considering the following surfaces which are both bounded by the Ampèrian circle:

- The surface S_1 is the flat solid disc bounded by C. Since the current I passes through S_1, in this case the current enclosed is $I_{enc} = I$.
- The surface S_2 is a hemisphere that is bounded by C, but which passes between the plates of the capacitor like a butterfly net. The current I does **not** pass through S_2.

However, we must get the same answer for the magnetic field when we apply Ampère's law regardless of whether we choose the surface S_1 or S_2: Our result must apply for any surface that is bounded by the path C. The **generalized form** of Ampère's law, $\oint_C \vec{\mathbf{B}} \cdot d\vec{\mathbf{s}} = \mu_0 I_{enc} + \epsilon_0 \mu_0 \frac{d\Phi_e}{dt}$, states that there is indeed a current, called **displacement current**, passing through S_2, where the displacement current equals $I_d = \epsilon_0 \frac{d\Phi_e}{dt}$. Why does the displacement current equal $I_d = \epsilon_0 \frac{d\Phi_e}{dt}$? According to Gauss's law, the electric flux between the parallel plates is $\Phi_e = \int_S \vec{\mathbf{E}} \cdot d\vec{\mathbf{A}} = \frac{q_{enc}}{\epsilon_0}$, such that $I_d = \epsilon_0 \frac{d\Phi_e}{dt} = \epsilon_0 \frac{d}{dt}\left(\frac{q_{enc}}{\epsilon_0}\right) = \frac{dq_{enc}}{dt}$. Recall that current is the rate of flow of charge: $I_d = \frac{dq_{enc}}{dt}$.

Partial Derivative Example. Given $f = 5x^3y^2$, find a partial derivative of f with respect to x, and also find a partial derivative of f with respect to y.

Solution. To find a partial derivative of f with respect to x, treat y as a constant.

$$\frac{\partial f}{\partial x} = \frac{\partial}{\partial x}(5x^3y^2) = 5y^2\frac{\partial}{\partial x}(x^3) = 5y^2(3x^2) = 15x^2y^2$$

To find a partial derivative of f with respect to y, treat x as a constant.

$$\frac{\partial f}{\partial y} = \frac{\partial}{\partial y}(5x^3y^2) = 5x^3\frac{\partial}{\partial y}(y^2) = 5x^3(2y) = 10x^3y$$

Note: We use the symbol ∂, which is a rounded version of the letter d, to distinguish partial derivatives from total derivatives.

Conceptual Example. Interpret the gradient, divergence, and curl operators.

Solution. The **gradient** ($\vec{\nabla}$) is a differential operator which applies partial derivatives to a scalar function to create a **vector** function. The gradient operator is defined as:

$$\vec{\nabla} = \hat{\mathbf{x}}\frac{\partial}{\partial x} + \hat{\mathbf{y}}\frac{\partial}{\partial y} + \hat{\mathbf{z}}\frac{\partial}{\partial z}$$

Recall that the **unit vectors** $\hat{\mathbf{x}}$, $\hat{\mathbf{y}}$, and $\hat{\mathbf{z}}$ point one unit along the $+x$-, $+y$-, and $+z$-axes, respectively. When the gradient operator acts on a scalar function f, we get:

$$\vec{\nabla}f = \hat{\mathbf{x}}\frac{\partial f}{\partial x} + \hat{\mathbf{y}}\frac{\partial f}{\partial y} + \hat{\mathbf{z}}\frac{\partial f}{\partial z}$$

The gradient is a sort of three-dimensional, multivariable derivative. Like an ordinary derivative, the gradient represents the **slope** of a tangent. The direction of the gradient at a given point on the function is along the direction in which the function increases the most (**the direction of greatest increase**) from that point.

Note: We use the symbol $\vec{\nabla}$ (called the **del** operator) to represent the gradient operator. The del symbol (∇) looks like an upside down uppercase delta (Δ).

The **divergence** operator ($\vec{\nabla}\cdot$) is a differential operator that uses the del operator ($\vec{\nabla}$) in a **scalar product** (we discussed the scalar product in Chapter 19). We apply the divergence to a vector function and obtain a **scalar** function as a result.

$$\vec{\nabla}\cdot\vec{\mathbf{F}} = \left(\hat{\mathbf{x}}\frac{\partial}{\partial x} + \hat{\mathbf{y}}\frac{\partial}{\partial y} + \hat{\mathbf{z}}\frac{\partial}{\partial z}\right)\cdot\left(F_x\hat{\mathbf{x}} + F_y\hat{\mathbf{y}} + F_z\hat{\mathbf{z}}\right) = \frac{\partial F_x}{\partial x} + \frac{\partial F_y}{\partial y} + \frac{\partial F_z}{\partial z}$$

The divergence of a vector field provides a measure of how much the field lines of a vector **diverge** (or spread out) from a given point.

The **curl** operator ($\vec{\nabla}\times$) is a differential operator that uses the del operator ($\vec{\nabla}$) in a **vector product** (we discussed the vector product in Chapter 19). We apply the curl to a vector function and obtain a different **vector** function as a result. **Note:** When finding a determinant that includes differential operators $\left(\text{like } \frac{\partial}{\partial x}\right)$, it's important to work out the determinant from **top to bottom** along each diagonal.

$$\vec{\nabla} \times \vec{\mathbf{F}} = \begin{vmatrix} \hat{\mathbf{x}} & \hat{\mathbf{y}} & \hat{\mathbf{z}} \\ \frac{\partial}{\partial x} & \frac{\partial}{\partial y} & \frac{\partial}{\partial z} \\ F_x & F_y & F_z \end{vmatrix} = \hat{\mathbf{x}} \begin{vmatrix} \frac{\partial}{\partial y} & \frac{\partial}{\partial z} \\ F_y & F_z \end{vmatrix} - \hat{\mathbf{y}} \begin{vmatrix} \frac{\partial}{\partial x} & \frac{\partial}{\partial z} \\ F_x & F_z \end{vmatrix} + \hat{\mathbf{z}} \begin{vmatrix} \frac{\partial}{\partial x} & \frac{\partial}{\partial y} \\ F_x & F_y \end{vmatrix}$$

$$\vec{\nabla} \times \vec{\mathbf{F}} = \left(\frac{\partial}{\partial y} F_z - \frac{\partial}{\partial z} F_y\right) \hat{\mathbf{x}} - \left(\frac{\partial}{\partial x} F_z - \frac{\partial}{\partial z} F_x\right) \hat{\mathbf{y}} + \left(\frac{\partial}{\partial x} F_y - \frac{\partial}{\partial y} F_x\right) \hat{\mathbf{z}}$$

The curl of a vector field expresses the **circulation** of the field lines.

Conceptual Example. Briefly describe the significance of each of Maxwell's equations in differential form (in vacuum).

Solution. Maxwell's equations in differential form apply the **divergence** and **curl** operators.

$$\vec{\nabla} \cdot \vec{\mathbf{E}} = \frac{\rho}{\epsilon_0} \quad \text{(Gauss's law in electricity)}$$

$$\vec{\nabla} \cdot \vec{\mathbf{B}} = 0 \quad \text{(Gauss's law in magnetism)}$$

$$\vec{\nabla} \times \vec{\mathbf{E}} = -\frac{\partial \vec{\mathbf{B}}}{\partial t} \quad \text{(Faraday's law)}$$

$$\vec{\nabla} \times \vec{\mathbf{B}} = \mu_0 \vec{\mathbf{J}} + \epsilon_0 \mu_0 \frac{\partial \vec{\mathbf{E}}}{\partial t} \quad \text{(generalized Ampère's law)}$$

The divergence ($\vec{\nabla} \cdot$) and curl ($\vec{\nabla} \times$) operators help to interpret these equations:

- Gauss's law in electricity, $\vec{\nabla} \cdot \vec{\mathbf{E}} = \frac{\rho}{\epsilon_0}$, states that the electric field lines tend to **radiate** away from (or towards) a pointlike charge, since the divergence of a vector field is nonzero when the field lines diverge.
- Gauss's law in magnetism, $\vec{\nabla} \cdot \vec{\mathbf{B}} = 0$, states that magnetic field lines **never diverge**. Rather, magnetic field lines tend to circulate (as shown by the curl in Ampère's law).
- Faraday's law, $\vec{\nabla} \times \vec{\mathbf{E}} = -\frac{\partial \vec{\mathbf{B}}}{\partial t}$, states that a changing magnetic field creates electric field lines that **circulate** (which could accelerate charges in a loop of wire, inducing a current in the loop).
- Ampère's law, $\vec{\nabla} \times \vec{\mathbf{B}} = \mu_0 \vec{\mathbf{J}}$, states that magnetic field lines tend to **circulate** around currents, since the curl of a vector field is nonzero when the field lines circulate. The generalized form of Ampère's law, $\vec{\nabla} \times \vec{\mathbf{B}} = \mu_0 \vec{\mathbf{J}} + \epsilon_0 \mu_0 \frac{\partial \vec{\mathbf{E}}}{\partial t}$, states that a changing electric flux associated with the displacement current ($I_d = \epsilon_0 \frac{d\Phi_e}{dt}$) also creates magnetic field lines that circulate.

Conceptual Example. In what sense is light an electromagnetic wave?

Solution. The figure below illustrates (in part) what this means. The graph below shows a light wave propagating to the right along the $+z$-axis. As the wave propagates to the right, the electric field ($\vec{\mathbf{E}}$) oscillates up and down and the magnetic field ($\vec{\mathbf{B}}$) oscillates into and out of the page.

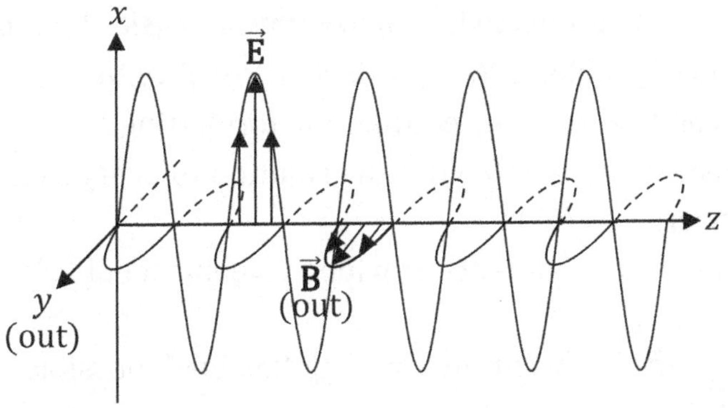

The **permittivity** of free space (ϵ_0) – which we encountered in Chapter 8 in the context of **electricity** with Gauss's law – and the **permeability** of free space (μ_0) – which we encountered in Chapters 26-27 in the context of **magnetism** with the law of Biot-Savart and Ampère's law – combine together to form the **speed of light** in vacuum (c) as follows.

$$c = \frac{1}{\sqrt{\epsilon_0 \mu_0}}$$

We can verify this by plugging in numbers. Recall that the **permittivity** of free space (ϵ_0) is related to **Coulomb's constant** (k) by $\epsilon_0 = \frac{1}{4\pi k}$. Also recall that **Coulomb's constant** is $k = 9.0 \times 10^9 \ \frac{\text{N}\cdot\text{m}^2}{\text{C}^2}$ and that the **permeability** of free space is $\mu_0 = 4\pi \times 10^{-7} \ \frac{\text{T}\cdot\text{m}}{\text{A}}$.

$$c = \frac{1}{\sqrt{\epsilon_0 \mu_0}} = \frac{1}{\sqrt{\frac{1}{4\pi k}\mu_0}} = \frac{1}{\sqrt{\frac{1}{4\pi(9 \times 10^9)}(4\pi \times 10^{-7})}} = \frac{1}{\sqrt{\frac{10^{-7}}{9 \times 10^9}}} = \frac{1}{\sqrt{\frac{10^{-7-9}}{9}}}$$

$$c = \frac{1}{\sqrt{\frac{10^{-16}}{9}}} = \sqrt{\frac{9}{10^{-16}}} = \sqrt{9 \times 10^{16}} = \sqrt{9}\sqrt{10^{16}} = 3.0 \times 10^8 \text{ m/s}$$

The speed of light in vacuum equals $c = 3.0 \times 10^8$ m/s to two (really, 3) significant figures.

The units work out as follows. Since ϵ_0's units are $\frac{\text{C}^2}{\text{N}\cdot\text{m}^2}$ (the reciprocal of k's units) and μ_0's units are $\frac{\text{T}\cdot\text{m}}{\text{A}}$, the units of $\epsilon_0\mu_0$ are $\frac{\text{C}^2\cdot\text{T}}{\text{N}\cdot\text{m}\cdot\text{A}}$. From the equation $F_m = ILB \sin\theta$, a Tesla can be expressed as $1 \text{ T} = 1 \ \frac{\text{N}}{\text{A}\cdot\text{m}}$, such that $\epsilon_0\mu_0$'s units become $\frac{\text{C}^2}{\text{m}^2\cdot\text{A}^2}$. Since $I = \frac{dq}{dt}$, an Ampère is $1 \text{ A} = 1 \ \frac{\text{C}}{\text{s}}$. With this, $\epsilon_0\mu_0$'s units are $\frac{\text{s}^2}{\text{m}^2}$, and $\frac{1}{\sqrt{\epsilon_0\mu_0}}$ has units of m/s.

WAS THIS BOOK HELPFUL?

A great deal of effort and thought was put into this book, such as:

- Breaking down the solutions to help make physics easier to understand.
- Careful selection of problems for their instructional value.
- Multiple stages of proofreading, editing, and formatting.
- Two physics instructors worked out the solution to every problem to help check all of the final answers.
- Dozens of actual physics students provided valuable feedback.

If you appreciate the effort that went into making this book possible, there is a simple way that you could show it:

Please take a moment to post an honest review.

For example, you can review this book at Amazon.com or BN.com (for Barnes & Noble).

Even a short review can be helpful and will be much appreciated. If you're not sure what to write, following are a few ideas, though it's best to describe what's important to you.

- Were you able to understand the explanations?
- Did you appreciate the list of symbols and units?
- Was it easy to find the information you were looking for?
- How much did you learn from reading through the examples?
- Would you recommend this book to others? If so, why?

Are you an international student?

If so, please leave a review at Amazon.co.uk (United Kingdom), Amazon.ca (Canada), Amazon.in (India), Amazon.com.au (Australia), or the Amazon website for your country.

The physics curriculum in the United States is somewhat different from the physics curriculum in other countries. International students who are considering this book may like to know how well this book may fit their needs.

GET A DIFFERENT ANSWER?

If you get a different answer and can't find your mistake even after consulting the hints and explanations, what should you do?

Please contact the author, Dr. McMullen.

How? Visit one of the author's blogs (see below). Either use the Contact Me option, or click on one of the author's articles and post a comment on the article.

www.monkeyphysicsblog.wordpress.com
www.improveyourmathfluency.com
www.chrismcmullen.wordpress.com

Why?

- If there happens to be a mistake (although much effort was put into perfecting the answer key), the correction will benefit other students like yourself in the future.
- If it turns out not to be a mistake, **you may learn something** from Dr. McMullen's reply to your message.

99.99% of students who walk into Dr. McMullen's office believing that they found a mistake with an answer discover one of two things:

- They made a mistake that they didn't realize they were making and learned from it.
- They discovered that their answer was actually the same. This is actually fairly common. For example, the answer key might say $t = \frac{\sqrt{3}}{3}$ s. A student solves the problem and gets $t = \frac{1}{\sqrt{3}}$ s. These are actually the same: Try it on your calculator and you will see that both equal about 0.57735. Here's why: $\frac{1}{\sqrt{3}} = \frac{1}{\sqrt{3}}\frac{\sqrt{3}}{\sqrt{3}} = \frac{\sqrt{3}}{3}$.

Two experienced physics teachers solved every problem in this book to check the answers, and dozens of students used this book and provided feedback before it was published. Every effort was made to ensure that the final answer given to every problem is correct.

But all humans, even those who are experts in their fields and who routinely aced exams back when they were students, make an occasional mistake. So if you believe you found a mistake, you should report it just in case. Dr. McMullen will appreciate your time.

VOLUME 3

If you want to learn more physics, volume 3 covers additional topics.

Volume 3: Waves, Fluids, Sound, Heat, and Light

- Sine waves
- Simple harmonic motion
- Oscillating springs
- Simple and physical pendulums
- Characteristics of waves
- Sound waves
- The decibel system
- The Doppler effect
- Standing waves
- Density and pressure
- Archimedes' principle
- Bernoulli's principle
- Pascal's principle
- Heat and temperature
- Thermal expansion
- Ideal gases
- The laws of thermodynamics
- Light waves
- Reflection and refraction
- Snell's law
- Total internal reflection
- Dispersion
- Thin lenses
- Spherical mirrors
- Diffraction
- Interference
- Polarization
- and more

ABOUT THE AUTHOR

Chris McMullen is a physics instructor at Northwestern State University of Louisiana and also an author of academic books. Whether in the classroom or as a writer, Dr. McMullen loves sharing knowledge and the art of motivating and engaging students.

He earned his Ph.D. in phenomenological high-energy physics (particle physics) from Oklahoma State University in 2002. Originally from California, Dr. McMullen earned his Master's degree from California State University, Northridge, where his thesis was in the field of electron spin resonance.

As a physics teacher, Dr. McMullen observed that many students lack fluency in fundamental math skills. In an effort to help students of all ages and levels master basic math skills, he published a series of math workbooks on arithmetic, fractions, algebra, and trigonometry called the Improve Your Math Fluency Series. Dr. McMullen has also published a variety of science books, including introductions to basic astronomy and chemistry concepts in addition to physics textbooks.

Dr. McMullen is very passionate about teaching. Many students and observers have been impressed with the transformation that occurs when he walks into the classroom, and the interactive engaged discussions that he leads during class time. Dr. McMullen is well-known for drawing monkeys and using them in his physics examples and problems, applying his creativity to inspire students. A stressed-out student is likely to be told to throw some bananas at monkeys, smile, and think happy physics thoughts.

Author, Chris McMullen, Ph.D.

PHYSICS

The learning continues at Dr. McMullen's physics blog:

www.monkeyphysicsblog.wordpress.com

More physics books written by Chris McMullen, Ph.D.:

- An Introduction to Basic Astronomy Concepts (with Space Photos)
- The Observational Astronomy Skywatcher Notebook
- An Advanced Introduction to Calculus-based Physics
- Essential Calculus-based Physics Study Guide Workbook
- Essential Trig-based Physics Study Guide Workbook
- 100 Instructive Calculus-based Physics Examples
- 100 Instructive Trig-based Physics Examples
- Creative Physics Problems
- A Guide to Thermal Physics
- A Research Oriented Laboratory Manual for First-year Physics

SCIENCE

Dr. McMullen has published a variety of **science** books, including:

- Basic astronomy concepts
- Basic chemistry concepts
- Balancing chemical reactions
- Creative physics problems
- Calculus-based physics textbook
- Calculus-based physics workbooks
- Trig-based physics workbooks

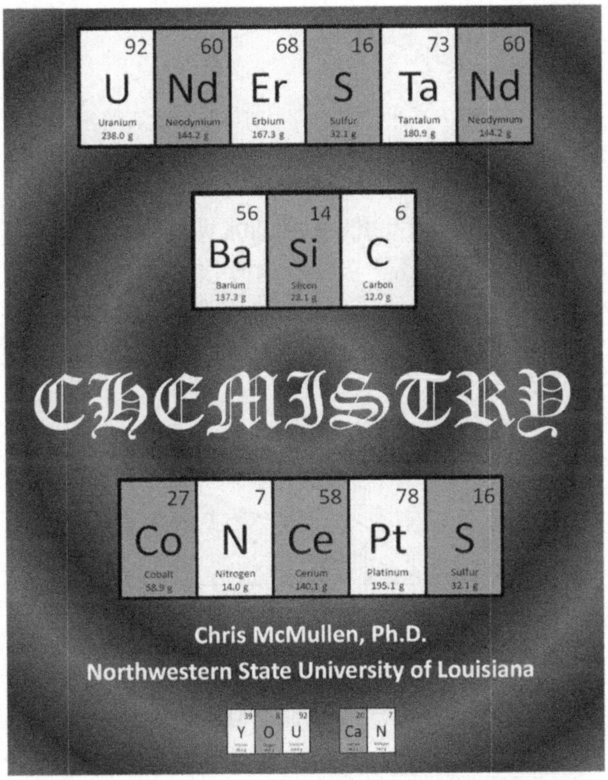

MATH

This series of math workbooks is geared toward practicing essential math skills:

- Algebra and trigonometry
- Fractions, decimals, and percents
- Long division
- Multiplication and division
- Addition and subtraction

www.improveyourmathfluency.com

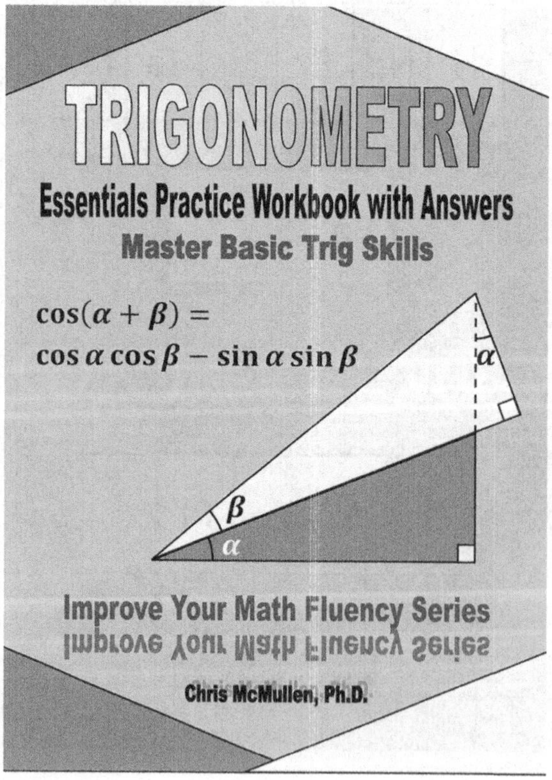

PUZZLES

The author of this book, Chris McMullen, enjoys solving puzzles. His favorite puzzle is Kakuro (kind of like a cross between crossword puzzles and Sudoku). He once taught a three-week summer course on puzzles. If you enjoy mathematical pattern puzzles, you might appreciate:

300+ Mathematical Pattern Puzzles

Number Pattern Recognition & Reasoning

- pattern recognition
- visual discrimination
- analytical skills
- logic and reasoning
- analogies
- mathematics

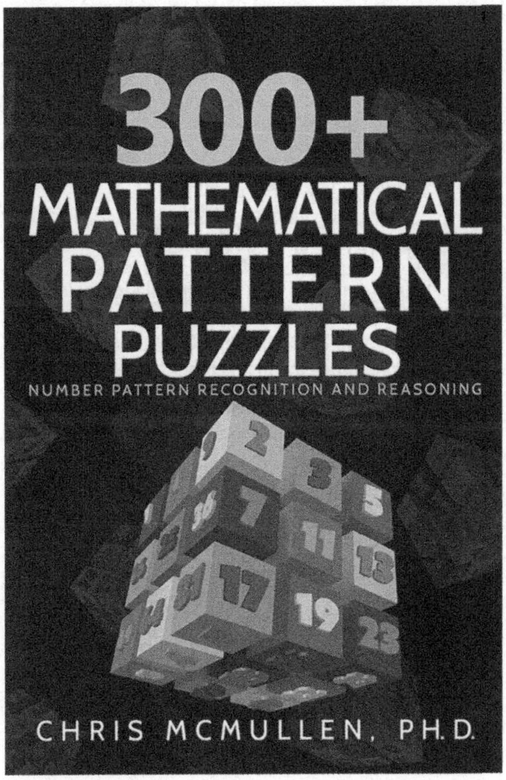

VErBAl ReAcTiONS

Chris McMullen has coauthored several word scramble books. This includes a cool idea called **VErBAl ReAcTiONS.** A VErBAl ReAcTiON expresses word scrambles so that they look like chemical reactions. Here is an example:

$$2\,C + U + 2\,S + Es \rightarrow S\,U\,C\,C\,Es\,S$$

The left side of the reaction indicates that the answer has 2 C's, 1 U, 2 S's, and 1 Es. Rearrange CCUSSEs to form SUCCEsS.

Each answer to a **VErBAl ReAcTiON** is not merely a word, it's a chemical word. A chemical word is made up not of letters, but of elements of the periodic table. In this case, SUCCEsS is made up of sulfur (S), uranium (U), carbon (C), and Einsteinium (Es).

Another example of a chemical word is GeNiUS. It's made up of germanium (Ge), nickel (Ni), uranium (U), and sulfur (S).

If you enjoy anagrams and like science or math, these puzzles are tailor-made for you.

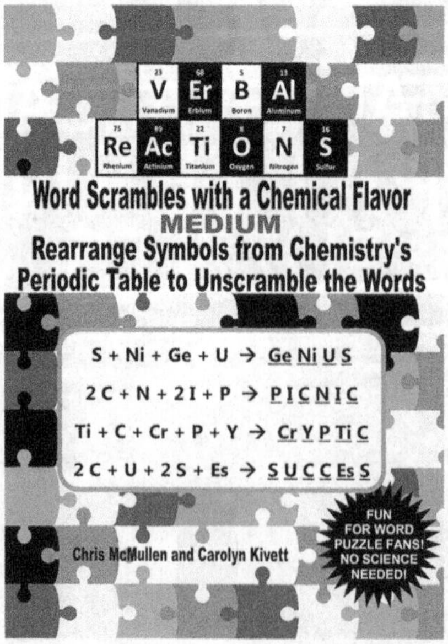

BALANCING CHEMICAL REACTIONS

$$2\ C_2H_6 + 7\ O_2 \rightarrow 4\ CO_2 + 6\ H_2O$$

Balancing chemical reactions isn't just chemistry practice.

These are also **fun puzzles** for math and science lovers.

Balancing Chemical Equations Worksheets

Over 200 Reactions to Balance

Chemistry Essentials Practice Workbook with Answers

Chris McMullen, Ph.D.

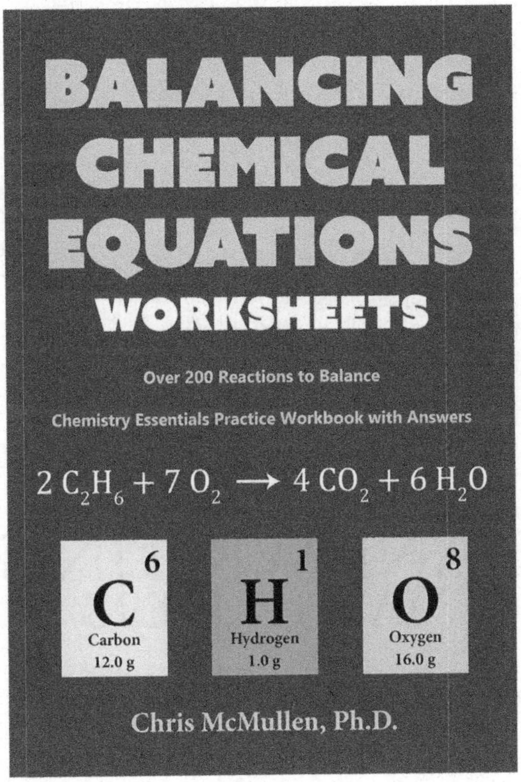

CURSIVE HANDWRITING

for... MATH LOVERS

Would you like to learn how to write in cursive?

Do you enjoy math?

This cool writing workbook lets you practice writing math terms with cursive handwriting. Unfortunately, you can't find many writing books oriented around math.

Cursive Handwriting for Math Lovers
by Julie Harper and Chris McMullen, Ph.D.

Made in the USA
Las Vegas, NV
19 February 2024

85976485R00223